48⁰⁰
20⁰⁰

D1014772

Fundamentals
of Ornithology

Fundamentals of Ornithology

SECOND EDITION

JOSSELYN VAN TYNE
Late Curator of Birds
University of Michigan Museum of Zoology

and

ANDREW J. BERGER
Professor of Zoology
University of Hawaii

A WILEY-INTERSCIENCE PUBLICATION

JOHN WILEY & SONS

New York • London • Sydney • Toronto

Copyright © 1959, 1976, by John Wiley & Sons, Inc.

All rights reserved. Published simultaneously in Canada.

No part of this book may be reproduced by any means,
nor transmitted, nor translated into a machine language
without the written permission of the publisher.

Library of Congress Cataloging in Publication Data:

Van Tyne, Josselyn, 1902–1957.
  Fundamentals of ornithology.

  "A Wiley-Interscience publication."
  Includes bibliographical references and index.
  1. Ornithology.   2. Birds—Classification.
I. Berger, Andrew John, 1915–   joint author.
II. Title.

QL673.V3   1975   598.2   75-20430
ISBN 0-471-89965-8

Printed in the United States of America

10 9 8 7 6 5 4 3 2 1

Preface to the Second Edition

The first edition of this book has served students for nearly 15 years. The rate of obsolescence of many chapters, however, has increased with each passing year so that a revised edition is in order. The objective remains the same as in the first edition: to present a readable and accurate text that covers the fundamentals of avian biology.

In a letter of October 23, 1887, Alfred Newton wrote: "We have more Natural History Journals than the country can afford, with the result that the numerosity is not only injurious to the Journals themselves but to Natural History itself, as it lowers the tone of the contributions." Newton would be bewildered, indeed, if he were faced with the number of journals that now contain papers dealing with birds. The references cited in this edition were taken from approximately 340 different journals! One of the most formidable tasks in preparing the second edition, therefore, was to decide which material should be included while staying within the framework of "fundamentals" and within a single volume. There is no chance that I will please all readers by my selections. The majority, but not all, of the references have been cited in the text; the others are included for the student who is interested in further reading on a particular subject. Although I have retained certain classic references, the advances in ornithology have been so great during the past 15 years that most of the books, monographs, and articles cited have appeared since 1958.

Some major changes in organization have been made. Chapter 11 (Social Relations) of the first edition has been deleted, and appropriate material has been incorporated where pertinent in other chapters. The chapter on breeding behavior has been expanded and reorganized to appear now as Chapters 10 and 11. A section on physiology has been added to Chapter 2. Space limitations made it necessary to delete the chapter on Ornithological Sources and the Glossary.

The number of illustrations has been increased from 254 in the first edition to

528. I am indebted to a number of people who have given permission to use previously published and unpublished illustrations; these are acknowledged in the legends for the appropriate figures. Illustrations without acknowledgment are mine. I am grateful to George Miksch Sutton for preparing the drawings of *Rhabdornis* and *Climacteris* for the two new families added to Chapter 13.

I express my special appreciation to colleagues who generously read the manuscripts of chapters in their areas of special research interests: Pierce Brodkorb, Causey Whittow, Mary H. Clench, Kenneth C. Parkes, Ernst S. Reese, Arthur N. Popper, David W. Johnston, Robert A. Kinzie, III, Frank Bellrose, and Eugene Eisenmann. John H. Ostrom kindly sent me manuscripts of papers that were in press. More than a decade ago, Oliver L. Austin, Jr., and Charles H. Rogers submitted suggestions for changes in the first edition. The comments and suggestions of all of these people have been very helpful to me, but I am solely responsible for errors in the text.

I am indebted to Sally Oshiro for her care in typing the manuscript, to Susan Monden for preparing illustrations, and to Diana M. Berger and Virginia B. Cleary for many hours of proofreading.

ANDREW J. BERGER

Honolulu, Hawaii
May 1975

From the Preface
to the First Edition

Josselyn Van Tyne asked me to promise to finish his textbook of ornithology in the event that he was unable to do so. I promised, and he made appropriate arrangements with the publisher to fulfill the contract in the event of his premature death.

I had read drafts of chapters written by Van; but, although it may appear inexplicable except to those who knew him well, we did not discuss the text in any detail after his operation. I am sure that during the summer of 1956 neither of us would admit that he might not live to see the book completed. During the autumn months he became too weak to do original work, and, perhaps for entirely different reasons, we rarely mentioned the book. It was not until mid-January that we made plans to meet at his office the following Tuesday in order to discuss the book again. The discussion did not take place, for Van did not return to the museum. He died on January 30, 1957. The text materials were turned over to me in March.

Van Tyne had written the chapters on Plumage and Molt, Migration, Bird Distribution, Ornithological Sources, and the long chapter on The Classification of World Birds by Families, which, in a sense (and for good reason), he considered *the book*. In addition, he had prepared manuscript pages for about one-half of two other chapters (Anatomy, Food and Feeding Habits) and he had written five pages on territory and eggs for the chapter on Breeding Behavior. I completed these and wrote the remaining chapters. For certain chapters (Paleontology, Voice and Sound Production, Food and Feeding Habits, Taxonomy and Nomenclature), there were either chapter outlines or lecture notes in the files, but for other chapters (Senses and Behavior, Flight, Social Relations) I found nothing in the files turned over to me except bibliography cards. Van Tyne had prepared on 3 by 5 cards about one-half of the definitions in the Glossary and had indicated many other terms that he intended to include. The Table of Contents (i.e., chapter topics and sequence) was planned by Van Tyne.

Van Tyne felt strongly that a student should become well grounded in the several basic branches of biology before specializing in ornithology. Birds of the World, as he taught it, was a graduate course intended for those who had at least an undergraduate degree in zoology. Van wrote that this book "should provide the background" for his graduate course. He anticipated that the book would fill many other needs: as a quick reference for information on all the families of birds; as a dictionary of ornithological terms; as a general reference for those interested in life history, taxonomy, and anatomy; as a summary of anatomical characters used in the classification of birds; and as a guide to ornithological literature.

The increase in knowledge during the past 50 years has been so great, both in ornithology and in biology in general, that most of the chapters included here might easily be expanded to book length, and a half dozen other subjects pertaining to birds might well receive the same treatment. When Van Tyne first conceived the outline for the book in 1946, he envisioned 21 chapters plus a glossary. It soon became evident, however, that the cost of publishing such a book would be prohibitive. Thus, instead of receiving full chapter treatment, some important subjects (e.g., physiology, genetics, ecology, study methods) had to be mentioned very briefly where pertinent in several chapters, and still other subjects (conservation, game management, population dynamics, museum techniques) not at all. Despite these necessary and intentional omissions, I believe that the book covers the fundamentals of bird study. Any graduate student worth his salt must gain a familiarity with the research literature and must acquire the habit of referring to it for additional information. Consequently, considerable emphasis has been placed on the list of references at the end of each chapter. Included in these lists are early "classic" works and a sampling of later papers that present different points of view. To be only partly facetious I may say that the Glossary should enable the student to translate the first paragraph of any of Robert Ridgway's "technical diagnoses."

A typical example of Van Tyne's attention to detail is seen in a folder labeled "Text: acknowledgment of assistance," which contained the names of those whom he planned to thank for assistance in various ways (reading manuscript, checking or loaning specimens, granting permission to use published illustrations, sending information on the habits of foreign birds, etc.): Dean Amadon, Paul H. Baldwin, William Beebe, Biswamoy Biswas, James P. Chapin, W. Powell Cottrille, Lee S. Crandall, David E. Harrower, Robert M. Mengel, Alden H. Miller, William H. Partridge, Frank A. Pitelka, Austin L. Rand, Robert W. Storer, Frederick H. Test, Alexander Wetmore. Had he been able to complete the book, Van Tyne would have expressed his deep appreciation to G. Reeves Butchart, a gifted editor who served as his highly valued editorial assistant for many years and until the very day of his death.

I wish to express my gratitude to the following, each of whom read one or more of the chapters that I wrote in part or entirely: John T. Emlen, Jr., Herbert Friedmann, Claude W. Hibbard, Hans P. Liepmann, Alfred M. Lucas, William A. Lunk, Harold Mayfield, Margaret M. Nice, George M. Sutton, Harrison B. Tordoff, Alexander Wetmore. In thanking these generous people, I do not in any way wish to imply that they agree with what I have said or with my method of presentation. One may always welcome suggestions, but, after study, decide not to follow them. In some instances I did just that. I wish to thank the several people who granted permission to use photographs or published illustrations; their names appear with their illustrations in the text. Special thanks are due George M. Sutton for his inimitable pen and ink drawings of world birds, and William L. Brudon and William A. Lunk for the remaining original illustrations. For many weeks of indispensable proofreading I am indebted to my wife and to Harriet Bergtold Woolfenden. To the McGregor Fund of Detroit, Michigan, I am grateful for a grant which enabled me to devote full time to indexing and publication details during the summer of 1958.

Josselyn Van Tyne had few peers among ornithologists. He was both a great ornithologist—in the modern sense—and a true scholar—in the oldest and best sense. For any ways in which I may have failed him, I offer my profound apologies to the spirit of Josselyn Van Tyne.

ANDREW J. BERGER

Ann Arbor, Michigan
March 1958

Artist's Preface

When I tackled the pen-and-ink drawings for this important work, I knew I would have trouble, for I had never seen alive, even in zoos, several of the birds that I would be obliged to illustrate. Confident that specimens, photographs, and field sketches would help, I went ahead despite my long-standing disapproval of copying, a feeling based partly on ethics and partly on the realization that photographs may distort badly. Occasionally my copying was restricted to one part of the bird figure—for example, the head of the Spotted Hemipode-quail (*Turnix ocellata*) on p. 643. Here I used a photograph by K. A. Hindwood that appeared in 1937 in *Emu* (Vol. 37, pl. 8). Occasionally I used a photograph considered by the photographer not good enough for reproduction as a halftone. An example is the Whimbrel (*Numenius phaeopus*); the photograph by Olin Sewall Pettingill, Jr., caught the fine bird in full stride, and I consider my copy one of the best drawings of the lot.

Other drawings that must be called out-and-out copies are the following: Owlet-frogmouth (*Aegotheles insignis*), based on O. Webb photo in *Emu*, Vol. 27, 1927, pl. 15; Indian Crested-swift (*Hemiprocne coronata*), based on E. H. N. Lowther photo in *J. Bombay Nat. Hist. Soc.*, Vol. 39, 1936; Cuban Tody (*Todus multicolor*), based on L. H. Walkinshaw photo in *Wilson Bull.*, Vol. 58, 1946; Scimitar-bill (*Rhinopomastus cyanomelas*), based on W. Hoesch photo in *Ornithol. Monatsber.*, Vol. 41, 1933, p. 34; Greater Honeyguide (*Indicator indicator*), based on J. P. Chapin photo in "Birds of the Belgian Congo," Part 2, (*Am. Mus. Nat. Hist. Bull.*, Vol. 75), 1939, pl. 21; Lyrebird (*Menura superba*), based on J. E. Ward photo in *Bull. N.Y. Zool. Soc.*, Vol. 42, p. 71, and on photo in *Emu*, Vol. 46, 1946–1947; Great Cuckoo-shrike (*Graiculus macei*), based on E. H. N. Lowther photo in *J. Bombay Nat. Hist. Soc.*, Vol. 41, 1939–1940; Orange-winged Treerunner (*Neositta chrysoptera*), based on P. A. Gilbert photo in *Emu*, Vol. 22, 1922; Paradise Flycatcher (*Terpsiphone paradisi*), based on E. H. N. Lowther photo in *J. Bombay Nat. Hist. Soc.*, Vol. 41, 1939–1940; White-browed Wood-swallow (*Artamus superciliosus*), based on A. J. Gwynne photo in *Emu*, Vol. 31, 1931; Paradise Whydah (*Steganura paradisea*), based on K. Plath photo in *Aviculture*, Vol. 2, 1930, p. 22.

The drawing of the Mexican Trogon (*Trogon mexicanus*) was based largely on a photograph taken by Olin Sewall Pettingill, Jr., but the three front toes of the drawing as it appeared in the first edition (1959) cannot be blamed on the photo. That mistake was purely my own. Before a new cut was made for this edition, I erased the feet, replacing the three front toes with two.

Before finishing the drawing of the Variegated Laughing-thrush (*Garrulax variegatus*), I watched certain living timaliids in zoos, consulted photographs of species believed to be closely related, and proceeded with a good specimen in hand.

For the following drawings I depended on photographs for certain important details: Fulmar (*Fulmarus glacialis*), photo by F. Darling in *Wild Country*, 1938, p. 11; Peruvian Diving-petrel (*Pelecanoides garnotii*), photo by L. E. Richdale in *Emu*, Vol. 43, 1943, opp. p. 44; Mallee Fowl (*Leipoa ocellata*), photo by E. Lewis in *Emu*, Vol. 40, 1940, pl. 22; Vulturine Guineafowl (*Acryllium vulturinum*), photo by H. Spang in *Aviculture*, May–June, 1941; Collared-hemipode (*Pedionomus torquatus*), photo by D. Le Souef in *Emu*, Vol. 15, 1916; African Finfoot (*Podica senegalensis*), photo by J. P. Chapin in "Birds of the Belgian Congo," Part 2, 1939, pl. VI; Painted-snipe (*Rostratula benghalensis*), photo in *J. Bombay Nat. Hist. Soc.*, Vol. 46, 1946–1947; Black Guillemot (*Cepphus grylle*), photo by A. Brook in Tate Regan's *Natural History*; Tibetan Sandgrouse (*Syrrhaptes tibetanus*), photo in *J. Bombay Nat. Hist. Soc.*, Vol. 46, 1946–1947; Wood Pigeon (*Columba palumbus*), photo by C. Reid in Tate Regan's *Natural History*; Pennant-winged Nightjar (*Semeiophorus vexillarius*), photo of dead bird by J. P. Chapin in "Birds of the Belgian Congo," Part 2, 1939, pl. XVI; Indian Roller (*Coracias benghalensis*), photo by O. and M. Heinroth in *Die Vögel Mitteleuropas*, Vol. 1, 1924, pl. 99; New Guinea Forest Butcherbird (*Cracticus cassicus*), photo by L. G. Chandler, *Emu*, Vol. 23, 1924; Magpie-lark (*Grallina picata*), photo by N. Chaffer, *Emu*, Vol. 29, 1930, pl. 26; Satin Bowerbird (*Ptilonorhynchus violaceus*), photo by J. Rawson, *Emu*, Vol. 15; Orange-bellied Leafbird (*Chloropsis hardwickii*), photo by E. H. N. Lowther, *J. Bombay Nat. Hist. Soc.*, Vol. 41, 1939–1940; White-breasted Dipper (*C. cinclus*), photo by O. and M. Heinroth, *Die Vögel Mitteleuropas*, Vol. 1, 1924, pl. X; Song Thrush (*Turdus ericetorum*), photo by G. S. C. Ingram in Tate Regan's *Natural History*; Blackcap (*Sylvia atricapilla*), photo by E. J. Hosking in Seton Gordon's *Wild Birds in Britain*, p. 14; White Wagtail (*Motacilla alba*), photo by A. H. Wilford in Tate Regan's *Natural History*; Rose-colored Starling (*Sturnus roseus*), photo by O. and M. Heinroth in *Die Vögel Mitteleuropas*, Vol. 1, 1924; Fire-breasted Flowerpecker (*Dicaeum ignipectum*), photo by S. A. Lawrence in *Emu*, Vol. 15, 1915; and Gray-breasted White-eye (*Zosterops lateralis*), photo by Mrs. A. S. Wilkinson in *Emu*, Vol. 31, 1931, p. 158.

A few of my delineations come close to being pure imagination, notably the drawing of the Brown Mesite (*Mesoenas unicolor*). This time I thought that certain persons with field experience in Madagascar might be able to help me, so, having made pencil sketches showing three different positions, I asked my friends Austin L. Rand and Jean Delacour to tell me frankly which of the three they considered best. Imagine my bewilderment on being told that any of them would do.

Several drawings may be poor to very poor, since I had nothing to work with save scientific skins and drawings by various artists, most of whom probably knew no more about the living birds than I did. These drawings are of the Crab-plover (*Dromas ardeola*), Cuckoo-roller (*Leptosomus discolor*), Crested Sharpbill (*Oxyruncus cristatus*), Velvet Asity (*Philepitta castanea*), Huia (*Heteralocha acutirostris*), and Iiwi (*Vestiaria coccinea*).

One drawing—that of the Chestnut-breasted Turco (*Pteroptochos castaneus*)—I based on a halftone illustration for an article by Frank M. Chapman: a drawing of "El Turco" by the great Louis Agassiz Fuertes (*Bird-Lore*, Vol. 21, 1919, p. 337).

Several of the drawings required from two to five attempts. The real *tour de force* of the series is the drawing of the Vulturine Guineafowl. Here the worst problem was the white spots, which are so geometrically arranged and so tiny that I had to work out a technique that would endow my original with beauty and authenticity and ensure clarity of reproduction when reduced. Even today, my eyes have a tendency to "swim around a bit" when I contemplate those dozens of tiny spots. The original drawing measures about 6.5 inches, from tip of bill diagonally to tip of tail.

GEORGE MIKSCH SUTTON

Norman, Oklahoma
May 1975

Contents

Fundamentals
of Ornithology

Temperature Conversion Table Centigrade to Fahrenheit

°C	°F	°C	°F
0	32	33	91.4
1	33.8	34	93.2
2	35.6	35	95.0
3	37.4	36	96.8
4	39.2	37	98.6
5	41.0	38	100.4
6	42.8	39	102.2
7	44.6	40	104.0
8	46.4	41	105.8
9	48.2	42	107.6
10	50.0	43	109.4
11	51.8	44	111.2
12	53.6	45	113.0
13	55.4	46	114.8
14	57.2	47	116.6
15	59.0	48	118.4
16	60.8	49	120.2
17	62.6	50	122.0
18	64.4		
19	66.2		

Interpolation Factors

°C	°F
20	68.0		
21	69.8	1	1.8
22	71.6	2	3.6
23	73.4	3	5.4
24	75.2	4	7.2
25	77.0	5	9.0
26	78.8	6	10.8
27	80.6	7	12.6
28	82.4	8	14.4
29	84.2	9	16.2
30	86.0	10	18.0
31	87.8		
32	89.6		

CHAPTER ONE

Paleontology

In order to appreciate fully the wide diversity in structure and habit of the world's birds today, one must be aware of what is known about the origin and early history of birds. It is unfortunate that most of the evidence of prehistoric life consists of fossilized bones, which hold little interest for some students. Nevertheless, a general knowledge of the main features of the bird skeleton is essential for understanding some of the differences between modern and ancient birds, as well as the similarities between the skeletons of ancient birds and those of their reptilian ancestors. In fact, the bird skeleton shows so many reptilian characters that T. H. Huxley characterized birds as "glorified reptiles." For example, birds and reptiles have a single occipital condyle and a single ear bone (columella), and most birds have uncinate processes on the ribs—a character found elsewhere only in certain reptiles (e.g., *Sphenodon, Euparkeria*). Some other characters shared by the two groups are nucleated red blood corpuscles, an egg-tooth on the upper jaw at hatching, an ambiens muscle, and eggs with large amounts of yolk.

An obvious feature which distinguishes birds from reptiles is the presence of feathers. Birds also are warm-blooded, and have only a right aortic arch (in contrast to the left aortic arch of mammals). Only birds and mammals have a true four-chambered heart.

1

THE ORIGIN OF BIRDS

Many predentate dinosaurs (order Ornithischia) had a pelvis superficially similar to that found in birds, and a few early writers believed that birds were descended from "birdlike" dinosaurs. However, these dinosaurs occurred too late (mostly in the Cretaceous) to fit into the ancestral line of birds, which are first known from the late Jurassic (Table 1).

Most authors state that both birds and dinosaurs are descended from primi-

Table 1.[a] **Distribution of Mesozoic Genera of Birds**

Stage	Inception[b]	Bird-bearing Formations and Genera
Upper Cretaceous		
Maestrichtian	72	Quiriquina beds: *Neogaeornis* Hornerstown marl: *Graculavus, Laornis, Telmatornis, Palaeotringa* Lance: *Lonchodytes, Torotix, Cimolopteryx, Ceramornis, Palintropus*
Campanian	81	Shell fragment limestone: *Parascaniornis* Claggett: *Coniornis* [Belly River: *Caenognathus*, avian?]
Santonian	84	Selma chalk: *Plegadornis*
Coniacian	88	Smoky Hill chalk: *Hesperornis, Baptornis, Apatornis, Ichthyornis*
Turonian	90	None
Cenomanian	110	[Dakota sandstone: tracks only, avian?]
Lower Cretaceous		
Albian	120	Upper Greensand: *Enaliornis*
Aptian	123 ±	None
Neocomian	135	Koonwarra claystones: feathers only Auxerre beds: *Gallornis*
Upper Jurassic		
Portlandian	138	Solnhofen limestone: *Archaeopteryx*
Kimmeridgian	148 ±	[Morrison: *Laopteryx*, avian?]
Oxfordian	135 ±	None
Callovian	158 ±	None

[a] Reproduced by permission of Pierce Brodkorb, from *Avian Biology*, Vol. 1, Academic Press, New York, 1971.
[b] Age at beginning of stage, in millions of years. The upper boundary of the Cretaceous is currently dated at 63 million years before the present.

tive, unspecialized thecodont reptiles belonging to the suborder Pseudosuchia (order Thecodontia). These were small carnivorous animals possessing many teeth set in sockets (i.e., thecodont) in the jaws. Their hindlimbs were longer than their front limbs, and in some forms, at least, the fifth hindtoe was reduced in length. However, no pseudosuchian yet described could well be *the* ancestor of the birds. Heilmann (1927) reconstructed a hypothetical "proavian" form intermediate between certain pseudosuchians (*Ornithosuchus* and *Euparkeria*) and the most primitive bird known.

Ostrom (1973), however, proposed that "*Archaeopteryx* must have been derived from an early or mid-Jurassic theropod" (a coelurosaur: Order Saurischia, suborder Theropoda). He pointed out that the earliest known fossils recognized as birds have only two avian osteological features (a furcula and possibly the position of the pubic bone). By contrast, several skeletal characters are shared by *Archaeopteryx* and coelurosaurs, and several of the latter possessed clavicles. Ostrom also commented that the absence of a clavicle has no phyletic significance because of its dermal origin; "it may well have been membranous (but not lost) in most theropods and thus not preservable." Ostrom concluded, therefore, that "the most likely origin of so many coelurosaurian features in *Archaeopteryx* is by direct inheritance from a small coelurosaurian ancestor."

Other kinds of clues to justify speculation on the origin of birds appear to be scarce. In his excellent study of fossil and recent bone tissue, Enlow (1956–1958), for example, emphasized that "in the history of bone tissue, a single 'evolutionary line' cannot be recognized. It is not possible to trace a precise series of progressive, increasingly complex developmental stages from extinct fish to modern mammal. It cannot be maintained that Haversian tissue is the evolutionary culmination of all bone development, or that the ontogeny of mammalian bone tissue recapitulates the phylogeny of the bone of lower vertebrate groups." Figure 1 shows the dense Haversian system (characteristic of human bone) in the rib of *Triceratops*, one of the horned dinosaurs of the Cretaceous period; Fig. 2 shows the rib of *Pteranodon* (the largest of the flying reptiles, with a wingspan of about 25 feet) and that of *Pinguinus impennis*, the recently exterminated Great Auk.

The Origin of Flight

One of the main points in the analysis of the immediate bird ancestor is the question of the origin of flight. Three main theories have been proposed: arboreal, cursorial, and predatory cursorial.

Marsh (1880), Osborn (1900), and Heilmann (1927) believed that the first birds were arboreal in habit. Their terrestrial ancestors became tree climbers before there was a great disparity in the relative lengths of the front and back

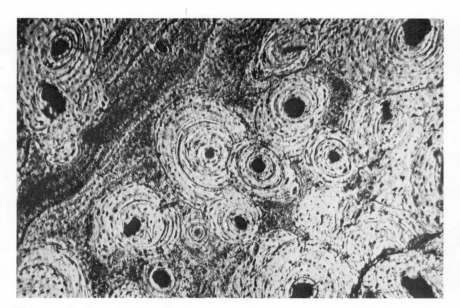

Figure 1. A thin section of the rib of *Triceratops*. (Courtesy of Donald H. Enlow and the editor of the *Texas Journal of Science*.)

limbs, although the bipedal gait had already resulted in some elongation of the metatarsals. Jumping from branch to branch favored the further elongation of the metatarsals and the development of a backward-directed hallux, which enabled the animals to secure a good grasp of branches. Now used for climbing, the forelimbs "preserved claws on their digits, remained large, and were not reduced as is commonly the case in cursorial animals which adopt the bipedal mode of progression" (De Beer, 1954). Each limb, therefore, became adapted to specialized and different functions. Böker (1927) suggested that the first birds flapped their forelimbs when jumping from branch to branch, but De Beer (1954) thought it probable "that simple gliding preceded flapping because *Archaeopteryx* had no carina and its pectoral muscles must have been feeble."

The fact seems to be, however, that we still do not know for certain that a carina was absent from the sternum of *Archaeopteryx*. Thus Parkes (1966) commented that "De Beer's drawing suggests to me the ossified base of a partly cartilaginous sternum, and, indeed, this possibility De Beer himself (1954: 41) briefly mentions. That birds with largely cartilaginous sterna are capable of adequate flapping flight is well illustrated by the precociously flying chicks of modern gallinaceous birds. I regard the sternal structure of *Archaeopteryx*, therefore, as an open question and would not use it as evidence

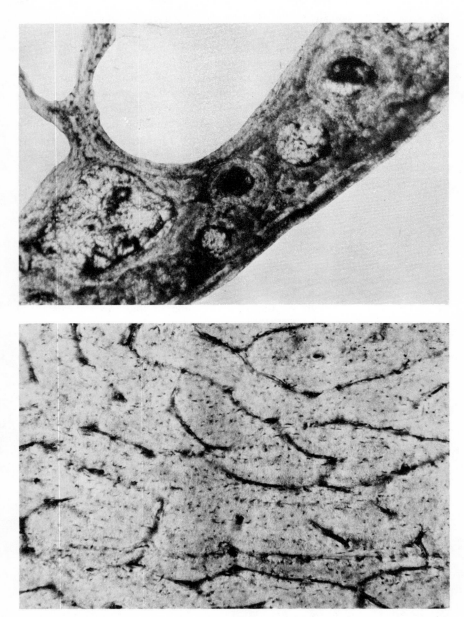

Figure 2. Bottom: Thin section of a rib of *Pteranodon*. Top: Thin section of a rib of *Pinguinus impennis*. (Courtesy of Donald H. Enlow and the editor of the *Texas Journal of Science*.)

either for or against the ability to fly." (Fig. 3). Moreover, after considering characters other than the carina, Yalden (1971) concluded that *Archaeopteryx* was capable of flapping flight, but that it was probably not long sustained. Heptonstall (1970) wrote that the "coracoids, humeri and sternum of *Archaeopteryx* indicate that it was a poor flier."

Beebe (1915) believed that flight in *Archaeopteryx* and its predecessors was limited to gliding and scaling, and he advocated a "Tetrapteryx" stage in the evolution of flapping flight. Beebe's hypothetical pre-Jurassic Tetrapteryx bird possessed not only an alar feather tract but also a "pelvic wing," both of which served merely as passive parachutes. Beebe thought that he found evidence in the embryos and nestlings of some modern birds of a recapitulation of such a four-winged stage. De Beer believed that Beebe placed too much emphasis on his "pelvic wing," that the "tibial quills" of *Archaeopteryx* actually were contour feathers, and that the feathers in question represented no more than the femoral feather tract in modern birds.

A cursorial origin of flight was postulated by Nopcsa (1907), who envisioned the development of flight in long-tailed, bipedal reptiles which flapped their forelimbs as they ran rapidly along the ground. The scales on the forelimb became elongated, and their posterior margins frayed out, in time evolving into feathers. Nopsca imagined three stages in the evolution of flight: (1) parachute or passive flight, (2) flight by flapping the wings or flight by force, and (3) soaring or flight by skill.

Ostrom (1974) contended that *Archaeopteryx* "was a very active, fleet-footed, bipedal, cursorial predator in which the hands, arms and pectoral arch were primarily adapted for seizing and holding small prey, as almost certainly was the situation in *Ornitholestes, Velociraptor, Deinonychus* and other small theropods." He added: "From the size of the head and mouth, and the shape of the teeth in *Archaeopteryx*, we may reason that the usual prey of *Archaeopteryx* consisted of relatively small animals, most probably large insects, and perhaps small lizards and mammals. Enlargement of the primitive contour feathers on the forelimb could have increased the efficiency of the forelimb as a prey-catching appendage by gradually converting it into a large snare, both arms becoming a trapping device or 'net' with which to corral or surround small prey so they could be more easily grasped in the mouth or hand." Ostrom believes that *Archaeopteryx* developed a metabolic rate higher than that of modern reptiles (i.e., was homoiothermic) and that feathers already existed in the reptilian ancestors, which is to say that feathers arose independently of flight, presumably in relation to temperature regulation. He discounts the significance of the backward-directed hallux as proof of arboreal habits for *Archaeopteryx*, stating that "the fact that it apparently was present in all carnivorous theropods, but never existed in equally bipedal but herbivorous ornithopods, indicates that the initial function of the reversed hallux was probably diet-related."

Figure 3. A new reconstruction of *Archaeopteryx* by Rudolf Freund. (Reproduced by courtesy of Kenneth C. Parkes, the Carnegie Museum, and the editor of *Living Bird,* 1966.)

THE FOSSIL RECORD OF BIRDS

The massive bones, dense plates, horn-cores, teeth, and spines of many vertebrates were admirably suited for preservation through the ages, assuming the requisite conditions for fossilization. By contrast, bird bones are fragile and many are hollow, and thus easily broken and fragmented. One may assume also that many ancient birds, like those of today, served as prey for carnivorous animals, whose digestive processes left little for preservation. In any event, relatively few complete fossilized skeletons of birds have been discovered. Numerous fragments of bones, however, have been found, usually the larger and denser ends of wing and leg bones or parts of the pectoral girdle; and many genera have been established on the basis of such fragments or on single bones. Several authors have discussed the problems involved in past attempts to relate inadequate fossil material to modern forms. Howard (1950:5), for example, pointed out that "taxonomy among fossil birds cannot be entirely coordinated with that of modern forms," and that the assignment of single or dissociated bones "to modern genera is to some extent a matter of convenience."

It is true that the fossil record of birds has been of little help in determining the evolutionary steps or the relationships of modern birds. "Nevertheless, paleornithology is still one of the least explored areas of avian biology, and hence it is one in which great advances may be made. It is still largely in the alpha stage of taxonomy, and the workers are few. Four authors are responsible for the description of one third, seven authors for one half, and fifteen authors for three fourths of the known 900-odd paleospecies. The reason for the imperfect state of our knowledge is not the alleged dearth of fossil birds—it lies in the scarcity of paleornithologists to study them" (Brodkorb, 1971).

Although comparatively few complete fossil bird skeletons have been discovered, some nearly complete skeletons have been found; and it probably is significant that, after careful analysis, the authors who studied them concluded that their specimens represented now-extinct families of birds. Wetmore (1925) examined a specimen of *Palaeospiza bella*, which had been described in 1878, and concluded that it belonged to a distinct family (Palaeospizidae), whose characters suggested that it "occupies a somewhat connecting position between the Mesomyodi and the acutiplantar Oscines." Howard (1957b) described a new bird, from the Miocene of California, which does not fit clearly into any family of living oscine birds. She named the specimen *Palaeoscinis turdirostris*, "an ancient oscine bird with a thrushlike beak" (Fig. 4).

Although Osborn (1900) and others believed that the avian ancestors lived during the Permian period of the Palaeozoic era or in the Triassic period of the Mesozoic era, the actual fossil record of birds begins millions of years later, in the Jurassic period. At least 902 extinct bird species are known, and "about

Figure 4. *Palaeoscinis turdirostris,* "an ancient oscine bird with a thrushlike beak," shown in type slab No. 1. (Courtesy of Hildegarde Howard and the Los Angeles County Museum.)

one tenth (857) of the 8656 neospecies of birds have also been reported as fossils. Thus the avian fossil record now includes some 1760 species" (Brodkorb, 1971). Some of the best-known forms are discussed in the following pages. The taxonomy of all fossil birds is given in Brodkorb's *Catalogue of Fossil Birds* (1963–1971).

The Mesozoic Era

1. Jurassic Period. Evidences of the oldest known bird were found in the Solnhofen or lithographic stone of Upper Jurassic age and are estimated to be about 138,000,000 years old. As of 1972, an isolated feather and portions of four skeletons had been discovered. *Archaeopteryx lithographica* was originally based on a feather, but by general consent the name is now applied to a skeleton found in 1861 in lithographic slate near Pappenheim, Bavaria; this specimen, which was described as a reptile and named *Griphosaurus prob-*

Figure 5. The 1877 specimen of *Archaeopteryx*. (Courtesy of Dr. Hermann Jaeger and the Museum für Naturkunde, Berlin.)

lematicus (Brodkorb, 1963), is now in the British Museum. A second skeleton (now in the Museum für Naturkunde in Berlin; Fig. 5) was found in 1877 about 10 miles distant from the site of the first one; this was named *Archaeornis siemensi*. A third specimen was discovered in 1958 in the same quarry that yielded the 1861 skeleton, but at a higher level in the stratum (Heller, 1959). In 1970 Ostrom (1970) found a specimen in the Teyler

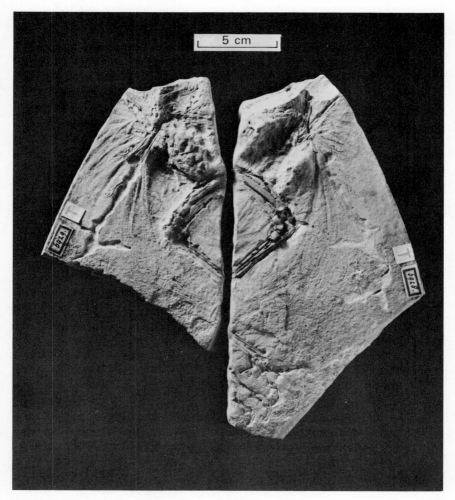

Figure 6. The counterpart slabs showing the Teyler Museum specimen of *Archaeopteryx*. (Courtesy of John H. Ostrom and the editor of *Science*; Vol. 170, October 30, 1970, pp. 537–538; copyright 1970 by the American Association for the Advancement of Science.)

Museum, Haarlem, Netherlands, that had been collected about 32 miles east of Pappenheim before 1857; at that time, it was described as a new species of pterosaur. Ostrom, however, recognized significant differences in skeletal elements and also observed faint impressions of wing feathers (Fig. 6). Of special interest were horny claws (not preserved with the other skeletons) on digits I and III of the wing (Fig. 7).

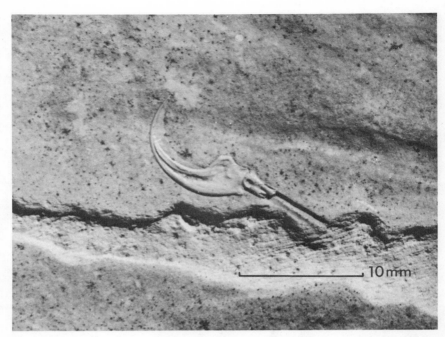

Figure 7. A greatly enlarged photograph of the terminal phalanx of digit III and its horny sheath from the left hand of the Teyler Museum specimen of *Archaeopteryx*; the scale is 10 mm long. (Courtesy of John H. Ostrom and the editor of *Science*; Vol. 170, October 30, 1970, pp. 537–538; copyright 1970 by the American Association for the Advancement of Science.)

Brodkorb (1971) wrote that more names have been proposed for the isolated feather and the skeletons of *Archaeopteryx* than there are specimens. According to Swinton (1960), "it is now clear that all these specimens are of the same genus and species, *Archaeopteryx lithographica* von Meyer, and that all of the specimens are of the same geological age, namely Kimmeridgian of the Upper Jurassic."

The skeleton of *Archaeopteryx* is more reptilian than avian; in fact, Lowe (1944) denied that *Archaeopteryx* was a bird. It was, Lowe suggested, an arboreal, climbing dinosaur that "takes its place not at the bottom of the avian phylum but at the top of the reptilian." Despite Lowe's assertions, however, the avian affinities of *Archaeopteryx* are now generally accepted. Its chief anatomical features may be summarized as follows.

The orbit is large, and a sclerotic ring of plates (Fig. 8) is present, as in modern birds and in certain early reptiles (e.g., pterosaurs). The palate has been described as schizognathus, but Simonetta (1960) said that nothing is left of the palate in the London specimen "and in the Berlin one it is so completely crushed as to make any attempt at identifying bones hopeless!" Both jaws possess teeth, which are set in sockets, and the mandible has a retroarticular process.

The vertebral column is composed of 49 or 50 vertebrae. There are 10 cervical vertebrae, 6 of which bear cervical ribs. Of the 19 or 20 trunk vertebrae, the caudal 5 are fused together to form a primitive synsacrum. Eleven pairs of "nonavian" ribs are unjointed, lack uncinate processes, and do not articulate with the sternum. A series of bony, dermal ribs (*gastralia*) has been preserved; such ribs were common in the ventral abdominal wall of early reptiles. There are 20 free caudal vertebrae; a pygostyle is not present. The vertebrae of *Archaeopteryx* have disclike or biconcave (*amphicoelous*) articular facets, as did those of many primitive reptiles. In modern birds the articular surfaces of the vertebral bodies are primarily saddle-shaped (*heterocoelous*).

The wing and leg are about equal in length. The hand contains three separate metacarpal bones. De Beer (1954) reported that there is a fusion of the third metacarpal with the carpal ossification to form a third carpometacarpal bone, but a carpometacarpus is not present. The two clavicles are fused to form a furcula. Whether or not the sternum had a carina is unknown.

The bones of the pelvic girdle (*ilium, ischium, pubis*) are separate elements, rather than being fused into a single pelvic bone. The two pubic bones are

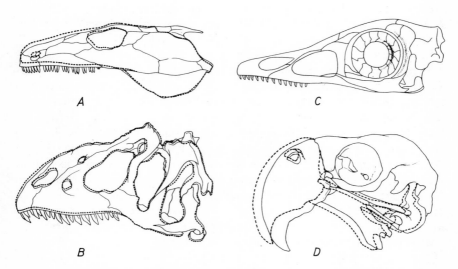

Figure 8. Diagrammatic drawings of the skulls of (*A*) a primitive labyrinthodont; (*B*) of *Antrodemus,* a Mesozoic carnosaur; (*C*) of *Archaeopteryx*; and (*D*) of a parrot to show kinesis. (Courtesy of Alberto M. Simonetta and the editor of *The Quarterly Review of Biology*.)

directed caudad, and they fuse to form a *pubic symphysis* (as in the pseudosu-chians and the Ostrich). The fibula is as long as the tibia, and the metatarsals are not fused to form a tarsometatarsus (i.e., there is a true heel joint). The short hallux is directed backward.

There seems to be little doubt that *Archaeopteryx* had a reptilelike brain, with some birdlike features. The cerebellum did not extend anteriorly to overlie the posterior part of the cerebral hemispheres, and the optic lobes were located dorsally (p. 51, Edinger, 1949).

The most important avian feature of the anatomy of *Archaeopteryx*, however, is the presence of feathers. Their structure is said to be typical of that of the feathers of modern birds. Authors have differed in their interpretation of the number of flight feathers, presumably because some of the feathers were double-struck during the process of fossilization. De Beer (1954) concluded that *Archaeopteryx* probably had six primaries and ten secondaries, and that the wing was eutaxic (p. 131), contrary to Steiner's (1918) theory that the wing was diastataxic. Swinton (1960:2) wrote that "there are six primaries and ten secondaries," although in the legend for his figure 4 he said: "Even in the earliest bird a satisfactory wing of nine primaries and fourteen secondaries exists"; at the 13th International Ornithological Congress in 1962, he stated that he then believed there were eight primaries. Saville (1957) also concluded that *Archaeopteryx* had eight primaries. There are similar conflicting state-ments in the literature concerning the number of tail feathers (rectrices), but De Beer said that there were 15 pairs "attached by ligaments to the lateral surface of their vertebrae." The first pair of rectrices was attached to the sixth free caudal vertebra. Contour feathers apparently were well developed.

In systems of classification, *Archaeopteryx* is placed in the subclass Ar-chaeornithes, meaning the ancestral birds; all other birds, in the subclass Neornithes, or true birds (Wetmore, 1951a). Most of the anatomical differences between *Archaeopteryx* and the true birds are associated with centralization of body weight and improvement of the basic structure for flight. Despite the many differences between *Archaeopteryx* and all other birds, however, it is dif-ficult to find many anatomical characters that hold without exception for the Neornithes themselves. This is true largely because of the extreme specializa-tion of the ratites. The true birds, however, have the tail much shortened, and the rectrices are arranged fanlike around the terminal caudal vertebrae, which typically, but not invariably, are fused to form a pygostyle. Uncinate processes are found on the ribs of all true birds except the Anhimidae. The bones of the wrist and hand (when present) are fused to form a carpometacarpus, the bones of the pelvic girdle are at least partially fused anteriorly, and a fused tarsometatarsus is present.

2. Cretaceous Period.   Nearly complete bird skeletons have been found in the marine Niobrara Cretaceous chalk beds of Kansas (Fig. 9). First described by

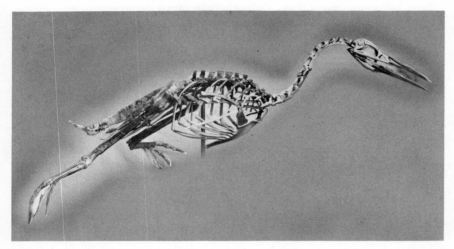

Figure 9. Skeleton of *Hesperornis regalis* from the Upper Cretaceous of Kansas. (Courtesy of Alexander Wetmore and the Smithsonian Institution.)

Marsh (1880 and earlier), these birds have long been spoken of as the Cretaceous "toothed birds" (Odontornithes; superorder Odontognathae in current systems of classification). Two orders (Hesperornithiformes and Ichthyornithiformes) and five families have been recognized.[1] Gregory (1952) studied the toothed jaw ascribed to *Ichthyornis* and concluded that it belonged to a small mosasaur (an aquatic lizard) and not to *Ichthyornis*; only fragments of an upper jaw had been described by Marsh, and these are "quite indeterminate as to what bone was represented as well as what animal." Hence it is not yet known whether the presence of teeth was a universal character of these Cretaceous birds; except for *Hesperornis*, complete jaws have not been found for the other genera placed in the superorder Odontognathae.

Hesperornis ("western bird") was a large (nearly 5 feet long), flightless, loonlike bird, highly specialized for swimming; *Ichthyornis*, a small, gull-like bird specialized for flying (Brodkorb said that the skeleton "shows considerable similarity to that of the medium-sized alcids such as *Cerorhinca monocerata* or *Lunda cirrhata*"). The striking divergence in the osteology of these two early birds reveals the wide radiation of birds during the interval (of unknown length) between the Jurassic and the Cretaceous periods, and also suggests a long history of development for these Cretaceous birds. Some differences between the two genera are given in Table 2. Many of the differences in structure between *Hesperornis* and *Ichthyornis* are obviously correlated with their

[1] Marsh erected the superorder Odontolcae for *Hesperornis*, in which the teeth were set in grooves in the jaws (the premaxillary bone of the upper jaw, however, was edentulous); in the superorder Odontotormae, established for *Ichthyornis*, the teeth were set in separate sockets.

**Table 2.   Some Osteological Differences between *Hesperornis* and
            *Ichthyornis***

Hesperornis	Ichthyornis
1. Clavicles are not fused.	1. Typical furcula is present.
2. Carina sterni is absent.	2. Carina is well developed.
3. Wing is composed of a vestigial humerus only.	3. Wing is well developed, including a carpometacarpus.
4. Vertebrae are heterocoelous.	4. Vertebrae are amphicoelous.
5. Teeth are present in both jaws.	5. Jaws are unknown.
6. The mandibular rami are united by ligament, rather than by a bony symphysis.	6. Jaws are unknown.

locomotor patterns, whereas other differences (e.g., type of vertebrae) cannot be explained adequately with information now available.

The following characters are shared by the two genera: a typical avian tarsometatarsus, a single-headed quadrate, and a small skull whose bones were fused nearly as well as those of modern birds. According to Edinger (1951), all evidence indicates that *Hesperornis* and *Ichthyornis* had typical avian brains, contrary to the conclusions reached by Marsh. Pearson (1972) discussed more fully the scant information available on the brains of these early birds.

Even older than *Hesperornis* and *Ichthyornis* are two fossils from the early Cretaceous: *Gallornis* from France and *Enaliornis* from England, plus two recently discovered feathers from claystones near Koonwarra, Victoria, Australia (Brodkorb, 1971). A third feather was described by Waldman (1970).

Other birds from the late Cretaceous include *Plegadornis antecessor*, a very small ibis from Alabama; *Coniornis altus*, a hesperornithid from Montana; and other genera from Wyoming, New Jersey, Chile, and Sweden. Information on these and other Cretaceous genera (e.g., *Baptornis, Apatornis*) may be found in the works of Wetmore (1956), Romer (1966), and Brodkorb (1971).

The Cenozoic Era

This era of geological history began some 60,000,000 years ago. Most modern orders of birds were well differentiated by the Eocene epoch. Many of the most completely preserved fossils from Eocene to Recent times have been much publicized, largely because of their great size; a bird with a head as big as that of a horse justifiably stimulates a man's imagination. Such gigantic birds, however, represent the ends of evolutionary series. This fact, coupled with the great hiatuses in the fossil record, means that the structures of these and many other fossil birds reveal little concerning their relationships to modern orders. By and large, most of the early Cenozoic fossil forms serve mainly to emphasize the early radiation of birds and the paucity of information about that early evolutionary history.

Fossil evidence of prehistoric life is recorded in many ways. The Miocene Lompoc diatomitic shales of California are noteworthy for reasons described by Miller and DeMay (1942; Fig. 10): "The bone [in bird fossils] has turned dark brown and completely fallen to powder. This material is dusted out with a camel's-hair brush and there is left an imprint of the skeleton etched in sepia on a white background. Most of the specimens show several of the bones in their proper anatomical relation, and many show nearly the complete skeleton."

1. The Tertiary Period. Drastic changes in the land and in climates occurred during the millions of years that form the Tertiary period. A tropical or subtropical climate was characteristic of north temperate regions during the Eocene epoch. "All the great mountain ranges of the world were made or remade during the Tertiary. There was a general rise of land which eventually

Figure 10. Lompoc Diatomite being sawn and split into blocks for commercial purposes. Fossil birds, marine mammals, and fish skeletons occasionally were exposed on cleavage planes. (Courtesy of the late Loye Miller.)

lifted the continents to their unusual high Pleistocene and present levels. . . . From the Mesozoic well into the Tertiary, the Tethys Sea extended across the whole of southern Europe and Asia. Europe was probably at times reduced to an archipelago. . . . North America was fairly constant in shape during the Tertiary except that shallow seas overlapped parts of what are now the southeastern coastal plain and the lower Mississippi Valley. But the surface was changed. The present Appalachians were raised and dissected mostly during the Tertiary, and the mountains of western North America were completely modified. . . .

"The climate of the Tertiary, at least north of the tropics, began much warmer than now and gradually cooled until the Pleistocene. Early in the Tertiary, what are now temperate types of forest extended northward into the arctic, and tropical or subtropical forest types extended far into the north-temperate zone" (Darlington, 1957 : 588–589).

Diatryma (Fig. 11) and its allies were terrestrial birds that lived and died during the Paleocene and Eocene. Their bones have been found in the United States (Wyoming, New Mexico, New Jersey), France, and Britain. *Diatryma steini* from the Lower Eocene of Wyoming was almost 7 feet tall. It had a massive head, holorhinal nostrils, greatly reduced wings and uncinate processes, a small hallux, and a very large pelvis. Taxonomists place *Diatryma* in or near the Gruiformes, a heterogeneous order to which several other large extinct birds have been assigned (e.g., *Phorusrhacos, Brontornis, Bathornis*) (Fig. 12).

Among the several diurnal raptores (Falconiformes) recorded from Eocene time in Europe and America (see Howard, 1950), one of the most interesting is *Neocathartes grallator*, a terrestrial vulture from Wyoming, with long legs, well-developed toes, and reduced wings (Fig. 13). Although able to fly, it evidently was mainly terrestrial and "apparently stood in the same relation to the species of the Cathartidae that the Secretary-bird, *Sagittarius serpentarius*, does to the true hawks and eagles" (Wetmore, 1955).

Feduccia (1973) described the first fossil bird that clearly showed a zygodactyl foot arrangement. He named this fossil from the Eocene of Wyoming *Primobucco kisterni*, stating that it "is probably most closely related to the primitive perching piciform birds of the families Capitonidae and Bucconidae" (Fig. 14).

The relationship of the penguins (Sphenisciformes) to other birds has puzzled ornithologists for a long time (Fig. 15). Penguins are "the most specialized of our living birds in the sense that they have departed far from the standard form adapted for flight that has controlled the development of other groups" (Wetmore, 1955). As long ago as 1883, Watson postulated that the ancestral penguins were flying birds, a view held also by many later ornithologists. Lowe (1933, 1939), however, denied that penguins were derived from flying ances-

Figure 11. Skeleton of *Diatryma giganteum* from the Lower Eocene of New Mexico. (Courtesy of the American Museum of Natural History.)

tors. In his opinion the evidence indicated that the penguins and the "Struthiones" represent two natural groups whose common ancestor left the main avian stem before flight had been attained. From that ancient, nonflying ancestor, the ratite birds specialized directly for a cursorial mode of existence, and the penguins specialized from the same ancestor for an aquatic life (1933:533). Simpson's studies (1946, 1957, 1970), however, leave little doubt that penguins are derived from flying ancestors, even though the earliest known fossil penguins already were flightless and specialized for an aquatic existence. All fossil penguins have been found within the present range of the group.

Howard (1957a) described *Osteodontornis orri*, a giant oceanic bird with a wingspread between 14 and 16 feet, from the Miocene of California (Fig. 16).

Figure 12. A restoration of *Phorusrhacos* (formerly *Phororhacos*), a giant flightless bird from the Miocene of Patagonia. (Courtesy of the American Museum of Natural History.)

Osteodontornis is exceptional because both jaws contain toothlike bony projections. These "bone teeth" are highly specialized elaborations of the jawbones (Fig. 17) and hence exhibit a similar histological pattern; they are in no way homologous with the teeth of *Hesperornis*. It is assumed that the "teeth" and the jaws were sheathed with a horny covering, as in other birds. (See Hopson, 1964.)

Pliocene fossils from Oregon were described by Brodkrob (1961), and others from Chihuahua, Mexico, by Howard (1966).

During the Tertiary period there was a great radiation among birds. "The early Tertiary essentially completed the development of the water-bird families and the nonpasserine forest dwellers. A second radiation of new families occurred in the Miocene, when the last of the specialized water birds appeared, and when the land birds exploited the expansion of alpine and xeric habitats. By the end of the Miocene all of the nonpasserine families were probably established, as well as most, if not all, of the passerines" (Brodkorb, 1971).

2. The Quaternary Period.   The birds discussed below lived during the Pleistocene epoch, when the avifauna was richer than that of today. Brodkorb (1971) estimated that the Pleistocene avifauna contained 4806 nonpasserines and 5847 passerines, or a total of 10,600 species (the number of species now living is about 8600). Brodkorb also estimated that "the total number of birds, past and present, is approximately 154,000 species."

Figure 13.   Restoration by Walter A. Weber of the terrestrial vulture *Neocathartes grallator* (Wetmore) "shown against a background of landscape designed to represent the ecological conditions of the late Eocene of Wyoming." (Courtesy of Alexander Wetmore and the Carnegie Museum.)

Figure 14. The holotype of *Primobucco kisterni*. (*A*) Flattened braincase; (*B*) right coracoid; (*C*) left humerus; (*D*) radius and ulna; (*E*) tibiotarsus; (*F*) tarsometatarsus; I–IV, digits I to IV. (Courtesy of J. Alan Feduccia.)

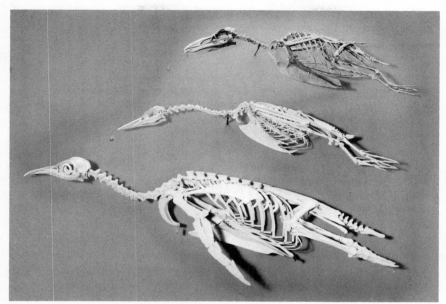

Figure 15. Upper: Great Auk (*Pinguinus impennis*). Middle: Common Loon (*Gavia immer*). Lower: Emperor Penguin (*Aptenodytes forsteri*). (Courtesy of the American Museum of Natural History.)

The Pleistocene was a time when "relatively brief advances and prolonged retreats of the continental ice cap altered the climate and the sea level. Elements of the flora and fauna shifted back and forth in response to fluctuating conditions. Many species and higher taxa became extinct, and the present avifauna was established. . . . Postglacial evolution has been mainly at the subspecific level" (Brodkorb, 1971).

Among the most remarkable fossil beds yet discovered are the La Brea "asphalt traps" of California (Figs. 18, 19). Miller and DeMay (1942) wrote that at the Rancho La Brea in Los Angeles 105 species of birds are represented by over 100,000 determinable bones. Mammals that are now extinct (saber-toothed tiger, wolf, horse, antelope), as well as birds, were entrapped in the asphalt pools (Fig. 20). "The bone, still clothed in its activating tissues, was plunged into liquid asphalt which soon penetrated its most minute structure and sealed it away from most of the destructive agencies, so that now after a thousand centuries it may be exposed to the air without traceable deterioration."

Another western site that has contributed to our knowledge of Pleistocene bird life is Fossil Lake, Oregon. Discovered in 1876, this desert fossil bed has

Figure 16. The Antolini quarry in Tepusquet Canyon, Santa Barbara County, California. The type specimen of *Osteodontornis orri* was taken from the tilted beds in the upper center of the photograph. (Courtesy of Hildegarde Howard and the Santa Barbara Museum of Natural History.)

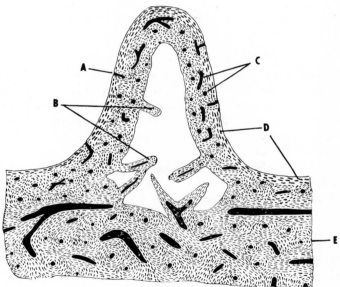

Figure 17. Reconstruction of a thin section through a "bone tooth" and jawbone of *Osteodontornis orri*. (*A*) Volkman canal; (*B*) bony trabeculae; (*C*) Haversian canals; (*D*) circumferential lamellae; (*E*) jawbone. (Courtesy of Hildegarde Howard and the Santa Barbara Museum of Natural History.)

Figure 18. An asphalt lens being excavated at Rancho La Brea by the University of California in 1910. (Courtesy of the late Loye Miller.)

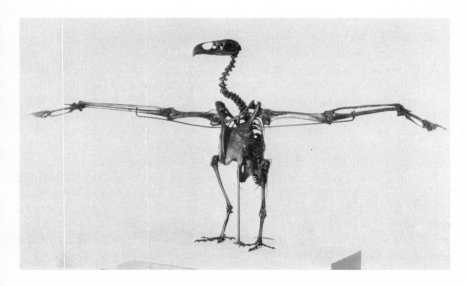

Figure 19. The Merriam Teratornis (*Teratornis merriami*) from the Pleistocene of Rancho La Brea, California. The wingspan of this bird was estimated to be 12 feet. (Courtesy of the Los Angeles County Museum of Natural History.)

25

Figure 20. A restoration of a scene from the Rancho La Brea tar pools in late Pleistocene times. The now-extinct animals shown are *Teratornis,* saber-tooth tigers, wolves, and horses. (From a painting by Charles R. Knight; courtesy of the Field Museum of Natural History.)

yielded some 2500 bones, representing 66 forms. Grebes, swans, geese, and ducks account for about four-fifths of the specimens. Howard (1946) concluded that the avifauna was "like that to be found living today about inland lakes of Oregon and California" (see also Wetmore's 1955 account of Fossil Lake).

Known fossil beds, however, are not limited to western North America. Wetmore (1931) reported on several widely separated areas in Florida where fossil birds had been discovered, and Hamon (1964) stated that at least 8000 specimens of birds' bones had been collected near Reddick, Florida (Fig. 21). Brodkorb (1959a, 1959b, 1963b, 1963c, 1964) discussed fossils from other localities in Florida, South Carolina, the Bahamas, and Barbados. Of significance is Brodkorb's (1963a) description of a giant flightless bird—"larger than the African ostrich and more than twice the size of the South American rhea"—from the late Pleistocene of Florida. He assigned *Titanis walleri*, a new genus and species, to the family Phorusrhacidae, which previously had been known only from the Oligocene to the early Pleistocene in South America.

The much-debated causes of extinction are multiple: e.g., overspecialization, climatic and environmental changes, competition with more adaptable forms, and the ascendancy of higher groups of animals. Two theories proposed during the early 1970s concerned warmbloodedness and thin egg shells as causes of the

Figure 21. One of the main fossil beds near Reddick, Florida, where Dr. Pierce Brodkorb and his students have collected some 8000 fossil bird bones. (Courtesy of J. Hill Hamon.)

extinction of dinosaurs. The first suggested that dinosaurs were endothermic animals that moved on land in the manner of modern elephants rather than being sluggish like lizards. Without an adequate means of controlling body temperature when the temperature dropped near the end of the Cretaceous Period, the animals died out. In regard to the second theory, intact dinosaur eggs and many fragments have been found in the French Pyrenees mountains, and the eggs of later times are said to have been significantly thinner than those from older strata. Be that as it may, some of the largest birds for which there are fossil records lived during the Pleistocene (and until Recent times) on Madagascar and New Zealand, large, isolated islands with a meager mammalian fauna and no large endemic carnivores. The elephantbirds, *Aepyornis* and *Mullerornis* (order Aepyornithiformes), of Madagascar and the moas, *Dinornis* and related genera,[2] of New Zealand became extinct after man reached these islands (Figs. 22, 23).

The largest species of *Dinornis* stood about 10 feet tall. The elephantbirds were smaller in stature (about Ostrich size) but had a massive build. Both orders are related to the ratite assemblage of flightless birds and may be

[2] See Archey (1941), Oliver (1955), and Brodkorb (1963–1971).

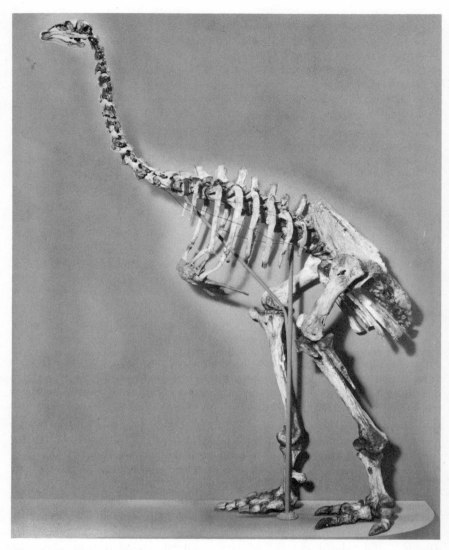

Figure 22. Skeleton of a small moa (*Euryapteryx*) from New Zealand. (Courtesy of the American Museum of Natural History.)

characterized, in part, as follows: the quadrate has a single articular condyle; the sternum is short and broad and lacks a keel; the scapula is fused with the coracoid; the wing is rudimentary (elephantbirds) or absent (moas); the ilium and ischium are not fused posteriorly; a pygostyle is wanting; most species have four toes (the hallux elevated), but some have only three. Archey (1941) wrote that some of the moas had a loop in the trachea. The feathers of the moas (e.g., *Megalapteryx*) lack barbicels and are soft and loose-webbed, as in living ratites; the aftershaft is well developed. Cream-colored eggs have been attributed to three different genera, and pale green fragments have been described. Deevey (1954) wrote that a female pelvis was found with a fully formed egg having a capacity of more than half a gallon. He also reported that "gizzard stones" ("some as large as a hen's egg") were found with a number of skeletons. Radiocarbon dating of presumed food materials suggested that *Dinornis* and *Euryapteryx* inhabited New Zealand as recently as A.D. 1300. Fisher (Fisher and Peterson, 1964) wrote that *Megalapteryx didina* may have survived until the eighteenth century, "possibly after 1773."

Figure 23. A restoration of the giant, wingless *Dinornis* of New Zealand. (From a painting by Charles R. Knight; courtesy of the Field Museum of Natural History.)

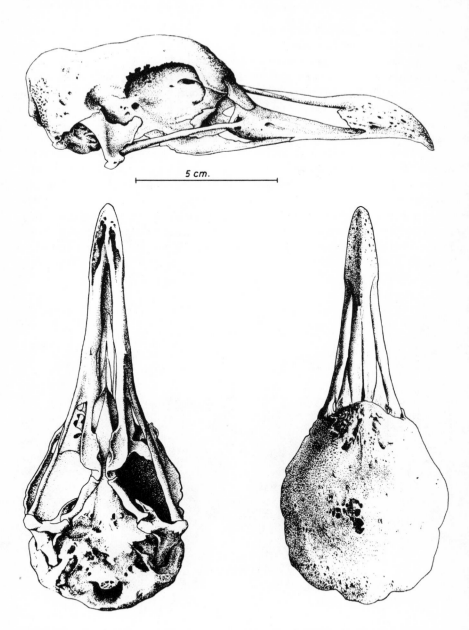

Figure 24. Three views of a skull of *Pezophaps solitarius*. Original drawings by Alberto M. Simonetta from a specimen in the British Museum. (Courtesy of Alberto M. Simonetta and the editor of *The Quarterly Review of Biology*.)

30

Aepyornis is noted for having laid the largest birds' eggs known: approximately 9.5 × 13 inches, with a capacity of over 2 gallons. It has been estimated that such an egg would hold the contents of 7 Ostrich eggs or 183 chicken eggs or more than 12,000 hummingbird eggs.

Edinger (1942) studied brain endocasts of four species of elephantbirds with special reference to the pituitary gland. She concluded that "the extraordinary size of the aepyornithid eggs and the loose luxuriant plumage of the large ratites now appear as signs of hyperpituitarism."

The Dodo (*Raphus cucullatus*) of Mauritius and the Solitaire (*Pezophaps solitarius*; Fig. 24) of Rodrigues Island are of interest even though they became extinct only some 300 years ago. Both have long been viewed as flightless pigeons and placed in a single family, but Storer (1970) proposed that each be placed in a separate monotypic family—the Raphidae and Pezophapidae, respectively. He reasoned that the distance between the islands in the Mascarene group would preclude the rafting of such large birds from one island to another, and therefore that the two species must have been independently derived from flying ancestors.

REFERENCES

Archey, G. 1941. The Moa, a study of the Dinornithiformes. *Bull. Auckland Inst. Mus.*, No. 1.

Ashmole, N. P. 1963. The extinct avifauna of St. Helena Island. *Ibis*, **103b**:390–408.

Beebe, C. W. 1915. A tetrapteryx stage in the ancestry of birds. *Zoologica*, **2**:39–52.

Böker, H. 1927. Die biologische Anatomie der Flugarten der Vogel und ihre Phylogenie. *J. Ornithol.*, **75**:304–371.

Bradbury, W. C. 1919. Some notes on the egg of *Aepyornis maximus*. *Condor*. **21**:97–101.

Brodkorb, P. 1959a. The Pleistocene avifauna of Arredondo, Florida. *Bull. Fla. State Mus.*, **4**, No. 9:269–291.

———1959b. Pleistocene birds from New Providence Island, Bahamas. *Ibid.* **4**, No. 11:349–371.

———1961. Birds from the Pliocene of Juntura, Oregon. *Quart. J. Fla. Acad. Sci.*, **24**:169–184.

———1963a. A giant flightless bird from the Pleistocene of Florida. *Auk*, **80**:111–115.

———1963b. Miocene birds from the Hawthorne Formation. *Quart. J. Fla. Acad. Sci.*, **26**:159–167.

———1963c. Fossil birds from the Alachua clay of Florida. *Fla. Geol. Surv.*, *Spec. Publ.* No. 2, paper No. 4, pp 1–11.

———1963–1971. Catalogue of fossil birds. *Bull. Fla. State Mus.*, **7**, 1963:179–293; **8**, 1964:195–335; **11**, 1967:99–220; **15**, 1971:163–266.

———1964. Fossil birds from Barbados, West Indies. *J. Barbados Mus. Hist. Soc.*, **31**:3–10.

———1971. Origin and evolution of birds. In *Avian Biology*, D. S. Farner and J. R. King, eds., Vol. 1, Academic Press, New York, pp. 19–55.

Cracraft, J. 1973. Continental drift, paleoclimatology, and the evolution and bigeography of birds. *J. Zool. London*, **169**:455–545.

Darlington, P. J., Jr. 1957. *Zoogeography: The Geographical Distribution of Animals.* John Wiley & Sons, New York.

De Beer, Sir Gavin. 1954. *Archaeopteryx lithographica.* British Museum (Natural History), London.

Deevey, Ed. S., Jr. 1954. The end of the moas. *Sci. Am.,* **190**:84–90.

Edinger, T. 1942. The pituitary body in giant animals fossil and living: a survey and a suggestion. *Quart. Rev. Biol.,* **17**:31–45.

————1949. Paleoneurology versus comparative brain anatomy. *Confina Neurol.,* **9**:5–24.

————1951. The brains of the Odontognathae. *Evolution,* **5**:6–24.

Enlow, D. H. 1956–1958. A comparative histological study of fossil and recent bone tissues. I, II, III. *Texas J. Sci.,* **8**:405–443; **9**:186–214; **10**:187–230.

Feduccia, A. 1973. A new Eocene zygodactyl bird. *J. Paleontol.,* **47**:501–503.

Fisher, J. 1967. Fossil birds and their adaptive radiation (pp. 133–154) and Aves (pp. 733–762). In *The Fossil Record, a Symposium with Documentation,* Geological Society, London.

—, and R. T. Peterson. 1964. *The World of Birds.* Doubleday & Co., New York.

Gregory, J. T. 1952. The jaws of the Cretaceous toothed birds, *Ichthyornis* and *Hesperornis.* *Condor,* **54**:73–88.

Hachisuka, M. 1953. *The Dodo and Kindred Birds.* H.F. & G. Witherby, London.

Hamon, J. H. 1964. Osteology and paleontology of the passerine birds of the Reddick, Florida, Pleistocene. *Fla. Geol. Surv., Geol. Bull.* No. 44, pp. 1–210.

Heilmann, G. 1927. *The Origin of Birds.* D. Appleton & Co., New York.

Heller, F. 1959. Ein dritter *Archaeopteryx*-fund aus den Solnhofener Plattenkalken von Langenaltheim/Mft. *Erlanger Geol. Abh.* **31**:3–25.

Heptonstall, W. B. 1970. Quantitative assessment of the flight of *Archaeopteryx.* *Nature,* **228**:185–186.

Holman, J. A. 1961. Osteology of living and fossil new world quails (Aves, Galliformes). *Bull. Fla. State Mus.,* **6**, No. 2:131–233.

Hopson, J. A. 1964. *Pseudodontornis* and other large marine birds from the Miocene of South Carolina. *Postilla,* **83**:1–19.

Howard, H. 1946. A review of the Pleistocene birds of Fossil Lake, Oregon. *Carnegie Inst. Wash. Publ.* No. 551, pp. 141–195.

————1950. Fossil evidence of avian evolution. *Ibis,* **92**:1–21.

————1957a. A gigantic "toothed" marine bird from the Miocene of California. *Santa Barbara Mus. Nat. Hist. Bull.* No. 1 (Geology).

————1957b. A new species of passerine bird from the Miocene of California. *Contrib. Sci.,* **9**:1–16 (Los Angeles County Museum).

————1962. Fossil birds. *Los Ang. County Mus., Sci. Ser.* 17, *Paleo.* No. 10.

————1966. Pliocene birds from Chihuahua, Mexico. *Contrib. Sci.* **94**:1–12.

Kaiser, H. E., and J. C. Bartone. 1968. A contribution to the comparative study of pachyostosis in recent birds. *Arch. Mex-Anat.,* **28**:43–59.

Lowe, P. R. 1933. On the primitive characters of the penguins, and their bearing on the phylogeny of birds. *Proc. Zool. Soc. London,* **1933**, Pt. 2:483–538.

————1939. Some additional notes on Miocene penguins in relation to their origin and systematics. *Ibis,* **1939**:281–296.

————1944. An analysis of the characters of *Archaeopteryx* and *Archaeornis.* Were they reptiles or birds? *Ibid.,* **86**:517–543.

Marsh, O. C. 1880. *Odontornithes: a Monograph on the Extinct Toothed Birds of North America*. U.S. Government Printing Office, Washington, D.C.

Miller, L. 1960. On the history of the Cathartidae in North America. *Novid. Columbianas*, 1:232–235.

———, and I. DeMay. 1942. The fossil birds of California. *Univ. Calif. Publ. Zool.*, 47:47–142.

Nopcsa, Baron F. 1907. Ideas on the origin of flight. *Proc. Zool. Soc. London*, 1907:223–236.

Oliver, W. R. B. 1955. *New Zealand Birds*, 2nd ed. A. H. & A. W. Reed, Wellington.

Osborn, H. F. 1900. Reconsideration of the evidence for a common dinosaur-avian stem in the Permian. *Am. Nat.*, 34:777–799.

Ostrom, J. H. 1970. *Archaeopteryx:* notice of a "new" specimen. *Science,* 170:537–538.

———1972a. *Pterodactylus crassipes* Meyer, 1857 (Aves): proposed suppression under the plenary powers. Z. N. (S) 1977. *Bull. Zool. Nomen.*, 29:30–31.

———1972b. Description of the *Archaeopteryx* specimen in the Teyler Museum, Haarlem. *Kon. Ned. Akad. Wet.*, Ser. B, 75:289–305.

———1973. The ancestry of birds. *Nature,* 242:136.

———1974. *Archaeopteryx* and the origin of flight. *Quart. Rev. Biol.*, 49:27–47.

Parkes, K. C. 1966. Speculations on the origin of feathers. *Living Bird,* 5th annual, pp. 77–86.

Pearson, R. 1972. *The Avian Brain*. Academic Press, New York.

Rich, P. V. 1972. A fossil avifauna from the Upper Miocene Beglia formation of Tunisia. *Notes Serv. Geol.*, 35:29–66.

Romer, A. S. 1966. *Vertebrate Paleontology*. University of Chicago Press, Chicago.

Saville, D. B. O. 1957. The primaries of *Archaeopteryx*. *Auk,* 74:99–101.

Selander, R. K. 1965. Avian speciation in the Quaternary. In *The Quaternary of the United States*, H. E. Wright, Jr., and D. G. Frey, eds., Princeton University Press, Princeton, N.J., pp. 527–542.

Simonetta, A. M. 1960. On the mechanical implications of the avian skull and their bearing on the evolution and classification of birds. *Quart. Rev. Biol.*, 35:206–220.

Simpson, G. G. 1946. Fossil penguins. *Bull. Am. Mus. Nat. Hist.*, 87:1–99.

———1957. Australian fossil penguins, with remarks on penguin evolution and distribution. *Rec. South Aust. Mus.*, 13:51–70.

———1970. Ages of fossil penguins in New Zealand. *Science,* 168:361–362.

Steiner, H. 1918. Das Problem der Diastataxie des Vogelflugels. *Jena Z. Naturwiss.*, 55:221–496.

Storer, R. W. 1970. Independent evolution of the Dodo and the Solitaire. *Auk,* 87:369–370.

Swinton, W. E. 1960. The origin of birds. In *Biology and Comparative Physiology of Birds*, A. J. Marshall, ed., Academic Press, New York.

Tchernov, E. 1968. *A Preliminary Investigation of the Birds in the Pleistocene Deposits of 'Ubeidiya*. Israel Academy of Science and Humanities, 38 pp.

Tyler, C., and K. Simkiss. 1959. A study of the egg shells of ratite birds. *Proc. Zool. Soc. London*, 133, Pt. 2:201–243.

Waldman, M. 1970. A third specimen of a Lower Cretaceous feather from Victoria, Australia. *Condor,* 72:377.

Wetmore, A. 1925. The systematic position of *Palaeospiza bella* Allen, with observations on other fossil birds. *Bull. Mus. Comp. Zool.*, 67:183–193.

——— 1931. The avifauna of the Pleistocene in Florida. *Smithson. Misc. Collect.*, 85, No. 2.

——— 1944. A new terrestrial vulture from the Upper Eocene deposits of Wyoming. *Ann. Carnegie Mus.*, 30:57–69.

————1951a. A revised classification for the birds of the world. *Smithson. Misc. Collect.,* **117,** No. 4.

————1951b. Recent additions to our knowledge of prehistoric birds, 1933–1949. *Proc. 10th Int. Ornithol. Congr.,* **1950:**51–74.

————1955. Paleontology. In *Recent Studies in Avian Biology,* Albert Wolfson, ed., University of Illinois Press, Urbana, pp. 44–56.

————1956. A check-list of the fossil and prehistoric birds of North America and the West Indies. *Smithson. Misc. Collect.,* **131,** No. 5, pp. 1–105.

————1959. Birds of the Pleistocene in North America. *Ibid.,* **138,** No. 4, pp. 1–24.

————1960. Pleistocene birds in Bermuda. *Ibid.,* **140,** No. 2, pp. 1–11.

Wheeler, W. H. 1960. The Uintatheres and the Cope-Marsh war. *Science,* **131,** No. 3408:1171–1176.

Williams, G. R. 1962. Extinction and the land and freshwater-inhabiting birds of New Zealand. *Notornis,* **10:**15–24.

Yalden, D. W. 1971. The flying ability of *Archaeopteryx. Ibis,* **113:**349–356.

CHAPTER TWO

Structure
and Function

The behavior of a bird is determined by the anatomy
and physiology of the several organ systems that form
its body. Moreover, a knowledge of embryology is essential to a full understanding of the anatomy of the
posthatching and adult bird, and separate college
courses are designed to deal with these interrelated
subjects. Obviously, in a single chapter only a concise
summary of the distinctive characteristics of birds can
be given.

THE MAJOR SYSTEMS

An anatomical system is composed of various tissues and
organs that serve a common function. The organization
of the body components into major systems facilitates
learning and is, therefore, the traditional approach to the
study of anatomy; but, as soon as one begins to think in
terms of functional anatomy, one is immediately led to a
consideration of the relationships among the different
systems.

The Integumentary System

The integumentary system includes the skin and all of its specializations: feathers, wattles, claws, horny spurs, dermal muscles, sensory nerve endings, and the oil gland.

It has been said that birds possess only a single skin gland, the oil gland. Lucas (1970), however, reported that the skins of the chicken, pigeon, and Eurasian Quail (*Coturnix coturnix*) contain numerous cells that release a fatty secretion.

Vestigial claws occur in the hands of many birds. These wing claws are horny sheaths that articulate with the distal end of the phalanx of one of the digits. Wing claws usually are found on digit I, but Fisher found them on digit II in *Opisthocomus*, newly hatched Anseriformes, and several other unrelated genera. Vestigial claws also have been found in the young of the African Fin-

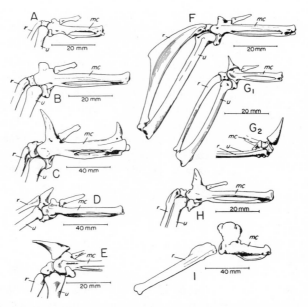

Figure 1. Bony structures in the region of the wrist of birds. (*A*) Red-legged Partridge. (*B*) European Oystercatcher. (*C*) Black-necked Screamer. (*D*) Spur-winged Goose. (*E*) Torrent Duck. (*F*) African Jaçana. (*G₁*) Jaçana. (*G₂*) Another view to show the curve of the spur. (*H*) Southern Lapwing. (*I*) Solitaire (an extinct bird). *A* and *B* show the "normal" process on metacarpal I; in *C, E, G,* and *H* this process is elongated into a spur (drawn with a horny sheath in *E* and *G*); *C* has an extra spur; *D* has the spur on a carpal bone; *F* has a thickened radius; *I* has a swollen knob on both metacarpal and radius. Abbreviations: r, radius; u, ulna; mc, metacarpal. (Courtesy of Austin L. Rand and the editor of *Wilson Bulletin*.)

foot (*Podica senegalensis*), as well as in adults of several passerine birds; examples are the Kiskadee Flycatcher (*Pitangus sulphuratus*), White-necked Crow (*Corvus leucognaphalus*), and Red-winged Blackbird (*Agelaius phoeniceus*).

Wing spurs occur in such birds as the Spur-winged Goose (*Plectropterus gambensis*), jacanas, some plovers, and two species of ducks (Fig. 1). Tarsal spurs are characteristic of certain gallinaceous birds (e.g., turkeys, some pheasants, francolins, partridges). Spurs differ from vestigial wing claws in that a spur consists of a projecting bony core and an outer covering of horn, a derivative of the skin. There presumably is an annual molt of the outermost layer of the horn in wing spurs (Rand, 1954; Weller, 1968).

The Skeletal System

The skeletal system forms an internal framework (*endoskeleton*) which protects and supports the soft structures of the body (e.g., brain, spinal cord, heart, lungs); it is composed of bone, cartilage, and ligament or gristle (Fig. 2). Bone contains living cells and undergoes continuous metabolic changes. Bones also serve as storage areas for calcium and phosphorus; they serve as the levers upon which muscles act; and certain bones, because of their red marrow cavities, are the primary blood-forming organs in the adult.

The avian skeleton is specialized for strength and for lightness. Strength and rigidity are attained by the architecture of the bones themselves, by the fusion of bones of the skull and synsacrum, and, in some birds, by the fusion of several of the dorsal vertebrae. Resorption of bone marrow and its replacement by extensions of air sacs decrease the weight of many of the bones in most bird families.

The *axial skeleton* is composed of the skull, vertebral column, hyoid elements, ribs, and sternum (Figs. 3, 4). The skull consists of the brain case (*cranium*) and the upper and lower jaws (*maxilla* and *mandible*). Typically, the main bones of the cranium are completely fused together in the adult bird.

The orbit and cranial cavity are large. The *lacrimal bone* (Fig. 5) may remain separate, or it may fuse with the other bones that form the anterior wall of the orbit. The *os opticum* is a small cancellous bone that supports the optic nerve at its exit from the eyeball. Typically horseshoe-shaped, the os opticum is thought to be of constant occurrence in the Falconidae, Trochilidae, Alcedinidae, Ramphastidae, Picidae, and Passeriformes; it is inconstant in the Ardeidae, Accipitridae, Phasianidae, Columbidae, and Psittacidae.

An anterior *sclerotic ring*, composed of from 10 to 18 overlapping platelike bones or ossicles, is found in the sclerotic coat of the eyeball. In owls, some hawks, and the Red-tailed Tropicbird (*Phaethon rubricauda*), this "ring" actually is a tubelike structure (Fig. 6). Congenital anomalies occur; only a single ossicle may develop in the eye of chickens homozygous for the mutant

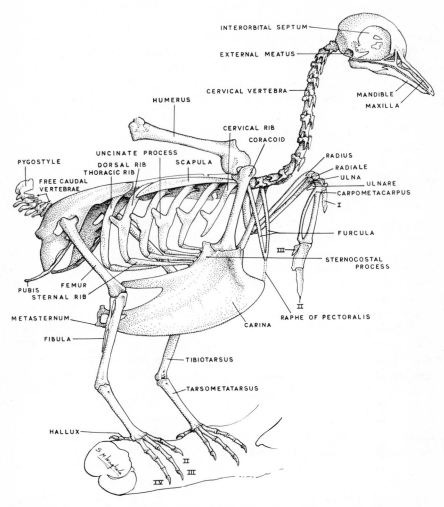

Figure 2. Lateral view of the skeleton of the Rock Dove (*Columba livia*). (By permission of J. C. George and A. J. Berger, from *Avian Myology*, Academic Press, copyright 1966.)

called "scaleless." (The nares, nasal bones, palate, and sternum are discussed on pp. 548–556.)

The regions of the vertebral column are designated as *cervical, dorsal, synsacral, caudal* (or *coccygeal*), and the *pygostyle*. The pygostyle is formed by the fusion of 4 to 7 embryonic vertebrae; a pygostyle is absent in most ratites

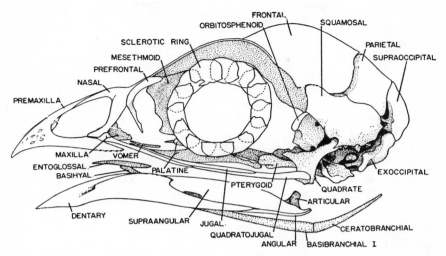

Figure 3. Lateral view of a 2- to 3-day-old chick head skeleton. (Courtesy of Malcolm T. Jollie and the editor of *Journal of Morphology*.)

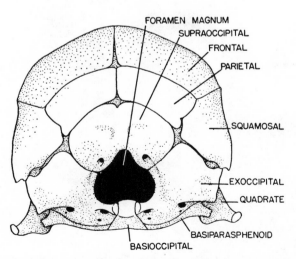

Figure 4. Posterior view of a 2- to 3-day-old chick skull. (Courtesy of Malcolm T. Jollie and the editor of *Journal of Morphology*.)

and in the Tinamidae. The number of vertebrae anterior to the pygostyle varies from about 33 to 58 in the several families of birds. The dorsal or back vertebrae (commonly 4 to 6 in number) are those that are connected by ribs to the sternum and that do not fuse with the synsacrum. In certain families (e.g., Podicipedidae, some Falconidae, Cracidae, Gruidae, Pteroclidae, Columbidae), 2 or more (commonly 3 to 5) of the dorsal vertebrae fuse to form an *os dorsale* (*notarium*); 1 or more free (unfused) dorsal vertebrae are interposed between the notarium and the synsacrum. The *synsacrum* is formed by the fusion of a series (10 to 23 bones) of thoracic, lumbar, sacral, and urosacral vertebrae; usually these vertebrae also fuse with the ilium. There are 4 to 9 free caudal vertebrae, 6 or 7 being most common. The numbers of cervical and dorsal vertebrae are relatively constant within a species, whereas the numbers of synsacral and free caudal vertebrae exhibit considerable individual variation.

The *true ribs* consist of a dorsal segment (vertebral rib) that has two articulations (*capitulum* and *tuberculum*) with a dorsal vertebra, and a ventral segment (sternal rib) that articulates with the sternocostal process of the sternum. There is, among the families of birds, much individual variation in the number of ribs that articulate with the sternum, and the number sometimes differs on the two sides of the same individual.

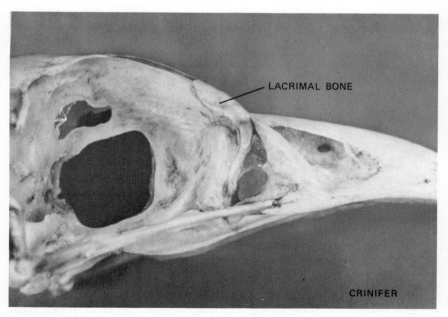

Figure 5. Lateral view of part of the skull of *Crinifer piscator* (Musophagidae) to illustrate the independent lacrimal bone.

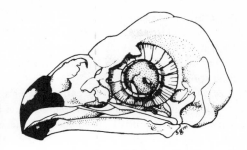

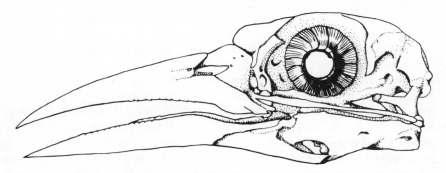

Figure 6. Examples of tubular sclerotic rings. Upper: Hawaiian Owl (*Asio flammeus sandwichensis*). Lower: Red-tailed Tropicbird (*Phaethon rubricauda*).

Thoracic ribs[1] may be present or absent. They can be identified when present because, by definition, they articulate dorsally with one of the anterior synsacral vertebrae rather than with a dorsal vertebra (see Fig. 2). A sternal

[1] This definition of thoracic ribs follows Gadow (1896, p. 849). In discussing dorsal and thoracic vertebrae and their ribs, he wrote that "unless clear definitions are strictly adopted, a promiscuous application of those terms will lead to much confusion. This remark applies with still greater force to the terms 'dorso-lumbar' and 'lumbar' vertebrae, which have a well-defined meaning in Mammals and in most Reptiles; but are absolutely inapplicable to Birds." Thomas H. Huxley defined the dorsal vertebrae as early as 1871 in his manual of the *Anatomy of Vertebrated Animals* when he wrote (p. 237): "The first dorsal vertebra is defined as such, by the union of its ribs with the sternum by means of a sternal rib; which not only as in the *Crocodilia,* becomes articulated with the vertebral rib, but is converted into complete bone, and is connected by a true articulation with the margin of the sternum."

segment may be present or absent: i.e., a thoracic rib may consist only of a vertebral segment that ends freely in the abdominal musculature, or it may have both a vertebral and a sternal segment; the latter may articulate with the sternum, or it may fuse with the sternal segment of the last true rib (or of the preceding thoracic rib).

The *appendicular skeleton* is composed of the upper and lower limbs and their supporting bony arches, the *pectoral* and *pelvic girdles*. Each pectoral (Latin *pectus,* breast) girdle is formed by three bones: *clavicle, coracoid,* and *scapula*. Superiorly, the three bones bound the *foramen triosseum,* through which the tendon of M.[2] supracoracoideus passes to its insertion on the humerus (see p. 387). The coracoid and scapula form the *glenoid fossa,* in which the head of the humerus articulates. In most birds (exceptions: some parrots, barbets, toucans), the clavicles fuse inferiorly to form the *furcula* (also *furculum*; wishbone); the area of fusion may be expanded into a *hypocleidium*. The clavicles are absent or rudimentary in the ratites and mesites, in some parrots, pigeons, and barbets, and in *Atrichornis* (Gadow, 1896:858; Glenny and Friedmann, 1954; Glenny and Amadon, 1955). In *Fregata* and *Opisthocomus* the furcula is ankylosed (fused) with the coracoids and the sternum. The coracoids and scapulae are fused in ratites.

The bones of a bird's wing are the *humerus, radius, ulna, two carpal bones,* the *carpometacarpus,* and, usually, *three fingers* or *digits* (Fig. 7). Anatomically, the *arm* is the most proximal segment of the forelimb, the part that extends from the shoulder to the elbow. The humerus is the supporting bone of the arm. The rounded head of the humerus articulates with the shallow glenoid fossa formed by the scapula and coracoid. This scapulocoracohumeral articulation is a ball-and-socket joint, which permits free rotation of the humerus. At its distal end, the humerus articulates with the ulna in a hinge joint; this joint permits movements of flexion and extension only (except in some hummingbirds; see George and Berger, 1966:360). The humerus is *pneumatic* in nearly all birds, i.e., the bone contains an air chamber that is an outgrowth of the clavicular air sac. The entrance (*pneumatic foramen*) to the air chamber lies in a depressed area (*pneumatic fossa*) on the palmar (ventral) surface near the proximal end of the humerus. In some birds there are two pneumatic fossae (Fig. 8; Berger, 1957; Bock, 1962).

The forearm extends from the elbow to the wrist; it contains two bones: the radius (on the thumb side) and the ulna. The adult bird has only two wrist or carpal bones. Other carpal bones fuse, during embryological development, with the hand (*metacarpal*) bones to form the carpometacarpus. No bird has more than three fingers. The reasons for calling these digits I, II, and III are given by George and Berger (1966:226). The alular digit (digit I) has one bony pha-

---

[2] *Musculus* before the name of a muscle is commonly abbreviated as M (plural, Mm).

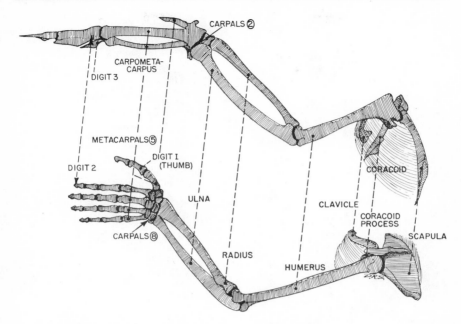

Figure 7. Homologies of the bones of the forelimb of a bird and man.

lanx in most birds; rarely there are two phalanges. The major digit (digit II) usually has two phalanges, but there are three in a few birds (e.g., *Gavia*, *Chen*). The minor digit (digit III) apparently always consists of a single phalanx.

The pelvic (Latin *pelveo,* a basin) girdle is formed by the partial fusion of three paired bones (*ilium, ischium, pubis*) and the synsacrum (Fig. 9). The pelvis is completely open ventrally in most birds; this facilitates the laying of relatively large eggs, although an open pelvis would not have been an essential phylogenetic development to permit egg-laying. The Ostrich (*Struthio*) has a *pubic symphysis* (as in mammals), and the rhea (*Rhea*) has an *ischiadic symphysis.*

The bones of the lower or hindlimb are the *femur, tibiotarsus, fibula, tarsometatarsus,* one independent *metatarsal bone,* and the *phalanges* of two or more toes. Figure 10 shows the considerable difference in the bony support for the lower limb between man and a bird.

The bird's thigh is nearly completely covered by contour feathers, and most of it is hidden by the skin of the body. Consequently, the thigh and the knee usually are not apparent in the living bird. For this reason, it sometimes is said that the bird's knee bends forward, or just opposite to the way that knee bends

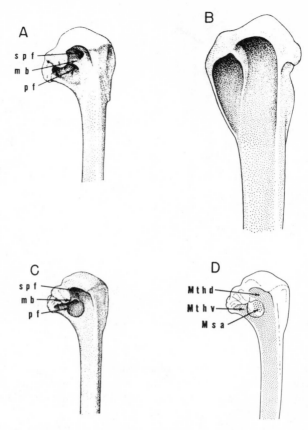

Figure 8. The right humerus of (*A*) *Turdus*, (*B*) *Larus*, (*C, D*) *Coccothraustes* to show degrees of development of the pneumatic fossae. Abbreviations: mb, medial bar of bone; pf, pneumatic fossa; spf, second pneumatic fossa; Msa, area of insertion of M. scapulohumeralis anterior; Mthd, Mthv, dorsal and ventral heads of origin of M. triceps humeralis. (Courtesy of Walter J. Bock and the editor of *Auk*.)

in man. This, of course, is not true. Because the thigh is concealed by feathers and skin, the feathered leg is the first obvious segment of the limb, and it is the bird's ankle joint that often is confused with the knee joint.

Anatomically, the leg (*crus*) extends from the knee to the ankle. In man, there are two bones in the leg: *tibia* and *fibula*. The ankle joint is between these two bones and the most proximal of the foot bones. In birds, however, the ankle joint is located *between* two rows of foot bones (*tarsal bones*), and, therefore, it is an *intertarsal joint*. During embryological development, the proximal tarsal bones fuse with the distal end of the tibia to form a *tibiotarsus*;

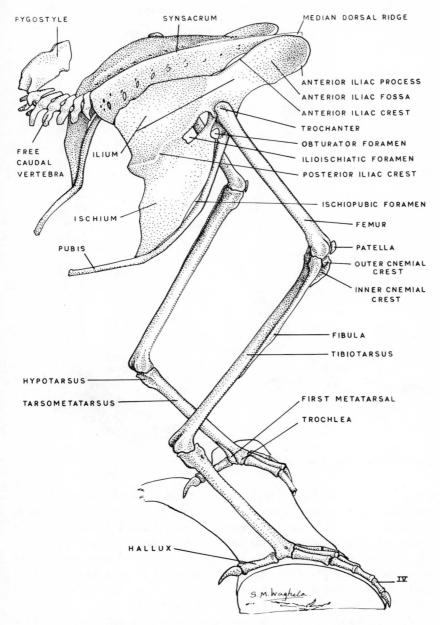

PYGOSTYLE

SYNSACRUM

MEDIAN DORSAL RIDGE

ANTERIOR ILIAC PROCESS

ANTERIOR ILIAC FOSSA

ANTERIOR ILIAC CREST

TROCHANTER

OBTURATOR FORAMEN

ILIOISCHIATIC FORAMEN

POSTERIOR ILIAC CREST

ISCHIOPUBIC FORAMEN

FEMUR

PATELLA

OUTER CNEMIAL CREST

INNER CNEMIAL CREST

FIBULA

TIBIOTARSUS

FIRST METATARSAL

TROCHLEA

FREE CAUDAL VERTEBRA

ILIUM

ISCHIUM

PUBIS

HYPOTARSUS

TARSOMETATARSUS

HALLUX

IV

S. M. Waghela.

Figure 9. An oblique view of the pelvic girdle and hindlimbs of the Rock Dove. (By permission of J. C. George and A. J. Berger, from *Avian Myology*, Academic Press, copyright 1966.)

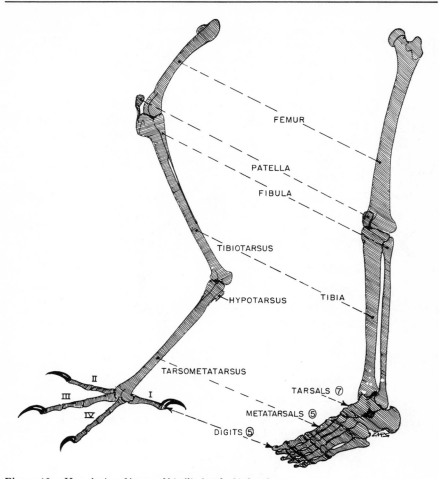

FEMUR

PATELLA

FIBULA

TIBIOTARSUS

HYPOTARSUS

TIBIA

TARSOMETATARSUS

II

III   I

IV

TARSALS ⑦

METATARSALS ⑤

DIGITS ⑤

Figure 10.   Homologies of bones of hindlimbs of a bird and man.

the distal tarsal bones fuse with the proximal ends of three *metatarsal bones,* which also fuse with each other, to form a *tarsometatarsus.* The tarsometa-tarsus is roughly comparable to the instep in man. Although the foot of the bird is composed of the tarsometatarsus and the phalanges of the toes, most birds walk on their toes only. A single, independent metatarsal bone supports the first toe or *hallux.* The fibula is poorly developed in most birds; rarely it extends as far as the distal end of the tibiotarsus: e.g., some penguins, the Osprey (*Pandion haliaetus*), the Eagle Owl (*Bubo bubo*).

In descriptions of birds, the tibiotarsus often is called the "tibia," and the tarsometatarsus, the "tarsus," but the abbreviated terms usually include the supporting bones, the muscles, and the skin with its covering scales and/or feathers. The bird crus forms the "drumstick" of roast chicken. The tarsus and toes are more or less completely covered by horny scales. The tarsus is wholly or partly feathered in some birds (e.g., the Golden Eagle, *Aquila chrysaetos,* grouse, and many owls), and in ptarmigan and the Snowy Owl (*Nyctea scandiaca*) even the toes are nearly completely covered by feathers.

A *sesamoid bone* (or cartilage) is one that develops in the tendon of a muscle and improves the angle of pull of that muscle. The largest sesamoid is the *patella* or kneecap; it is found in the patellar tendon, which passes in front of the knee joint and inserts on the tibiotarsus. The *os humeroscapulare* (*scapula accessoria*) is a sesamoid found in the capsule of the shoulder joint in many families; this bone is exceptionally well developed in *Otus asio, Tauraco leucotis, Procnias nudicollis, Dendroica kirtlandii,* and some genera of cuckoos (e.g., *Crotophaga, Guira, Cuculus, Morococcyx*) but not others (e.g., *Tapera*).

The Muscular System

The muscles are the organs of the body that produce motion. Most of the "fleshy" structure of the body consists of *striated muscle* whose action is consciously controlled. The integumentary, respiratory, circulatory, digestive, and urogenital systems contain *smooth* or *nonstriated muscle* whose action is involuntary. *Cardiac* or *heart muscle* is unique because of its *inherent rhythmicity.* In other words, the heartbeat originates within the muscle itself; the nerves to the heart regulate, but do not initiate, the heartbeat—in fact, the heart of the embryo begins to beat before any nerves have grown to it.

Two different kinds of striated muscles occur in birds: "red" and "white." Red muscle fibers contain myoglobin (an oxygen-transporting protein), have a richer blood supply, are smaller in diameter, and have their more abundant nuclei located more peripherally than do white fibers. White muscle fibers lack myoglobin. George (George and Berger, 1966), however, has shown that there is also a third type of fiber, and that the muscles in different bird species may contain one, two, or all three fiber types. Although the pectoralis (breast) muscle of the chicken is called "white" meat and that of the pigeon is called "red" meat, both muscles contain mixtures of red and white fibers (Fig. 11). The general color and texture of the muscle in each of the two species is due to the predominance of either white or red fibers. George also found that the pectoralis muscles of the Ruby-throated Hummingbird (*Archilochus colubris*) and of the House Sparrow (*Passer domesticus*) contain only red fibers. He found no species, however, in which the muscle consisted only of white fibers. The biochemical properties of the fiber types and of wing and abdominal muscles are discussed in detail by George, particularly in relation to energy requirements and migration.

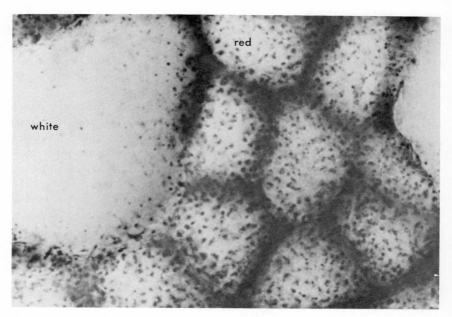

Figure 11. Photomicrograph of a transverse section of M. pectoralis of the Rock Dove, showing the distribution and localization pattern of succinic dehydrogenase in the two main types of muscle fibers. The narrower red fibers show a markedly greater deposition of diformazan granules than do the broader white fibers. ×1173. (Courtesy of J. C. George and the editor of *The Quarterly Journal of Microscopical Science*.)

The long Latin names of muscles frequently discourage the beginning student, but a familiarity with these can be quickly acquired if the names of certain classes of muscles are first learned separately. A few muscle names are arbitrary or fanciful; an example is the sartorius, so called because *sartor* is the Latin name for tailor, and in man this muscle helps rotate the leg for sitting in tailor fashion. But the longer muscle names are compact descriptions of the muscles themselves, one word often defining the type of motion (extensor), a second the part of the body which the muscle moves (digitorum), and a third its relative size (longus, brevis). Thus *Mm. extensor digitorum longus* and *extensor digitorum brevis* are two muscles (one long and one short) that extend the toes. Pairs of muscles also are named because of their relative size (major, minor) or their relative positions: superior, inferior; medialis, lateralis; profundus, superficialis. Some names tell where the muscles arise and insert: coracobrachialis (origin: coracoid; insertion: brachium = arm). The name of a muscle may describe its shape (rhomboideus), its structure (semimembranosus), or its location (brachialis).

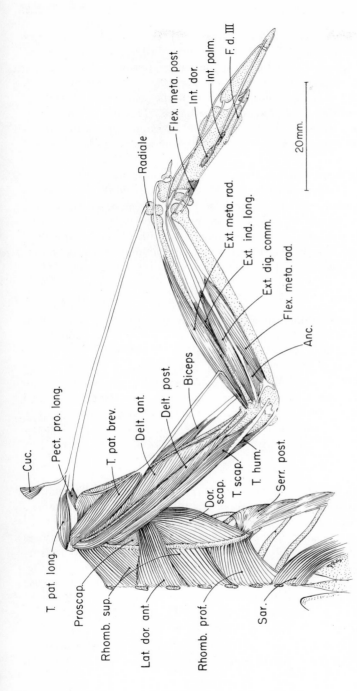

Figure 12. Dorsal view of wing muscles of the extinct Bourbon-crested Starling (*Fregilupus varius*). (Courtesy of Andrew J. Berger and the American Museum of Natural History.)

49

Many of the appendicular muscles of birds are characterized by having very long tendons of insertion (Fig. 12). The fleshy bellies of such muscles, therefore, occupy a proximal position near the center of gravity. The long tendons of some muscles cross two or more joints and act on each joint crossed. For example, M. flexor hallucis longus arises from the distal end of the femur; its tendon inserts on the distal phalanx of the hallux. This muscle, therefore, aids in flexing the tibiotarsus at the knee, aids in extending the tarsometatarsus, and flexes the tarsometatarsal-phalangeal joint and the interphalangeal joint of the first toe.

Functionally one cannot divorce the muscular system from the skeletal and nervous systems: nerves stimulate the muscles to contract, and the muscles then move bones. A complete understanding of functional myology requires a knowledge of a complicated set of factors: the origin and insertion of a muscle, its internal architecture (i.e., the arrangement of its fibers), the biochemical properties of the fibers, and the structure of the joint around which the muscle produces movement. (See Bock, 1968; Gans and Bock, 1965; Cracraft, 1971.)

The Nervous System

The world exists for an animal because it has a nervous system. This system perceives stimuli from the environment (both external and internal) and, by effecting adjustments to these stimuli, integrates all bodily functions.

The functional unit of the nervous system is the *neuron,* a cell highly specialized for conducting nerve impulses to other nerve cells, to muscles, or to glands. Although individual nerve cells are microscopic in size, some axons and dendrites in large birds, such as the Ostrich, are about 4 feet in length.

The brain and spinal cord form the *central nervous system* (CNS). Cranial and spinal nerves, with their associated ganglia, compose the *peripheral nervous system* (PNS).

The Central Nervous System.   Birds' brains are, in many respects, reptilian in form. The highly developed *cerebral cortex* (Latin, bark or shell) of mammals is lacking. The mammalian cortex covers completely the outer surface of the cerebral hemispheres; in man, the cortex contains so many folds and fissures that only about a third of the total cortex is visible. By contrast, the outer surface of the bird cerebrum (two cerebral hemispheres) is relatively smooth (Fig. 13), and most of it consists primarily of *basal ganglia* (such as the *corpus striatum,* or striated body), the oldest part of the vertebrate forebrain. The bird forebrain contains the olfactory lobes, hippocampus, thalamus, striatum, and connecting tracts. Most of it, however, is formed by the striatum; this consists of three parts (*archistriatum, paleostriatum, neostriatum*) that also occur in reptiles and mammals, plus a fourth part, the *hyperstriatum,* that is unique to

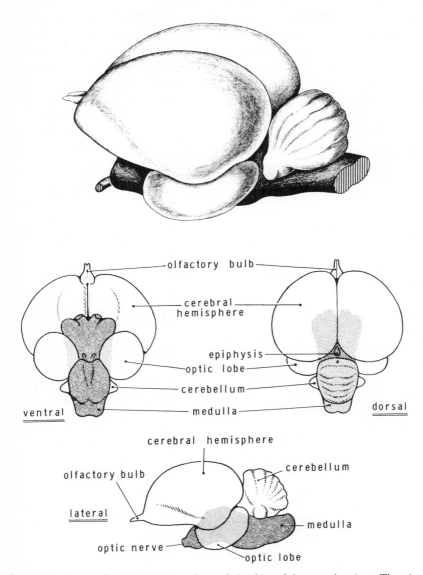

Figure 13. A generalized bird brain to show relationships of the several regions. The stippled area in the lower three drawings forms the "brain stem rest" (diencephalon, basal portion of mid-brain, and medulla) of Portmann. (Modified by permission after Portmann.)

birds (Fig. 14). The striatum is overlain only by a thin covering of white matter and cortexlike tissue, and authorities disagree as to what, if any, part of the avian cerebral hemisphere is homologous to the neocortex of mammals. Thus Cobb (1960:406) wrote: "In my own preparations, I cannot find the 'cortex' as described by Stingelin and Craigie, when 'cortex' is taken to mean a peripherally placed coating (pallium) of cells arranged in layers. It is difficult for me to make out definite layers in any of the structures of the bird's brain except in the optic lobes of the midbrain and in the hippocampus." Adopting Cobb's definition of "cortex," Van Tienhoven (1969) also concluded that a cortex is

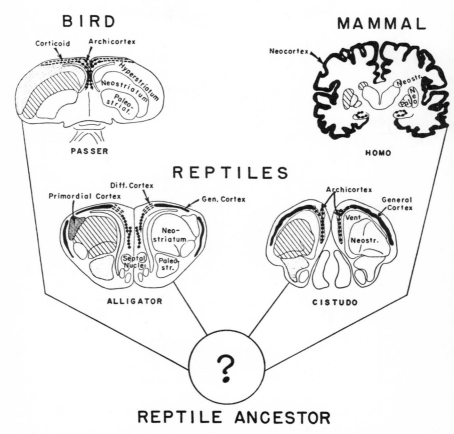

Figure 14. Diagrammatic frontal sections of brains of man (*Homo*), House Sparrow (*Passer domesticus*), alligator, and turtle, showing relationships of corticoid layers in the bird to the cortex in mammals, and the relative sizes and positions of the several parts of the striatum in the three classes. (Courtesy of Stanley Cobb and the editor of *Perspectives in Biology and Medicine*.)

absent in birds. Two parts of the hyperstriatum (hyperstriatum dorsale and hyperstriatum accessorium) form the swollen dorsomedial surface of each hemisphere in many birds; this area of the hemisphere, which is limited laterally by the vallecula (a fissure), was named the *Wulst* by Stingelin. The extent of the Wulst varies in different birds (Stingelin and Senn, 1969), but Stingelin wrote (*in litt.*, 1972): "Authors agree that part of the dorsal cortex of the reptilian brain is the forerunner of the Neocortex. Since the Wulst in birds is, as compared with reptiles, a new and progressing structure in the same position as the dorsal cortex (between hippocampus and the rudimentary paleocortex), I regard the whole Wulst (in a broad sense) as homologous to the mammalian Neocortex."

Stettner and Matyniak (1968) reported simple experiments with Bobwhite (*Colinus virginianus*) that suggested the Wulst was involved in learning, and Van Tienhoven (1969) and Pearson (1972) summarized other experiments on the hyperstriatum. Van Tienhoven concluded that these experiments demonstrated that "hyperstriatal lesions have no effect on body temperatures, food and water intake, respiratory and cardiac rates, and cardiac conditioning," and that often, in fact, "hyperstriatal lesions show no obvious effects; rather sophisticated methods are required to find their effects." Hence the precise function of the hyperstriatum remains to be delineated in its finer details.

Portmann (1955) developed a *cerebral index* to show the weight of the cerebral hemispheres in relation to the weight of the "brain stem rest," which he defined as consisting of the diencephalon, the basal portion of the midbrain (but not the optic lobes), and the medulla oblongata (see Fig. 13). A high index number, which indicates that the species has large cerebral hemispheres, occurs in unrelated birds: e.g., a hawk (*Buteo buteo*), 9.87; a parrot (*Ara ararauna*), 28.02; an owl (*Bubo bubo*), 15.07; and a crow (*Corvus corax*), 18.95. Portmann and Stingelin (1961) report that "evolution of the avian brain to higher degrees of complexity is always combined with a tendency to concentrate the mass of higher centers in the front part of the hemisphere," although, in different birds, the specialization may occur in the dorsal or basal part of the hemisphere.

It has been suggested that high index numbers are an indication of higher intelligence or adaptability, even though Portmann and Stingelin pointed out that "these indexes are quantitative expressions of the relative mass development and cannot therefore be taken as a measure of psychic complexity or intelligence. They give instead the first objective basis for any determination of the level of brain evolution and permit a deeper understanding of morphological and ethological considerations. Together with the morphological analysis, the index method allows an appreciation of evolutionary trends in different groups." Cobb (1960) remarked that "merely showing the size of the hemisphere is of little interest without describing its component parts," and, it may be added, without having a firm knowledge of the functioning of the avian brain. (See also Pearson, 1972.)

Interesting differences among birds occur in the relative position of the brain within the skull. In the Double-crested Cormorant (*Phalacrocorax auritus*), for example, the axis of the brain (a line from the center of the olfactory bulb to the center of the medulla) forms a 15° angle with the axis of the bill. An extreme specialization is found in the American Woodcock (*Philohela minor*), in which "the *foramen magnum* actually points somewhat forward, and the brain is indeed 'an arc around the back of the orbit'" (Cobb, 1960; Fig. 15). Dubbledam (1968) discussed in detail four basic differences in brain axes

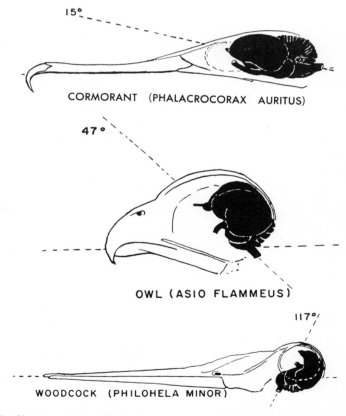

Figure 15. Variation of the cerebral axis compared to the bill axis in three different birds. The brain-bill angle is measured in degrees. (Courtesy of Stanley Cobb and the editor of *Perspectives in Biology and Medicine*.)

among birds correlated with corresponding differences in skull shape. (The olfactory bulbs and the sense of smell are discussed in Chapter 4.)

The *diencephalon* is the posterior part of the forebrain; it consists of an *epithalamus, thalamus,* and *hypothalamus.* The hypothalamus (the floor of the diencephalon) is concerned with regulation of the autonomic nervous system. Groups of cell bodies (called "centers") are responsible for temperature regulation, for feelings of hunger, thirst, and satiety, and (in man) for such physical expressions of emotion as increased heart rate, blushing, dryness of the mouth, and sweaty hands. All of these feelings and physiological changes are due to the operation of the hypothalamus and the autonomic nervous system. The hypothalamus also has a critical influence on the secretions of the pituitary gland and, through it, on the activity of other endocrine glands.

The nerve connections between the thalamus and other parts of the brain are presumed to be related to the stereotyped behavior patterns of birds. The *pineal gland* is a diverticulum from the epithalamus (the roof of the diencephalon). Quay and Renzoni (1963) described the histology and probable secretory activity of cells in the pineal glands of 24 species of passerine birds, and they described experiments with adult female House Sparrows that produced an increase in nuclear size after an increase in photoperiod. Wight (1971) reported that "in the fowl, there is both morphological and biochemical evidence of the effect of light" on the activities of the pineal gland. The possible hormonal activities of the gland are inadequately known; Wight reviewed the evidence for possible interrelations of the pineal gland to other endocrine glands. (See also Quay, 1972a; Quay and Renzoni, 1967.)

The *optic lobes* are the most prominent parts of the bird midbrain. Rather than forming the roof (*tectum*) of the midbrain as in other animals, however, the optic lobes occupy a position on the sides of the brain. Moreover, they occupy a lower position in the more advanced groups (e.g., Passeriformes); this is presumed to be an example of neurobiotaxis—the phylogenetic movement of nerve cell bodies toward the stimuli. The Kiwi (*Apteryx australis*) of New Zealand, which has small eyes and is most active at night, has small optic lobes.

The anterior part of the embryological hindbrain (*rhombencephalon*) becomes the *cerebellum* (Latin diminutive for brain), which is concerned with the sense of balance or equilibrium, in birds especially in relation to flight. Hence the cerebellum is highly developed (with a small ventricle) in most birds but is relatively small (with a large ventricle) in *Apteryx,* a flightless bird; the cerebellum is highly developed in hummingbirds. The relative sizes of the several folia in different birds are correlated generally with their locomotor patterns (Larsell, 1967). Cerebellar hemispheres (the neocerebellum of mammals) are not present in birds.

The most posterior and primitive part of the brain in all vertebrate animals is the *medulla oblongata.* Although the bird medulla is similar to that in reptiles, there is an increase in cell bodies that is related to the development of the syrinx and to incoming nerve fibers from the ear, including those from the semicircular canals for the sense of balance. A *pons* (Latin, bridge) is absent both in reptiles and in birds.

The *spinal cord* lies in a protective neural (or vertebral) canal formed by the bones of the vertebral column. The neurons in the spinal cord conduct impulses both to and from the brain, and they function in intersegmental reflexes. The bird spinal cord extends throughout the length of the vertebral column, from the *foramen magnum* (Latin, great hole) in the base of the skull into the neural canal in the caudal vertebrae in the tail. Enlargements occur in the regions of the spinal cord that accommodate the large numbers of nerves supplying the wings (*cervical enlargement*) and the legs (*lumbosacral enlargement*). The relative sizes of the two enlargements depend on the locomotor habits of the bird.

A unique feature of the avian spinal cord is the presence in the lumbosacral region of a *rhomboid sinus* (*intumescentia lumbalis*). This sinus (a cavity) results from the lateral deflection of the dorsal column of white matter in the spinal cord due to an increase in numbers of incoming sensory (afferent) nerve fibers from the leg and is thought to exhibit its best development in birds with powerful legs. The cavity contains a gelatinous substance and myelinated nerve fibers; the roof is a glycogen depot and is called the *glycogen body*. The function appears to be unknown, but the deposition of glycogen may be under hormonal control.

The Peripheral Nervous System.   *Cranial nerves* arise from the brain; in addition to its name, each nerve traditionally is referred to by a Roman numeral (Table 1). Each name tells something about the nerve; e.g., the olfactory nerve (Latin *olor,* a smell) carries fibers for the sense of smell; the oculomotor nerve (Latin *oculus,* eye) conveys motor fibers to four of the muscles that move the eyeball; the trochlear nerve supplies a muscle whose tendon passes through a trochlea (pulley) before inserting on the eyeball; the trigeminal nerve has three main branches; the glossopharyngeal nerve sends its branches to the tongue (Greek *glossa*) and the pharynx; the vagus nerve is a "wandering" nerve (Latin *vago,* to wander), so named because its branches supply structures in the head, neck, chest, and abdomen (see Watanabe, 1960).

Birds have only 11 readily demonstrated (i.e., by gross dissection) cranial nerves. Portmann and Stingelin (1961) state that the spinal accessory nerve of mammals "is merely a side branch" of the vagus nerve in birds. Rogers (1965), however, studied histological sections of chick embryos through 17 days of incubation and concluded that the bird "has a true spinal accessory nerve." Gross dissections indicated that the nerve innervates the anterior portion of M. cutaneous colli lateralis (part of M. cucullaris); this muscle, however, appears to be a dermal one, and we know of no evidence to suggest that it is homologous with the mammalian trapezius (George and Berger, 1966:271).

Table 1. Cranial Nerves in Birds

Number	Name	Function
I	Olfactory	Smell; small in most birds.
II	Optic[a]	Vision; large in most birds.
III	Oculomotor	Motor to medial, superior, and inferior rectus, and the inferior oblique ocular muscles, and to the levator of the upper eyelid; parasympathetic to eye (via ciliary ganglion), and (?) to Harder's gland.
IV	Trochlear	Motor to superior oblique muscle.
V	Trigeminal	Sensory to eye, oral cavity, bill, etc.; motor to certain jaw muscles and to Mm. tensor periorbita and depressor palpebrae.
VI	Abducens	Motor to lateral rectus muscle and to muscles of the nictitating membrane.
VII	Facial	Motor to muscles associated with the hyoid arch (e.g., depressor mandibulae, mylohyoideus posterior, sphincter colli); sensory (including taste) from palate, also part of nasal cavity, orbit, etc.; parasympathetic to salivary glands.
VIII	Acoustic	Hearing and equilibration.
IX	Glossopharyngeal	Sensory (including taste) from oral cavity, pharynx, larynx, etc.; motor to M. keratomandibularis (geniohyoideus); parasympathetic to salivary glands.
X	Vagus	Sensory (including taste) and motor (parasympathetic) for viscera of pharynx, neck, and thoracoabdominal cavity; motor to M. cucullaris. Recurrent laryngeal branch (? bulbar XI) is the motor supply to muscles of the larynx.
XI	Spinal accessory	Nature and function uncertain; see p. 56.
XII	Hypoglossal	Motor to muscles of tongue and syrinx.

[a] This is not a true nerve but is a fiber tract of the brain.

The number of *spinal nerves* varies in different birds (e.g., 39 pairs in *Columba,* 51 pairs in *Struthio*). Each nerve has a single dorsal root but several ventral rootlets. The *ventral rami* of several (usually 4 to 6) of the lower cervical nerves, and sometimes the first thoracic, form a *brachial plexus* whose branches distribute primarily to the wing. A variable number of lumbosacral nerves take part in the formation of three plexuses (*lumbar, sacral, pudendal*) that supply the lower extremity, the pelvis, and the caudal region (see Fisher and Goodman, 1955; Baumel, 1958).

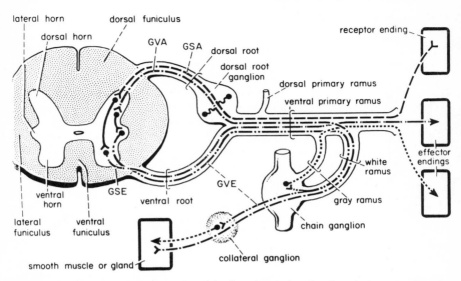

Figure 16. Diagrammatic cross section of the thoracic spinal cord and a spinal nerve to show the relationships among nerve components and ganglia of the sympathetic portion of the autonomic nervous system in man. (By permission of Andrew J. Berger, from *Elementary Human Anatomy*, John Wiley & Sons, copyright 1964.)

The Autonomic Nervous System. The subdivision of the nervous system into parts (central, peripheral, autonomic) is a pedagogical technique to facilitate the learning of a mass of information. In reality, all parts form one elaborate functional system, and the neurons of the autonomic system are located in both the CNS and the PNS (Fig. 16).

The autonomic nervous system is composed of motor fibers that distribute to glands, cardiac muscle, and smooth muscle in the viscera, blood vessels, and skin. It is an involuntary system that regulates the functions over which we have little or no conscious control; it consists of two-neuron chains composed of *preganglionic* and *postganglionic* neurons. Two anatomical or physiological parts of the autonomic nervous system are recognized: (1) *craniosacral* or *parasympathetic* components of cranial nerves III, VII, IX, X, XI, and also the sacral portion of the spinal cord, and (2) a *thoracolumbar* or *sympathetic* portion, which has its origin in cervical, thoracic, and upper lumbar segments of the spinal cord in birds. Visceral organs are innervated by both sympathetic and parasympathetic fibers; there is no parasympathetic supply to the limbs or the surface of the body. In general, parasympathetic innervation controls vegetative activities; e.g., it slows the heart rate, constricts the pupil of the eye, increases muscular contractions in the stomach and intestine, and empties the urinary and digestive systems. The sympathetic system prepares the animal for "fight or flight" responses; it is an emergency system that accelerates the heart rate, dilates the pupil, increases the flow of blood to skeletal muscles, and decreases digestive activities.

The Endocrine System

The endocrine system consists of widely separated ductless glands (Fig. 17). These glands secrete *hormones*, which diffuse directly into the blood stream and thence are carried to all parts of the body to stimulate or regulate the activities of other glands called *target organs*. Hence hormones, like the nervous system, integrate physiological processes.

The pituitary, thyroid, parathyroid, and adrenal glands are exclusively endocrine glands. The pancreas, duodenum, stomach, and gonads are both endocrine and exocrine glands. As an endocrine gland, the pancreas secretes hormones (*insulin, glucagon*) that pass into the blood stream and are important in carbohydrate metabolism; as an exocrine gland, the pancreas secretes digestive enzymes that reach the duodenum via pancreatic ducts. The secretory activity of the pancreas is controlled by parasympathetic nerves and by a hormone produced in the wall of the duodenum. The ovaries and testes secrete exocrine products (ova and spermatazoa, respectively) and endocrine products (sex hormones).

The *pituitary gland* (or *hypophysis*) is suspended by a stalk (the *infundibulum*) from the ventral surface of the hypothalamus. The hormones secreted

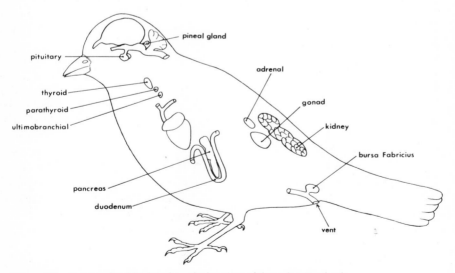

Figure 17. A generalized bird to show the locations of the endocrine glands.

by the hypophysis affect nearly all of the other endocrine glands. Its two lobes have different embryological origins and different functions.

The *anterior lobe* (*adenohypophysis*) is derived from ectodermal cells in the roof of the embryonic oral cavity. This lobe secretes *ACTH* (*adreno-corticotropic hormone*), *TSH* (*thyrotropin*: thyroid-stimulating hormone), and three gonadotropic hormones: FSH, LH, and *prolactin*. A gonadotropic hormone effects changes in the *primary sex organs* (*gonads*) and in other structures related to reproduction. FSH (follicle-stimulating hormone) stimulates the growth or maturation of the follicle cells surrounding individual ova in the female, or the seminiferous tubules and the interstitial cells in the testes of the male. LH (luteinizing hormone) causes ovulation; in some weaverfinches (Ploceidae) it effects changes in bill color and plumage. Prolactin is a *lactogenic* (milk-producing) *hormone* in mammals; it stimulates the secretion of *pigeon milk* in the crop of pigeons and doves. Prolactin also causes broodiness behavior, and it is involved in the development of incubation patches and in the inhibition of further secretions by the ovaries and testes; also, it may either stimulate or inhibit certain kinds of sexual behavior, such as bowing and cooing in pigeons.

The *posterior lobe* (*neurohypophysis*) of the pituitary gland arises embryologically from the floor of the diencephalon of the brain. This lobe has at least two hormones: an *antidiuretic hormone,* which aids in conserving water by its action on the kidney tubules; and *oxytocin,* which stimulates contraction of the oviduct in egg-laying.

The *thyroid gland* consists of two dark red lobes located at the base of the neck near the junction of the common carotid and subclavian arteries. *Thyroxine* (a hormone containing iodine) is produced in response to TSH from the anterior pituitary gland. Thyroxine plays a role in general metabolic processes, growth, and sexual development. It increases the number of red blood corpuscles, increases the heart rate, and is essential for pigmentation and the normal growth of feathers; in addition, it is presumed to be involved in the molting of feathers. Although injections of thyroxin into some passerine birds produce "migratory restlessness," the precise relationship between thyroxin and migration remains unknown.

Four *parathyroid glands* develop in the embryo, but because of fusion the number varies in the adult. The usual location of the glands is near the posterior poles of the paired thyroid glands; other locations are common, however, and accessory parathyroid tissue sometimes is found embedded in the thyroid gland or in the ultimobranchial bodies. The parathyroids produce *parathormone,* which controls the metabolism of calcium and phosphorus.

Located a short distance posterior to the parathyroid glands are the *ultimobranchial glands*; these are pink because of their rich blood supply. The glands secrete *calcitonin*; although little is known about this hormone in birds, "there is now a growing body of opinion that an important function of calcitonin may be to protect the skeleton from excessive resorption" (Simkiss and Dacke, 1971).

The *adrenal glands* (Latin *ad,* towards; *ren,* kidney) typically are paired glands that lie in the pelvis just anterior to the kidneys; they may, as a matter of individual variation, be fused into a single mass. Variation in color (orange, yellow, cream, pink, gray, reddish brown) may be due partly to diet (Hartman and Albertin, 1951). Each gland consists of two functionally unrelated parts that are named, because of the structure of the human kidney, the *cortex* (shell) and the *medulla* (pith). In birds, however, the two kinds of tissue are intermingled; the cortical material is called *interrenal tissue,* and the medullary tissue is termed *chromaffin* (colored) *tissue.*

The interrenal tissue arises from mesodermal cells near the developing kidneys in the embryo. The steroid secretions, which are essential to life, of this tissue fall into three main groups: *mineralocorticoids,* which aid in controlling the proper levels of sodium and potassium in the blood and in extracellular fluids; *glucocorticoids* (e.g., *cortisone*), which regulate the concentration of blood sugar, promote the storage of glycogen in the liver, and stimulate the conversion of body protein to sugar; and *sex hormones,* which consist of male hormones (*androgens*) and female hormones (*estrogens* and *progesterone*), both of which are secreted in both sexes. In addition to these three main groups of hormones, secretions of the interrenal tissue are believed to be important in enabling an animal to adjust to stresses in the environment, a process referred to as the *general adaptation syndrome*; secretions of the pituitary gland also are important in regulating the physiological responses to stress.

The chromaffin tissue has its embryonic origin in the same type of ectodermal cells that form the cells of the sympathetic portion of the autonomic nervous system. In essence, therefore, this means that the chromaffin tissue is composed of highly modified postganglionic neurons of the sympathetic nervous system; consequently, it receives preganglionic neurons but does not contain postganglionic neurons, and it is not under the control of the secretions of the pituitary gland.

The chromaffin tissue secretes *adrenalin* and in some birds, at least, a closely related hormone called *noradrenalin.* As is true of the sympathetic portion of the autonomic nervous system, adrenalin and noradrenalin prepare the animal for emergency situations: they produce an increase in heart rate and blood pressure, a rise in blood sugar (which is then available for increased muscular action), and a decrease in the activity of the muscles of the digestive tract.

The gonads are important endocrine glands. The *interstitial cells* of the testes secrete the male sex hormone called *testosterone* (an androgen; from the Greek word for man). The ovaries secrete *estrogens* (from the Greek word for a strong desire), such as *estradiol* and *progesterone*; inconclusive evidence suggests that they may also secrete some male hormone.

There is an intimate relationship among the sex hormones secreted by the pituitary gland, the interrenal tissue, and the gonads. In addition to their effect on each of the other glands, these hormones have a profound influence on the development and maintenance of secondary sex characters, on physiological processes, and on behavior.

The *thymus gland* consists of several nodules of lymphoid tissue widely scattered in the neck, from the angle of the jaw to the upper thorax. Although no hormone has yet been isolated from the thymus gland, there may be a seasonal increase in the size of the gland after sexual maturity has been reached, suggesting a possible hormonal function. Freeman (1971) discusses the relationship of thymectomy to the production of lymphocytes and a lowering of resistance to certain diseases.

The *bursa of Fabricius* is a lymphoepithelial gland that lies dorsal to the cloaca and opens into the roof of the proctodeum. It generally atrophies after hatching but is useful as a criterion of age in some groups of birds (e.g., gallinaceous birds). Although the wall of the bursa is composed primarily of lymphoid tissue, "at present it seems likely that the bursa produces secretions concerned with immunological competence, the activity of the adrenal 'cortex' and possibly the thyroid" (Freeman, 1971:576).

Endocrines, Autonomic Nerves, and Hypothalamus

It should be clear from the preceding discussion that both the endocrine system and the nervous system are concerned with the initiation and regulation of body functions, and that the actions of the two systems are very closely integrated. Indeed, the intimate relationship between nerve tissue and endocrine tissue is demonstrated by reference to the posterior lobe of the pituitary gland, which has its embryonic origin in the floor of the brain, and the adrenal medullary tissue, which arises from the same kind of cells that develop into the neurons of the sympathetic portion of the autonomic nervous system.

As already pointed out, the neuron centers in the hypothalamus effect the ultimate control and coordination of the autonomic nervous system. The hypothalamus also has a critical role in the secretory activity of the pituitary gland and, consequently, on nearly all of the other endocrine glands. This control results from the presence in the hypothalamus of neurosecretory cells. The secretory products (containing one or more hormones) of some groups of cells are conducted by nerve tracts to the posterior lobe of the pituitary gland. Neurohumors produced in other regions of the hypothalamus are carried to the anterior lobe by means of a *hypophysial portal system of veins*[3] (Figs. 18, 19).

The distinguished research by Benoit and Assenmacher (1959, 1970) on

[3] A portal system is one that begins and ends in capillary networks. Thus the hypophysial portal system begins in capillaries in the hypothalamus and ends in capillaries in the anterior lobe of the pituitary gland.

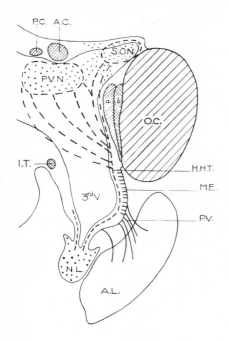

Figure 18. Schematic illustration of the hypothalamohypophysial neurosecretory pathway. A.C., anterior commissure; A.L., anterior lobe; H.H.T., hypothalamohypophyseal tract; I.T., infundibular tract; M.E., median eminence; N.L., neural lobe; O.C., optic chiasma; P.C., pallial commissure; P.V., portal veins; P.V.N., paraventricular nucleus; S.O.N., supraoptic nucleus; 3rd V., third ventricle; a, dorsal supraoptic decussation; b, ventral supraoptic decussation. (Courtesy of Jacques Benoit and Ivan Assenmacher.)

ducks has shown that one stimulus for increased seasonal activity of the hypothalamic secretory cells comes from the effects of changing photoperiod on the retina of the eye. The impulses evoked (primarily by orange and red radiations) in the sensory cells of the retina are conducted to the brain and, in part, to the hypothalamus itself, where they stimulate the secretory cells (Fig. 20). Moreover, these authors presented evidence that photoreceptors are present within the hypothalamus itself, and that these receptors were stimulated when light was applied through a quartz stick placed in the orbit even after both eyeballs had been removed.

What all of these complex interrelationships mean is that a bird is what it is because of its nervous and endocrine systems. The bird's anatomical development, its physiological processes, and its specific behavior patterns are the result of the interplay between the nervous and endocrine systems. Basically, the bird does what it does because it has little or no control over the physiological and neurophysiological processes that take place within its body; its brain, unlike that of man, has not evolved so that a cerebral cortex can modify the basic drives resulting from the interaction of hormones and neurophysiological processes. This is the basis for the stereotyped behavior patterns characteristic of birds.

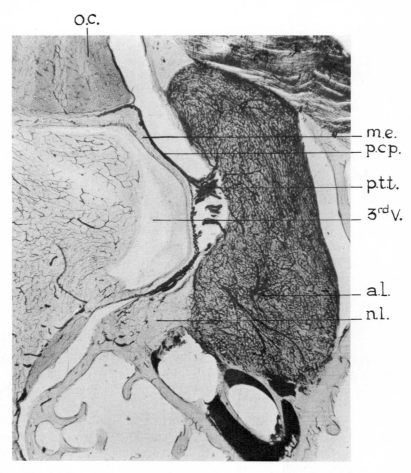

O.C.

m.e.
p.c.p.

p.t.t.

3rdV.

a.l.
n.l.

Figure 19. Median sagittal section through the hypophysial complex. Celloidin, 50 μ; a previous intracardiac injection of India ink was performed before autopsy. a.l., anterior lobe; m.e., median eminence; n.l., neural lobe; o.c., optic chiasma; p.c.p., primary capillary plexus; p.t.t., portotuberal tract; 3rd V., third ventricle. Magnification: ×26. (Courtesy of Jacques Benoit.)

The Circulatory System

The circulatory system transports food materials and oxygen to the cells and the waste products of metabolism away from the cells; it also carries hormones, antibodies, and leucocytes. Thus it aids the nervous system in integrative processes, and it is important in protecting the animal against disease and infection, in repairing injuries, and in controlling body temperature. It is composed of the *blood-vascular system* and the *lymphatic system*.

The Blood-Vascular System. Blood is composed of *plasma, corpuscles,* and *thrombocytes.* Plasma carries the ultimate products of digestion (amino acids, glucose, glycerol, fatty acids), nitrogenous waste products (urea, uric acid, creatinine) of metabolism, hormones, antibodies, carbon dioxide, inorganic salts, and plasma proteins (albumins, globulins, fibrinogen).

Several types of white corpuscles (*leucocytes*) aid in combating infections. Red corpuscles (*erythrocytes*) and *thrombocytes* are nucleated in birds and generally oval in shape. Thrombocytes are homologous to blood platelets in mammals; they play an important role in blood coagulation.

The heart lies in the anterior part of the body cavity deep to the sternum (Fig. 21), and there is enclosed in a protective fibrous sac, the *pericardium.* Birds, like mammals, have a four-chambered heart, divided into right and left halves by *interatrial* and *interventricular septa.* Venous blood from all parts of the body enters the "right heart," passing from its *atrium* through a right atrioventricular opening into the right *ventricle*; contraction of the right ventricle pumps the blood through the *pulmonary trunk* to the lungs. This path of blood through the right side of the heart and lungs is the *pulmonary circuit.*

Four *pulmonary veins* conduct oxygen-rich blood from the lungs to the left

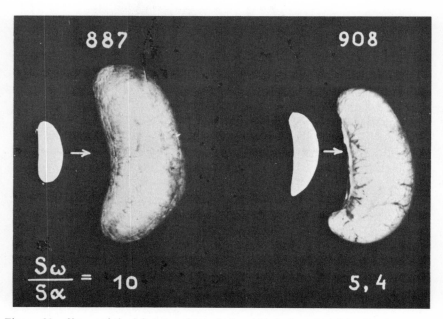

Figure 20. Shapes of the left testes of two ducks which had received light directly onto the hypothalamus. In white, drawings of the testes before the experiment, compared with photographs taken at the time of autopsy. The extent of stimulation is proportional to the increase in testicular surface during the experiment. Magnification: ×1. (Courtesy of Jacques Benoit.)

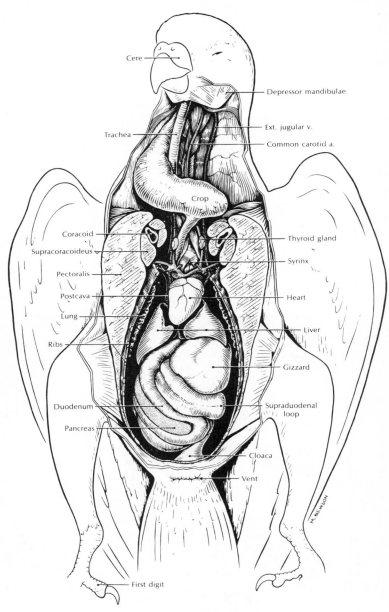

Figure 21. The viscera of the Budgerigar (*Melopsittacus undulatus*) *in situ*; the sternum and abdominal wall have been removed. (By permission of Howard E. Evans, from Petrak's *Diseases of Cage and Aviary Birds,* Lea and Febiger, copyright 1969.)

atrium, and thence to the left ventricle. Contraction of the left ventricle pumps the blood through the *right aortic arch,* whose branches carry blood to all parts of the body. This is the *systemic circuit* (Fig. 22). In correlation with the greater work necessary to pump the blood throughout the body, the left ventricle has a much thicker wall than the right ventricle.

Coronary arteries (usually two in number) arise from the first part of the aorta and pass over the surface of the heart, sending branches into the muscle. *Cardiac veins* carry the blood from the heart muscle back to the right atrium.

The innate rhythmicity of cardiac muscle results from the highly specialized nature of this type of muscle. The *sinoatrial node* in the wall of the right atrium initiates the heartbeat and, therefore, is referred to as the "pacemaker" of the heart. It stimulates contraction of the right atrium, and this in turn stimulates the *atrioventricular node,* located in the bottom of the interatrial septum. An *atrioventricular bundle* leads to the ventricular muscle.

Birds are unique in having only a *right systemic aortic arch.* The aortic arch gives rise to right and left *brachiocephalic arteries.* Each of these bifurcates to form *subclavian* and *common carotid arteries.* The numerous branches of the subclavian artery supply the wing, sternum, ribs, and associated muscles. The common carotid arteries supply the head and neck.

The aorta gives rise to three unpaired branches in the abdomen (Fig. 23). A *celiac artery* supplies the liver, pancreas, spleen, proventriculus, gizzard, duodenum, and lower part of the esophagus. A *cranial* (anterior) *mesenteric*

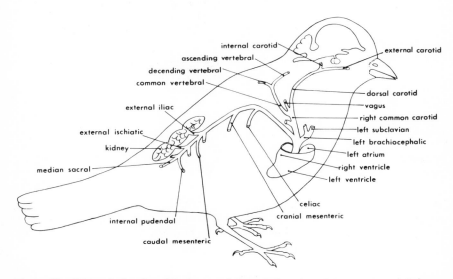

Figure 22. Schematic drawing of the heart and the main branches of the aorta; most of the segmental branches have been omitted. Terminology after Baumel (1964).

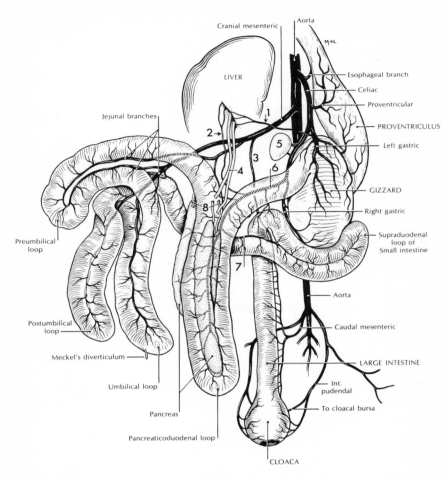

Figure 23. The arterial supply of the digestive tract. 1 and 2, Hepatic arteries; 3, colic artery; 4, bile ducts; 5, spleen; 6, pancreaticoduodenal artery; 7, ileocolic artery; 8, pancreatic arteries. (By permission of Howard E. Evans, from Petrak's *Diseases of Cage and Aviary Birds,* Lea and Febiger, copyright 1969.)

artery carries most of the blood to the small intestine and ceca. A *caudal* (posterior) *mesenteric artery* supplies the large intestine and cloaca; in some birds this branch arises from the right internal pudendal artery.

The aorta gives off a series of segmental branches to the body wall (e.g., *lumbar, lateral sacral,* and *lateral coccygeal arteries*) and visceral branches to the abdomen and pelvis (e.g., *adrenal, renal, internal spermatic, ovarian arteries*). Right and left *external iliac arteries* pass downward and laterally between the cranial and middle lobes of the kidneys. Paired *external ischiatic arteries* follow a similar course but emerge between the middle and caudal divisions of the kidneys. The external iliac artery gives rise to the *femoral artery,* which supplies the bones and muscles of the thigh; the external ischiatic artery is the main artery to the lower limb, and it has more than 20 named branches. After giving off paired *internal pudendal arteries* (which supply the cloaca, bursa of Fabricius, and muscles of the pelvic wall), the aorta continues as the *median sacral artery*; it becomes the median *coccygeal artery* at the level of the coccygeal vertebrae.

Venous tributaries.   By tradition one speaks of *tributaries* of veins and *branches* of arteries. Veins have much thinner walls than arteries and are larger in diameter. Large veins also possess valves; these prevent blood from flowing backward in the system.

In general, veins run with arteries and a vein has the same name as the corresponding artery. One exception is that blood is returned from the head and neck in *jugular* (Latin *jugulum,* throat) *veins,* whereas carotid arteries supply the head and neck. Blood from the head, neck, and wings drains into either a right or left *anterior vena cava* (precaval vein). Blood from the legs, tail, and posterior body wall drains into a single *posterior vena cava* (postcaval vein). Both venae cavae empty into the right atrium of the heart.

Venous blood from the digestive tract and related glands flows through vessels corresponding to the branches of the celiac, cranial, and caudal mesenteric arteries. All of these terminate in the *hepatic* (Greek *hepatos,* liver) *portal vein,* which conducts blood directly to the liver. A portal system was defined (footnote 3) in relation to the hypophysial portal system as one that begins and ends in capillaries. The hepatic portal vein begins in capillary networks in the walls of the stomach, intestine, pancreas, and spleen; it ends in a capillary network in the liver. Hence the products of digestion are carried directly to the liver in the hepatic portal vein. Blood leaves the liver through two large *hepatic veins*; these drain into the posterior vena cava just before it enters the heart.

A modified *renal portal circulation* is found in birds (Fig. 24). "Birds have the anatomical basis for a renal portal system but have the ability to switch it on or off according to requirements" (Akester, 1971). A unique *renal portal valve,* containing smooth muscle, occurs at the junction between the external iliac vein and the renal vein. Consequently, venous blood can bypass a kidney, and the pattern of flow of blood through the two kidneys may differ during the same time span. (For other details of the vascular system, see Glenny, 1955; Baumel and Gerchman, 1968.)

The Lymphatic System.   This system is composed of a specialized form of connective tissue and a series of vessels that conduct lymph from nearly all parts of the body back to the venous system. Blood plasma diffuses through capillary

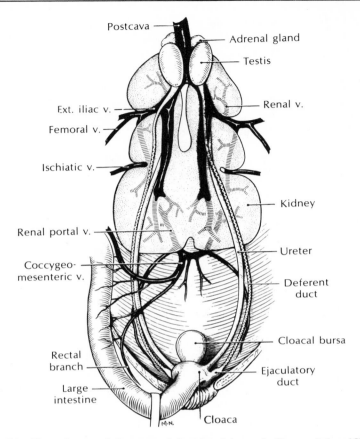

Postcava

Adrenal gland

Testis

Ext. iliac v.

Renal v.

Femoral v.

Ischiatic v.

Kidney

Renal portal v.

Ureter

Coccygeo-
mesenteric v.

Deferent
duct

Cloacal bursa

Rectal
branch

Ejaculatory
duct

Large
intestine

Cloaca

Figure 24. The male urogenital system and the veins of the trunk. (By permission of Howard E. Evans, from Petrak's *Diseases of Cage and Aviary Birds,* Lea and Febiger, copyright 1969.)

walls into interstitial spaces between cells, which are bathed by this tissue fluid. Some of the fluid passes back into the venous end of the capillary network and thence into veins; some of the tissue fluid, however, enters lymphatic capillaries and then is called *lymph.* Lymphatic capillaries lead into progressively larger channels which drain into the venous system—usually into jugular, precaval, and hypogastric veins.

In mammals, the products of fat digestion enter *lacteals* in the intestinal villi, from which they pass through an *intestinal lymph trunk* to a *thoracic duct;* this drains into the venous system in the neck. The breakdown products of protein and carbohydrate metabolism are carried directly to the liver via the hepatic portal vein. Central lacteals are absent in the villi in birds (Hill, 1971), and large amounts of fatty acids reach the liver through the hepatic portal vein. Right and left thoracic ducts are present in birds and apparently receive some of the absorbed fat; these ducts also receive lymph from the thorax, abdomen, pelvis, and lower limb. The thoracic ducts may drain into the right and left jugular veins or into the right and left anterior venae cavae.

True lymph nodes have been described for only a few species of birds (e.g., Anatidae, *Fulica, Larus*; Biggs, 1957); these consist of a central sinus enveloped by dense lymphoid tissue with germinal centers. Payne (1971:1027) suggested that "the avian node may therefore be a less efficient filter of lymph than the mammalian node." Discrete patches of lymphoid tissue are found in the thymus gland, spleen, wall of the intestine, ceca, and bursa of Fabricius (see Payne, 1971).

Pulsating *lymph hearts,* located near the first segmental coccygeal vein, have been found in the embryos of all birds studied. Although they usually atrophy after hatching, they presumably persist throughout life in *Struthio, Casuarius,* Anatidae, Laridae, Ciconiidae, and Passeriformes.

The Respiratory System

Oxygen is the fuel used by cells for carrying on chemical reactions involving the "burning" or oxidation of food materials with a consequent release of energy and two waste products: water and carbon dioxide. *Internal respiration* consists of the oxidative processes that occur within cells. *External respiration* involves the passage of oxygen from atmospheric air into the blood capillaries in the lungs and, at the same time, the release of carbon dioxide through the walls of blood and respiratory capillaries in the lungs.

The respiratory system is composed of *two lungs,* several air sacs, and a series of channels that conduct air to and from them. In sequence, these channels are the *nasal cavities, pharynx, larynx, trachea,* and two *bronchi* (singular, *bronchus;* from the Greek word for windpipe).

The two *nasal cavities* are separated by a *nasal septum,* which supports, on each side, three thin *conchae* (Greek, shell), also called *turbinates* (scroll-shaped), covered by mucous membrane (Fig. 25). The mucus traps dust in the air, and the membranes aid in warming the air. The endings of olfactory neurons are embedded in the membrane of the posterior concha. Two *internal nares* lead from the nasal cavities into the roof of the oral cavity; usually the nares open through a single slit between two *palatine folds* (Latin *palatum,* roof of the mouth).

The *pharynx* (Greek, throat) begins at the posterior border of the tongue. It is a space where the digestive and respiratory systems cross. The nasal cavities lie above the oral cavity, whereas the trachea lies ventral to the esophagus in the neck; consequently, the two paths cross in the pharynx.

Figure 25. A section through the nasal chambers of the Broad-tailed Hummingbird (*Selasphorus platycercus*) between the olfactory bulb and the external nares. C, choana; D, diplöe space; E, olfactory epithelium; G, salivary gland; H, Harder's gland; L, lower mandible; M, mouth cavity; MC, middle concha; O, olfactory chamber; R, respiratory chamber; S, nasal septum; T, tongue. ×12. (Courtesy of Stanley Cobb and the editor of *Archives of Neurology*.)

The *larynx* (Greek, gullet) does not have a primary sound-producing function in birds but serves as a valve to regulate the flow of air between the pharynx and the trachea (and to prevent the passage of food into the trachea). The bird larynx is composed of several small cartilages (*cricoid, procricoid, arytenoid*), muscle, and a lining of mucous membrane. Two *laryngeal folds* bound the *glottis,* the entrance to the trachea.

The *trachea* is held open, for the free passage of air, by a series of cartilaginous or bony rings. Usually elliptical or round in cross section, the trachea is compressed dorsoventrally in parrots. In most birds the trachea follows a relatively straight course from the glottis into the thorax, but in some it is greatly elongated and looped. These loops may be subcutaneous in the cervical, thoracic, or abdominal regions (e.g., in *Anseranas, Tetrao,* male of *Ortalis, Rostratula, Manucodia, Phonygammus*). The trachea may be looped within the thorax (male of *Ibis ibis*) or within the sternum (*Olor columbianus, Grus canadensis*; Fig. 26); in some cranes and swans the trachea winds through the entire length of the carina of the sternum. In the guineafowl *Guttera plumifera* the trachea of both sexes is looped into the furcula. In different species of a given genus, or in male and female of the same species, the trachea may be either coiled or straight.

The trachea may contain one (screamers, males of some ducks, some cotingas) or two (some ducks) dilatations between the larynx and the bronchi, in addition to the *tracheal bulla* (*bulla ossea*), an enlargement at its inferior end in ducks. These specializations modify the sounds produced by the syrinx. Inferiorly the trachea divides to form right and left *extrapulmonary primary bronchi*. In most birds (some exceptions: *Ciconia,* swallows) at least the first few bronchial rings are incomplete, with only a membrane completing the ring (a semiring) on the inner side. After the bronchi enter the lungs, they are called right and left *intrapulmonary primary bronchi*; each of these, in the domestic chicken, gives rise to four series of *secondary bronchi,* which, in turn, give rise to *tertiary bronchi* (Figs. 27, 28). In describing this branching pattern, King and Molony (1971) point out that anastomosing channels between two of the secondary bronchi and their tertiary branches form "one great functional unit

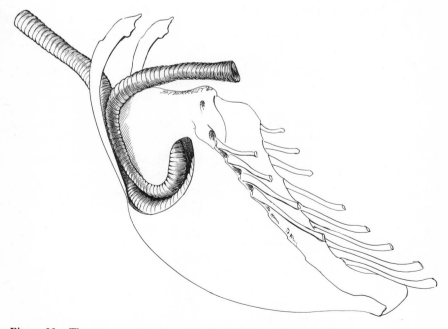

Figure 26. The sternum and trachea of the Sandhill Crane (*Grus canadensis*).

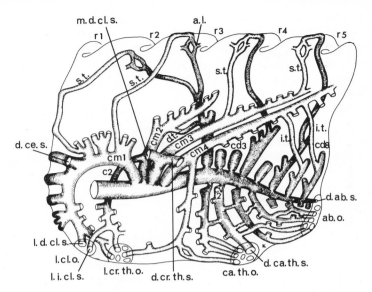

Figure 27. Ventromedial view of the right lung of an adult chicken drawn from casts and dissections, to show the primary bronchi and main secondary bronchi, and examples of tertiary bronchi and their connections with air sacs. a.l., anastomotic line caused by embryonic fusion of the primordial tertiary bronchi; ab.o., ostium of abdominal air sac; ca.th.o., ostium of caudal thoracic sac; cd, caudodorsal secondary bronchi; cm, craniomedial secondary bronchi; c2, circumflex branch of cm2; cv, caudoventral secondary bronchi; d.ab.s., direct connection of abdominal sac; d.ca.th.s., direction connection of caudal thoracic sac; d.ce.s., direct connection of cervical sac; d.cr.th.s., direct connection of cranial thoracic sac; d.t., a deep tertiary bronchus; i.t., tertiary bronchi of intermediate depth; l.cl.o., lateral ostium of clavicular sac; l.cr.th.o., lateral ostium of cranial thoracic sac; l.d.cl.s., lateral direct connection of clavicular sac; l.i.cl.s., lateral indirect connection of clavicular sac; m.d.cl.s., medial direct connection to clavicular sac; r1–5, five impressions of vertebral ribs 2 to 6; s.t., superficial tertiary bronchi. Cranial is toward the left and dorsal toward the top of drawing. (Courtesy of A. S. King and the editor of *International Review of General and Experimental Zoology*.)

comprising about two-thirds of the lung." Numerous openings in the wall of each tertiary bronchus lead into *atria,* which are interconnected with *air* or *respiratory capillaries*. It is through the walls of the latter that oxygen and carbon dioxide exchange takes place.

Thin-walled air sacs develop embryologically from the bronchial tree, and they occupy most of the space between visceral organs in the adult. Different birds may have three or four paired air sacs and one or two unpaired sacs. The paired sacs are *cranial thoracic, caudal thoracic, abdominal,* and, in a few birds, *cervical*; paired sacs frequently are asymmetrical in shape. The unpaired air sacs are the *median clavicular* (or interclavicular) and *median cervical* (Fig.

29). The cervical air sac sends diverticulae into the bones of the vertebral column; an extension of the clavicular air sac enters the humerus through its pneumatic foramen; diverticulae of the abdominal air sacs pneumatize the synsacrum and femur. The bones most frequently pneumatized are the humerus, femur, ribs, and vertebrae (Fig. 30), but in screamers and hornbills (Anhimidae and Bucerotidae) even the pygostyle and the phalanges of the fingers and toes are pneumatic. In some birds (e.g., swifts, many passerines) the cranial bones are pneumatized by extensions from the nasopharyngeal chambers.

Flying birds have very high oxygen requirements, and their respiratory system has evolved to accommodate these demands. A bird in flight may require as much as 21 times more oxygen than when at rest (Tucker, 1968). King and Molony (1971:152) explained that the increase in oxygen is "provided by a massive increase in the rate and depth of respiration, a concurrent increase in cardiac output, and probably an increase in the diffusing capacity of the lungs." However, more is involved than an increase in respiratory rate,

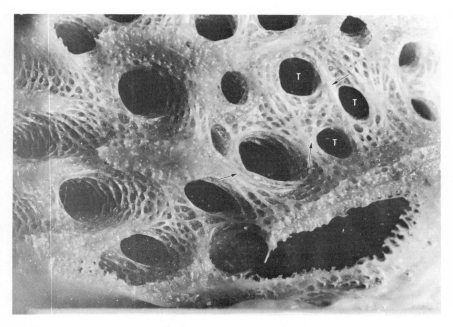

Figure 28. Wall of dissected chicken lung fixed *in situ* by intratracheal perfusion with 5% formalin, showing the interior of a secondary bronchus and the openings into many tertiary bronchi. The arrows point to bronchial muscle around the openings into tertiary bronchi. (Courtesy of A. S. King, A. F. Cowie, and the editor of *Journal of Anatomy*.)

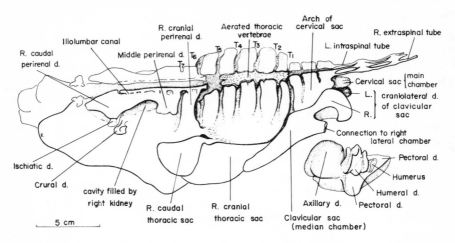

Figure 29. Drawing of the right side of a cast of the lungs and air sacs of an adult chicken. The right lateral chamber of the clavicular air sac has been detached (lower right). T1–7, grooves for the seven vertebral ribs; d., diverticulum; L., left; R., right. (By permission of A. S. King, from *The Anatomy of Domestic Animals*, W. B. Saunders, copyright 1972.)

namely, the path of airflow through the lungs and air-sac system and the nature of the oxygen-carbon dioxide exchange. Bretz and Schmidt-Nielsen (1971) demonstrated that there is a unidirectional flow of air through most of the lung during both inspiration and expiration. During inspiration air flows through the primary bronchi both to the posterior parts of the two lungs and directly to the posterior air sacs (caudal thoracic and abdominal), but it flows indirectly to the anterior air sacs via secondary and tertiary bronchi, during which oxygen is given off and carbon dioxide is picked up. Consequently, the concentration of carbon dioxide is greater in the anterior sacs (6 or 7 percent) than in the posterior sacs (3 or 4 percent). During expiration oxygen-rich air from the posterior sacs flows into tertiary bronchi of the lungs, whereas air from the anterior sacs takes the most direct path via craniomedial secondary bronchi to the primary bronchi and trachea. However, enough air from the anterior air sacs remains in the trachea "to ensure the right concentration of carbon dioxide in the posterior sacs after the next inhalation." Schmidt-Nielsen (1971) described the implications of this unidirectional airflow with respect to the maintenance of an adequate concentration of carbon dioxide in the air and in the blood for the regulation of breathing movements. (See also Duncker, 1971.) Electromyography has shown that at least a dozen different muscles are involved in respiration in nonflying birds (King and Molony, 1971).

The lungs are enclosed in serous-lined right and left *pleural sacs*. A *broncho-*

pleural membrane (pulmonary aponeurosis) stretches obliquely across the chest ventral to the lung. The membrane actually is a double-layered structure: the dorsal layer is parietal pleura, and the ventral layer is the wall of the cranial thoracic air sac. Four slips of striated muscle (*costopulmonary muscle*) pass laterally from the bronchopleural membrane to insert near the midpoint of several true ribs. Lying ventral to the air sac is the *bronchoperitoneal membrane* (oblique septum), also a double-layered structure: its dorsal component is air-sac wall, and its ventral component is parietal peritoneum. A sheet of smooth muscle (about 1 cm wide) forms much of the medial border of the bronchoperitoneal membrane. Birds do not have a thoracic diaphragm (a muscular partition that separates the thoracic and abdominal cavities in mammals).

The Digestive System

This system is specialized to take in food, to secrete enzymes that break up food into simple compounds that can be absorbed through the walls of the intestine, and to eliminate nondigestible waste products. The digestive system is a tube that extends throughout the body from the bill-tip to the *vent,* the external

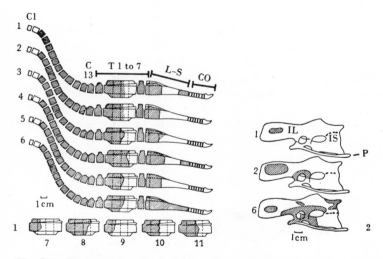

Figure 30. Left: Diagrams of vertebral columns of six specimens of the domestic chicken and of the second to fifth dorsal vertebrae in five additional specimens to show the pneumaticized bones. Right: Diagrams of left halves of the pelvic girdles of the three specimens, showing the pneumatized areas (hatched). Dotted lines show the approximate boundaries between the ilium (IL) and ischium (IS); P, pubis; C1–13, cervical vertebrae; T1–7, dorsal vertebrae; L-S, lumbosacral vertebrae; CO, coccygeal vertebrae, including the pygostyle. (Courtesy of A. S. King and the editor of *Acta Anatomica.*)

opening of the *cloaca*. Three main subdivisions are the mouth (oral cavity), throat (pharynx), and alimentary canal (esophagus, proventriculus, ventriculus, small intestine, and large intestine).

The bill is the usual food-getting organ of birds, although some (e.g., hawks, owls, the Osprey) catch prey with their feet. The *maxilla* is, by definition, the upper jaw and the *mandible* is the lower jaw in vertebrate animals. Because of the absence of teeth and other specializations in birds, however, both upper and lower jaws traditionally are referred to as mandibles. The upper jaw is formed primarily of two bones (*premaxillae* and *nasal bones*) and, in many birds, is capable of limited movement at a *nasofrontal hinge* (between the nasal bones and the frontal bones); this movement is called *kinesis* (Bock, 1964; Zusi, 1967).

On the basis of the relative lengths and the relationships of the two mandibles, different types of bills have been named epignathous, hypognathous, fissirostral, conirostral, etc. (Coues, 1903:105). Sexual dimorphism in shape (see *Heteralocha*, p. 755) or color and seasonal changes in color (e.g., *Sturnus vulgaris, Passer domesticus*) of the bill occur.

The tongue, adapted to feeding habits, exhibits great diversity in form and structure. Among different birds it serves as a probe, a sieve, a capillary tube, a brush, a rasp, etc. Consequently, in shape the tongue may be rectangular, cylindrical, lanceolate, spoon-shaped, flat, cupped, grooved, tubular, or bifid; it may be fleshy, horny, spined, "feathery," or brush-tipped (Scharnke, 1932). The tongue is very small (rudimentary) in gannets, pelicans, ibises, spoonbills, storks, and some kingfishers. The tongues of some woodpeckers and hummingbirds are very long and can be protruded for some distance; in the wryneck *Jynx ruficollis* the tongue is nearly two-thirds the length (exclusive of tail) of the bird. In such birds, the posterior elements of the hyoid (the support for the tongue) curve upward to follow the contour of the skull (Fig. 31).

The bones of the two mandibles are covered by a horny sheath, the *rhamphotheca*; it is subdivided into the *rhinotheca*, the covering of the upper mandible, and the *gnathotheca*, the covering of the lower mandible. A median

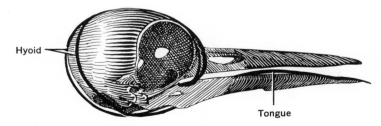

Figure 31. Skull of a Hairy Woodpecker (*Dendrocopus villosus*) to show the relations of the hyoid apparatus to the tongue and the skull.

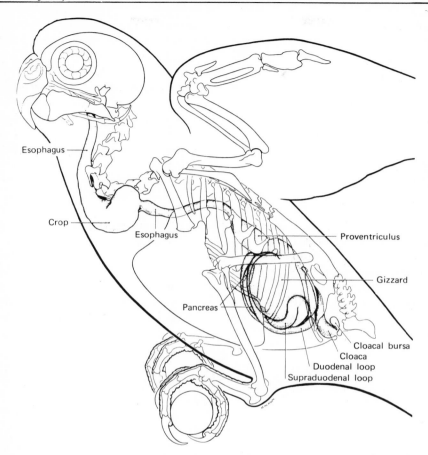

Figure 32. The digestive tract of the Budgerigar in relation to the skeleton. (By permission of Howard E. Evans, from Petrak's *Diseases of Cage and Aviary Birds,* Lea and Febiger, copyright 1969.)

ridge and a pair of lateral ridges on the palatal surface of the rhinotheca form the "horny palate" of Sushkin (1927) and Beecher (1951b). Caudal to the horny palate, *palatal folds* bound a palatal fissure in the roof of the pharynx. The internal nares open into the roof of the pharynx anteriorly; posteriorly, two auditory tubes from the middle ear open into it, usually into a single, median fossa. The glottis lies in the floor of the pharynx.

Dorsal to the glottis, the pharynx leads into the *esophagus,* which conveys food from the pharynx to the stomach (Fig. 32). In some birds (e.g., Galliformes, Thinocoridae, Pteroclidae, Columbidae, Psittacidae, and some members of the Drepanididae, Icteridae, Zosteropidae, and Fringillidae) there is a permanent dilatation (*crop*) in the esophagus; in some species the esophagus can be dilated to form a temporary crop. In *Opisthocomus* the crop is unusually large and muscular, forming a double loop which impinges on the pectoral muscles and the carina of the sternum. Esophageal diverticulae may be used for food storage, and in some birds they may play a role in courtship behavior.

The stomach of most birds is divided into a glandular *proventriculus* and a muscular *ventriculus* or *gizzard*. The proventriculus secretes mucus, a strong acid, and digestive enzymes; in some birds (e.g., petrels, cormorants, herons, gulls, terns, some hawks, some woodpeckers), it also serves as a temporary storage organ. The gizzard is lined by a series of hard, sometimes leathery, ridges. Birds with well-developed gizzards typically eat *grit* (quartz and other hard objects). The heavy muscular wall of the gizzard, its internal ridges, and grit all aid in grinding and mixing food with digestive enzymes. In cormorants, anhingas, storks, and herons, a third chamber, the *pyloric stomach*, is located between the gizzard and the duodenum.

In certain tanagers (*Chlorophonia* and *Tanagra*) the gizzard is reduced to a thin, membranous area lying between the proventriculus and the duodenum (Fig. 33). In nectar-eating hummingbirds, honeyeaters, and sunbirds (Tro-

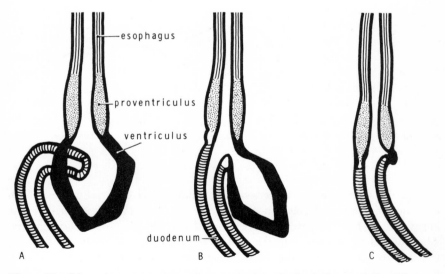

Figure 33. Schematic representation to show progressive reduction in the ventriculus of fruit-eating birds. (*A*) Ventriculus of a primitive insect-eating flowerpecker (Dicaeidae); (*B*) ventriculus of a specialized flowerpecker that eats both fruit and animal food; (*C*) rudimentary ventriculus of a euphonia (Thraupidae). (After Desselberger, 1932.)

chilidae, Meliphagidae, Nectariniidae) the openings to the esophagus and the duodenum are adjacent to each other rather than being at the opposite ends of the stomach. This specialization is carried further in certain fruit-eating flower-peckers (Dicaeidae), in which the gizzard is a blind diverticulum from the lowermost end of the proventriculus, so that the latter leads directly into the duodenum (Desselberger, 1932).

The small intestine is composed of a short, U-shaped *duodenum* and a long, coiled *ileum*. The pattern of coiling of the intestine differs considerably among birds and has been used taxonomically. The pancreas lies between the two limbs of the duodenum, and the pancreatic and bile ducts open into it. The final stage of digestion occurs in the ileum, and it is there that most of the breakdown products of digestion are absorbed across the mucosal lining of the intestinal wall.

The large intestine in birds consists of the *rectum* (the term "colon" is inappropriate). The junction between small and large intestine is marked by the presence of a pair of ceca in most birds (see p. 576). The rectum usually is a short, straight tube, but in *Opisthocomus, Chauna,* and struthious birds it is long and coiled. The functions of the rectum apparently are limited to the movement of intestinal contents, their temporary storage, and the reabsorption of water from them. The rectum empties into the *coprodeum*, the innermost of three compartments of the *cloaca* (Latin, sewer). The coprodeum is continuous with the *urodeum*, into which the ureters from the kidneys and the genital ducts from the gonads drain (Fig. 34). The urodeum continues as the *procto-deum*; it opens to the exterior of the body through the *vent* (erroneously called "anus," which is the external opening of the digestive tract only, as in most mammals). The bursa of Fabricius opens into the roof of the proctodeum, and a *copulatory organ* is found in its floor in some birds.

The *liver, pancreas,* and *salivary glands* are accessory glands derived embryologically from the digestive tube (the liver and pancreas from endoderm, the salivary glands from stomodeal ectoderm); they retain their connection to it by means of ducts, through which their exocrine secretions reach the lumen of the gut.

The primary function of salivary glands in most birds presumably is to moisten food, and birds possess several groups of these glands: palatine, angular, sublingual, maxillary, mandibular, sphenopterygoid, cricoarytenoid, and esophageal. They are said to be absent in the Pelecaniformes. Most avian salivary glands contain mucus-secreting cells, but serous cells (which presumably secrete an enzyme) have been reported in seed-eating birds, especially fringillids. Woodpeckers and wrynecks have a highly specialized *glandula picorum* (Latin *picus,* woodpecker), which secretes a sticky fluid effective in trapping ants and other insect food on the tongue. Mandibular salivary glands are hypertrophied in the Gray Jay (*Perisoreus canadensis*). Their mucous

DORSAL

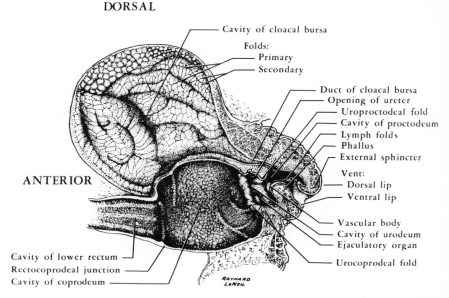

Cavity of cloacal bursa
Folds:
— Primary
— Secondary

Duct of cloacal bursa
Opening of ureter
Uroproctodeal fold
Cavity of proctodeum
Lymph folds
Phallus
External sphincter
Vent:
Dorsal lip
Ventral lip

Vascular body
Cavity of urodeum
Ejaculatory organ
Urocoprodeal fold

ANTERIOR

Cavity of lower rectum
Rectocoprodeal junction
Cavity of coprodeum

Figure 34. Midsagittal section of the cloaca and bursa of Fabricius of the chicken. (By permission of Alfred M. Lucas, from *Diseases of Poultry,* Iowa State University Press, copyright 1965.)

secretion presumably coats the tongue, "thereby making it sticky and suitable for probing into crevices for insects and under the scales of cones for seeds during the winter when other food is scarce" (Bock, 1961).

The Chimney Swift (*Chaetura pelagica*), European Common Swift (*Apus apus*), House Swift (*Apus affinis*), and members of the tropical Pacific cave swiftlet genus *Collocalia* have salivary glands that secrete an adhesive substance used in nest-building; in *Collocalia* the entire nest is composed of dried saliva (Fig. 35).

Indicative of its many functions, the liver is the largest of the viscera. Its relative size varies both with diet and with age. Of the two major lobes, the right is larger than the left in most birds, but there is much individual variation not only in the relative sizes of the two lobes but also in their shapes (Lucas and Denington, 1956). A series of peritoneal ligaments (falciform, coronary, gastrohepatic, etc.) connects the liver with adjacent structures.

The liver has many functions not related to digestion proper; it stores sugars as glycogen, stores fats, is active in the synthesis of glycogen and proteins, forms uric acid, produces heparin, removes foreign substances from the blood,

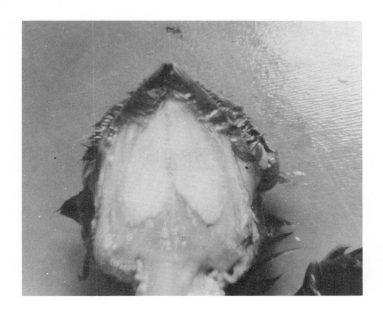

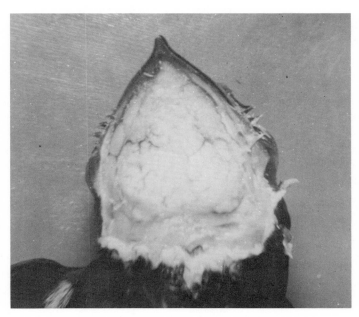

Figure 35. Developed and undeveloped salivary glands of the Indian House Swift (*Apus a. affinis*); Baroda, India, August 26, 1964. This swift builds its nest of grasses and feathers, cementing the materials together with saliva.

83

and produces bile. Bile has two functions: neutralizing acidity and emulsifying fats.

A hepatic or bile duct leads from each lobe of the liver directly into the duodenum. Many birds have a *gall bladder*, a vesicle for the storage and concentration of the bile secreted by the liver. The gall bladder is attached by peritoneum to the inferior surface of the right lobe of the liver and usually is oval or saccular in form. It is very long and tubular ("intestiniform") in the Capitonidae, Indicatoridae, Ramphastidae, and some Picidae. The gall bladder is said to be absent in the Trochilidae and in one species of the genus *Falco* (*F. peregrinus*). It is present in some genera of the Columbidae, Psittacidae, Cuculidae, and Picidae, but absent in others. In certain instances (e.g., *Rhea, Mergus, Grus, Cuculus*), absence of the gall bladder seems to be a matter of individual variation.

The pancreas lies in a two-layered fold (*mesoduodenum*) that extends between the two limbs of the duodenum. The pancreas arises from three pancreatic buds in the embryo, but usually is a single-lobed or bilobed organ, exhibiting, in the adult, considerable individual variation in configuration. Its endocrine functions were discussed earlier. The pancreas also secretes pancreatic juice, a mixture of enzymes that aid in the digestion of proteins, carbohydrates, and fats. The enzymes reach the duodenum through two or three pancreatic ducts. (See Langslow and Hales, 1971.)

The Urogenital System

"Urogenital" is a compound word formed from "urinary" (Latin *urina*, urine) and "genital" (Latin *gignere*, to beget). The genital system is the reproductive system. The urinary system is also called the *excretory system* because it excretes urine. The two systems are discussed together because of their intimate adult and embryological relationships; for example, ducts that arise in relation to the developing kidneys persist but are used in the adult for the passage of the male germ cells.

The Urinary System. The urinary system consists of paired *kidneys* and *ureters*. The kidneys lie in depressions between transverse processes of synsacral vertebrae and, hence, are situated deep in the pelvis against the vertebral column; the entire irregular area occupied by the kidneys is called the *renal depression*. Avian kidneys usually are described as irregular-shaped structures that are composed of three interconnected lobes in most birds, but some hornbills and kingfishers and many passerine birds have only two obvious divisions, and *Apteryx* has five. Moreover, numerous other variations in lobation and shape occur throughout the orders of birds. Part of the variation appears to be directly related to the shape of the pelvis (Kuroda, 1963), which in turn determines the shape of the renal depression. The posterior lobes often are fused along their medial borders, especially in passerine birds. Among different species, the weight of the two kidneys varies from about 0.5 to 2.6 percent of total body weight.

A tubular ureter conducts urine from each kidney to the urodeum (see Fig. 24), where the whitish, semisolid urine is mixed with the feces. In the male, each ureter passes posteriorly parallel to the ductus deferens; in the female, the ureters lie dorsal to the oviducts. Adult birds do not have a urinary bladder.

The nitrogenous wastes of metabolism are synthesized in both the liver and the kidney. References on the avian kidney state that the primary constituent of bird urine is the relatively insoluble uric acid, with small amounts of urea and creatinine. However, after studying dried urine from various American and Australian birds with both polarizing and electron microscopes, Folk (1969) concluded that "bird urine has a varied composition, and x-ray analysis shows that it does not consist largely of uric acid." Analysis by ultraviolet spectrophotometry revealed that "the droppings contained the urate radical but did not reveal the mode of combination. Thus the chemical composition of the white product of the bird's cloaca remains unknown."

The urine excreted in the kidney is concentrated as it passes through the *renal tubules* because much of the water is resorbed by them. The ability to excrete concentrated urine is said to be related to the number of Henle's loops in the kidney (e.g., House Finch, *Carpodacus mexicanus*). Consequently, the ratio of medulla to cortex differs in different birds. Additional water resorption may take place in the cloaca, although that organ probably plays a more important role in the reabsorption of sodium.

The Reproductive System. The reproductive system is composed of the *primary sex organs* or gonads (Greek *gonē*, seed) and the *secondary* or *accessory sex organs*. Unlike other systems, the reproductive system is in a state of reduced physiological activity during part of the year. An increase in the secretion of gonadotropic hormones by the pituitary gland with the approach of the breeding season stimulates the growth and development of the gonads. The testes may increase from several hundred times to nearly a thousand times their nonbreeding size, and similar changes occur in the ovaries and oviducts (Figs. 36, 37).

The male reproductive system. The male gonads are the *testes* (singular, testis). These produce male germ cells, *spermatozoa*. The two testes vary in shape from oval to vermiform. Either the right or the left testis may be larger; asymmetry in size and shape has been reported for several species (McNeil, 1964).

In mammals the testes are divided into many compartments by connective tissue septa, and there are many independent seminiferous tubules. In birds,

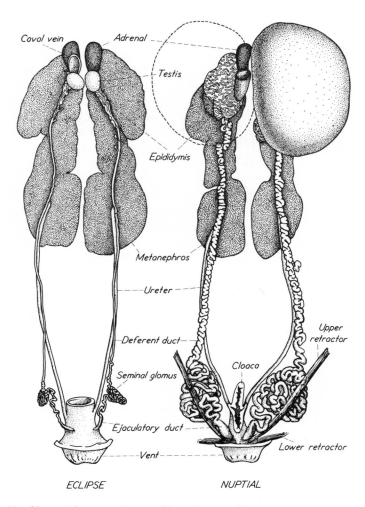

Caval vein Adrenal

Testis

Epididymis

Metanephros

Ureter

Deferent duct

Upper
retractor

Seminal glomus

Cloaca

Ejaculatory duct

Lower retractor

Vent

ECLIPSE NUPTIAL

Figure 36. Urogenital organs of a male House Sparrow (*Passer domesticus*) in eclipse and in nuptial condition; the latter was induced out of season by 17 daily injections of 0.1 cc of pregnant mare serum. The retractor muscles are not shown in the left figure. (By permission of the late Emil Witschi, from *Biology and Comparative Physiology of Birds,* Vol. 2, Academic Press, copyright 1961.)

septa are absent and the seminiferous tubules form a more or less single, continuous network (Bailey, 1953). In addition to spermatogonia and their derivatives, the seminiferous tubule contains *Sertoli cells*. In a study of the rat testis, Elftman (1963) concluded that the Sertoli cells function primarily in nourishing the maturing germ cells; it has been suggested (Marshall, 1961) that in birds Sertoli cells may also secrete a hormone. The primary endocrine function of the testis, however, results from the presence of *Leydig cells*, which secrete testosterone, in the loose interstitial connective tissue around the seminiferous tubule. Pigment cells (*melanoblasts*) occur in the interstitial tissue in many birds. The melanoblasts become widely separated when the testes enlarge at the onset of the breeding season; they become aggregated when the testes shrink after the breeding season. Consequently, the testes of some birds (e.g., *Sturnus vulgaris, Leiothrix lutea*), are nearly black during the nonbreeding season but paler during the breeding season (Serventy and Marshall, 1956). In other birds, the testes vary in color (flesh-colored, pearl, yellow,

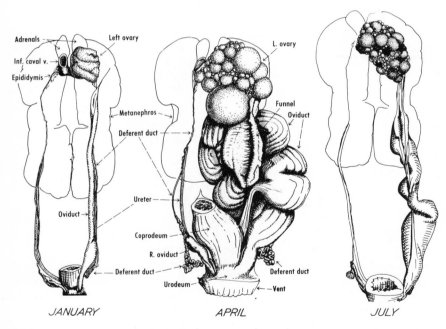

Figure 37. Urogenital organs of a female House Sparrow. January: eclipse condition. April: early nuptial condition; note enlargement of the deferent ducts and right oviduct. July: regression stage in an incubating female taken from the nest on July 27; the ovary contains several empty follicles and degenerating eggs. (By permission of the late Emil Witschi, from *Biology and Comparative Physiology of Birds,* Vol. 2, Academic Press, copyright 1961.)

orange), partly because of seasonal changes in the chemical composition of the Leydig cells; examples of birds that have orange or orange-yellow postnuptial testes are the White-headed Stilt (*Himantopus leucocephalus*), Magpie-lark (*Grallina cyanoleuca*), and Western Thornbill (*Acanthiza inornata*).

The accessory sex organs are the *ductus deferens* and, in some birds, a *cloacal penis*. A ductus deferens emerges from the medial border of each testis; it carries sperm to the urodeum. In gross appearance during the nonbreeding season, each ductus is a small tube with a slight expansion (*ampullary duct*) at its caudal end. With the onset of the breeding season, however, each ductus becomes highly coiled and enlarged, and the region of the ampullary duct is then called a *seminal sac* or *seminal vesicle*. The hypertrophy of the right and left seminal sacs is so great in passerine birds that it causes a marked protrusion of the cloacal region. The *cloacal protuberance* thus formed makes it possible to identify male birds and to determine the stage of their reproductive cycle (Salt, 1954; Wolfson, 1954; Fig. 38). Because the so-called seminal vesicle in birds is simply a specialized part of the ductus deferens, it is not homologous with the seminal vesicles of mammals, nor does it appear to be important for sperm storage, even though the temperature in the cloacal protuberance has been reported to be lower than body temperature (this is reminiscent of the lower temperature in the scrotum of mammals). In writing about the chicken, Lake (1957) commented that "the common conception that the swollen distal portion of the vas deferens serves as a storage organ for spermatozoa in the fowl appears unlikely since the increased size is mainly accounted for by increased musculature and connective tissue in the subepithelial layers. The whole vas serves as a transitory storage place in the absence of an extensive ductus epididymis." Additional work on passerine birds is needed (see Riddle, 1927; Middleton, 1972).

The ampullary duct, or the seminal sac, narrows distally, and the ductus deferens passes through the wall of the urodeum as the *ejaculatory duct*, which consists of erectile tissue similar to that in the mammalian penis.

A *copulatory organ* (*cloacal penis*) occurs in the ventral wall of the proctodeum in some birds and presumably aids in the transfer of sperm to the female cloaca during copulation. A cloacal penis is well developed in the Struthionidae, Casuariidae, Dromiceiidae, and Apterygidae, and can be protruded and retracted by special muscles. In the Rheidae and the Anatidae, the two halves are specialized by "being spirally twisted and being reversible like the finger of a glove" (Gadow, 1896:91). A smaller, less elaborate penis has been found in the Tinamidae, the Galliformes (especially the Cracidae), many Ciconiiformes, and the Burhinidae. A "phalloid organ" has also been described and illustrated for *Bubalornis* (Ploceidae) by Sushkin (1927).

Spermatozoa are transferred into the female cloaca during copulation. The transfer is accomplished by muscular contractions when the male and female

Figure 38. The cloacal protuberance of an adult male Swamp Sparrow (*Zonotrichia georgiana*). Anterior is toward the left, and the bird is lying on its back. The cloacal opening is not visible, but it is surrounded by the anal tuft of feathers; usually the opening is centrally located in the protuberance. Measurements of the protuberance were as follows: anterior wall, 6.2 mm; posterior wall, 6.7 mm; largest diameter, 7.0 mm. Photographed June 23, 1952. (Courtesy of Albert Wolfson.)

cloacae are pressed together, aided by the cloacal penis when present. The number of sperm ejaculated varies among the species. Estimates have been made of 200 million per ejaculation in some pigeons and as many as 3.5 billion in domestic fowl (there are about 500 million sperm per ejaculation in man). In his study of passerine birds, Wolfson (1960) noted that the ejaculate was "a tiny drop of thick material, not quite 1 mm. in diameter and gray in colour. . . . It proved to be a dense mass of sperm which were highly motile and moved radially. The semen was viscous and tended to dry out rapidly." Wolfson also noted that "a small papilla is extruded from the cloacal protuberance when the proctodeum is squeezed." Closer examination revealed that the papilla was formed by two folds and that these virtually formed a tube when pressed together. Wolfson thought it possible, therefore, that the tubelike papilla might actually be "inserted into the everted opening of the oviduct" during copulation.

Coil and Wetherbee (1959) described a *cloacal gland*, embedded in the dorsal lip of the cloaca in male Eurasian Quail (*Coturnix coturnix*). This quail also has a small penis, and the secretions of the gland presumably serve as a lubricant during copulation.

The female reproductive system. The female gonads are the *ovaries*. They produce hormones and *ova* (singular, *ovum*; Latin, egg). The word "egg" obviously has two meanings: (1) it is the ovum or female reproductive cell; and (2) it is the hard-shelled bird egg that contains the "white" (*albumen*) and the "yellow" (*yolk*) part. The yolk is stored in the ovum, so that this single cell of a domestic hen's egg accounts for about 32 percent of the total hard-shelled egg. Most female birds have only a single (left) functional ovary, but in some hawks (especially the genera *Accipiter, Circus, Falco*) 50 percent of the females have paired ovaries.

A rudimentary right gonad may persist as a potential ovary, testis, or ambisexual organ. Destruction of the left ovary by disease (or surgery) may thus result in sex reversal in an individual with a testicular rudiment (Witschi, 1961). Marshall and Serventy (1956) described a Western Magpie (*Gymnorhina dorsalis*) from Australia that was a phenotypic male but had a left ovotestis in which ova occupied about three-fourths of the organ, whereas the remainder had spermatogonia.

The female secondary sex organs are the *oviducts* and, in some birds, a *cloacal clitoris*. Typically, only the left oviduct is functional; the right oviduct is either absent or rudimentary (Fig. 39). Different regions of the oviduct are specialized for different functions. The *infundibulum* is the expanded, open, ciliated proximal end of the oviduct. Ova are secreted into the body cavity (as in mammals), but they soon enter the infundibulum and begin their passage downward. Albumen is added to the ovum in the *magnum*; the shell membranes, in the *isthmus*. The several layers of the calcareous shell and its pigments are added in the *shell gland* ("uterus"). Approximately 24 hours is required for an ovulated egg to become a shelled egg ready for laying; for most of this time (18 to 20 hours) the egg rests in the shell gland. A muscular sphincter separates the shell gland from the "*vagina*," the terminal portion of the oviduct, which opens into the urodeum.

Sperm have a longer viable life in the avian cloaca and oviduct than in the mammalian uterus and oviduct (exceptions are found in armadillos and bats). The duration of viability of avian sperm varies among species and apparently is not correlated with the number of copulations in some birds. Thus a hen turkey may lay as many as 15 fertile eggs after a single copulation, and even after 30 days 83 percent of the eggs laid may be fertile. In their study of domestic Mallards (*Anas platyrhynchos*), however, Elder and Weller (1954) found that

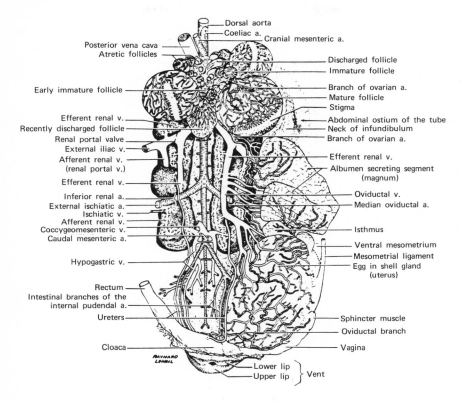

Dorsal aorta
Coeliac a.
Cranial mesenteric a.
Posterior vena cava
Atretic follicles
Discharged follicle
Immature follicle
Early immature follicle
Branch of ovarian a.
Mature follicle
Stigma
Efferent renal v.
Abdominal ostium of the tube
Recently discharged follicle
Neck of infundibulum
Renal portal valve
Branch of ovarian a.
External iliac v.
Afferent renal v.
Efferent renal v.
(renal portal v.)
Albumen secreting segment
Efferent renal v.
(magnum)
Inferior renal a.
Oviductal v.
External ischiatic a.
Median oviductal a.
Ischiatic v.
Afferent renal v.
Coccygeomesenteric v.
Isthmus
Caudal mesenteric a.
Ventral mesometrium
Mesometrial ligament
Hypogastric v.
Egg in shell gland
(uterus)
Rectum
Intestinal branches of the
internal pudendal a.
Ureters
Sphincter muscle
Oviductal branch
Cloaca
Vagina
Lower lip
Vent
Upper lip

Figure 39. A ventral view of the female urogenital system of the chicken. (By permission of Alfred M. Lucas, from *Diseases of Poultry,* Iowa State University Press, copyright 1965.)

the number of fertile eggs laid after separation from the drakes decreased from 64 percent the first week to 3 percent the third week (last fertile egg laid on the 17th day), and that the hatchability of fertile eggs decreased from 73 percent to zero during that period. By contrast, some single-egg layers, such as the Griffon Vulture (*Gyps fulvus*) of Europe and Africa, may copulate frequently for about 1 month before egg-laying. Such extended periods of copulatory behavior, however, probably serve to maintain the pair-bond rather than being essential for fertilization.

After deposition in the female cloaca (or oviduct?), the sperm cells move toward the infundibulum of the oviduct. Fertilization must, of course, take place in the upper end of the oviduct before the albumen, membranes, and shell are secreted around the ovum. By definition, fertilization occurs when the nuclei of a sperm cell and an ovum come together to form a single cell, called a *zygote*. Cell division and growth of the zygote continue as it passes down the oviduct, but development stops after the egg is laid unless incubation begins. The significance of fertilization is that the union of the sperm and egg nuclei brings together the paternal and maternal genetic material in the genes, and this determines the entire development of the new individual.

SELECTED PHYSIOLOGICAL TOPICS

Bird physiology deals with energy requirements, heat production, maintenance of body temperature under different climatic and environmental conditions (i.e., avian bioenergetics), water and salt balance, and the peculiarities of the respiratory and circulatory systems. There is a vast literature on these subjects; only the broadest outlines can be given here.

Metabolism

In both birds and mammals, the metabolic rate per unit weight increases as body weight decreases. Decrease in size is limited by the potential for mobilizing energy. Kendeigh (1970) wrote that "lower rates of standard metabolism in nonpasserine species permit attainment of both lower and greater sizes than in passerine species," and that "a given biotype will support a larger biomass of large than of small birds but a larger number of individual small birds. Evolution toward small size in birds has occurred, in spite of the greater energy demands involved on the individual and on the biotype, apparently because small size has enabled species to exploit niches which would otherwise remain unoccupied."

Energy to produce some kind of work is obtained by the oxidation of an animal's food: carbohydrates, fats, and proteins. Each type of food yields a specific amount of energy when oxidized. This energy can be measured directly because of its caloric value (direct calorimetry), or indirectly from the amount of oxygen utilized in oxidizing the food and the amount of carbon dioxide produced (indirect calorimetry). In most research on birds the latter method has been used. Food energy is expended in several ways: as energy absorbed from the gut (*digestible energy*), in the production of fecal and urinary wastes, in all anabolic processes (*metabolizable energy*), and in the production of heat (the heat increment or specific dynamic action of the diet). Most studies omit urinary nitrogen excretion because of technical difficulties in collecting the urine (King and Farner, 1961).

A study of oxygen consumption in Anna's Hummingbird (*Calypte anna*) and Allen's Hummingbird (*Selasphorus sasin*) showed that these small birds

utilized from 10.7 to 16.0 cc of oxygen per gram of weight per hour when resting. In hovering flight, however, oxygen consumption increased to 85 cc per
gram of body weight per hour in Allen's Hummingbird, and to 68 cc per gram
of body weight per hour in Anna's Hummingbird. At night, when the birds became torpid and their body temperatures decreased almost to the air temperature, oxygen consumption declined to less than 3 cc of oxygen per gram of body
weight per hour. Similar studies have been conducted on a wide variety of
birds: e.g., other species of hummingbirds, the Papuan Frogmouth (*Podargus
ocellatus*), crossbills (*Loxia* sp.), the White-crowned Sparrow (*Zonotrichia leucophrys gambelii*), the Ostrich, emus, and rheas (Crawford and Lasiewski,
1968; Dawson and Tordoff, 1964; King, 1964; Lasiewski, 1964; Lasiewski *et
al.*, 1967).

Respiratory Rate

The respiratory rates of birds are inversely related to body weight; some are
lower than those of mammals with comparable body weights (Table 2). The
resting rate of the Ostrich may be only 5 or 6 breaths per minute, and the
breathing rate of domestic turkeys and chickens has been recorded as 16 to 18
breaths per minute, similar to the rate in man. In smaller birds, however, rates
vary from 45 breaths per minute in the Cardinal to 113 in the Orange-cheeked
Waxbill (*Estrilda melpoda*). Respiratory rates of sleeping Black-capped
Chickadees (*Parus atricapillus*) varied from 65 per minute when the air
temperature was 11°C to 95 per minute when the temperature was raised to
32°C. By contrast, respiratory rates as slow as 8 breaths per minute have been
recorded in torpid nestlings of the European Swift (*Apus apus*). The nestlings
of this species enter a torpid state and are not fed during periods of very bad
weather.

Oxygen requirements and respiratory rates rise during flight; increased
respiratory movements increase the rate at which excess heat can be lost with
the expired air, and they also increase evaporative water loss. Respiratory
movements in a flying House Sparrow may increase from 50 to 212 breaths per
minute. Respiratory rates of a flying Budgerigar vary from about 200 to 270
breaths per minute, and the rate is lower at 35 than at 20 or at 45 km per hour
(Tucker, 1968; Fig. 40).

It is often stated that the lungs, because of their close attachment to the rib
cage, must expand and contract with each wing-beat cycle, and breathing does
coincide with the wing beats in pigeons: inhalation occurs on each upstroke,
and exhalation on each downstroke. A study of respiratory movements was
made by encasing a bird's bill with a thin rubber balloon and then monitoring
the flight with a motion picture camera. Analysis of the film revealed that
breathing in flight frequently is irregular and is not always synchronized with

Table 2. Respiratory Rates of Resting Birds[a]

Species	Body Weight (g)	Respiratory Rate (breaths/min)
Struthio camelus	100,000	5
S. camelus	90,000	6
Rhea americana	21,700	8.5
Dromiceius novae-hollandiae	38,300	7.1
Gavia immer	2,123	6.8
Pelecanus erythrorhynchos	7,500	6.3
P. occidentalis	3,130	8.0
Phalacrocorax auritus	1,340	15.0
Anser sp.	3,425	7.6
Anas platyrhynchos	785	19.0
A. platyrhynchos (Pekin)	2,608	17.0
Cathartes aura	2,000	9.2
Buteo buteo	658	18
Polyborus tharus	350	15
Gallus domesticus	5,200	13
Excalfactoria chinensis	42.7	68
Larus argentatus	930	19
L. canus	388	26
Columba livia	315	29.1
C. livia	317	28
C. livia	382	26
Melopsitticus undulatus	38.2	69
Geococcyx californianus	284.7	28.9
Strix aluco	350	24
Podargus strigoides	675	22
Colaptes auratus	112	26
Empidonax flaviventris	10.2	74

the wing beats. It was learned that, with an average number of wing beats of 242 per minute, the air-breathing rate of a flying Western Gull (*Larus occidentalis*) was 81 times per minute, whereas, with an average number of wing beats of 645 per minute in the Lesser Scaup (*Aythya affinis*), the air-breathing rate was 140 times per minute. A Red-tailed Hawk (*Buteo jamaicensis*) completed 13 wing beats without breathing at all. Moreover, it was discovered that the intake of air may occur on either the upstroke or the downstroke of the wing (Tomlinson, 1963). Berger *et al.* (1970) studied the relationship of respiration to wing beats in 10 species and concluded that "respiration of birds

Table 2. (Continued)

Species	Body Weight (g)	Respiratory Rate (breaths/min)
Cyanocitta cristata	77.1	49
Corvus corone	339	20
C. frugilegus	341	25
C. monedula	240	26
C. monedula	246	60.2
Parus atricapillus	12	64
Troglodytes aedon	11	83
Dumetella carolinensis	29.4	57
Toxostoma rufum	59.2	30
Turdus merula	92	48
T. philomelos	81	94
T. migratorius	69.5	36.5
Hylocichla mustelina	30.5	43
Erithacus rubecula	20	97
Sturnus vulgaris	74	92
Passer domesticus	29.0	57
P. domesticus	24.6	59
Estrilda troglodytes	6.9	95
E. melpoda	6.9	113
Serinus canaria	17	108
S. canaria	16	57
Cardinalis cardinalis	40	45
Zonotrichia melodia	20	63
Pyrrhula pyrrhula	23	42
Loxia curvirostra	39.1	70

[a] Reproduced by permission from Calder (1968).

in flight is usually co-ordinated with wing beats, but the co-ordination is not obligatory. Respiration synchronous with wing beats (1 : 1 co-ordination) was found only in pigeons and crows; the other species exhibited one of 11 other types of co-ordination. Quails, ducks and pheasants, birds with relatively high wing beat frequencies (with relatively small wings), showed a 5 : 1 co-ordination." The same authors also found that the type of coordination sometimes varied during flight.

Deep-diving birds face special problems. It is assumed that buoyancy is reduced by decreasing the reservoir of air in the air-sac system. Diving birds

Figure 40. Budgerigar flying at 35 km per hour in a wind tunnel while wearing a cellulose acetate mask. (Courtesy of Vance A. Tucker and the editor of *Journal of Experimental Biology*.)

usually have a greater blood volume and a higher hemoglobin concentration than terrestrial birds; there is a decrease in blood flow to the muscles during a dive as well as a general lowering of metabolic rate. Although most diving birds spend only a short time under water, Common Loons (*Gavia immer*) can stay submerged for at least 15 minutes, and loons have been caught accidentally in fish nets laid 180 feet deep. Kooyman *et al.* (1971) found that most dives made by Emperor Penguins (*Aptenodytes forsteri*) lasted less than 1 minute, but they recorded one of more than 18 minutes, and they estimated that one bird dove to a depth of more than 800 feet.

Heart Size and Heart Rate

The hearts of birds are relatively larger than those of most mammals of comparable size. In birds, there is a tendency toward an inverse ratio between body size and heart size. In the Ostrich and the Sandhill Crane (*Grus canadensis*), for example, the heart weighs less than 1 percent of the total body weight; in hummingbirds it may constitute as much as 2.75 percent of the body weight. The males of some species seem to have larger hearts than the females, and there may be an increase in the ratio between heart weight and body weight among species that inhabit higher elevations or latitudes, but there are exceptions. Dunson (1965) found that both heart weight and lung weight were significantly higher in montane Robins than in lowland Robins. Johnson and Lockner (1968) reported that mountain-inhabiting White-tailed Ptarmigan (*Lagopus leucurus*) had a much lower ratio of heart size to body weight than did two other species of lowland ptarmigan. Brush (1966), however, found that "there is no correlation between heart ratio and body weight for a wide array of birds of different ecological and taxonomic relationships." Brush showed that "birds of similar body weight but different heart size employed different combinations of capacity (stroke volume) and intensity (heart rate) to adjust to increased cardiovascular loads. These differences were not predictable from either heart size or heart rate alone. Further, birds with relatively larger hearts possess a temperature independent zone of heart rate not seen in species with lower heart ratios." Hence heart size is an index of stroke volume, and therefore of the resting cardiac output, but, under conditions of a greater demand for blood and oxygen, the important factors are an increase in both heart rate and oxygen extraction.

Heart rate tends to increase with decreasing body size (Table 3). The resting heart rate of the Ostrich and of cassowaries is about 70 beats per minute, approximating the rate in man. The heart beat is considerably faster in most other birds, reaching its highest rate in the smallest and most active birds, such as hummingbirds. Heart rate is dependent not only on the age and size of the bird but also on its activity, its body temperature, and the air temperature. The 3-day-old nestling House Wren has a heart rate of 121 beats per minute at an air temperature of 21°C, but this increases to 320 beats per minute at 32°C and to 411 beats per minute at 38°C.

The flapping flight of falcons requires more energy and a faster heart rate than the soaring flight of vultures and buteo hawks. Falcons have relatively larger hearts than the soaring members of their order. In the Greater Black-backed Gull (*Larus marinus*) and in the Mallard, however, the increased cardiac output during flight is due to a greater stroke volume rather than to a drastic increase in the number of beats per minute.

Blood pressure tends to be higher in birds than in mammals. Some comparisons with the mean arterial pressure of 100 for man are as follows: Common Pigeon, 135; Robin, 118; Starling, 180. Extreme fright in birds may increase the blood pressure so much that the aorta or thin-walled atria may rupture, resulting in death. Postmortem examinations have revealed such conditions in several birds: Baldpate (*Anas americana*), Cardinal, and Field Sparrow (*Spizella pusilla*). (*See* Walkinshaw, *1945;* Dilger, *1955*).

Body Temperature

Body temperatures among birds range from about 37.0°C (98.6°F) to 44.6°C (112.3°F); see Table 4. In one study the average resting temperature of 311

Table 3.   Heart Rates of Resting Birds[a]

Species	Body Weight (g)	Heart Rate (beats/min)
Pelecanus erythrorhynchos	7500	150
Anser sp.	3420	113
Anas platyrhynchos (Pekin)	2670	118
Cathartes aura	2000	132
Lophortyx californicus	138	250
Larus argentatus	930	218
Columba livia	382	166
Zenaida macroura	130	135
Z. macroura	91.4	120
Phalaenoptilus nuttallii	40	210
Chordeiles minor	72.5	180
Archilochus colubris	4	615
Colaptes auratus	112	230
Empidonax flaviventris	10.2	545
Cyanocitta cristata	77.1	307
Parus atricapillus	12	480
Troglodytes aedon	11	450
Cinclus mexicanus (fledgling)	40	370
Dumetella carolinensis	28.9	427
Toxostoma rufum	59.2	303
Turdus migratorius	69.5	328
Hylocichla mustelina	30.5	363
Passer domesticus	28	350
Serinus canarius	16	514
Cardinalis cardinalis	40	375
Pipilo erythrophthalmus	40	445
Spizella passerina	12	440
Zonotrichia melodia	20	450

[a] Reproduced by permission from Calder (1968).

passerine species was 40.6°C (105.1°F), and about 40.1°C (104.2°F) for 90 charadriiform species. In such nocturnal species as the Kiwi, owls, nighthawks, and some seabirds, the body temperature is higher at night, when the birds are active, than in the daytime. The Kiwi is a primitive bird, and it has both a lower and a greater fluctuation in its body temperature (Farner *et al.*, 1956).

Although adult birds are able to maintain a high body temperature under varying environmental conditions, newly hatched young have imperfect

Table 4. Body Temperatures for Birds of Various Orders[a]

Order	Number of Species Sampled	Range[b] (°C)
Sphenisciformes (penguins)	6	37.0–38.9
Struthioniformes (Ostrich)	1	39.2
Casuariiformes (cassowaries and emus)	4	38.8–39.2
Apterygiformes (Kiwi)	3	37.8–39.0
Tinamiformes (tinamous)	1	40.5
Gaviiformes (loons)	1	39.0
Podicipediformes (grebes)	4	38.5–40.2
Procellariiformes (albatrosses, shearwaters, petrels, and allies)	13	37.5–41.0
Pelicaniformes (tropicbirds, pelicans, frigatebirds, and allies)	9	39.0–41.3
Ciconiiformes (herons, storks, ibises, flamingos, and allies)	12	39.5–42.3
Anseriformes (screamers, swans, geese, and ducks)	28	40.1–43.0
Falconiformes (vultures, hawks, and falcons)	12	39.7–42.8
Galliformes (megapodes, curassows, pheasants, and Hoatzin)	22	40.0–42.4
Gruiformes (cranes, rails, and allies)	7	40.1–41.5
Charadriiformes (shorebirds, gulls, auks, and allies)	39	38.3–42.4
Columbiformes (sandgrouse, pigeons, and doves)	5	40.0–43.3
Cuculiformes (cuckoos and plantain eaters)	2	41.9–42.3
Strigiformes (owls)	9	39.2–41.2
Caprimulgiformes (goatsuckers, Oilbird, and allies)	5	37.6–42.4
Apodiformes (swifts and hummingbirds)	25	35.6–44.6
Coraciiformes (kingfishers, motmots, rollers, bee-eaters, and hornbills)	1	40.0
Piciformes (woodpeckers, jacamars, toucans, and barbets)	10	39.0–43.0
Passeriformes (perching birds)	101	39.2–43.8

[a] Reproduced by permission from Dawson and Hudson (1970).

[b] The values presented are the extremes of the means for the active phase of the daily cycles of the species sampled. Data for species on which only single determinations have been made have been excluded from this summary.

temperature control mechanisms. This is true for both the nearly naked and helpless newly hatched altricial bird (Fig. 41) and the down-covered precocial species (Fig. 42).

The following factors are important in the development of an efficient temperature-regulating system in altricial birds, especially: the growth of feathers or down; a decrease in the body surface area relative to bulk; an increase in nervous and hormonal control, resulting in increased heat produc-

Figure 41. Two Willow Flycatchers (*Empidonax traillii*) less than 12 hours old; Ann Arbor, Michigan.

tion; and the development of functional air sacs. Among many passerine species, all of these changes appear to take place within about 1 week after hatching, and one study revealed that nestling Field Sparrows and Chipping Sparrows (*Spizella passerina*) were able to maintain body temperature effectively when 7 days old (Dawson and Evans, 1957). Nestling Cactus Wrens (*Campylorhynchus brunneicapillus*), however, which do not leave the nest until they are 20 days old, do not develop thermogenic homeostasis until the 13th day after hatching (Ricklefs and Hainsworth, 1968). By contrast, a study of the Killdeer (*Charadrius vociferus*), a precocial species, showed that the newly hatched downy young could maintain body temperatures at air temperatures between 23 and 40°C, although they could not do so at lower temperatures. The efficiency of the temperature control mechanism increased rapidly during the first 10 days, but did not reach the efficiency of the adult system until the young birds were 27 days old. Dawson *et al.* (1972) found that Laughing Gull (*Larus atricilla*) chicks less than 32 hours old had moderately effective temperature control at ambient temperatures between 21.7 and 41.2°C; the chicks were able to dissipate 131 percent of their heat production by panting at temperatures between 44.5 and 45.2°C, thereby maintaining a body temperature averaging 2.3°C below the highest ambient temperature. However, the chicks were unable to maintain body temperatures when the ambient temperature fell below 20°C. These authors pointed out that "considerably more variation in thermoregulation capacities exists among birds at hatching than is implied by the convention of classifying species alternatively as 'precocial' or 'altricial.' . . . Thus, among nominally precocial birds, newly hatched ducklings

Figure 42. A newly hatched Common Tern (*Sterna hirundo*) panting from heat stress; Detroit, Michigan.

can regulate body temperature more effectively in the cold than either gallinaceous birds or gulls of comparable age. In one case this appears correlated with differences in distribution of the species considered, hatchlings of ducks breeding at higher latitudes having better control of body temperature at cool T_a's than those of ducks breeding at somewhat lower latitudes."

The newly hatched altricial bird is helpless and spends most of its time sleeping. Its primary need is protection from the environment. In temperate zones, the basic need may be heat, as provided by the brooding adult, or it may be protection from the sun (Fig. 43). In extremely hot environments (as in desert areas or on oceanic lava or coral islands), the greatest need often is protection from the burning sun, and in these species one of the adults typically stands over the egg or young bird to shield it from the sun (Fig. 44).

Both water (evaporative water loss) and heat (evaporative cooling) are discharged during respiration. Excess heat also may be dissipated by means of *gular fluttering*. An increased blood supply to the thin membranes of the gular pouch effects a continuous release of heat as the pouch is fluttered rapidly (Fig. 45). Gular flutter compensates for the absence of sweat glands in birds. It is

Figure 43. A female Red-Winged Blackbird shielding nestlings from the sun. (Courtesy of Samuel A. Grimes.)

Figure 44. A Sooty Tern (*Sterna fuscata oahuensis*) shielding its egg from the sun; East Island, French Frigate Shoal, Hawaiian Islands, March 23, 1966.

practiced by boobies, pelicans, cormorants, herons, owls, nighthawks, and other, unrelated species (Bartholomew *et al.*, 1968). Young pelicans also may walk into shallow water to lower body temperature during the hottest part of the day (Bartholomew and Dawson, 1953). Increased blood flow through bare areas of the head and legs also facilitates heat loss. Conversely, special *arteriovenous shunts*, plus a decrease in blood flow, reduce the amount of heat loss in cold weather.

Feathers provide an excellent insulating mechanism for conserving heat, and birds typically "fluff up" their feathers on cold wintry days and nights. They also frequently perch on one leg, drawing the other up into the breast and belly feathers, and they may tuck their bills into the scapular feathers as in sleep. The legs and feet of some birds, such as the Herring Gull (*Larus argentatus*), are less sensitive to cold (because of special anatomical adaptations) than are those of most birds (Chatfield *et al.*, 1953).

In 1946 Edmund C. Jaeger discovered that some birds actually hibernate. He found a Poor-will (*Phalaenoptilus nuttallii*) in a crevice in a rock in the Chuckawalla Mountains of southeastern California (Fig. 46). The bird was in such a torpid state that it could be picked up and handled without awaking. Its metabolic rate had slowed so much that no heartbeat or respiratory movements could be detected. The normal temperature for this species is about 41°C. In the hibernating bird, the temperature varied from about 17 to 19°C over a period of time. Later experimental studies on the Poor-will revealed that individuals could be forced into a torpid state by lowering the air temperature below 19°C. It also was calculated that 10 grams of stored fat would enable a bird to hibernate as long as 100 days (Jaeger, 1949; Bartholomew *et al.*, 1962).

Torpid states have been discovered in several species of swifts and hummingbirds and in the European Nightjar (*Caprimulgus europaeus*), Common Nighthawk, Lesser Nighthawk (*Chordeiles acutipennis*), and Speckled Mousebird (*Colius striatus*). Although capable of entering torpid states at other times, incubating female hummingbirds maintain their nighttime temperatures, a necessity in order to maintain a proper temperature for the developing embryos (Bartholomew *et al.*, 1957). Lasiewski and Thompson (1966) described apparent torpidity in Violet-green Swallows (*Tachycineta thalassina*) in California.

Figure 45. A group of young White Pelicans (*Pelecanus erythrorhynchus*) at the Salton Sea, California, rookery during the middle of a warm June day, showing the characteristic gular fluttering. (Courtesy of George A. Bartholomew.)

Figure 46. Close-up of a Poor-will in its hibernation crypt in the face of a granite rock; Chuckawalla Mountains, Riverside County, California. On November 15, 1964, The Nature Conservancy established the Edmund C. Jaegar Nature Sanctuary, an area of 160 acres that includes the site where Dr. Jaeger found the first hibernating poorwill. (Courtesy of Edmund C. Jaeger and the editor of *Condor*.)

Water and Salt Regulation

Maintenance of the correct proportions of salts and water within the cells, in the intercellular spaces, and in the blood is of critical importance to the survival of the individual cell and of the animal itself. Although the complicated regulatory mechanisms that maintain the internal environment within normal limits are beyond the scope of this chapter, certain aspects deserve brief mention.

Reptiles, birds, and mammals develop in a watery environment, even though reptiles and birds grow in hard-shelled eggs and mammals in a uterus (Fig. 47), and these three classes are called *amniotes* because of the fetal membranes characteristic of their embryological development. At an early age, each embryo becomes enveloped in a closed *amniotic sac* containing a watery amniotic fluid, which bathes the embryo. Embryonic cells also grow downward to surround the yolk, thus forming a second fetal membrane, the *yolk sac*. The yolk is slowly digested and carried to the growing embryo or fetus by veins. The yolk sac and its remaining contents are drawn through the umbilical opening into the body cavity a day or two before the bird hatches. Passerine birds do not need food immediately after hatching, and a precocial species may have enough food material in the yolk sac to live for 4 days without eating. The third fetal membrane is the *allantois*; it develops as a pouch from the floor of the hindgut, rapidly grows outward through the umbilical opening, and eventually lines the entire inner surface of the shell membrane of the egg. The allantois is both urinary bladder and functional respiratory organ during the fetal period of development. The allantois is richly supplied with blood vessels, and oxygen diffuses through the eggshell into the vessels and carbon dioxide passes from them out through the shell. The nitrogenous wastes of protein digestion are deposited in the allantois, which dries up near hatching time, thus leaving the wastes outside the fetus. Adult birds do not have a urinary bladder, but the proximal part of the allantois remains as the functional bladder in mammals.

The posthatching and the adult bird, as well as other amniotes, face a problem of water conservation. The skin, with its appendages (feathers, scales), not only protects the bird but also is essential for preventing an excessive loss of body fluids. Birds lose very little water in excretion because most of the water filtered through the kidneys is reabsorbed and thus conserved. However, birds may lose a considerable amount of water in respiration, especially when high air temperatures require evaporative cooling. In a study of 12 species of desert birds, Bartholomew and Dawson (1953) found that the birds lost more water than was produced by the metabolism of their food, even at moderate air temperatures, and that the loss was much greater for small than for large birds. In order to occupy desert habitats, therefore, the birds must have access to water or to foods that contain large amounts of water. Desert-inhabiting pigeons and doves, which subsist largely on dry seeds, may fly many miles each

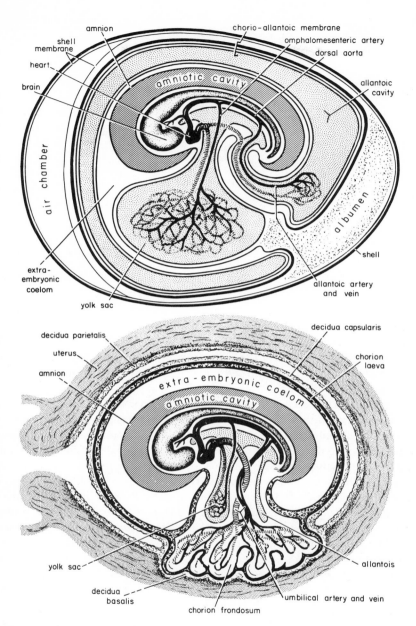

Figure 47. Schematic representation of the fetal membranes and their relationships in a bird (above) and in a human female (below). The *decidua* is the name given to the lining (endometrium) of the pregnant uterus. The *decidua basalis* and the *chorion frondosum* (a specialization of the embryo) together form the placenta. (By permission of Andrew J. Berger, from *Elementary Human Anatomy*, John Wiley & Sons, copyright 1964.)

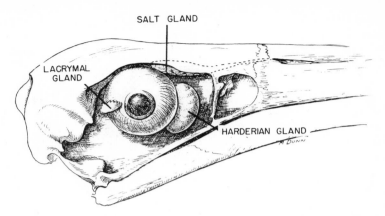

Figure 48. The orbital region of a Brown Pelican (*Pelecanus occidentalis*) showing the location of major glands. (Courtesy of Knut Schmidt-Nielsen and the editor of *Auk*.)

day to obtain water. Bartholomew and Dawson (1954) reported that captive Mourning Doves (*Zenaida macroura*) kept at an air temperature of 39°C drank four times as much water per day as those kept at 23°C. Doves held without water for 24 hours at the higher temperature might lose 15 percent of their weight but could regain this weight within a few minutes after water was provided.

A series of interrelated physiological processes maintains a concentration of salts in the blood within certain narrow limits which vary in different animals. A pathological imbalance of the salt-regulating mechanism can result in a drastic increase in the concentration of blood salts, subsequent dehydration, and death if the imbalance is prolonged or extreme. Excess salts are excreted by the kidneys and by sweat glands in mammals; sweat glands are lacking in birds. For landbirds in general, however, a source of fresh water is not a problem, and the excretory capacity of the kidneys is adequate for maintaining a proper balance of salts and electrolytes. Of special interest is the discovery that certain geographical races of the Savannah Sparrow (*Passerculus sandwichensis*) have the ability to survive on salt water far better (and longer) than other terrestrial birds that have been studied; this was strikingly true for *P.s. beldingi*, an inhabitant of salt marshes and marine beaches. This race was able to ingest an amount of salt equal to more than 2 percent of the body weight per day. These birds excrete a highly concentrated urine, and their kidneys contain more Henle's loops per unit area than are found in the kidneys of most other passerine species (the House Finch being an exception). Cade and Bartholomew (1959) pointed out that "the unusual ability of savannah sparrows to withstand dehydration and their capacity to drink sea water preadapts members of this species to occupy the deserts of western North America, which are of relatively recent development" (see also Poulson and Bartholomew, 1962; Poulson, 1965).

Seabirds, however, not only survive without fresh water but also drink sea water. At the same time, the kidneys secrete a urine only about half as salty as sea water. Schmidt-Nielsen and his co-workers were the first to study this problem experimentally, using such birds as penguins, gulls, cormorants, and pelicans. Schmidt-Nielsen and Fange (1958) summarized the problem as follows.

"In order to profit from the ingestion of sea water it is necessary for an animal to excrete salts in a concentration at least as high as that in the water ingested. The elimination of the salt would otherwise require an additional amount of water which would be taken from the body tissues. Therefore, if the organism cannot excrete a highly concentrated salt solution, the drinking of sea water only will lead to a progressive dehydration or a harmful accumulation of salt.

"The bird kidney is able to excrete salts in a concentration only about one-half that found in sea water. Hence, if the bird kidney should excrete the salts from a given amount of sea water, it would be necessary to produce twice as much urine as the amount of water ingested. Thus, the kidney is not able to keep a marine bird in a favorable water balance if it drinks sea water."

Schmidt-Nielsen showed that excess salts in seabirds are excreted by a pair of *salt glands* (long called *nasal glands*) located in the orbits (Fig. 48). Salt glands reach their highest development in marine birds, and the size of the glands is directly correlated with the extremeness of the marine habitat. It is this provision for extrarenal salt excretion that makes it possible for marine birds to survive on salt water. Still unstudied is salt metabolism in certain landbirds that inhabit oceanic islands.

Cooch (1964) investigated the hypothesis that "ducks, gulls, grebes, and shorebirds inhabiting the Great Plains of North America must possess a functional salt gland in order to cope with the highly alkaline and saline waters of many prairie sloughs and lakes." Although most of his experiments were conducted with ducks, he concluded that "water birds inhabiting the Great Plains of North America possess a salt gland which may function as effectively as that of marine birds." Hughes (1970) reported that birds with functional salt glands have heavier kidneys than birds with nonfunctional glands. Carpenter and Stafford (1970) found that the nasal secretions of the American Coot and the Guam Rail (*Rallus owstoni*) were insufficient to permit these birds to drink sea water, and they discovered that nasal secretion occurred only when "sodium concentrations in the blood reached levels significantly above the sodium levels in non-secreting birds." Hence they postulated the presence of sodium receptors in these birds in contrast to the osmoreceptors in other species with functional salt glands.

REFERENCES

Akester, A. R. 1971. The blood vascular system. In *Physiology and Biochemistry of the Domestic Fowl*, Vol. 2, Academic Press, New York, pp. 783–839.

Bailey, R. E. 1953. Accessory reproductive organs of male fringillid birds: seasonal variations and response to various sex hormones. *Anat. Rec.* **115**:1–19.

Bartholomew, G. A., Jr., and W. R. Dawson. 1953. Respiratory water loss in some birds of southwestern United States. *Physiol. Zool.,* **26**:162–166.

———, and ——— 1954. Temperature regulation in young pelicans, herons, and gulls. *Ecology,* **35**:466–472.

———, T. R. Howell, and T. J. Cade. 1957. Torpidity in the White-throated Swift, Anna Hummingbird, and Poor-will. *Condor,* **59**:145–155.

———, J. W. Hudson, and T. R. Howell. 1962. Body temperature, oxygen consumption, evaporative water loss, and heart rate in the Poor-will. *Ibid.,* **64**:117–125.

———, R. C. Lasiewski, and E. C. Crawford, Jr. 1968. Patterns of panting and gular flutter in cormorants, pelicans, owls, and doves. *Condor,* **70**:31–34.

———, and C. H. Trost. 1970. Temperature regulation in the Speckled Mousebird, *Colius striatus. Ibid.,* **72**:141–146.

Baumel, J. J. 1958. Variation in the brachial plexus of *Progne subis. Acta Anat.,* **34**:1–34.

——— 1964. Vertebral-dorsal carotid artery interrelationships in the pigeon and other birds. *Anat. Anz.,* **114**:113–130.

———, and L. Gerchman. 1968. The avian intercarotid anastomosis and its homologue in other vertebrates. *Am. J. Anat.,* **122**:1–18.

Beecher, W. J. 1951a. Adaptations for food-getting in the American blackbirds. *Auk,* **68**:411–440.

——— 1951b. Convergence in the Coerebidae. *Wilson Bull.,* **63**:274–287.

Benoit, J., and I. Assenmacher. 1959. The control by visible radiations of the gonadotropic activity of the duck hypophysis. In *Recent Progress in Hormone Research*, Academic Press, New York, pp. 143–164.

——— 1970. *La Photoregulation de la Reproduction chez les Oiseaux et les Mammiferes.* Editions du Centre National de la Recherche Scientifique, Paris.

Berger, A. J. 1957. On the anatomy and relationships of *Fregilupus varius*, an extinct starling from the Mascarene Islands. *Bull. Am. Mus. Nat. Hist.,* **113**:225–272.

——— 1969. Appendicular myology of passerine birds. *Wilson Bull.,* **81**:220–223.

Berger, M., J. S. Hart, and O. Z. Roy. 1970. Respiration, oxygen consumption and heart rate in some birds during rest and flight. *Z. Vergl. Physiol.,* **66**:201–214.

Biggs, P. M. 1957. The association of lymphoid tissue with the lymph vessels in the domestic chicken (*Gallus domesticus*). *Acta Anat.,* **29**:36–47.

Bock, W. J. 1961. Salivary glands in the gray jays (*Perisoreus*). *Auk,* **78**:355–365.

——— 1962. The pneumatic fossa of the humerus in the Passeres. *Ibid.,* **79**:425–443.

—— 1964. Kinetics of the avian skull. *J. Morphol.* **114**:1–42.

—— 1968. Mechanics of one- and two-joint muscles. *Am. Mus. Novit.*, No. 2319:1–45.

Boyd, J. C., and W. J. L. Sladen. 1971. Telemetry studies of the internal body temperatures of Adélie and Emperor penguins at Cape Crozier, Ross Island, Antarctica. *Auk,* **88**:366–380.

Bretz, W. L., and K. Schmidt-Nielsen. 1971. Bird respiration: flow patterns in the duck lung. *J. Exp. Biol.,* **54**:103–118.

Brush, A. H. 1966. Avian heart size and cardiovascular performance. *Auk,* **83**:266–273.

Cade, T. J., and G. A. Bartholomew. 1959. Sea-water and salt utilization by Savannah Sparrows. *Physiol. Zool.,* **32**:230–238.

Calder, W. A. 1968. Respiratory and heart rates of birds at rest. *Condor,* **70**:358–365.

Carpenter, R. E., and M. A. Stafford. 1970. The secretory rates and the chemical stimulus for secretion of the nasal salt glands in the Rallidae. *Condor,* **72**:316–324.

Chatfield, O., C. P. Lyman, and L. Irving. 1953. Physiological adaptation to cold of peripheral nerve in the leg of the Herring Gull (*Larus argentatus*). *Am. J. Physiol.,* **172**:639–644.

Cobb, S. 1960. Observations on the comparative anatomy of the avian brain. *Perspect. Biol. Med.,* 3 (Spring, 1960): 383–408.

—— 1962. Notes on the brain of the hummingbird. *Arch. Neurol.,* **6**:57–62.

Coil. W. H., and D. K. Wetherbee. 1959. Observations on the cloacal gland of the Eurasian Quail, *Coturnix coturnix. Ohio J. Sci.,* **59**:268–270.

Cooch, F. G. 1964. A preliminary study of the survival value of a functional salt gland in prairie Anatidae. *Auk,* **81**:380–393.

Coues, E. 1903. *Key to North American Birds,* 5th ed. Dana Estes & Co., Boston.

Cracraft, J. 1968. The lacrimal-ectethmoid bone complex in birds: a single character analysis. *Am. Midl. Nat.,* **80**:316–359.

—— 1971. The functional morphology of the hind limb of the domestic pigeon, *Columba livia. Bull. Am. Mus. Nat. Hist.,* **144**:171–268.

Crawford, E. C., Jr., and R. C. Lasiewski. 1968. Oxygen consumption and respiratory evaporation of the emu and rhea. *Condor,* **70**:333–339.

Curtis, E. L., and R. C. Miller. 1938. The sclerotic ring in North American birds. *Auk,* **55**:225–243.

Dawson, W. R., and F. C. Evans. 1957. Relation of growth and development to temperature regulation in nestling Field and Chipping Sparrows. *Physiol. Zool.,* **30**:315–327.

——, and J. W. Hudson. 1970. Birds. In *Comparative Physiology of Thermoregulation,* Vol. 1. Academic Press, New York, pp. 223–310.

——, ——, and R. W. Hill. 1972. Temperature regulation in newly hatched Laughing Gulls (*Larus atricilla*). *Condor,* **74**:177–184.

——, and H. B. Tordoff. 1964. Relation of oxygen consumption to temperature in the Red and White-winged crossbills. *Auk,* **81**:26–35.

Desselberger, H. 1932. Ueber den Verdauungskanal nektarfressender Vögel. *J. Ornithol.,* **80**:309–318.

Dilger, W. C. 1955. Ruptured heart in the Cardinal (*Richmondena cardinalis*). *Auk,* **72**:85.

Dubbledam, J. L. 1968. *On the Shape and Structure of the Brainstem in Some Species of Birds: An Architectonic Study.* Meppel, Krips Repro. Industrieweg, 107 pp.

Duncker, H. R. 1971. The lung air sac system of birds. A contribution to the functional anatomy of the respiratory apparatus. *Ergeb. Anat. Entwickl. Gesch.,* **45**, 171 pp.

Dunson, W. A. 1965. Adaptation of heart and lung weight to high altitude in the Robin. *Condor,* **67**:215–219.

Elder, W. H. 1954. The oil gland of birds. *Wilson Bull.,* **66**:6–31.

———, and M. W. Weller. 1954. Duration of fertility in the domestic mallard hen after isolation from the drake. *J. Wildl. Manage.,* **18**:495–502.

Elftman, H. 1963. Sertoli cells and testis structure. *Am. J. Anat.,* **113**:25–34.

Farner, D. S., N. Chivers, and T. Riney. 1956. The body temperatures of the North Island Kiwis. *Emu,* **56**:199–206.

Fisher, H. I. 1940. The occurrence of vestigial claws on the wings of birds. *Am. Midl. Nat.,* **23**:234–243.

——— 1955. Major arteries near the heart in the Whooping Crane. *Condor,* **57**:286–289.

———, and D. C. Goodman. 1955. The myology of the Whopping Crane, *Grus americana. Ill. Biol. Monogr.* No. 24, 127 pp.

Folk, R. L. 1969. Spherical urine in birds: Petrography. *Science,* **166**:1516–1519.

Frank, G. H. 1954. The development of the chondrocranium of the Ostrich. *Ann. Univ. Stellenbosch,* Ser. A, No. 4:179–248.

Freeman, B. M. 1971. The endocrine status of the bursa of Fabricius and the thymus gland. In *Physiology and Biochemistry of the Domestic Fowl,* Vol. 1, Academic Press, New York, pp. 575–587.

Gadow, H. 1896. In *A Dictionary of Birds,* A. Newton and H. Gadow. Adam and Charles Black, London.

Gans, C. and W. J. Bock. 1965. The functional significance of muscle architecture—a theoretical analysis. In *Reviews of Anatomy, Embryology, and Cell Biology,* Springer-Verlag, Berlin, pp. 115–142.

Gaunt, A. S. 1969. Myology of the leg in swallows. *Auk,* **86**:41–53.

George, J. C., and A. J. Berger. 1966. *Avian Myology.* Academic Press, New York, 500 pp.

Glenny, F. H. 1955. Modifications of pattern in the aortic arch system of birds and their phylogenetic significance. *Proc. U.S. Natl. Mus.,* **104**:525–621.

———, and D. Amadon. 1955. Remarks on the Pigeon, *Otidiphaps nobilis* Gould. *Auk,* **72**:199–203.

——— and H. Friedmann. 1954. Reduction of the clavicles in the Mesoenatidae, with some re-marks concerning the relationship of the clavicle to flight-function in birds. *Ohio Jr. Sci.,* **54**:111–113.

Hamilton, T. H., and C. Teng. 1965. Sexual stabilization of Müllerian ducts in the chick embryo. In *Organogenesis,* R. L. DeHaan and Heinrich Ursprung, eds., Holt, Rinehart and Winston, New York, pp. 681–700.

Hartman, F. A., and R. H. Albertin. 1951. A preliminary study of the avian adrenal. *Auk,* **68**:202–209.

———, and K. A. Brownell. 1961. Adrenal and thyroid weights in birds. *Ibid.,* **78**:397–422.

Hill, K. J. 1971. The structure of the alimentary tract. In *Physiology and Biochemistry of the Fowl,* Vol. 1, Academic Press, New York, pp. 1–23.

Hill, W. C. O., and C. J. Skead. 1952. On terminal claws on the manual digits in Ardeiform birds. *Ibis,* **94**:62–67.

Hudson, G. E. 1937. Studies on the muscles of the pelvic appendage in birds. *Am. Midl. Nat.,* **18**:1–108.

———, R. E. Parker, J. Vanden Berge, and P. J. Lanzillotti. 1966. A numerical analysis of the modifications of the appendicular muscles in various genera of gallinaceous birds. *Ibid.*, **76**:1–73.

Hughes, M. R. 1970. Relative kidney size in nonpasserine birds with functional salt glands. *Condor,* **72**:164–168.

Jaeger, E. C. 1949. Further observations on the hibernation of the Poor-will. *Condor,* **51**:105–109.

Johnson, O. W. 1974. Relative thickness of the renal medulla in birds. *J. Morphol.,* **142**:277–284.

Johnson, R. E., and F. R. Lockner. 1968. Heart size and altitude in ptarmigan. *Condor,* **70**:185.

Jollie, M. T. 1957. The head skeleton of the chicken and remarks on the anatomy of this region in other birds. *J. Morphol.,* **100**:389–436.

Karten, H. J. 1969. The organization of the avian telencephalon and some speculations on the phylogeny of the amniote telencephalon. *Ann. N.Y. Acad. Sci.,* **167**:164–179.

———, and W. Hodos. 1967. *A Stereotaxic Atlas of the Brain of the Pigeon (Columba livia).* John Hopkins Press, Baltimore.

Kendeigh, S. C. 1961. Energy of birds conserved by roosting in cavities. *Wilson Bull.,* **73**:140–147.

——— 1970. Energy requirements for existence in relation to size of bird. *Condor,* **72**:60–65.

——— 1972. Energy control of size limits in birds. *Am. Nat.,* **106**:79–88.

King, A. S., and V. Molony. 1971. The anatomy of respiration. In *Physiology and Biochemistry of the Domestic Fowl,* Vol. 1, Academic Press, New York, pp. 93–169.

King, J. R. 1964. Oxygen consumption and body temperature in relation to ambient temperature in the White-crowned Sparrow. *Comp. Biochem. Physiol.,* **12**:13–24.

———, and D. S. Farner. 1961. Energy, metabolism, thermoregulation and body temperature. In *Biology and Comparative Physiology of Birds,* Vol. 2, Academic Press, New York, pp. 215–288.

Klemm, R. D. 1969. Comparative myology of the hind limb of procellariiform birds. *South. Ill. Univ. Monogr., Sci. Ser.* 2, 269 pp.

Kooyman, G. L., C. M. Drabek, R. Elsner, and W. B. Campbell. 1971. Diving behavior of the Emperor Penguin, *Aptenodytes forsteri. Auk,* **88**:775–795.

Kuroda, N. 1963. A fragmental observation on avian kidney. *Misc. Rept. Yamashina's Inst. Ornithol. Zool.,* **3**:280–286.

Lake, P. E. 1957. The male reproductive tract of the fowl. *J. Anat.,* **91**:116–129.

Langslow, D. R., and C. N. Hales. 1971. The role of the endocrine pancreas and catecholamines in the control of carbohydrate and lipid metabolism. In *Physiology and Biochemistry of the Domestic Fowl,* Vol. 1, Academic Press, New York, pp. 521–547.

Larsell, O. 1967. *The Comparative Anatomy and Histology of the Cerebellum from Myxinoids through Birds.* University of Minnesota Press, Minneapolis.

Lasiewski, R. C. 1964. Body temperature, heart and breathing rate, and evaporative water loss in hummingbirds. *Physiol. Zool.,* **37**:212–223.

———, W. R. Dawson, and G. A. Bartholomew. 1970. Temperature regulation in the Little Papaun Frogmouth, *Podargus oscellatus. Condor.* **72**:323–338.

———, and H. J. Thompson. 1966. Field observation of torpidity in the Violet-green Swallow. *Ibid.,* **68**:102–103.

———, W. W. Weathers, and M. H. Bernstein. 1967. Physiological responses of the Giant Hummingbird, *Patagonia gigas. Physiol. Zool.,* **23**:797–813.

Lewis, R. A. 1967. "Resting" heart and respiratory rates of small birds. *Auk,* **84**:131–132.

Lucas, A. M. 1970. Avian functional anatomic problems. *Fed. Proc.,* **29**:1641–1648.

———, and E. M. Denington. 1956. Morphology of the chicken liver. *Poultry Sci.,* **35**:793–806.

———, and C. Jamroz. 1961. *Atlas of avian hematology. U.S. Dept. Agric. Monogr.* No. 25, vi + 271 pp.

McNeil, R. 1964. Un cas inusite d'asymetrie testiculaire chez le *Crotophaga ani* L. *Rev. Univ. Zulia, Kasmera,* **1**:273–287.

Malinovský, L. 1965. Contribution to the comparative anatomy of the vessels in the abdominal part of the body cavity in birds. III. Nomenclature of branches of the a. coeliaca and of tributaries of the v. portae. *Folia Morphol.,* **13**:252–264.

Marshall, A. J. 1961. Reproduction. In *Biology and Comparative Physiology of Birds,* Vol. 2, Academic Press, New York, pp. 169–213.

———, and D. L. Serventy. 1956. A case of intersexuality in *Gymnorhina dorsalis. Emu,* **56**:207–210.

Middleton, A. L. A. 1972. The structure and possible function of the avian seminal sac. *Condor,* **74**:185–190.

Nishida, T. 1963. Comparative and topographical anatomy of the fowl. X. The blood vascular system of the hind-limb in the fowl (in Japanese). *Jap. J. Vet. Sci.,* **25**:93–106.

Payne, L. N. 1971. The lymphoid system. In *Physiology and Biochemistry of the Domestic Fowl,* Vol. 2, Academic Press, New York, pp. 985–1037.

Pearson, R. 1972. *The Avian Brain.* Academic Press, New York.

Portmann, A. 1955. Die postembryonale Entwicklung der Vögel als Evolutions-problem. *Acta XI Congr. Int. Ornithol.,* **1954**:138–151.

———, and W. Stingelin. 1961. The central nervous system. In *Biology and Comparative Physiology of Birds,* Vol. 2, Academic Press, New York, pp. 1–36.

Poulson, T. L. 1965. Countercurrent multipliers in avian kidneys. *Science,* **148**:389–391.

———, and G. A. Bartholomew. 1962. Salt balance in the Savannah Sparrow. *Physiol. Zool.,* **35**:109–119.

Quay, W. B. 1967. Comparative survey of the anal glands in birds. *Auk,* **84**:379–389.

——— 1972a. Infrequency of pineal atrophy among birds and its relation to nocturnality. *Condor,* **74**:33–45.

——— 1972b. Integument and the environment: glandular composition, function, and evolution. *Am. Zool.,* **12**:95–108.

———, and A. Renzoni. 1963. Comparative and experimental studies of pineal structure and cytology in passeriform birds. *Riv. Biol.* **56**:363–407.

———, and ——— 1967. The diencephalic relations and variably bipartite structure of the avian pineal complex. *Ibid.,* **60**:9–75.

Rand, A. L. 1954. On the spurs on birds' wings. *Wilson Bull.,* **66**:127–134.

Richardson, F. 1972. Accessory pygostyle bones of Falconidae. *Condor,* **74**:350–351.

Ricklefs, R. E., and F. R. Hainsworth. 1968. Temperature regulation in nestling Cactus Wrens: the development of homeothermy. *Condor,* **70**:121–127.

Riddle, O. 1927. The cyclical growth of the vesicula seminalis in birds is hormone controlled. *Anat. Rec.,* **37**:1–11.

Rogers, K. T. 1965. Development of the XIth or spinal accessory nerve in the chick. *J. Comp. Neurol.,* **125**:273–285.

Romanoff, A. L. 1960. *The Avian Embryo.* Macmillan Co., New York, 1305 pp.

Salt, W. R. 1954. The structure of the cloacal protuberance of the Vesper Sparrow (*Pooecetes gramineus*) and certain other passerine birds. *Auk*, **71**:64–73.

Scharnke, H. 1932. Ueber den Bau der Zunge der Nectariniidae, Promeropidae und Drepanididae nebst Bemerkungen zur Systematik der blutenbesuchen Passeres. *J. Ornithol.*, **80**:114–123.

Schmidt-Nielsen, K. 1971. How birds breathe. *Sci. Am.*, December 1971:72–79.

———, and R. Fange. 1958. The function of the salt gland in the Brown Pelican. *Auk*, **75**:282–289.

Serventy, D. L., and A. J. Marshall. 1956. Factors influencing testis coloration in birds. *Emu*, **56**:219–221.

Simkiss, K., and C. G. Dacke. 1971. Ultimobranchial glands and calcitonin. In *Physiology and Biochemistry of the Domestic Fowl*, Vol. 1, Academic Press, New York, pp. 481–488.

Stettner, L. J., and K. A. Matyniak. 1968. The brain of birds. *Sci. Am.*, **218**:64–76.

Stingelin, W. 1956. Studien am Vorderhirn von Waldkauz (*Strix aluco* L.) und Turmfalk (*Falco tinnunculus* L.). *Rev. Suisse Zool.*, **63**:551–660.

———, and D. G. Senn. 1969. Morphological studies on the brain of Sauropsida. *Ann. N.Y. Acad. Sci.*, **167**:156–163.

Sushkin, P. P. 1927. On the anatomy and classification of the Weaverbirds. *Bull. Am. Mus. Nat. Hist.*, **57**:1–32.

Tiemeier, O. W. 1950. The os opticus of birds. *J. Morphol.*, **86**:25–46.

Tomlinson, J. T. 1963. Breathing of birds in flight. *Condor*, **65**:514–516.

Tucker, V. A. 1968. Respiratory exchange and evaporative water loss in the flying Budgerigar. *J. Exp. Biol.*, **48**:67–87.

Vanden Berge, J. C. 1970. A comparative study of the appendicular muscles of the order Ciconiiformes. *Am. Midl. Nat.*, **84**:289–364.

Van Tienhoven, A. 1969. The nervous system of birds: a review. *Poultry Sci.*, **48**:10–16.

——— 1970. Neurological approach to photosexual relationships in birds. In *La Photorégulation de la Reproduction chez les Oiseaux et les Mammifères*, Éditions du Centre National de la Recherche Scientifique, Paris, pp. 259–280.

Walkinshaw, L. H. 1945. Aortic rupture in Field Sparrow due to fright. *Auk*, **62**:141.

Watanabe, T. 1960. Comparative and topographical anatomy of the fowl. VII. On the peripheral course of the vagus nerve in the fowl. *Jap. J. Vet. Sci.*, **22**:152–154.

Weller, M. W. 1968. Plumages and wing spurs of Torrent Ducks, *Merganetta armata*. *Wildfowl*, **19**:33–40.

Wight, P. A. L. 1971. The pineal gland. In *Physiology and Biochemistry of the Domestic Fowl*, Vol. 1, Academic Press, New York, pp. 549–573.

Witschi, E. 1961. Sex and secondary sexual characters. In *Biology and Comparative Physiology of Birds*, Vol. 2, Academic Press, New York, pp. 115–168.

Woolfenden, G. E. 1961. Postcranial osteology of the waterfowl. *Bull. Fla. State Mus.*, **6**:1–129.

Wolfson, A. 1954. Notes on the cloacal protuberance, seminal vesicles, and a possible copulatory organ in male passerine birds. *Bull. Chicago Acad. Sci.*, **10**:1–23.

——— 1960. The ejaculate and the nature of coition in some passerine birds. *Ibis*, **102**:124–125.

Zusi, R. 1967. The role of the depressor mandibulae muscle in kinesis of the avian skull. *Proc. U.S. Natl. Mus.*, **123**, No. 3606:1–28.

CHAPTER THREE

Plumage and Molt

Feathers, unlike the skin structures of vertebrates other than birds, have become highly diversified and specialized in form, color, and arrangement. Feathers are peculiar to birds, and this characteristic alone distinguishes them from all other animals.

There are several opposing views as to the origin of feathers. One suggests that the earliest birds had a covering of scales with a few feather filaments scattered among them. As birds became warm-blooded, the filaments became of survival value as an insulating layer and in the course of evolution became more abundant and longer, that is, more downlike. These theoretical simple filaments of primitive birds are called *cryptoptiles*.

A second theory proposes that the first feathers were pennaceous, similar to the contour feathers of modern birds and, in fact, like those preserved in the fossil specimens of *Archaeopteryx*. To proponents of this theory, down feathers were derived secondarily during the long evolution of birds.

Regal (1975) argued that elongation of body scales on reptiles resulted as an adaptive response to excessive solar heat, and that the first feathers evolved directly by subdivisions of such elongated scales. He doubted that feathers evolved in response to a need to conserve body heat.

If one accepts as a fact that *Archaeopteryx* was, indeed, the "first bird," there are two schools of thought as to the primary function of the early pennaceous feathers: (1) they evolved in relation to the development of warm-bloodedness, or (2) they evolved in relation to the development of flying ability. Parkes (1966) presents arguments in favor of the latter proposal. Ostrom (see Chapter 1) believes that feathers existed in the reptilian ancestors of *Archaeopteryx,* which already was homoiothermous, and, therefore, that feathers arose independently of flight; *Archaeopteryx* used the hand and forearm feathers to aid in capturing prey.

Whatever their origin, feathers provide a very light weight, durable, and flexible covering that retains innumerable pockets of air. The result is one of the most efficient natural insulating materials known to man. The rate of heat transmission through this body covering is regulated by changes in the angle at which the shinglelike feathers stand out from the body of the bird. Under conditions of cold the feathers (each of which has a number of muscles attached to its base from various directions; see Lucas and Stettenheim, 1972) may stand out from the body in a manner that results in the thickest possible layer of insulated air spaces about the bird; in extreme heat the feathers may be pressed against the body so that the insulation is reduced to a minimum (if there is a breeze, however, the bird may fluff its feathers).

So much confusion exists in the terminology of natal downs that the older terms probably should be abandoned (an exhaustive study of the subject is needed). The following discussion is given only as a necessary historical background for the student delving into the older literature; very few authors now use these terms.

Pycraft (1907) proposed that the feathers of modern birds be named as follows.

1.   Neossoptiles (nestling feathers)
     a.   Protoptiles[1]
     b.   Mesoptiles
2.   Teleoptiles (adult feathers)

Two successive coats of nestling down occur in penguins, in most owls, and in some other groups. Most of the higher birds have but one coat of nestling down, and some authors believed that in them the first, or protoptile, coat has persisted and that the second, or mesoptile, coat has been suppressed. Pycraft believed that the protoptile, not the mesoptile, coat is suppressed in most birds.

Percival (1942) later studied down feathers in a series of birds and reached interpretations different from those of earlier workers. He pointed out that

[1] These terms were coined by Pycraft (1907:11); his spellings: "protoptyle" and "mesoptyle."

there are two distinct types of down: (1) the rachis of the first down feather is continuous with that of the second as well as with the rachis of the definitive feather; and (2) the down lacks a rachis, and "there is continuity of individual barbs from first down to second, and on the removal of the constricting epitrichial sheath the barbs can be separated without further breaking of tissue."

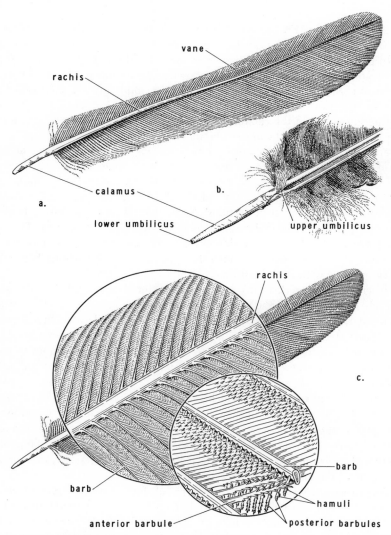

Figure 1. A typical flight feather and the nomenclature of its parts. (*a*) General view. (*b*) Detail of the base of the feather. (*c*) Detail of the vane.

Percival suggested several new terms, which have rarely been used: *haploptile,* single down; *diploptile,* double down; *pseudoprotoptile* and *pseudomesoptile,* the two parts of the double down.

THE FEATHER

The typical feather (Fig. 1) has a long, tapering central axis, supporting on each side a row of small branches, or barbs (*rami*), set at an angle inclined toward the tip of the feather. These branches form the web or vane (*vexillum*) of the feather. The central shaft is composed of two main parts: (1) a hollow, cylindrical basal section called the *calamus*; (2) a solid, angular shaft called the *rachis* (or *rhachis*).

At the lower end of the calamus is a small hole, the *lower umbilicus*; at the upper end another hole, the *upper umbilicus*.

Branching from each barb, but in the same plane, are two rows (one on each side) of still smaller branches, or *barbules*, set at an angle inclining toward the tip of the barb. The barbules on the two sides of the barb are very unlike; the *anterior barbules* (which point toward the tip of the feather) are flat basally and have a series of small projections (*barbicels*), whereas the barbicels along the middle portion of the under side are hooked (then called *hamuli*). The *posterior barbules* (on the opposite side of the barb) are also flat basally but have no hamuli; they form ridges on which the opposite anterior barbules hook, thus making the vane with its extraordinary lightness and strength.

Feathers are horny (keratin) structures, produced by papillae in the skin. The early development of the feather is quite similar to that of a scale. Strongly vascular dermal tissue, covered with a thin layer of epidermis, first forms a cone-shaped structure. Then (unlike a developing scale) the epidermal layer sinks inward and forms a follicle, out of which the papilla continues to grow and develop into a feather. At least once a year the feather is molted and a new feather grows from the same papilla.

Contour feathers are those which form the contour or outline of the body. Typically they have a large, firm vane, but the base of the feather is commonly plumulaceous (downy). A few of the contour feathers on the wing (remiges) and tail (rectrices) have become highly specialized for flight (or display); these include the largest and most highly developed of all feathers. Typical contour feathers grow only in the feather tracts (pterylae).

Semiplumes (Figs. 2*b*, 3) are a type of loose-webbed (plumulaceous) contour feather, with a definite rachis but no hamuli and therefore no firm vane. These feathers occur especially at the margins of the feather tracts and in the apteria, and are usually overlain by the general mass of typical contour feathers. There is a complete intergradation between semiplumes and firm-

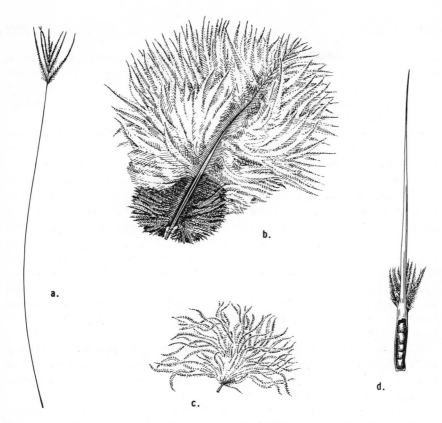

Figure 2. (*a*) Filoplume, *Lophura edwardsi* (×2). (*b*) Semiplume, Curassow (×¾). (*c*) Down, *Grus* sp. (×1). (*d*) Bristle: an eyelash, *Circus c. hudsonius* (after Chandler, 1914) (×16).

Figure 3. A feather typical of the malar and interramal regions of the Hawaii Amakihi (*Loxops v. virens*; Drepanididae).

121

webbed (pennaceous) feathers, but the fully developed "semiplume" type of feather is such a well-marked entity that it deserves a special name. Semi-plumes apparently serve to insulate, to provide flexibility at constricted points (particularly about the base of moving parts), and in the case of water birds to increase buoyancy.

Filoplumes (Fig. 2a) are very specialized, hairlike feathers, always asso-ciated with contour feathers. At present, two distinct kinds of feathers seem to be included under this name. One type nearly always has a minute vane (or a vestige of one) at the tip; filoplumes of this kind occur in groups of one to many about the base of a contour feather and appear to grow from the same papilla (Morlion, 1971). They are usually completely covered by the surround-ing feathers. The second type of filoplume lacks any vane whatever and often extends like a long hair beyond the contour feathers. This type occurs most frequently on the nape and upper back and may be readily seen on close exami-nation, for example, of any fresh-plumaged American Robin (*Turdus migra-torius*) or Lavender Fire-finch (*Estrilda caerulescens*). Notable instances of this second type of filoplume occur among the bulbuls (Pycnonotidae), the extreme condition being found in the Hairy-backed Bulbul (*Hypsipetes criniger*) of Malaysia (Fig. 4).

Filoplumes of at least the first type occur among the contour feathers of most birds (except the Ostrich and other ratites, and perhaps some pelicans). The function of filoplumes is unknown, but it may be associated with sensory nerve endings in the skin; in a few cases filoplumes may be decorative.

Nitzsch (1867:15) applied the term "filoplume" to the white filamentous feathers on the head and neck of some cormorants (some later authors even in-cluded the ornamental white plumes on the flanks of some birds in this cate-gory), but these feathers do not seem to come within the modern definition of filoplumes, for on close examination a rather extensive, though vestigial, web may be seen along the whole shaft. Certain other feathers superficially resemble

Figure 4.   Filoplumes on the nape and back of the Hairy-backed Bulbul (*Hypsipetes criniger*). (After Delacour, 1943.)

filoplumes but actually differ in structure. The "bristles" of the Bristle-thighed Curlew (*Numenius tahitiensis*) and the long, frontal plumes of the Spangled Drongo (*Dicrurus h. hottentottus*), for example, are true contour feathers (with a webbed base) whose outer portions have become bare and hairlike.

Bristles, which occur about the mouth, nostrils, and eyes (Fig. 2*d*) of many birds, are modified (vaneless and nearly vaneless) stiff feathers. Chandler (1914:360) has shown that in the Marsh Hawk (*Circus cyaneus*) these range, with all intergrades, from loosely woven feathers with some vane basally and a prolonged naked shaft to a simple, bare shaft. Therefore we conclude that they are not filoplumes, as some authors have thought, but specialized contour feathers.

The bristles about the nostrils seem to sift the air drawn in there; the long bristles (rictal bristles or vibrissae) about the mouth of many birds probably enlarge the effective gape of the mouth and thus facilitate the capture of flying insects. Chandler commented that the various bristles about the face of the Marsh Hawk are less easily ruffled, worn, and soiled than ordinary contour feathers; they may also have some sensory functions (see Lederer, 1972).

Bristles that form definite mammal-like eyelashes occur in some birds of widely scattered groups (Fig. 5): e.g., the Ostrich (Struthionidae), hornbills (Bucerotidae), many cuckoos (Cuculidae), the Hoatzin (Opisthocomidae), seriamas (Cariamidae), and some hummingbirds (Trochilidae). In the latter, the eyelashes may be black (*Eugenes fulgens, Archilochus colubris, Calypte costae*), brownish gray (*Calypte anna, Selasphorus platycercus, S. rufus, S. sasin*), or metallic green (*Popelairia conversii*).

Other examples of greatly modified feathers are the following: scalelike feathers in the Scale-feathered Cuckoo (*Phoenicophaeus cumingi*) and the Curl-crested Aracari (*Pteroglossus beauharnaesii*); bristlelike loral feathers in *Nyctibius* and *Podargus*; hairlike feathers on the neck of the Adjutant Stork (*Leptoptilus crumeniferus*); stiff feathers on the back of the head of the Crowned Crane (*Balearica pavonina*); many stiff, hairlike feathers in a pectoral tuft in turkeys; waxlike feather tips in waxwings (*Bombycilla*); and "wires" in some birds-of-paradise (Paradisaeidae).

Down feathers (plumules) (Fig. 2*c*) of adult birds are usually concealed beneath the contour feathers. They are small and soft (without a vane), with rachis very short or even vestigial. Down feathers are not confined to the pterylae (except in the tinamous); they are usually rather widely distributed on the bird, but they may be confined to the apteria or be absent entirely. Their principal function seems to be insulation. Down feathers are especially well developed in water birds. Research on comparative structure in down feathers has been scattered, fragmentary, and much obscured by contradictory statements, noncomparable techniques, and confused terminology. No generalizations can safely be made in advance of a critical and sweeping review.

Figure 5. An 11-week-old Secretarybird (*Sagittarius serpentarius*); note the well-developed eye-lashes and the specialized crest feathers. (Courtesy of W. T. Miller.)

Powder downs are much-modified body feathers which retain little of the definite structure of the normal feather. A typical powder down from an American Bittern (*Botaurus lentiginosus*) may have a calamus several centimeters long. In other species the calamus may barely protrude from the skin. From the calamus there extend a number of long, silky filaments, among which the powder occurs. The most strongly developed powder downs, such as those found in herons, grow continuously from the base and disintegrate at the tip. Part of the keratin of the growing powder-down feather, instead of forming barbs, is given off as a very finely divided powder made up of minute scalelike particles. This fine, waxy, powdery substance produces a characteristic bloom on the plumage of the bird, and it may protect the feathers from water.

Although powder downs are usually massed in solid, paired tracts on the ventral and sometimes the dorsal parts of the body, they may be scattered throughout the plumage, as in parrots. According to Schüz (1927), powder downs are lacking, or nearly so, in ratite birds, but are present to some degree in most other birds. The extreme development of powder downs occurs in the tinamous, herons, mesites (Mesoenatidae), the Cuckoo-roller (Leptosomatidae), some cotingas (Cotingidae), and wood-swallows (Artamidae) (Lowery and O'Neill, 1966).

Bock and Short (1971) discussed a "tuft of bristly feathers" located in the middle of the back of Asian woodpeckers of the genus *Hemicircus*. The tuft contains a viscid material "having a peculiar resinous odour." No cutaneous glands that could be a source of the material were found, and the authors concluded that "the most probable hypothesis is that the resinous secretion that coats the feathers . . . is produced by 'fat-quills,' which are modified powder feathers."

The **aftershaft** is part of the complete primitive feather, which consists of two shafts, an outer one (the feather we ordinarily see) and an inner one, the aftershaft or *afterfeather* (also called hyporachis or hypoptilum). In emus (Fig. 6) the two are nearly equal, but in most other birds the aftershaft is much reduced in size or may be absent. The presence of an aftershaft is a primitive condition, but its reduction or loss has not occurred uniformly during the evolution of higher types of birds. Miller (1924) summarized the different degree of loss of the aftershaft as follows.

1. Only two families, the cassowaries (Casuariidae) and emus (Dromiceiidae), have aftershafts nearly equal to the main feather in size.

2. Many tinamous (Tinamidae) and most of the higher gallinaceous birds (Tetraonidae, Phasianidae) have large aftershafts, often nearly as long as the main feather but considerably narrower and more plumulaceous. The aftershaft in these groups is unquestionably still of real importance in insulation.

3. Many herons, hawks, parrots, and others have aftershafts that are nearly or quite as long as those in Group 2, but the aftershaft is still less like the main feather and usually has a very feeble rachis.

4. The toucans, woodpeckers, and most songbirds (Oscines) have still less aftershaft, and the aftershaft itself has very little rachis.

5. The curassows (Cracidae) and the Hoatzin (Opisthocomidae) have the aftershaft represented only by a tuft of barbs that lack a central shaft.

6. The ducks, American vultures, and owls have only a vestigial fringe of tiny filaments to represent the aftershaft.

7. The aftershaft has been entirely lost by pigeons, hornbills, and most of the primitive passerine birds (Dendrocolaptidae, Formicariidae, etc.).

In general, the aftershaft is never found on the rectrices or larger remiges but commonly occurs on all of the other contour feathers of any bird that has the structure at all. In a few species, however, the aftershaft is found only on certain body feathers.

It has been suggested that there may be some correlation between low temperatures and the occurrence of a large aftershaft (presumably of value as insulation) in such northern groups as the grouse (Tetraonidae), but for the

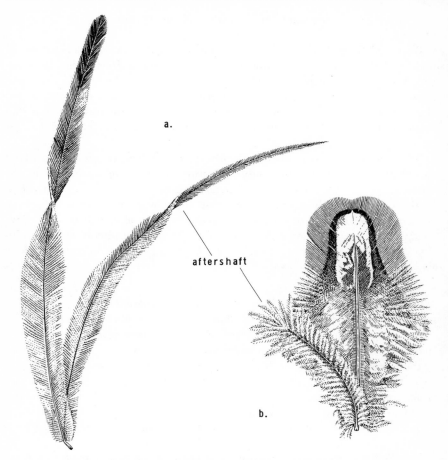

aftershaft

Figure 6. Feathers with well-developed aftershaft. (*a*) Emu; the aftershaft is nearly equal to the main feather (×¾). (*b*) Pheasant (*Phasianus*) back feather (×1).

most part the afershaft seems to be a nonadaptive character and for that very reason is of considerable value in taxonomy.

REMIGES

Primaries

The flight feathers attached to the hand (manus) are called primaries (Fig. 7). Their number is very constant within most bird groups. Flying birds have 12,

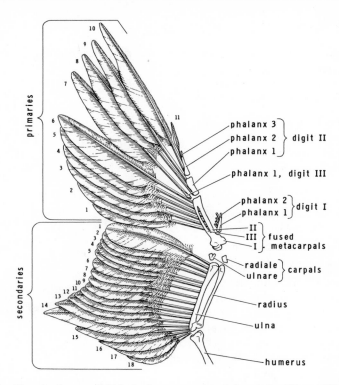

Figure 7. The wing bones and flight feathers as seen from below. (After Wray, 1887b; drawn by W. A. Lunk.)

11, 10 or 9 primaries. Among primitive flightless birds some (rhea and Ostrich) have an increased number of remiges on the hand (serving as purely decorative plumes), whereas others have a reduced number.

The outermost primary (12th, 11th, or 10th) is reduced in size in many species, sometimes so much so that it is difficult to recognize. This small outer primary is called the "remicle" (little remex) (Wray, 1887b; Miller, 1924). Although passerine birds have 10 primaries, the outermost (the 10th) is so short (a remicle) in certain families that these are said to have 9 "functional primaries"—hence the phrase "9-primaried oscines." As a general rule, all of the species of a family have the same number of "functional" primaries; the Warbling Vireo (*Vireo gilvus*), however, has a remicle, whereas the Philadelphia Vireo (*Vireo philadelphicus*) has a relatively long 10th primary.

Some recorded primary numbers are as follows.

Flying Birds

12,   grebes (Podicipedidae), except *Rollandia micropterum*, a flightless
       species
       storks (most) (Ciconiidae)
       flamingos (Phoenicopteridae)

11,   many families, including
       herons (Ardeidae)
       ducks (Anatidae)
       gulls (Laridae)

10,   the majority of birds

"9,"   honeyguides (Indicatoridae)
       some Old World passerines, viz.:
       white-eyes (Zosteropidae)
       some New World passerines, viz.:
       swallows (Hirundinidae)
       larks, in part (Alaudidae)
       pipits (Motacillidae)
       vireos, in part (Vireonidae)
       Hawaiian honeycreepers (Drepanididae)
       American wood-warblers (Parulidae)
       troupials (Icteridae)
       Swallow-tanager (Tersinidae)
       tanagers (Thraupidae)
       Plush-capped Finch (Catamblyrhynchidae)
       finches (Fringillidae)

Flightless Birds

3,   cassowaries (*Casuarius*)

4,   Kiwi (*Apteryx*)

7,   emus (*Dromiceius*)

8,   Laysan Rail (*Porzanula palmeri*)

9,   Henderson Island Rail (*Nesophylax ater*)

12,   rheas (*Rhea*)

16,   Ostrich (*Struthio*)

Primaries are referred to individually by number, and the count should always be made from the inner (proximal) primary to the outer (distal) one because, when primaries are lost in the course of evolution, they are lost from the outer end of the series. Thus homologous primaries in various families and genera will have the same number only if the count is made from the inner end of the series. Unfortunately, some ornithologists, especially in the Old World, ignore this fundamental reason for numbering the primaries outward and call the outermost (and most easily found) primary "number 1." The student must watch for this pitfall when reading papers that refer to primaries by number.

The several primaries are typically of different relative lengths, and these relationships are extraordinarily consistent in any given kind of bird (Eisenmann, 1969).

The proximal primaries are attached to the fused metacarpal bones and are referred to as "metacarpal primaries"; the others, attached to the second finger, are called "digital primaries." The number of metacarpal primaries is apparently a rather fundamental character (rheas, grebes, storks, and flamingos have seven; most other birds have six). Nearly all authors have assumed that the primitive number of metacarpal primaries was seven and that most birds have lost one. Stresemann (1963b), however, takes the opposite view that there were only six metacarpal primaries in early birds. A prime reason for his belief is that individual birds rarely have one more primary than is typical for the species; this may occur bilaterally or in only one wing. In both types, Stresemann found that the added feather was a metacarpal primary, not a digital primary.

Stresemann also differs in his interpretation of the remicle. He believes that, rather than being a vestigial primary feather, its original function, as in *Archaeopteryx,* was as a covert to the terminal claw on digit II. With the evolutionary loss of the claw in nearly all birds, the feather remained but has no covert because it itself was a covert.

Secondaries ("Cubitals"; Latin *cubitus,* elbow, ulna)

The flight feathers attached to the ulna of the forearm are called secondaries (Fig. 8). They have a much greater range in number (6 to 32) than the primaries; the number seems to be related to the length of the forearm. The smallest number (6 to 7) is found in hummingbirds, which combine very small body size and short forearm with great development of the manus of the wing. Although most passerine birds have 9 secondaries, the number varies from 9 to 14 (Stephan, 1965). Other recorded numbers are as follows.

10,	Bobwhite (*Colinus*)	11,	woodpeckers (Picidae)
	Blue Coua (*Coua*)		Hoatzin (*Opisthocomus*)

12, Cuban Trogon (*Priotelus*) kingfishers (*Alcedo*)

13, bee-eaters (*Merops*) rollers (*Coracias*)

14, Great Horned Owl (*Bubo*) Passenger Pigeon (*Ectopistes*)

15, Cooper's Hawk (*Accipiter cooperii*)

16, Wild Turkey (*Meleagris*) Prairie Falcon (*Falco mexicanus*)

17, Red-tailed Hawk (*Buteo jamaicensis*)

18, Turkey Vulture (*Cathartes*)

19, Black Vulture (*Coragyps*) Mallard (*Anas platyrhynchos*)

20, Osprey (*Pandion*)

21, King Vulture (*Sarcoramphus*)

22, California Condor (*Gymnogyps*)

25, Andean Condor (*Vultur*)

32, Wandering Albatross (*Diomedea*)

Secondaries are numbered from the outermost (next to the primaries) inward to the elbow. As explained under "Diastataxy and Eutaxy," many species have a gap in the row of secondaries just proximal to the fourth, but this gap should

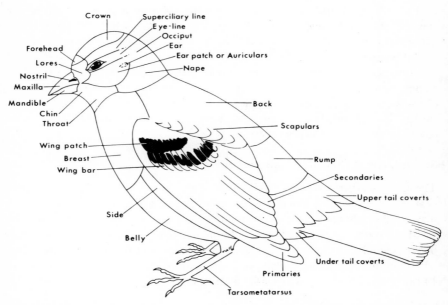

Figure 8. Topography of a bird, showing the regions used in descriptions of plumages of birds. (By permission of Andrew J. Berger, from *Hawaiian Birdlife,* University Press of Hawaii, copyright 1972.)

be ignored in numbering the secondaries, for we now believe that "diastataxic" and "eutaxic" species differ only in the presence or absence of a gap, not of a feather.

Carpal Remex

On the dorsal surface of the wing, in the gap between the primaries and secondaries, many birds have an additional remex, the "carpal remex," with a major upper covert ("carpal covert"). The carpal remex seems to be disappearing in the course of evolution. It appears in some gallinaceous birds and in some gulls as a strong pennaceous feather half as long as the adjacent secondaries—to which series it seems to belong. Some groups, such as the woodpeckers, retain it only as a small feather without a covert. Frequently we find the covert without the remex (e.g., in the hornbill *Aceros undulatus,* in the shrike *Lanius,* and in passerine birds generally), or a small covert with an even smaller carpal remex (Fig. 9).

Diastataxy and Eutaxy[2]

In a great many species the fifth secondary (counting inward from the wrist) seems to be missing, but its usual position is clearly marked by a gap and a major covert above it (Fig. 10). Steiner (1918) studied "diastataxy" (literally, "arranged with a gap"), and decided that the secondaries are derived from two

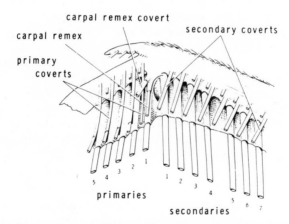

Figure 9. The carpal remex and the diastataxic condition of the secondaries in the Painted Snipe (*Rostratula*). (Adapted from Lowe, 1931; drawn by W. A. Lunk.)

[2] The old terms "aquincubital" (for diastataxic) and "quincubital" (for eutaxic) are to be avoided, since it is now agreed that a gap not a feather, is present or absent.

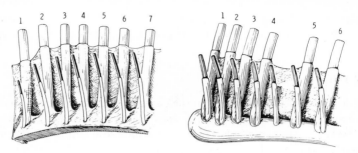

Figure 10. Eutaxic and diastataxic arrangements of the major coverts and secondaries (dorsal view). Left: Pheasant (eutaxic). Right: Golden Eagle (diastataxic). (After Wray, 1887b.)

embryonic series, one on the lower and the other on the upper surface of the wing. During embryonic development the under row moves dorsally and comes to form a continuous series with the upper row. At the point of junction of the two series (between secondaries 4 and 5), an extra greater upper covert is retained (but without a secondary to match it). The arrangement without a gap is termed "eutaxic."

Stephan (1970), however, found no evidence of a dislocation of rows of feathers or any evidence that diastataxy is the result of a reduction or disappearance of a secondary feather. He stated that the diastataxic condition results from the addition of an additional transverse row of coverts. He also concluded that "no evidence could be given for a diastataxic origin of the eutaxic wing. The eutaxic and diastataxic wings are of eutaxic and diastataxic origin, respectively." Moreover, "diastataxy cannot be established for *Archaeopteryx*. At the present state it may be supposed that *Archaeopteryx* was eutaxic."

Nevertheless, many of the more primitive groups of birds are diastataxic; all passerine birds and some other families, such as bee-eaters, motmots, and trogons, are eutaxic. In a number of other orders and families both conditions are found. For example, some hummingbirds are eutaxic and others diastataxic; the American Woodcock (*Philohela*) is eutaxic, but, as far as is known, all other members of the Charadriiformes (shorebirds, gulls, etc.) are diastataxic (Miller, 1924).

Alula

The group of feathers borne by the thumb (alular digit) is called the alula (Fig. 11). There may be only two quills (e.g., hummingbirds), but the number varies from five to seven in certain cuckoos, the peafowl (*Pavo*), trumpeter (*Psophia*), and a touraco (*Tauraco*). In most passerines the number is three or four, but six are recorded in the Lyrebird (*Menura*) and five in two other large Australian passerines (*Gymnorhina* and *Struthidea*).

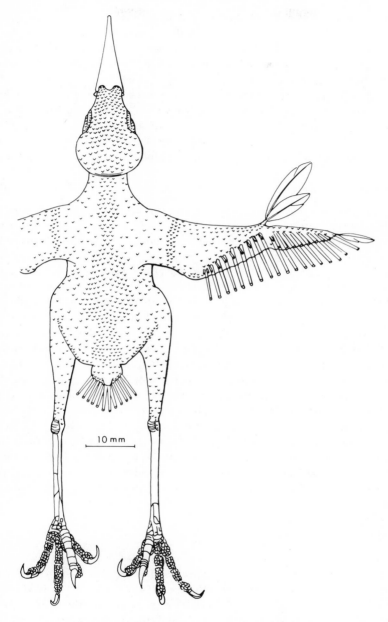

Figure 11. Dorsal feather tracts of the Apapane (*Himatione sanguinea*; Drepanididae), showing the two alula feathers and the remicle (10th primary).

The relative lengths of alula quills vary; in all passerines and in most other birds the distal one is the longest, but in the cuckoo *Tapera,* for example, the next-to-distal one is the longest. *Tapera* is called "four-winged" because of the very large and conspicuously displayed alula, and a strongly developed alula is, in fact, characteristic of a number of cuckoos.

Alula quills should be numbered from the innermost to the outermost, since in the known cases of reduction in number within a taxonomic group it is the outermost quills that are lost, and this order agrees with the order of molt. Unfortunately, the practice has not been standardized, and such an eminent student of ptilosis as Miller (1924) numbered alula quills in the opposite order.

Tertiaries (or "Tertials")

Long ago, Wray (1887b:344) proposed that the name "tertial" or "tertiary" be abandoned. Unfortunately, Stone (1896) and Dwight (1900) revived this term and applied it to the proximal three secondaries of passerine birds. (Since Stone and Dwight did not recognize any vestigial innermost secondary, they listed six secondaries and three "tertials" for all the passerine birds whose molt they described.) Because these inner secondaries are often sharply differentiated in ducks, Kortright (1942) yielded to the temptation and labeled them "tertials" in his figures of the wing, but admitted in the accompanying text that they merely "pass under the name of tertials"! It is true that a group of inner secondaries are often differently colored and shaped and behave differently in molt; but if we attempt to count them separately from the other secondaries, we are completely lost when making comparisons with species in which these feathers do not differ morphologically. Therefore modern students of pterylosis have usually followed the practice advocated here and have termed "secondaries" all feathers attached to the ulna.

Some birds, especially large, long-winged forms, have well-developed flight feathers attached to the humerus; these may be called tertiaries, although in some birds they seem to be continuous with upper wing coverts rather than with secondaries. Fisher (1942:32) recorded 13 tertiaries in the Andean Condor (*Vultur*).

Reduction in Remiges

It is debatable that the general trend in the evolution of birds has been toward a reduction in the number of primaries and secondaries (see Stresemann, 1963b). The large numbers of secondaries in some modern birds obviously seem to be correlated with the bird's pattern of flight and, therefore, with the length of the forearm. Thus, frigatebirds and albatrosses, which are adept at soaring, have very long forearms and large numbers of secondaries; the forearm in hummingbirds, however, is very short, and these birds have the smallest number of secondaries. Among flying birds the greatest number of remiges occurs in certain long-winged birds, such as the Wandering Albatross, *Diomedea exulans* (10 primaries + 32 secondaries) and the Andean Condor, *Vultur gryphus* (11 + 25). The minimum number of functional primaries in a flying bird is 9, and that is found in certain families such as the Fringillidae and Icteridae, which on the basis of other evidence are commonly considered to represent the highest point in avian evolution. (The only birds that have fewer than 9 primaries are certain flightless rails and ratite birds with obviously very degenerate wings.)

Wing Coverts ("Tectrices")

Typically, each wing quill (remex) is covered at its base on the upper side of the wing by a greater ("major") covert (Fig. 12). Passing from the greater coverts toward the leading edge of the spread wing, we find successive rows of

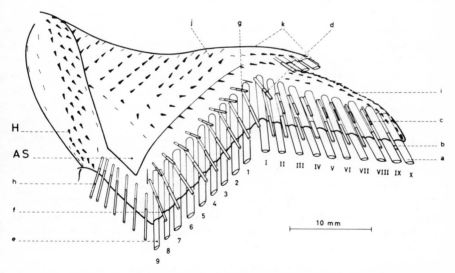

Figure 12. Dorsal view of the feather tracts of the wing of Vieillot's Black Weaver (*Ploceus nigerrimus*). a, I–X, primaries; b, greater primary coverts; c, middle primary coverts; d, carpal covert; e, 1–9, secondaries; f, greater secondary coverts; g, middle secondary coverts; h, tertiary coverts; i, carpometacarpal coverts; j, marginal coverts; k, alula (feathers and coverts); H, humeral feather tract; AS, dorsal alar apterium. (Courtesy of Maria Morlion and the editor of *Verhandelingen van de Koninklijke Vlaasmse Academie voor Wetenschappen, Letteren en Schone Kunsten van Belgie.*)

smaller and smaller feathers: one row of middle ("median") coverts, several rows of lesser ("minor") coverts, and an indefinite number of marginal coverts. The rows of coverts above the primaries are all referred to as "primary coverts"; those above the secondaries are called "secondary coverts." For details, see Lucas and Stettenheim (1972).

The coverts on the under side of the wing are arranged similarly, but the feathers are less fully developed and some rows, especially among the lesser coverts, are incomplete. In passerine birds and woodpeckers, greater secondary under coverts are lacking or are rudimentary.

All of the remiges overlap laterally in a uniform way: the inner vane of each feather is overlapped (as considered from above) by the next proximal feather. All greater coverts, above and below, overlap in the same way as the remiges, and therefore, following Bates (1918), we say that these coverts have a "conforming" overlap (i.e., are like the remiges); the opposite type is called "contrary" overlap.[3] Many of the rows or parts of rows of smaller coverts have, in various combinations in different species and genera, partly a contrary and partly a conforming overlap. Bates showed that in birds of strong flight the method of overlap is clearly adaptive, a contrary overlap (which seals the interstices between the remiges more effectively) appearing regularly in such birds, especially in the areas of greatest strain.

In 1843 the Swedish ornithologist Sundevall (1886) pointed out that some or all of the greater and median coverts on the under surface of the wing are "reversed," i.e., their concave surface faces outward. Wray (1887b) was the first to see that this was explained by their origin. These coverts, now on the under surface of the wing, were originally on the upper surface and have retained their primitive orientation. A study of their embryological history shows that they begin to develop on the dorsal side of the wing and then are carried to the ventral side as the remiges of the adjoining row become larger and larger. The dorsal origin of these under coverts is most strikingly demonstrated in such a bird as the great Military Macaw (*Ara militaris*), which has on the upper side of the wing bicolored coverts (blue-green on the exposed surface and dull olive beneath). On turning over the "reversed" under coverts, one finds that they too are a bright blue-green dorsally, but in their position under the wing the bright surface is concealed and the dull olive surface exposed.

[3] In the early literature, overlap is described as "distal" or "proximal," but Goodchild (1886) called one type of overlap "distal," whereas Gadow (1893:949–952) and others used the term "distal" to denote the opposite type of overlap. Bates pointed out this conflicting usage and the added problem that arises when we wish to discuss both upper and under coverts; he therefore proposed the terms "conforming" and "contrary."

FEATHERS OF THE TAIL

Rectrices

The flight feathers of the tail are called rectrices. Normally they are even in number. Ornithologists identify them numerically, counting from the center outward.

Most birds have 12 rectrices (6 pairs); a few have 10 (hummingbirds, swifts, most cuckoos, most of the motmots, the toucans, etc.); some have but 8 (the cuckoo genus *Crotophaga,* some rails, some grebes); and at least three species, all very small birds, have only 6 (a timaliid, *Pnoepyga,* of Asia; a sylviid, *Stipiturus,* of Australia; and a furnariid, *Sylviorthorhynchus,* of southern South America). Adolf Portmann wrote that adult males of the Marvelous Spatuletail (*Loddigesia mirabilis*), a hummingbird from Peru, have only 4 rectrices; younger males have 10. Many examples of an increase over the usual 12 rectrices can be cited.

14,	guineafowl (*Numida*) Scaled Quail (*Callipepla*)	20,	♂ Peafowl (*Pavo*)
16,	ptarmigan (*Lagopus*) Silver Pheasant (*Lophura*) Lyrebird (*Menura*)	16–22,	Blue Grouse (*Dendragapus*)
		24,	White Pelican (*Pelecanus erythrorhynchos*)
18,	♀ Peafowl (*Pavo*) Prairie Chicken (*Tympanuchus*) Ring-necked Pheasant (*Phasianus*)	26,	Pintail Snipe (*Capella stenura*)
		32,	White-tailed Wattled Pheasant (*Lobophasis bulweri*)

Tail feathers vary in number within bird groups more than wing feathers do. For example, the Rhinocryptidae as a family have 12 tail feathers (but one genus, *Pteroptochos,* has 14); the Furnariidae vary from 6 to 12 or 14, with the number varying even within a single genus; the Cuculidae characteristically have 10 (but the genus *Crotophaga* has 8); the Dicruridae as a family have 10 (but the genus *Chaetorhynchus* has 12). The female peafowl (*Pavo*) has 2 tail feathers fewer than the male (which has a tremendous "train" of tail coverts, supported by 20 rectrices). There is also some individual variation. The Japanese have bred domestic chickens for exceedingly long, continuously growing tail feathers, and roosters are known to attain tails over 30 feet long.

Tail Coverts

The rectrices are overlain by the greater upper coverts, with lesser upper coverts above them. Clark (1918) investigated the numbers and arrangement of the greater coverts. In many birds he found that each rectrix had a greater upper covert; however, in a few groups (especially swimming birds) there were more coverts than rectrices, and in the majority of bird species, including all of the passerine birds that he examined, the greater upper coverts were fewer in number than the rectrices. Usually it was the central pair of rectrices that lacked greater upper coverts. Miller (1931:128) and others have found a few passerine birds that have equal numbers of rectrices and major upper coverts. The greater under tail coverts are often less clearly defined than the greater upper coverts; their number has been recorded in but few species, and it may be the same as the number of rectrices, or several more or less.

NUMBER OF CONTOUR FEATHERS

Since the flight feathers and their coverts have been found to be regularly arranged and definite in number according to the kind of bird, it should not surprise us that the whole covering of the bird is made up of a rather definite number of feathers according to kind. Few feather counts have been made on significant numbers of birds, however, so that counts should be considered approximations (see Clench, 1970).

The smallest number of feathers (940) was recorded for a Ruby-throated Hummingbird (*Archilochus colubris*), and the largest (25,216) for a Whistling Swan (*Olor columbianus*). Brodkorb (1955) counted 7182 feathers on a female Bald Eagle (*Haliaetus leucocephalus*) and noted that the feathers weighed 677 grams, which was more than twice the weight of the skeleton (272 grams). This Bald Eagle also had fewer feathers than were found in a Barred Owl (*Strix varia*) and a Screech Owl (*Otus asio*). Markus (1963, 1965) examined 11 specimens of the Laughing Dove (*Streptopelia senegalensis*) and found numbers of feathers ranging from 3900 to 4390, with an average of 4192 feathers per bird. He gave the following numbers for three species of barbet: Crested Barbet (*Trachyphonus vaillantii*), 2653 and 2904; Black-collared Barbet (*Lybius torquatus*), 3014; Yellow-fronted Tinker Barbet (*Pogoniulus chrysoconus*), 2210. In number of feathers, passerine birds (omitting the most diminutive species) range between about 1500 and 3000 (in one case, 4900). Clench (1970) noted that "species resident in cold northerly regions tended to have more heavily feathered pterylae than those that migrate south for the winter. This trend seemed also to be generally true of most passerines that breed in temperate Connecticut: e.g., six migratory species of Parulidae averaged 114 feathers in 9 saddle rows, compared to two specimens of the similar-sized but usually resident *Parus atricapillus* which had an average of 151 in the same number of rows." She found "no significant variation between individuals of the same species, and no effect on a specimen by its sex, plumage stage, etc. The same kinds of differences between juveniles and adults are also seen in other passerines."

PTERYLOSIS OF ADULT BIRDS

The feathers of the Ostrich, penguins, South American screamers (Anhimidae), and mousebirds (Coliiformes) are distributed over the body in ·an almost continuous fashion, but all other birds have their contour feathers restricted to definite tracts (*pterylae*), leaving spaces (*apteria*) between. Some apteria have bare skin, others are sparsely feathered with semiplumes, and still others (especially in waterfowl) are densely filled with downs. In life, the apteria are entirely covered by the overlapping contour feathers from adjoining pterylae.

Ornithologists have used two primary methods for studying the elaborate patterns formed by pterylae: clipping the feathers close to the skin, leaving only short stubs; and removing the skin and reversing it, tracing the patterns formed by the feather papillae on the inside of the skin. Clench (1970), however, made "soft" x-rays of the tracts after the skin had been prepared as a flat mount (Fig. 13). Her method has the advantage of providing specimens in which there is minimal disturbance of the pterylae in both the dorsal and the ventral tracts. Her study demonstrated that very little variation occurs in the pattern of the tracts in different individuals of the same species provided that they are of comparable age; there are, however, significant differences between the tracts of adults and those of nestlings and juveniles of a species.

The feather tracts are classified and named according to their anatomical location. The major tract names used in discussing the pterylae are as follows.

Capital Tract. This consists of the upper half of the head, from the base of the bill posteriorly to the base of the skull, where it continues as the spinal tract. The capital tract is commonly subdivided into a number of "regions."

Spinal or Dorsal Tract. The spinal pteryla extends along the back from the base of the skull to the base of the tail, ending at the oil gland and upper tail coverts. This tract may be divided into cervical, interscapular, dorsal, and pelvic regions; it is a tract that varies greatly among the families and genera of birds. Compare Fig. 14 (shrike) with Fig. 15 (cuckoo).

Humeral Tract (also scapulohumeral). This narrow tract, beginning at the shoulder, extends down across the dorsal surface of the base of the wing to the region of the elbow; it includes the scapular feathers. This tract shows little variation in form.

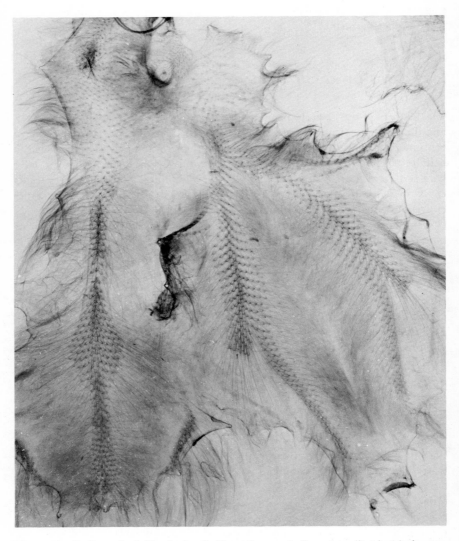

Figure 13. Radiograph of skin of a female House Sparrow in first winter (first basic) plumage, collected November 20, 1960. The dorsal tract is on the left; two horns of the ventral tract on the right. (Courtesy of Mary H. Clench.)

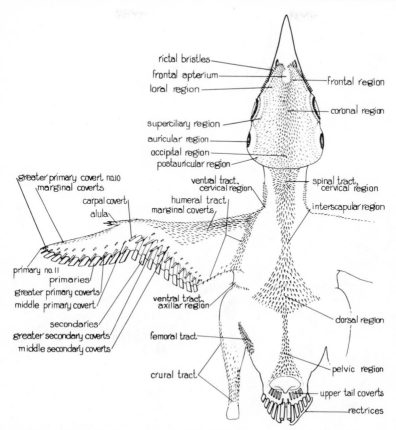

Figure 14. Pterylography of the shrike *Lanius ludovicianus* (dorsal view). (Courtesy of the late A. H. Miller.)

Femoral Tract.   This narrow, rather uniform tract, extends from the region of the femur along the flank toward the tail.

Crural Tract.   The sparsely feathered crural tract (inner and outer) covers the leg.

Alar Tract.   This consists of the feathers of the wing exclusive of the humeral tract.

Caudal Tract.   The rectrices, their coverts, the anal circlet, and the feathers on the oil gland (the last are lacking in passerines and many other birds) constitute this tract.

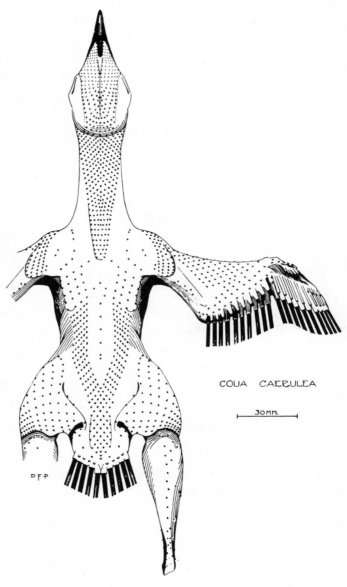

COUA CAERULEA

├──── 30MM ────┤

DFP

Figure 15. Dorsal view of the feather tracts of a cuckoo (*Coua caerulea*); drawn by David F. Parmelee. (Courtesy of Andrew J. Berger and the editor of *Wilson Bulletin*.)

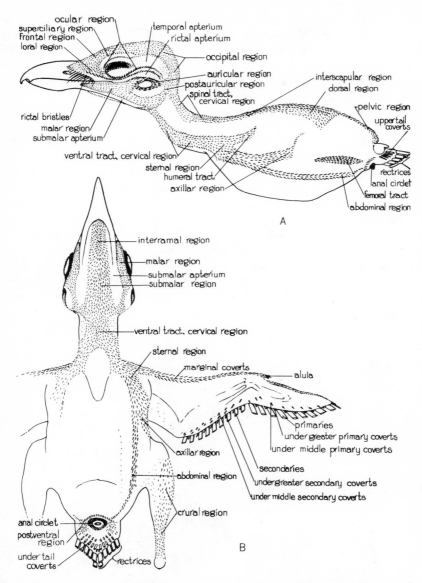

Figure 16. Pterylography of the shrike *Lanius ludovicianus*: (*A*) lateral view; (*B*) ventral view. (Courtesy of the late A. H. Miller.)

Ventral Tract. This tract consists of the feathers of the under parts exclusive of the humeral and femoral tracts. Some authors consider it to extend from the base of the bill to the cloaca; others believe that the ventral chin and neck feathering are part of the capital tract, the ventral tract beginning on the anterior breast. This tract is usually forked. It may be divided into interramal, submalar, cervical, sternal, axillar, and abdominal regions (Fig. 16).

PTERYLOSIS OF NESTLING BIRDS

The newly hatched bird may be naked, may contain a relatively small number of natal down feathers in the regions of later feather tracts (prepennae), or may be almost completely covered with a thick coat of down of both prepennae and preplumulae (Figs. 17, 18). Nestlings of the first two types are said to be *psilopaedic* (also gymnopaedic); those covered with a full coat (in pterylae and in apteria) of down are called *ptilopaedic*. Ptilopaedic young are characteristic of ducks, gallinaceous birds, and some seabirds. Most passerine birds (an exception is *Procnias averano*) and some nonpasserine species have psilopaedic young.

Figure 17. A naked, newly hatched Purple Sunbird (*Nectarinia asiatica*) and an egg; Baroda, India, February 25, 1965.

Figure 18.   A newly hatched White Tern (*Gygis alba*); Laysan Island, Hawaiian Islands, March 30, 1966.

Wetherbee (1957) described and illustrated the pattern of distribution of the prepennae in a large number of passerine species (Fig. 19). Despite conflicting statements in the literature, Wetherbee believes that natal down is fully developed in the egg and that there is no growth of it after the bird hatches. Downs that develop after hatching always grow from follicles that produce only downs from then on; the downy coat developed by young pelicans, cormorants,

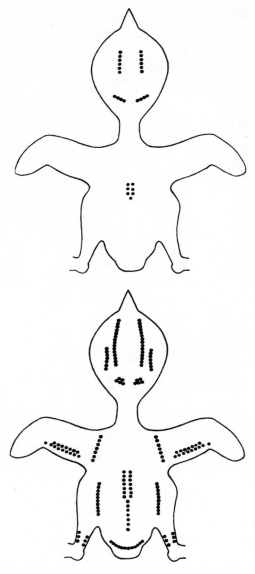

Figure 19. Top: Basic neossoptile pterylosis of newly hatched wrens (Troglodytidae). Bottom: Basic neossoptile pterylosis of newly hatched mockingbirds, thrashers, and catbirds (Mimidae). (Courtesy of David K. Wetherbee and the American Museum of Natural History.)

Figure 20. A 19-day-old nestling Chestnut-collared Swift. (Courtesy of Charles T. Collins and the editor of *Condor*.)

boobies, etc., after hatching represents precocious growth of the body downs of the juvenal plumage (Parkes and Clark MS.).

Collins (1963) described an interesting condition in the Chestnut-collared Swift (*Cypseloides rutilus*): the nestlings are naked at hatching but grow a dense covering of downy *semiplumes* by the time they are about 19 days old (Fig. 20).

PLUMAGES AND MOLTS

Beginning with the first plumage, worn at hatching or acquired soon after it, birds pass through one or more series of molts[4] and plumages before they attain the fully adult plumage. Thereafter, in all healthy birds all feathers are re-placed at least once a year (except the flight feathers of some very large birds, such as *Aquila*, in which more than 2 years may be required for a complete re-placement of the wing feathers); in some species much of the plumage is re-placed two or even three times during a year.

A single annual molt (the postnuptial) appears to be characteristic of most species. Many have also a second molt (the prenuptial), which is in some cases complete (e.g., in the Sharp-tailed Sparrow, *Ammospiza caudacuta*), but in most birds partial (e.g., in the crowned sparrows *Zonotrichia*); ptarmigan (*Lagopus*) have three or even four molts per year (Amadon, 1966).

Watson (1963) presented evidence that appears to justify his conclusion that "in the repeated molt cycle of many, if not all birds, the dropping of the old generation of feathers is brought about by the initiation of growth in the new generation which pushes the old feathers passively out of the follicles. Molt in birds is consequently a single growth process actively concerned only with the production of the new generation of feathers. This new growth causes the passive loss of the old generation of feathers" (Figs. 21, 22). Consequently, this normal method of molt has been termed "active molt." Not all loss of feathers is accomplished by this method, however; exceptions include the simultaneous loss of flight feathers, the loss of feathers in the development of brightly colored bare skin or wattles for display, the development of brood patches, certain cases of abnormal loss of feathers (Van Tyne, 1943; Michener and Michener, 1946), and "frightmolt" (Juhn, 1957).

The series of plumages of the typical bird may be outlined as follows:

Natal (or Nestling) Down

This plumage is found on the majority of birds at hatching. In some groups (e.g., the pelicans and their relatives) the young are hatched naked, but after some days they develop a downy covering to the whole body. Other groups (e.g., woodpeckers, kingfishers, and a few passerine birds) have completely naked young; their first plumage, when it finally grows out, consists of feathers very much like those of adults. In still other groups (penguins, petrels, most owls, and a few other birds), the first coat of down is succeeded by a second nestling down coat. Down feathers are lost over a period of time as the contour feathers grow; they are attached to the tip of the growing juvenal feather and are pushed out, whereupon they either wear or break off. The growing contour

---

[4] Obviously, change of plumage by "molt" involves two processes: the loss of the old feathers and the growth of new ones. Ornithologists formerly discriminated clearly between the two processes in their terminology, calling them "ecdysis" (putting off) and "endysis" (putting on). Unfortunately, these terms are now little used, and when ornithologists use the term "molt" they sometimes are referring to the loss of feathers only, and at other times to the double process of loss and replacement of feathers.

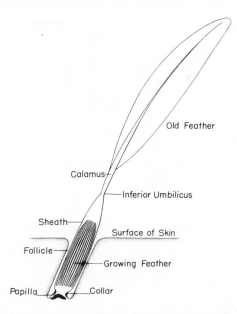

Figure 21. Diagram of feather structures during molt; sheath and follicle below the inferior umbilicus are shown in longitudinal section. (Courtesy of George E. Watson and the editor of *Science*; Vol. 136, January 4, 1963, pp. 50–51; copyright 1963 by the American Association for the Advancement of Science.)

feather in its sheath is called a "blood quill." The enveloping sheath first splits at the tip (Fig. 23); the definitive feather assumes its typical shape after the sheath is completely removed, in part because of the preening activities of the young bird.

Juvenal Plumage

The first plumage of typical contour feathers was named the "juvenal" plumage by Dwight (1900). It is usually worn for only a short time. Sutton (1935) showed that many young sparrows begin to molt into the next plumage before the juvenal rectrices have completed their growth. Although composed of the adult type of feather, the juvenal plumage is often easily distinguished from true adult plumages by the softer, looser-textured character of the body feathers and the different shape of the flight (especially tail) feathers. Passerine birds develop the juvenal plumage before they leave the nest. As the juvenal plumage grows in, it carries out on the feather tips remnants of natal down.

Figure 22. A juvenal flank feather (arrow) of a living Chukar Partridge (*Alectoris chukar*) pushed out of follicle still attached to tip of incoming feather sheath. Other feathers already breaking sheaths. Photographer's fingers at bottom. (Courtesy of George E. Watson and the editor of *Auk*.)

Figure 23. Incoming primaries and alula feathers in the right wing of a 38-day-old Redhead (*Aythya americana*). Photographed at the Delta Waterfowl Research Station, Manitoba, Canada, July 31, 1962.

First Winter (First Basic) Plumage

This plumage is much like the adult winter plumage but is frequently distinguishable by duller colors or immature types of pattern; it is acquired by a partial or complete postjuvenal molt (Fig. 24). A number of species retain at least the flight feathers of the juvenal stage through the first winter plumage. This plumage may be acquired very early. For example, Sutton (1948) described a young Warbling Vireo (*Vireo gilvus*) which had acquired an apparently complete first winter plumage by July 28. By contrast, Selander and Giller (1960) reported that only 2 percent of male Brown-headed Cowbirds (*Molothrus ater*) and only 28 percent of female Red-winged Blackbirds (*Agelaius phoeniceus*) undergo a complete postjuvenal molt.

First Nuptial (First Alternate) Plumage

In many species this plumage is like, or nearly like, the adult nuptial plumage; it is acquired by the wearing off of contrasting tips and edges of the body feathers (hence is not really a new plumage) or by a (usually incomplete) prenuptial molt. The name of this plumage is somewhat misleading, for birds of

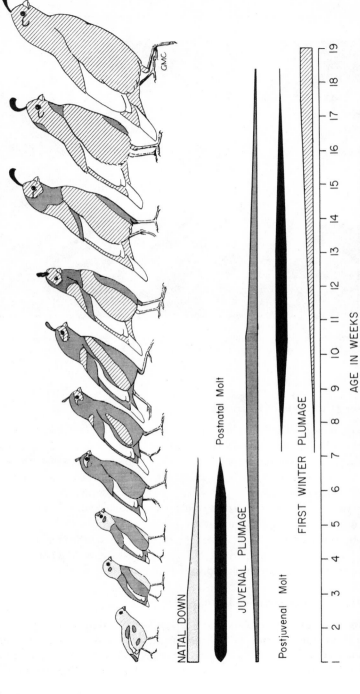

Figure 24. Development of body plumage of California Quail (*Lophortyx californicus*). Stippling indicates areas bearing natal down; vertical hatching, areas covered by juvenal feathers; oblique hatching, areas covered by first winter plumage. (Courtesy of Ralph J. Raitt, Gene M. Christman, and the editor of *Condor*.)

many species (especially large birds) are not yet ready to breed when they assume the first nuptial plumage. The nuptial plumage of the common Starling (*Sturnus vulgaris*) is a familiar example of the effects of wear; other examples of this type are the male House Sparrow and the Snow Bunting (*Plectrophenax nivalis*).

Second Winter (Second Basic) Plumage

In most species of birds this plumage is indistinguishable from that of adults of any age; in this case it may be called simply "adult winter plumage." It is acquired by a complete postnuptial molt—the complete annual molt of virtually all birds.

Humphrey and Parkes (1959) proposed a new terminology for plumages and molts because of their belief that "molts should be named in terms of the incoming generation of feathers," and as an indication of homologies between the molts and plumages of different groups of birds without semantic dependence on other phenomena such as breeding or north temperate seasons. They suggested that the postjuvenal molt be renamed the *first prebasic molt*; the first prenuptial molt, the *first prealternate molt*; etc. (Table 1). The new terms were criticized by some authors (e.g., Miller, 1961; Stresemann, 1963a [reply by Humphrey and Parkes, 1963]; Amadon, 1966) and supported by others (e.g., Watson, 1963; Clench, 1970).

Table 1. Terminologies for Plumages and Molts

Humphrey and Parkes (1959)	Dwight (1900)
Natal down	Natal down
Prejuvenal molt	Postnatal molt
Juvenal plumage	Juvenal plumage
First prebasic molt	Postjuvenal molt
First basic plumage	First winter plumage
First prealternate molt	First prenuptial molt
First alternate plumage	First nuptial plumage
Second prebasic molt	First postnuptial molt
Second basic plumage	Second winter plumage
Second prealternate molt	Second prenuptial molt
Second alternate plumage	Second nuptial plumage
Third prebasic molt	Second postnuptial molt
Etc.	Etc.

TIME OF MOLT

Most species of migratory birds complete the postnuptial molt before they start on their fall migration, but swallows, some tyrant-flycatchers, bee-eaters (Meropidae), shearwaters (*Puffinus*), and some Palaearctic Sylviidae normally postpone the molt until they reach their winter quarters.

Molt periods are closely synchronized with the reproductive cycle and with the seasons of the year. The increase and decrease of daylight appears to be the chief environmental factor involved in timing molt in the Temperate Zone. In tropical areas the major factor is often rainfall patterns. Ashmole (1963) discussed "a uniquely flexible molt program" of Sooty Terns (*Sterna fuscata*) on tropical islands.

TIME REQUIRED FOR ATTAINING ADULT PLUMAGE

There are great differences among species in the age at which the adult plumage is acquired, and bird banders are constantly finding new plumage characters to distinguish age classes of birds. In many small birds, a plumage much like that of adults follows the juvenal plumage (i.e., at about 3 months of age). The adult nuptial plumage of the Bobolink (*Dolichonyx oryzivorous*) is acquired by an extensive (or complete) prenuptial molt during the late winter or early spring of the bird's first year (at an age of about 8 months). The Rose-breasted Grosbeak (*Pheucticus ludovicianus*) molts into the adult plumage in August of its second year (when about 14 months old). The Herring Gull (*Larus argentatus*) usually attains adult plumage with its fourth winter plumage.

VARIATION IN EXTENT OR OCCURRENCE OF MOLT

Whistler (1941) has shown that there may be marked differences, apparently of adaptive significance, between the molts of two otherwise very similar subspecies. The European form of Short-toed Lark (*Calandrella cinerea*), for example, has no spring (prenuptial) molt, but an Asiatic subspecies, *dukhunensis*, which lives under very windy desert conditions causing excessive abrasion of the plumage, has a very extensive body molt in spring. In the Phainopepla (*Phainopepla nitens lepida*), Miller (1933) found marked differences in the postjuvenal molt between different geographical components of a single subspecies. The nonmigratory population of the Colorado Desert of California molted more fully than the population of the cooler coastal belt, which began nesting later and thus had less time to complete a molt before starting its fall migration (see Selander and Giller, 1960).

ORDER OF MOLT IN THE INDIVIDUAL

Except in penguins (in which great patches of feathers are shed in a seemingly irregular pattern), molt in birds is a surprisingly orderly process. A complete molt commonly begins with the loss of the innermost primaries on both sides (for some exceptions, see Stresemann, 1963c). Then, usually in regular succession, the other primaries drop out as their predecessors are being replaced, so that there is only a small gap at any one time, and the bird thus retains the power of flight. When the molt of the primaries is about half completed, the secondaries usually begin to molt, starting with the outermost and progressing inward. Soon after the molt of the wing feathers is under way, molt begins in the tracts of body feathers. The tail feathers, like the wing feathers, are usually molted serially (by pairs), so that here, too, a continuously useful organ of flight is maintained. In most birds the molt begins at the center (or next to it) and progresses outward—an order of molt which Beebe (1916) termed *centrifugal*; but the molt is *centripetal* (progresses from outer feathers inward) in colies, certain toucans, woodpeckers (Friedmann, 1930), and a few passerines. Some pheasants have a centrifugal, some a centripetal, tail molt; in others the molt starts at an intermediate point and progresses both outward and inward.

SIMULTANEOUS MOLT OF FLIGHT FEATHERS

Striking exceptions to the rule that remiges are molted gradually and serially occur in some unrelated birds that have one thing in common: a habitat (usually aquatic) where they can elude their enemies for a few weeks without resorting to flight. These birds drop all their remiges at once. Examples of birds molting thus are waterfowl (Anatidae), most rails (Fig. 25), many alcids (Alcidae), and loons (Gaviidae). The Dipper (*Cinclus*) has a very compressed molt which leaves the bird flightless for a period.

The Mayrs (1954) found that small owls usually lose all of their tail feathers at once, whereas large owls (e.g., *Tyto, Bubo, Strix*), which pursue larger and speedier prey and probably cannot secure adequate food without strong powers of flight, molt their tail feathers gradually. Woodpeckers (Picidae) and creepers (Certhiidae) maintain a supporting tail without interruption by molting the long central feathers only after the other rectrices have been replaced.

The most remarkable of all simultaneous molts occurs in females of a number of hornbills (Bucerotidae), which molt all their wing and tail feathers

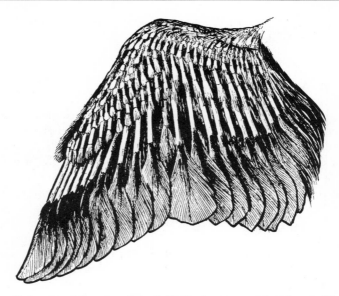

Figure 25. Under side of the wing of a rail (*Gallinula chloropus*) in full molt. The incoming feathers are all of about the same age. (After Grant, 1914:302.)

at once (and in some species the body feathers also) while they are walled up in the nest with their eggs and are being fed entirely by the male.

Although no passerine species is known to molt all of the flight feathers simultaneously, some insectivorous species that nest at high latitudes undergo a postnuptial molt in such a short period of time that they may lose their ability to fly. Haukioja (1971) presented evidence to suggest this phenomenon in several species (e.g., Willow Warbler, *Phylloscopus trochilus,* and Bluethroat, *Luscinia svecica*) in Finnish Lapland (69.5° N., 27° E.).

ECLIPSE PLUMAGE

It has long been known that the males of most Northern Hemisphere ducks assume an "eclipse" plumage in midsummer. Brooks (1938) called attention to eclipse plumages in the males of some Southern Hemisphere ducks and in some female ducks in both hemispheres.

When the females begin incubating their eggs, the males flock by themselves, usually in secluded marshes, and molt rapidly into a very inconspicuous, female type of plumage[5] which is retained, in most species, for only about a month,

[5] The eclipse plumage and posteclipse molt are beautifully illustrated by T. M. Shortt (in Kortright, 1942, especially col. pls. 6, 11, 14, 17, 21, and 29).

after which the birds assume the full winter plumage—which is also the nuptial plumage. The Blue-winged Teal and Shoveller retain the eclipse plumage for several months; the Ruddy Duck, until the following March or April. In molting into the eclipse plumage the male ducks suddenly (actually in a matter of hours) lose all their flight feathers and are flightless until the new feathers have grown in. These new flight feathers are retained through the following year, whereas the body feathers are replaced at the end of the eclipse period. The eclipse plumage may be prevented by castration or by keeping experimental birds under abnormally low temperatures through the summer.

Eclipse plumages also have been described among bee-eaters (Meropidae), cuckoo-shrikes (Campephagidae), sunbirds (Nectariniidae), and weaverbirds (Ploceidae). Any dull winter plumage can be (and has been) called an eclipse plumage.

COLORS

Some of the colors (biochromes) in bird feathers are due to pigmentation; others (schemochromes) are the effect of feather structure, either wholly (as in white feathers) or in part. As Fox (1953), who suggested these two terms, pointed out, "the production of conspicuous schemochromic colors depends upon underlying light-absorbing deposits of biochromic material" (i.e., on natural pigments—usually one or more melanins).

Schemochromes (Structural Colors)

The simplest case of a schemochrome is whiteness. The color white in a bird is the result of the reflection of all elements (wavelengths) of the white light striking the feather. "Chalky" and other nonlustrous whites are produced by a close-packed fibrous structure; lustrous whites, by a porous, air-filled structure (in the latter case, immersion in balsam renders the feather transparent instead of white).

Blue in feathers is apparently never the result of blue pigment. Noniridescent blues in feathers are caused by the structure of the barbs: a turbid, porous layer without pigment overlies a dark layer colored by granules of melanin. The colorless layer reflects blue light, but other wavelengths pass through the layer and are absorbed by the dark pigment below. The cuticular sheath of the barb, above the reflecting layer, is transparent and lacks pigment. Such noniridescent blue feathers (as in parrots, kingfishers, Blue Tits, etc.) are the result of Tyndall scattering, and the color does not change with the angle of vision.

Noniridescent green is, in rare cases, the result of green pigmentation, but in the great majority of noniridescent-green-feathered birds the color results from the "blue structure" described above. In this case, however, the cuticular sheath of the barb is pigmented, either with yellow carotenoids (producing pure greens) or with melanins (producing olive greens).

Purple colors sometimes result from Tyndall scattering plus red pigment. In the purple feathers on the head of the Blossom-headed Parakeet (*Psittacula cyanocephala*) "the barbs produce Tyndall blue while the barbules contain a red pigment. In other birds, for instance, certain rollers (*Coracias, Eurystomus*) and kingfishers (*Halcyon*), the purple is produced in the barbs by Tyndall blue over a background of reddish brown instead of the usual black" (Fox and Vevers, 1960).

Iridescent colors are the result of interference of light by thinly laminated structures in the barbules. The iridescent colors in hummingbirds were analyzed by Greenewalt *et al.* (1960a, 1960b). They stated that "interference as the cause of iridescent colors in birds was first proposed by Altum in 1857 and, while there have been doubting Thomases in the intervening years, nearly all subsequent investigators confirm his findings. Today, interference as the cause of iridescence is universally accepted." In hummingbirds, they found minute (about 2.5 microns in length) colored platelets in the form of a mosaic in the barbules; interstices between platelets were dark. "The barbule proper then has a surface which is 15–20 microns wide and 100 microns long, divided by diagonal boundary lines into a series of cells which look like parallelograms, each cell made up of a mosaic of 100 or more beautifully colored elliptical platelets."

Biochromes (Pigment Colors)

Most of the pigments in feathers belong to three groups: melanins, carotenoids, and porphyrins.

Melanins. The most common bird pigments are melanins, which range from dull yellow through red-brown to dark brown and black. They are synthesized by birds and mammals and occur in feathers and skin in granular form. Pinkish brown, fawn, and fawn-gray colors (as in the Chaffinch, *Fringilla coelebs*, and the Marsh Tit, *Parus palustris*) are due to the presence of phaeomelanin and/or eumelanin. Rollin (1962) described mutations that produce fawn-colored or cinnamon plumage in the Budgerigar, Starling, Zebra Finch (*Poephila guttata*), and canaries.

Carotenoids. These pigments usually produce yellow, orange, or red and occur in diffused form in the feathers and skin of birds. Birds are dependent on plants for most of their carotenoid pigments. Some of these pigments are modified after ingestion, but the bird cannot synthesize them.

A number of carotenoids have been identified with some exactness (Fox and Vevers, 1960).

Lutein (in the Old-world oriole *Oriolus*, the cotingas *Ampelion* and *Rupicola*,
   the wagtail *Motacilla*, and the weaverbird *Ploceus*)
Astaxanthin (in the shrike *Laniarius*)
Zeaxanthin (in the cotinga *Rupicola*)
Rhodoxanthin (in the cotinga *Phoenicircus* and the fruit pigeon *Ptilinopus*)
Picofulvin (in the woodpeckers *Picus* and *Piculus*)
Canary-xanthophyll (in the carduelines *Serinus, Chloris,* and *Spinus*)
Alpha-carotene (in the woodpecker *Colaptes*)

Olson (1970) reported that crimson, violet, and deep maroon colors in certain broadbills (Eurylaimidae), cotingas (Cotingidae), and manakins (Pipridae) were the result of heavy concentrations of carotenoid pigments plus modifications of the barbs of the feathers (Fig. 26). "The barbs are flattened and usually lack barbules. The internal structure of the barbs differs from [that

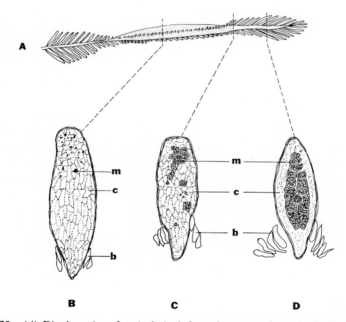

Figure 26. (*A*) Distal portion of a single barb from the crown of a male *Rupicola rupicola*; shaded portion is maroon, and unshaded portions are orange. (*B–D*) Cross sections of barb; dotted lines show approximate levels of the sections. All sections shown ventral side uppermost. m, medulla; c, cortex; b, barbule. (Courtesy of Storrs L. Olson and the editor of *Condor*.)

of] schemochromatic feathers in that the medulla is reduced or absent, there are distinct cortical cells which contain the pigments, and a thickened exterior cuticle is present, giving the barbs a glossy appearance."

Porphyrins. These are nitrogenous compounds synthesized by all animals. They are readily detected by their red fluorescence and are very common among birds. Some porphyrins reported from feathers are the following.

Turacoverdin, a green pigment found only in touracos (family Musophagidae).
Turacin, a red pigment found only in touracos.
Coproporphyrin III, a pink pigment found in the bustards *Lophotis* and
    Lissotis (family Otididae), owls (Strigidae), and 11 other orders of birds.

Some parrots (e.g., *Kakatoe, Melopsittacus, Psittacula*) contain fluorescent pigments, whereas in other parrots (e.g., *Amazona, Lorius*) the pigments are nonfluorescent.

Melanism is an abnormally dark color of the plumage due to an increase in the amount of melanin pigment. This condition is common in hawks and herons and has been recorded in a number of other species: e.g., the Herring Gull, Common Snipe (*Capella gallinago*), Pileated Woodpecker (*Dryocopus pileatus*), Hairy Woodpecker (*Dendrocopus villosus*), Black-capped Chickadee, and Ovenbird *Seiurus aurocapillus*.

Erythrism, or the presence of an excessive amount of reddish brown color, is uncommon, although it occurs with some regularity in the Prairie Chicken (*Tympanuchus cupido*) and the Bobwhite (*Colinus virginianus*). The red-plumaged Least Bittern, once considered a distinct species ("*Ixobrychus neoxenus*"), is an erythristic phase. A reddish color in certain ducks, geese, and cranes is a ferric oxide stain from iron-bearing water or mud and is not to be confused with erythrism. The feathers of House Sparrows and other birds that live in industrial areas may become darker because of an accumulation of soot. These are examples of *adventitious* coloring.

Schizochroism results from the absence of one of the pigments normally present in the plumage. Such birds commonly have the normal plumage pattern of the species but an abnormally pale, washed-out appearance. Storer (1952) figured Dovekies (*Alle alle*) and Murres (*Uria lomvia*) of this sort. California Quail (*Lophortyx californicus*) and Mourning Doves (*Zenaida macroura*), described by others under the terms "dilute" and "pale mutant," are probably examples of schizochroism. The blue strain of the common green Budgerigar (*Melopsittacus undulatus*) is a familiar instance, but this condition is rare in wild birds. A blue Yellow-headed Parrot (*Amazona ochrocephala*) was reported from the New York Zoological Society (*Bulletin,* 1941, **44:**157). In this species, yellow and brown pigments produce a green plumage; when the yellow pigment is lacking, the plumage is blue.

Xanthochroism, an abnormally yellow coloration, is very rare in the wild state. Saunders (1958) described a yellow mutant Evening Grosbeak (*Hesperiphona vespertina*) in which "the head, neck, back and entire underparts were a brilliant, clear yellow." Xanthochroism sometimes is seen in captive parrots, in which it apparently results from loss of the dark pigment that, with the physical structure of the feather and yellow pigment, produces the normal green color; these yellow birds are called "lutinos." Rollin (1962) described a Blue Tit (*Parus caeruleus*) that combined abnormal white and yellow plumage, so that the bird was "not an albino or a lutino."

Albinism is the abnormal absence of color; it results from the absence or the diminution of melanin (Figs. 27, 28). A *complete albino* lacks color in all struc-

Figure 27. An albinistic Willow (Traill's) Flycatcher (*Empidonax traillii*), one of two such birds found in a nest near Ann Arbor, Michigan (Berger, 1956).

Figure 28. An albino Red-shouldered Hawk (*Buteo lineatus*). (Courtesy of Heinz Meng.)

tures derived from the skin and in the iris. In addition to having white feathers, therefore, a complete albino has pale-colored tarsi, a pale bill, and pink eyes (because the iris pigment is absent that ordinarily conceals the red color of the blood in the eyeball).

A photograph of a pink-eyed, albino **Red-necked Grebe** (*Podiceps grisegena*) appeared in the June-July 1955 issue of *Nature Magazine*, and Weller (1959) described two juveniles of the same species and summarized other records on albinism in grebes. Bergtold (1909) told of three "pure albino robins" from a nest in Denver, Colorado, during the summer of 1905 and of two the following summer. Dexter (1957) reported one complete albino **Robin**; Lincoln (1958) described a complete albino **Purple Martin** (*Progne subis*).

Partial albinism is much more common, and the tendency for mutation to albinism appears to be present far more frequently in some bird families than in others. A large number of examples in North America have been described among species of the Tyrannidae: **Eastern Kingbird** (*Tyrannus tyrannus*), Scissor-tailed **Flycatcher** (*Muscivora forficata*), **Eastern Phoebe** (*Sayornis phoebe*), **Willow** (Traill's) **Flycatcher** (*Empidonax traillii*), **Alder Flycatcher** (*E. alnorum*), and **Eastern Wood-pewee** (*Contopus virens*).

FUNCTIONS OF COLOR AND PATTERN IN BIRDS

Our knowledge and theories of the uses that color may serve in birds are outlined below.

1. Uses of color unrelated to the vision of other animals.
   a. Prevention of wear in feathers. Averill (1923) pointed out that melanin-bearing feathers (or parts of feathers) resist wear much better than white feathers (Fig. 29). The wing tips are especially subjected to wear, and strong-flying birds that are much on the wing usually have black wing tips. An albino Robin raised by Berger went through several apparently normal molts in captivity, but the wing and tail feathers were unusually brittle. In the space of 5 or 6 weeks after the new feathers were fully grown, the tips of all of the flight feathers would be broken. In time the bird lost such a large part of each of the wing and tail feathers that it was able to fly only a few feet along the ground. The feathers of a normal-plumaged nest mate of the albino, however, did not break but remained in excellent condition from one molt to the next.
   b. Insulation from injurious light and extremes of temperature. Mei-

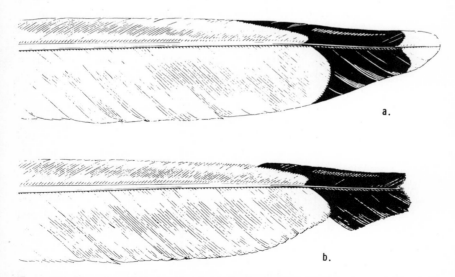

Figure 29. The wear-resisting quality of melanin pigment in feathers is shown in a primary of a Ring-billed Gull (*Larus delawarensis*) in fresh, winter condition (*a*) and in worn, late-June condition (*b*). (Drawn by W. A. Lunk.)

nertzhagen (1951) suggests that the pinkish buff so characteristic of desert birds is especially valuable in this connection.

    c.   Absorption of needed light and heat (Heppner, 1970).

    d.   In the case of eye-lines, a predatory function. Ficken *et al.* (1971) suggested that eye-lines "serve as lines of sight in tracking and capturing swifly moving prey. . . . A line drawn through the center of the eye to the farthest extent of the eye-line usually bisects the eye-line. Further, the eye-line is directed toward the point where prey would be expected" (Fig. 30).

2.   Uses of color related to the vision of other animals.

    a.   Advertisement.

        (1)   A warning label on inedible prey species. Two African bustards (*Afrotis afra* and *Eupodotis rüppelli*) are, respectively, conspicuously and cryptically colored; the conspicuous one, Meinertzhagen (1951) testifies, is (to man at least!) most inedible. ("Aposematic" colors or patterns.)

        (2)   A means (especially in gregarious birds) of attracting the attention of other individuals of the same species. Many flocking birds, such as crows and flamingos, seem to be free of any need of cryptic coloration for protection from enemies, and we are probably justified in thinking that their conspicuous colors are useful in enabling individual birds to correlate their movements with those of the flock. In certain coots and other rails the downy young are brilliantly colored about the head, and this seems to assist them in securing the attention of the parent bird bringing food. ("Episematic" colors or patterns.) White outer tail feathers (as in larks, meadowlarks, juncos) and white rump patches (as in flickers) are called "signal marks" and, sometimes, "deflective patterns."

        (3)   A means of attracting the attention of (and, perhaps, stimulating) members of the other sex. ("Gamosematic" and "epigamic" colors or patterns.) Both bizarre feather patterns and courtship behavior occur, especially in species in which the males and females come together only for copulation (e.g., many birds-of-paradise and hummingbirds).

    b.   Concealment from other animals. Many species of birds are "cryptically" marked, some matching their backgrounds so well that when motionless they are difficult or impossible to detect from a short distance. Many owls, frogmouths (Podargidae), nightjars (Caprimulgidae), and other ground-nesting birds constitute excellent examples of cryptic coloration (Figs. 31, 32). A special aspect of cryptic coloration is designated by the term "countershading." The

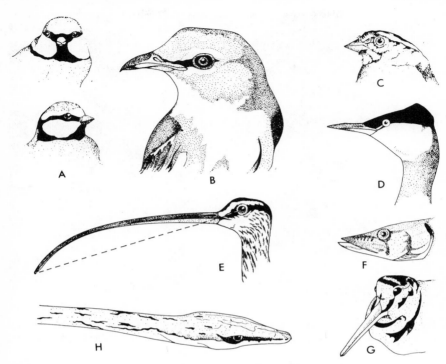

Figure 30. (*A*) Simple eye-line of the partially insectivorous Blue Tit (*Parus caeruleus*). This is the most common type of eye-line in vertebrates. The eye-line in this species is wider than it is in other avian examples, but it was illustrated because it was the only species for which we could find both frontal and lateral photographs. (*B*) Tracing of close-up photograph showing the combined eye circle and eye-line of the Yellow-throated Vireo (*Vireo flavifrons*). (*C*) Raised yellow feathers above the eye-line in a White-throated Sparrow (*Zonotrichia albicollis*) which may cast light along the line of sight. (*D*) Red-necked grebe (*Podiceps grisegena*), showing eye-line slanting downward. Such a line also occurs in some other piscivorous birds. (*E*) Long-billed Curlew (*Numenius americanus*), showing direction of eye-line forward of center of pupil to bill tip. (*F*) Teardrop mark of the pickerel (*Esox americanus*) associated with downward dashes at prey. (*G*) Rearward-pointed eye-line of the European Woodcock (*Scolopax rusticola*), probably used in the sighting of predators coming from behind; associated with 360° vision plus front and back binocularity. (*H*) Head of the arboreal vine snake (*Oxybelis aeneus*), showing eye-line and groove. (Courtesy of Robert W. Ficken and the editor of *Science*; Vol. 173, September 3, 1971, pp. 936–939; copyright 1971 by the American Association for the Advancement of Science.)

essential point of this concept, presented by Abbott H. Thayer in *Auk* in 1896 and elaborated by his son Gerald Thayer (1909), is that an animal with pale or white under parts is less conspicuous than a uniformly shaded one because the white surface counteracts the revealing shadows cast by a solid object.

Figure 31. Protective or cryptic coloration. Spotted Nightjar (*Eurostopodus guttatus*) of Australia on its nest. (Courtesy of *Australian Museum Magazine* and Norman Chaffer, photographer.)

Figure 32. Protective coloration. A female Painted Sandgrouse (*Pterocles indicus*) incubating eggs. (Courtesy of Shivrajkumar of Jasdan, India.)

Figure 33. The disruptive plumage pattern of the Semipalmated Plover is very conspicuous against a plain background (*B*), but actually makes the bird hard to discover in its normal habitat of stony shore (*A*). (Courtesy of the Smithsonian Institution; W. A. Weber, artist.)

"Disruptive" patterns are a special category of cryptic marking. The pattern breaks the familiar bird outline and causes the eye of man or other predator to "pass over" the bird without recognizing its true nature. The White-headed Fruit Pigeon (*Leucotreron cincta*) of Malaysia, whose pure white head and neck are sharply demarcated from a black and green body, is said to be very difficult to detect as it sits motionless in jungle trees (Cott, 1940:197). Friedmann pictured the Semipalmated Plover (*Charadrius semipalmatus*) as an example of this phenomenon (Fig. 33).

"Mimicry" in the restricted sense of advantageous resemblance to other birds has rarely been described by ornithologists. The young of some cuckoos (*Clamator* and *Eudynamis*) which are raised in the nests of crows and magpies have a blackish nestling plumage much like that of the host young with which they are raised and unlike the plumage of their parents or of related young cuckoos (Jourdain, 1925:657).

REFERENCES

Amadon, D. 1966. Avian plumages and molts. *Condor,* **68**:263–278.

Ashmole, N. P. 1963. Molt and breeding in populations of the Sooty Tern, *Sterna fuscata. Postilla,* **76**:1–18.

Averill, C. K. 1923. Black wing tips. *Condor,* **25**:57–59.

Bartlett, L. M. 1953. A white Kingbird in Pelham. *Bull. Mass. Audubon Soc.,* **37**:216.

Bates, G. L. 1918. The reversed under wing-coverts of birds and their modifications, as exemplified in the birds of West Africa. *Ibis,* **1918:**529–583.

Becker, R. 1959. Die Strukturanalyse der Gefiederfolgen von *Megapodius freyc. reinw.* und ihre Beziehung zu der Nestlingsdune der Hühnervögel. *Rev. Suisse Zool.,* **66:**411–527.

Beebe, C. W. 1916. Notes on the birds of Pará, Brazil. *Zoologica,* **2:**54–106.

Bennett, W. W. 1934. An abnormal Little Flycatcher. *Condor,* **36:**24–27.

Berger, A. J. 1953. The pterylosis of *Coua caerulea. Wilson Bull.,* **65:**12–17.

——— 1956. Two albinistic Alder Flycatchers at Ann Arbor, Michigan. *Auk,* **73:**137–138.

Bergtold, W. H. 1909. Albino Robins. *Auk,* **26:**196–198.

Bock, W. J., and L. L. Short, Jr. 1971. "Resin secretion" in *Hemicircus* (Picidae). *Ibis,* **113:**234–236.

Brodkorb, P. 1951. The number of feathers in some birds. *Quart. J. Fla. Acad. Sci.,* **12,** for 1949 (1951):241–245.

——— 1955. Number of feathers and weights of various systems in a Bald Eagle. *Wilson Bull.,* **67:**142.

Brooks, A. 1938. Eclipse in ducks of the Southern Hemisphere. *Auk,* **55:**272–273.

Chandler, A. C. 1914. Modifications and adaptations to function in the feathers of *Circus hudsonius. Univ. Calif. Publ. Zool.,* **11:**329–376.

Clark, H. L. 1918. Tail-feathers and their major upper coverts. *Auk,* **35:**113–123.

Clench, M. H. 1970. Variability in body pterylosis, with special reference to the genus *Passer. Auk,* **87:**650–691.

Collins, C. T. 1963. The "downy" nestling plumage of swifts of the genus *Cypseloides. Condor,* **65:**324–328.

Cott, H. B. 1940. *Adaptive Coloration in Animals.* Methuen & Co., London.

Delacour, J. 1943. A revision of the genera and species of the Family Pycnonotidae (Bulbuls). *Zoologica,* **28:**17–28.

Dexter, R. W. 1957. Observations on three albino American Robins. *Wilson Bull.,* **69:**185–186.

Dwight, J., Jr. 1900. The sequence of plumages and moults of the passerine birds of New York. *Ann. N.Y. Acad. Sci.,* **13:**73–360.

Dyke, J. 1971. Structure and colour-production of the blue barbs of *Agapornis roseicollis* and *Cotinga maynana. Z. Zellforsch.,* **115:**17–29.

Eisenmann, E. 1969. Wing formula as a means of distinguishing Summer Tanager, *Piranga rubra,* from Hepatic Tanager, *P. flava. Bird-Banding,* **40:**144–145.

Ficken, R. W., P. E. Matthiae, and R. Horwich. 1971. Eye marks in vertebrates: aids to vision. *Science,* **173:**936–939.

Fisher, H. I. 1942. The pterylosis of the Andean Condor. *Condor,* **44:**30–32.

Fox, D. L. 1953. *Animal Biochromes and Structural Colours.* Cambridge University Press, England.

Fox, H. M., and G. Vevers. 1960. *The Nature of Animal Colours.* Macmillan Co., New York.

Friedmann, H. 1930. The caudal molt of certain coraciiform, coliiform, and piciform birds. *Proc. U.S. Natl. Mus.,* **77,** Art. 7:1–6.

Gadow, H. 1893. In A. Newton and H. Gadow, *A Dictionary of Birds.* Adam and Charles Black, London.

Goodchild, J. G. 1886. Observations on the disposition of the cubital coverts in birds. *Proc. Zool. Soc. London,* **1886:**184–203.

Grant, C. H. B. 1914. The moults and plumages of the Common Moorhen (*Gallinula chloropus* Linn.). *Ibis,* **1914:**298–304. (See also 528–532, 652–654.)

Greenewalt, C. H., W. Brandt, and D. D. Friel. 1960a. The iridescent colors of hummingbird feathers. *Proc. Am. Phil. Soc.,* **104:**249–253.

———, ———, and ——— 1960b. Iridescent colors of hummingbird feathers. *J. Opt. Soc. Am.,* **50:**1005–1013.

Haukioja, E. 1971. Flightlessness in some moulting passerines in northern Europe. *Ornis Fenn.,* **48:**101–116.

Heppner, F. 1970. The metabolic significance of differential absorption of radiant energy by black and white birds. *Condor,* **72:**50–59.

Humphrey, P. S., and K. C. Parkes. 1959. An approach to the study of molts and plumages. *Auk,* **76:**1–31.

———, and ——— 1963. Comments on the study of plumage successions. *Ibid.,* **80:**496–503.

Jourdain, F. C. R. 1925. A study on parasitism in the Cuckoos. *Proc. Zool. Soc. London,* **1925:**639–667.

Juhn, M. 1957. "Frightmolt" in a male Cardinal. *Wilson Bull.,* **69:**108–109.

Kischer, C. W. 1963. Fine structure of the developing down feather. *J. Ultrastruct. Res.,* **8:**305–321.

Kortright, F. H. 1942. *The Ducks, Geese and Swans of North America.* American Wildlife Institute, Washington, D.C.

Lederer, R. J. 1972. The role of avian rictal bristles. *Wilson Bull.,* **84:**193–197.

Ligon, J. D. 1964. Albinism in the Scissor-tailed Flycatcher. *Wilson Bull.,* **76:**98.

Lincoln, F. C. 1958. An albino Purple Martin. *Auk,* **75:**220–221.

Lowe, P. R. 1931. On the relations of the Gruimorphae to the Charadriimorphae and Rallimorphae. . . . *Ibis,* **1931:**491–534.

Lowery, G. H., Jr., and J. P. O'Neill. 1964. A new genus and species of cotinga from eastern Peru. *Auk,* **83:**1–9.

Lucas, A. M., and P. R. Stettenheim. 1972. Avian anatomy: Integument. I, II. *Agric. Handbook* No. 362, Washington, D.C.

Markus, M. B. 1963. The number of feathers in the Laughing Dove *Streptopelia senagalensis* (*Linnaeus*). *Ostrich,* **34:**92–94.

——— 1965. The number of feathers on birds. *Ibis,* **107:**394.

Mayr, E. and M. Mayr. 1954. The tail molt of small owls. *Auk,* **71:**172–178.

Meinertzhagen, R. 1951. Desert colouration. *Proc. 10th Int. Ornithol. Congr.,* **1950:**155–162.

Michener, H. and J. R. Michener. 1946. Loss of feathers at times other than the normal molt. *Condor,* **48:**283–284.

Middleton, A. L. A. 1969. The moult of the European Goldfinch near Melbourne, Victoria. *Emu,* **69:**145–154.

Miller, A. H. 1928. The molts of the Loggerhead Shrike, *Lanius ludovicianus* Linnaeus. *Univ. Calif. Publ. Zool.,* **30:**393–417.

——— 1931. Systematic revision and natural history of the American Shrikes (Lanius). *Ibid.,* **38:**11–242.

——— 1933. Postjuvenal molt and the appearance of sexual characters of plumage in *Phainopepla nitens. Ibid.,* **38:**425–446.

——— 1961. Molt cycles in equatorial Andean sparrows. *Condor,* **63:**143–161.

Miller, W. DeW. 1924. Further notes on ptilosis. *Bull. Am. Mus. Nat. Hist.*, **50**:305–331.

Morlion, M. 1971. Vergelijkende Studie van de Pterylosis in Enkele Afrikaanse Genera van de Ploceidae. *Verh. K. Vlaam. Acad. Wet. Lett. Schone Kunsten Belg.* **33,** No. 119: Part 1, Text, 327 pp.; Part 2, tables, 256 pp.

Naik, N. L., and R. M., Naik. 1969. On the plumages and moults of some Indian starlings. *Pavo,* **7**:57–73.

Nitzsch, C. L. 1867. *Pterylography* (translated by W. S. Dallas; edited by P. L. Sclater). Ray Society, London.

Olson, S. L. 1970. Specializations of some carotenoid-bearing feathers. *Condor,* **72**:424–430.

Parkes, K. C. 1966. Speculations on the origin of feathers. *Living Bird,* **1966**:77–86.

Percival, E. 1942. The juvenile plumage of some birds and an interpretation of its nature. *Trans. Proc. Roy. Soc. N. Z.,* **72**:6–20.

Pycraft, W. P. 1907. On some points in the anatomy of the Emperor and Adélie penguins. *Natl. Antarct. Exped. 1901–1904, Zool.,* Vol. 2.

Regal, P. J. 1975. The evolutionary origin of feathers. *Quart. Rev. Biol.,* **50**:35–66.

Rollin, N. 1962. Abnormal white, yellow and fawn plumages. *Bull. Brit. Ornithol. Club,* **82**:83–86.

Rutschke, E. 1960. Untersuchungen über Wasserfestigkeit und Struktur des Gefieders von Schwimmvögeln. *Zool. Jahrb.,* **87**:441–506.

Saunders, A. A. 1958. A yellow mutant of the Evening Grosbeak. *Auk,* **75**:101.

Schüz, E. 1927. Beitrag zur Kenntnis der Puderbildung bei den Vögeln. *J. Ornithol.,* **75**:86–224.

Selander, R. K., and D. R. Giller. 1960. First-year plumages of the Brown-headed Cowbird and Redwinged Blackbird. *Condor,* **62**:202–214.

Snyder, L. L., and H. G. Lumsden. 1951. Variation in *Anas cyanoptera. Occas. Pap. Roy. Ontario Mus. Zool.,* No. 10.

Steiner, H. 1918. Das Problem der Diastataxie des Vogelflügels. *Jena Z. Naturwiss.,* **55**:221–496.

Stephan, B. 1965. Die Zahl der Armschwingen bei den Passeriformes. *J. Ornithol.,* **106**:446–458.

——— 1970. Eutaxie, Diastataxie und andere Probleme der Befiederung des Vogelflügels. *Mitt. Zool. Mus. Berlin,* **46**:339–437.

Stone, W. 1896. The molting of birds with special reference to the plumages of the smaller land birds of Eastern North America. *Proc. Acad. Nat. Sci. Phila.,* **1896**:108–167.

Storer, R. W. 1952. A comparison of variation, behavior, and evolution in the sea bird genera *Uria* and *Cepphus. Univ. Calif. Publ. Zool.,* **52**:121–222.

Stresemann, E. 1963a. The nomenclature of plumages and molts. *Auk,* **80**:1–8.

——— 1963b. Variation in the number of primaries. *Condor,* **65**:449–459.

——— 1963c. Taxonomic significance of wing molt. *Proc. XIII Int. Ornithol. Congr.,* **1**:171–175.

Sundeval, C. J. 1886. On the wings of birds. *Ibis,* **1886**:389–457.

Sutton, G. M. 1935. The juvenal plumage and postjuvenal molt in several species of Michigan sparrows. *Cranbrook Inst. Sci. Bull.* No. 3.

——— 1948. The juvenal plumage of the Eastern Warbling Vireo (*Vireo gilvus gilvus*). *Univ. Mich. Mus. Zool. Occas. Pap.* No. 511.

Test, F. H. 1940. Effects of natural abrasion and oxidation on the coloration of flickers. *Condor,* **42**:76–80.

Thayer, G. H. 1909. *Concealing-coloration in the Animal Kingdom.* Macmillan Co., New York.

Van Tyne, J. 1943. Abnormal feather loss by Cardinals. *Wilson Bull.,* **55**:195.

Vaurie, C. 1971. *Classification of the Ovenbirds (Furnariidae).* H. F. & G. Witherby, London.

Voitkevich, A. A. 1966. *The Feathers and Plumage of Birds.* October House, New York.

Watson, G. E. 1963. The mechanism of feather replacement during natural molt. *Auk,* **80**:486–495.

Weller, M. W. 1959. Albinism in *Podiceps grisegena* and other grebes. *Auk,* **76**:520–521.

——— 1967. Notes on plumages and weights of the Black-headed Duck, *Heteronetta atricapilla.* *Condor,* **69**:133–145.

Wetherbee, D. K. 1957. Natal plumages and downy pteryloses of passerine birds of North America. *Bull. Am. Mus. Nat. Hist.,* **113**:339–436.

Whistler, H. 1941. Differences of moult in closely allied forms. *Ibis,* **1941**:173–174.

Woolfenden, G. E. 1967. Selection for a delayed simultaneous wing molt in loons (Gaviidae). *Wilson Bull.,* **79**:416–420.

Wray, R. S. 1887a. On the structure of the barbs, barbules, and barbicels of a typical pennaceous feather. *Ibis,* **1887**:420–423.

——— 1887b. On some points in the morphology of the wings of birds. *Proc. Zool. Soc. London,* **1887**:343–357.

CHAPTER FOUR

Senses and Behavior

Behavior is the overt expression of the coordinated life processes of an animal, including the means by which it maintains its relation with the environment (Emlen, 1955). In order to deal with coordinated life processes, one must become familiar with the physical equipment that effects this coordination. The sense organs are an indispensable part of that equipment, and we elect to discuss them here rather than in Chapter 2.

THE SENSE ORGANS

Vision

The bird eye is so large that the two eyeballs nearly touch each other in the median plane of the skull, and they may equal or exceed the weight of the brain, from which the eyes develop in the embryo. The eyeballs of some hawks and owls are larger than those of man. Coulombre and Crelin (1958) wrote that the growing embryonic eye exerts a critical mechanical influence on the morphogenesis of the chick's skull, and that the alignment of the axes of the upper beak and the head is dependent, in part, on equal expansion of the two eyes. Berger and Howard (1968) reported the bilateral absence of the eyes in an American Robin and a concomitant absence of well-defined bony orbits (Fig. 1). Lord (1956a) told of congenital eye defects (cola-

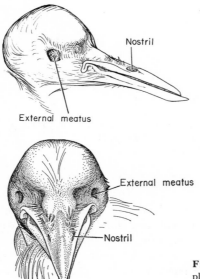

External meatus

Nostril

External meatus

Nostril

Figure 1. Lateral and frontal views of the plucked head of an eyeless Robin; note the deformity of the upper mandible. (From Berger and Howard, 1968.)

bomas) in one eye of a Rough-legged Hawk (*Buteo lagopus*) and one of a Red-tailed Hawk (*Buteo jamaicensis*). These adult hawks managed to survive even though each had but a single functional eye.

The eyes of most birds (exceptions: penguins, cormorants, pelicans, gulls, hornbills) have very limited mobility, and those of owls are immovably fixed in their orbits. The six oculomotor muscles are present but much reduced in size, and in a few birds they are functionless. What the bird lacks in eye mobility is compensated for by increased neck mobility (e.g., some owls are said to be able to rotate their heads through an arc of at least 270°). This set of circumstances may be significant in certain behavior studies because complex *reflex eye movements* tend to be replaced by *reflex neck movements*.

Birds have a well-developed *nictitating membrane*, or third eyelid, as do frogs, reptiles, and some mammals (Sandoval, 1965). The *semilunar fold,* a half-moon-shaped structure at the medial angle of the human eye, is thought to be a remnant of the nictitating membrane of lower animals. The cornea of the human eye is kept moist by the frequent blinking of the eyelids, and blindness can result if the muscles of the eyelids are paralyzed so that the cornea becomes excessively dry. Birds rarely close their eyes except in sleep, and it is the nictitating membrane that passes from front to back (or medial to lateral in owls) across the cornea and keeps it moist. According to Goodge (1960), however, the American Dipper blinks its upper eyelid between 40 and 50 times per minute.

Two muscles operate the nictitating membrane (Fig. 2). Musculus quadratus arises from the superior surface of the eyeball and appears to insert on the tendon of insertion of M. pyramidalis, but in many birds M. quadratus actually forms a sling through which the tendon of the pyramidalis muscle passes. Musculus pyramidalis arises from the inferior surface of the eyeball; its tendon inserts in the inferior margin of the nictitating membrane. Contraction of the two muscles draws the translucent nictitating membrane over the cornea. When the two muscles relax, elastic tissue in the membrane returns it nasal-ward to its resting position. Adult birds may draw the membrane across the eye when feeding their young.

Each of the three layers of the avian eyeball exhibits special features, some of which are found also in lizards and other reptiles. The outer layer is composed of a posterior fibrous *sclera* and an anterior transparent *cornea*. The cornea is opaque during early development, but there is full transparency by the time a chick embryo is 19 days old (Coulombre and Coulombre, 1958). Embedded in the sclera are a cup-shaped scleral cartilage, a sclerotic ring, and, in some birds, an os opticum (Gemminger's ossicle); the early development of the first two has been described by O'Rahilly (1962).

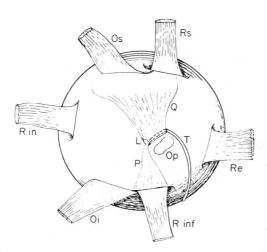

Figure 2. Posterior surface of the right eyeball of the House Sparrow with the rectus muscles reflected. Re, rectus externus; R in, rectus internus; R inf, rectus inferior; Rs, rectus superior; Oi, inferior oblique; Os, superior oblique; Op, optic nerve; P, pyramidalis; Q, quadratus; L, loop of quadratus muscle for passage of tendon (T) of pyramidalis muscle. (After Slonaker, by permission of the editor of *Journal of Morphology*.)

The middle or vascular tunic is composed of a thick *choroid,* a complex *ciliary body,* and a thin *iris.* In addition to blood vessels, the choroid may contain muscle fibers (either striated or smooth); the iris has striated sphincter and dilator fibers. The inner or nervous tunic is termed the *retina.* Its light-perceptive cells are called *rods* and *cones.* The rods (containing rhodopsin) are highly sensitive to light, and in nocturnal birds are present in greater numbers than cones; rods are greatly reduced in numbers in diurnal birds and may be absent in some. Cones are responsible both for sharp images (visual acuity) and for color vision. The cones of diurnal birds may contain red, orange, yellow, or colorless oil droplets, and pale-pigmented droplets are present in some nocturnal species; the functions of these oil droplets, however, are poorly known. Birds that have a retina containing only rods are said to have a scotopic retina and *scotopic* (night) *vision*; those that have only cones have *photopic vision.*

The *area centralis* is a modified region of the retina that provides the greatest visual acuity. In shape it may be round or rectangular, or there may be two round areae connected by a band; in some birds it is absent. The *fovea,* a depression within the area centralis that serves to magnify images (Fig. 3), may be pitlike, troughlike, or absent; some birds have two foveae (e.g., hawks, king-

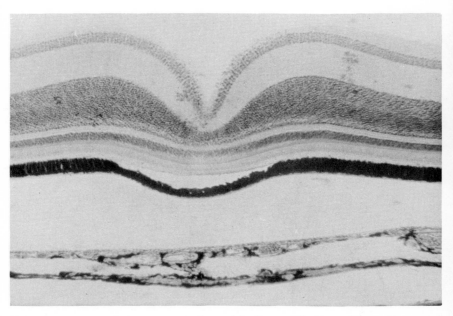

Figure 3. A section of the retina at the fovea of the Dipper (*Cinclus mexicanus*). (Courtesy of William R. Goodge and the editor of *Journal of Morphology.*)

fishers, hummingbirds, swallows; see Lord, 1956b; Duijm, 1959; Pearson, 1972).

Several theories have been offered to explain the avian pecten (*pecten oculi*), a pigmented and folded vascular structure projecting dorsally from the head of the optic nerve. Walls (1942:658) concluded that the pecten oculi is a supplemental nutritive device, and the need for a large or small pecten "seems to depend solely upon the rate-of-living of the sensory retina." Thus a large pecten is needed by very active diurnal birds in which there is a high retinal temperature. The nutritional role of the pecten is now generally accepted (O'Rahilly and Meyer, 1961). Evidence obtained from electron microscope studies "suggests that the *pecten oculi* operates as a marginal apparatus for the maintenance of avian intraocular pressure control under certain physiologic conditions and that the ciliary epithelium is the steady contributor to the aqueous humor" (Seaman and Himelfarb, 1963). The same authors postulate that "the pigmented intervascular tissue of the *pecten oculi* is probably the energy depot and source for the fluid transport occurring in the neighboring pectineal capillary endothelial cells."

The intrinsic or intraocular eye muscles of birds and reptiles (especially lizards) differ from those of other vertebrates in being composed of striated rather than smooth muscle. The striated fibers presumably effect the rapid accommodation or focusing of images on the retina that is necessary in fast-flying birds. Goodge (1960) pointed out that "the divisions of the avian ciliary musculature are, to a large degree, artificial separations due mainly to the positions of the ciliary nerve." These named portions of the ciliary muscle are as follows: Crampton's muscle, Brücke's muscle, Müller's muscle, and the temporal ciliary muscle (Fig. 4). The precise action of the ciliary muscle on the lens and the entire mechanism of accommodation in birds are still debatable. It is known that accommodation is accomplished by adjustments of the lens and changes in the curvature of the cornea, although the process undoubtedly is not the same in all birds.

Interesting modifications correlated with habits are seen in the eyes of certain amphibious birds (e.g., loons, cormorants, diving ducks, auks). In such birds, the transparent cornea is reduced in size; in some the nictitating membrane has a specialized central window that acts as a contact lens when it overlies the cornea; Crampton's muscle is reduced or absent; and Brücke's muscle is hypertrophied. The eyes of certain kingfishers (*Alcedo*) are so constructed that the birds apparently see clearly in air and water even though there is no elaborate accommodation mechanism to effect changes as the kingfishers plunge from air into water. The lens and ciliary body are asymmetrical in shape, and each eye has two foveae, one of which is thought to function for aerial vision, the other for underwater vision. Goodge (1960) described the eye of the American Dipper, a species that feeds under water and presumably has equally good vi-

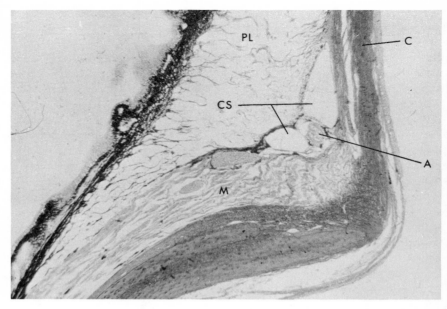

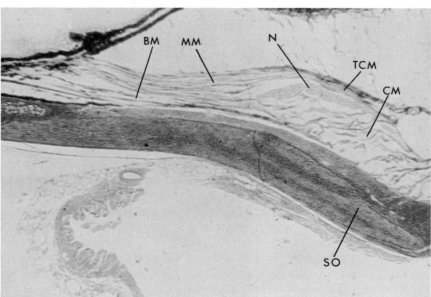

Figure 4. Upper: Section of the eye of the Dipper in the region of the corneoscleral junction and canal of Schlemm. A, artery; C, cornea; CS, canal of Schlemm; M, ciliary muscle; PL, pectinate ligaments. ×70.20. Lower: Ciliary musculature of the Dipper. BM, Brücke's muscle; CM, Crampton's muscle; MM, Müller's muscle; N, branch of ciliary nerve; SO, scleral ossicle; TCM, "temporal ciliary muscle." ×74.10. (Courtesy of William R. Goodge and the editor of *Journal of Morphology*.)

sion in air and in water. The only specialization he found was a much larger ciliary muscle (2.5 times greater in cross section than that of the Robin or Varied Thrush, *Ixoreus naevius*). He concluded that "the much better development of the iridial sphincter in *Cinclus* than in non-aquatic species suggests that the curvature of the lens is changed by the pressure this musculature exerts upon it, and that in *Cinclus*, it makes possible a greater power of accommodation."

Birds have excellent vision, "no sharper but a good deal faster than that of man" (Pearson, 1972). Small passerine birds, however, may have visual acuity that is three times less than that of man. Pumphrey (1961) concluded that the visual acuity of falconiform birds does not exceed three times the acuity of man. Similarly, Shlaer (1972) suggested that the resolution of the eye of the Golden Eagle (*Aquila chrysaetos*) is between 2.0 and 2.4 times and that of the Martial Eagle (*Spizaetus bellicosus*) is between 3.0 and 3.6 times better than the resolution of the human eye.

Owls have a binocular field of 60 to 70°; hawks, up to 50° or more. Most granivorous birds, on the other hand, have binocular fields considerably less than 25°. With the exception of the nocturnal *Strigops*, parrots have the smallest binocular fields (most often between 6 and 10°) yet found in birds. Only in certain penguins (*Spheniscus*) are the eyes so placed in the skull that there is no overlap in the visual fields of the two eyes, i.e., they have monocular vision in that each eye sees a different picture; but their field of vision may be as great as 340°. Two interesting specializations in the location of the eyes are seen in bitterns and in the Woodcock (*Philohela minor*). When a bittern "freezes" with bill pointed skyward, the eyes can be turned ventrally so that the bird has a binocular field of vision parallel with the ground. The Woodcock's eyes are set so far back on the head that the "posterior binocular field probably is much wider than the anterior" (Walls, 1942:295).

Although the iris is dark (black, brown) in the majority of birds, there are some striking exceptions. Examples of red-eyed birds are the Wood Pigeon (*Columba palumbus*), Wood Duck (*Aix sponsa*), some bee-eaters (Meropidae), Phainopepla (*Phainopepla nitens*), Bronzed Cowbird (*Molothrus aeneus*), and Rufous-sided Towhee (*Pipilo erythrophthalmus*). A number of species have yellow eyes: e.g., King Vulture (*Sarcoramphus papa*), Stonecurlew (*Burhinus oedicnemus*), many owls, Collared Toucan (*Pteroglossus torquatus*), and some juncos. The Acorn Woodpecker (*Melanerpes formicivorus*) has white eyes. The eye of the cormorant *Phalacrocorax carbo* varies from blue-green to dark emerald-green; the eyes are turquoise-blue in some bowerbirds (Ptilonorhynchidae); downy young White Terns, the Galapagos Flightless Cormorant (*Phalacrocorax harrisi*), and the Ariel Toucan (*Ramphastos ariel*) have blue eyes. Some albino birds (e.g., Jackdaw) are said to have blue eyes (Weir, 1891).

Sexual differences in eye color are found in a number of species: for example, scaups, mergansers, Wood Duck, Redhead, Spectacled Eider, White-winged and Surf scoters, Bronze Grackle, Boat-tailed Grackle, and Bushtit. Some species exhibit age differences in the color of the iris: it is brown in immature Jungle Babblers (*Turdoides striatus*) but white in adults.

The pupil of the eye is round or circular in most birds; it is elliptical in a few species: Black Skimmer (*Rynchops nigra*) and Collared Toucan, and in pepper-shrikes (*Cyclarhis*).

Eyeshine in certain mammals and other animals is produced by a tapetum in the eye. Although birds lack such a structure, the eyes of a number of nocturnal species reflect the artificial light of a flashlight or headlamp and appear to glow or shine. Van Rossem (1927) listed 14 species that exhibit such eyeshine. Among those whose eyeshine is brilliant orange-red, glowing pink, or dark red are the Double-striped Thick-knee (*Burhinus bistriatus*), Killdeer (*Charadrius vociferus*), Barred Owl (*Strix varia*), Common Potoo (*Nyctibius griseus*), Chuck-will's-widow (*Caprimulgus carolinensis*), Whip-poor-will (*C. vociferus*), Little Nightjar (*C. parvulus*), Poor-will (*Phalaenoptilus nuttallii*), and Lesser Nighthawk (*Chordeiles acutipennis*). According to Van Rossem, a pale green eyeshine also had been reported for the Lesser Nighthawk, which suggests "a seasonal change, possibly correlated with sexual activity." The Wedge-tailed Shearwater (*Puffinus pacificus chlororhynchus*) of the Hawaiian Islands has a pale orange eyeshine; the Black-footed (*Diomedea nigripes*) and Laysan (*Diomedea immutabilis*) albatrosses have a silvery white eyeshine when viewed at close range at night (Berger, 1972). Van Rossem wrote that "under favorable circumstances and with a good light, a Whip-poor-will's eye can be seen for over 100 yards, and the eyes of Giant Goatsuckers and Thick-kneed Plovers for twice that distance."

Hearing and Balance

The ear evolved in fishes primarily as an organ for maintaining equilibrium in the watery environment. The added function of hearing developed in the higher vertebrates. Hearing is keen in birds and is second in importance only to vision for them. As in mammals, the auditory system consists of three parts, although the external ear flap or auricle (Latin *auricula,* external ear) is absent in birds. The external ear consists only of the *external acustic canal* (or meatus) in the skull; in most birds (exceptions: vultures, the Ostrich), the meatus is concealed by feathers (the *auriculars*). Sound waves enter the external meatus and strike the ear drum (tympanic membrane; Latin *tympanum,* a drum), which forms part of the boundary between the external ear and the middle ear (Fig. 5). The middle ear (tympanic cavity) is a small space in the skull, bounded by bone except for the tympanic membrane and the opening into the Eustachian

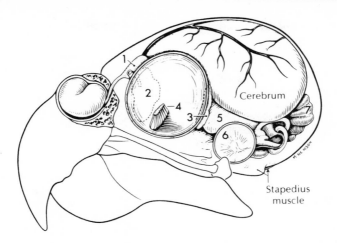

Figure 5. Lateral view of the head of a Budgerigar with the roof of the skull removed, the fundus of the eye *in situ,* and the inner ear exposed. 1, Nasal gland; 2, Harder's gland; 3, lacrimal gland; 4, pecten; 5, optic lobe; 6, tympanic membrane; 7, cerebellum. (By permission of Howard E. Evans, from Petrak's *Diseases of Cage and Aviary Birds,* Lea and Febiger, copyright 1969.)

(pharyngotympanic) tube, which leads from the middle ear to the pharynx. A single, partly cartilaginous ear bone (*columella*) transmits vibrations from the tympanic membrane to the membrane of the oval window in the lateral wall of the internal ear. The movements of the latter membrane produce comparable movements in the fluids of the chambers of the internal ear, which, in turn, stimulate highly specialized nerve endings (Fig. 6).

The *internal ear* consists of a complicated series of fluid-filled membranous channels (*labyrinths*) in the bone of the skull; these membranous channels also are surrounded by fluid. Three parts (*sacculus, cochlea,* and *lagena*) are concerned solely with hearing, whereas two other parts (*utricle* and three *semicircular canals*) are concerned with balance or the orientation and movement of the head in space. The cochlea is responsible for transforming sound-wave stimuli to nerve impulses, which are conducted to the brain by the cochlear nerve. The lagena is a specialized portion of the cochlea that contains minute calcium carbonate "ear stones" (*otoliths*). The lagena is presumed to be responsive to low frequencies, the rest of the cochlea to high frequencies, and Beecher (1951) proposed that the lagena and other otolith organs form a "gravity system, capable of reporting changes in head position." The length and configuration of the cochlea vary among birds, and there is evidence of a positive correlation between length of cochlea and complexity of song (Pumphrey, 1961). One study revealed that in a series of birds the range of

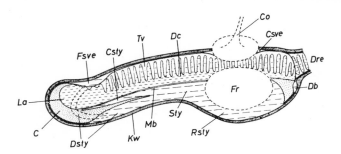

Figure 6. A schematic longitudinal section through the inner ear of birds. C, cartilage; Co, columella; Csty, cavum scalae tympani; Csve, cisterna scalae vestibulae; Db, ductus brevis; Dc, ductus cochlearis; Dre, ductus reuniens; Dsty, ductus scalae tympani; Fr, fenestra rotunda; Fsve, fossae scalae vestibuli; Kw, bony wall; La, lagena; Mb, membrana basilaris; Rsty, recessus scalae tympani; Sty, scala tympani; Tv, tegmentum vasculosum. (By permission of J. Schwartzkopff.)

hearing varied from a low of 40 cycles per second (in the Budgerigar, *Melopsittacus undulatus*) to 29,000 cycles per second (in the Chaffinch, *Fringilla coelebs*), with each species having its own total range of hearing. There are also hearing differences between young and adult birds. Downy young chickens are primarily responsive to the low-pitched calls of the adult hen, whereas the hen is especially sensitive to the high-pitched "peeping" notes of recently hatched chicks.

Experiments have shown that certain birds use hearing, in a way similar to that characteristic of bats, to avoid obstacles when flying in the dark. This echolocation is well developed in the Oilbird (*Steatornis caripensis*) of Trinidad and parts of South America. These birds roost and nest in dark caves. In flight, an Oilbird utters a continuous series of very short pulses of sound which bounce back from the cave walls, "informing" the bird of its position in relation to the walls. This extremely rapid response in a flying bird is much greater than the response possible in the human ear. The Cave Swiftlet (*Callocalia brevirostris*) also uses echolocation in flying to and from its nest. Medway and Wells (1969) discussed the taxonomic implications of ability or inability to echolocate in the genus *Collocalia*.

Most owls are nocturnal in habits, and some species are noted for having both a greatly increased size and an asymmetry in the location of right and left external ear canals (Fig. 7). Studies of Barn Owls show that these nocturnal birds locate their prey in total darkness with surprising accuracy (Konishi, 1973a, 1973b). Some owls have a fleshy flap or operculum along the anterior margin of the ear opening. The "ears" of eared owls and the "horns" of horned owls are simply feathers growing from the top of the head and have no relation to the true ears. Pesquet's Parrot (*Psittrichas fulgidus*) of New Guinea

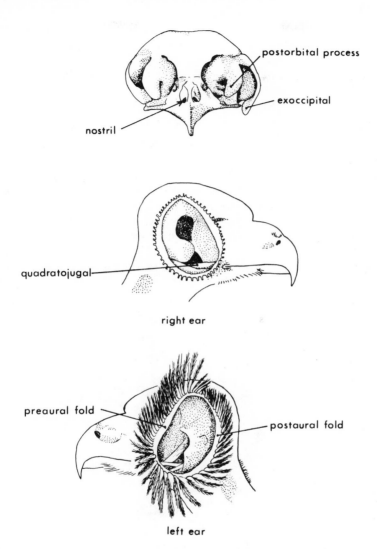

Figure 7. Three views of the skull of Tengmalm's Owl (*Aegolius funereus*), showing the asymmetry in the skull and the external auditory canals. (After Pycraft, 1898).

is notable because there is a striking asymmetry in the position of the external ear openings in the skin, but not in the positions of the bony canals in the skull (Stonor, 1939). The outer end of the meatus typically is located posterior to the orbit, but in the Common Snipe and the Woodcock the opening lies ventral to the orbit.

Olfaction

The debate over the olfactory powers of birds has continued ever since John James Audubon wrote that vultures did not find carrion or freshly killed animals if they were covered, but that the birds very quickly discovered a picture of a partly dissected sheep and tried to eat the picture. Bang (1964) remarked that "with few exceptions . . . laboratory tests on bird olfaction have either been done on birds with little olfactory equipment . . . or have been predicated on the types of odors to which human receptors are known to respond, or both" (see also Michelsen, 1959; Tucker, 1965).

As a result of a fine study, however, Stager (1964) could conclude that

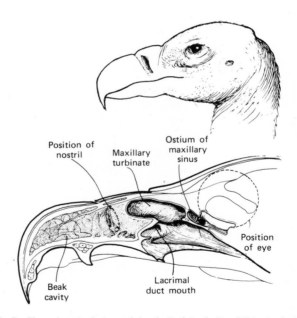

Figure 8. Left: Profile and sagittal views of the skull of the Indian White-backed Vulture (*Gyps bengalensis*). Upper right: Sagittal section of the skull of the Falkland Islands Turkey Vulture (*Cathartes aura falandica*). Lower right: Profile and sagittal sections of the skull of the Egyptian Vulture (*Neophron p. percnopterus*). (Courtesy of Betsy Garrett Bang and the editor of *Journal of Morphology.*)

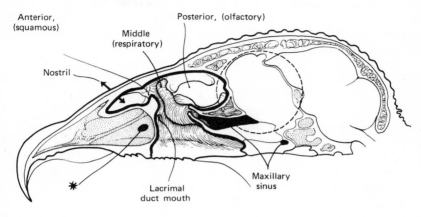

Turbinates, or conchae:

Anterior, (squamous)

Middle (respiratory)

Posterior, (olfactory)

Nostril

Lacrimal duct mouth

Maxillary sinus

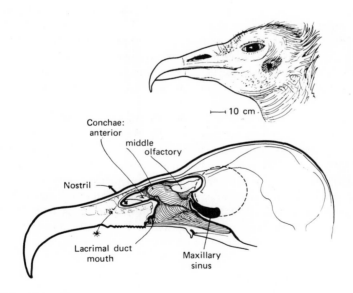

⊢—— 10 cm

Conchae: anterior

middle

olfactory

Nostril

Lacrimal duct mouth

Maxillary sinus

Figure 8 (*Continued*)

"among the cathartine vultures, the turkey vulture (*Cathartes aura*) possesses and utilizes a well-developed olfactory food-locating mechanism. The king vulture (*Sarcoramphus*) of tropical America, although its behavior is little known, appears, on the basis of present ethological and morphological data, also to utilize olfaction in its location of food. . . . There is no evidence, either ethological or morphological, to indicate that olfaction plays more than a minor, if any, role in food location by *Coragyps*, *Gymnogyps*, and *Vultur*. There likewise were no data to indicate that the Old World vultures employ any sense other than vision in the location of food." Stager added that "the turkey vulture appears to be the most successful of all cathartine vultures in the New World today, and that this success can be attributed largely to its demonstrated olfactory acuity. Olfaction, assisted by a specialized type of foraging flight, enables the turkey vulture to seek and locate food in a greater variety of terrain types than are available to the other members of the family Cathartidae."

Bang's (1964) equally outstanding study compared the olfactory apparatus of several species of vultures (Fig. 8). She verified the "enormous development of the olfactory organ of the Turkey Vulture." Among its specializations, the Turkey Vulture has very large olfactory bulbs, olfactory epithelium, and nostrils. Nasal glands and Harderian glands also are well developed in vultures; their functions are incompletely known, but Bang remarks that "the carrion-feeding birds have presumably developed pretty effective antibiotic resources in the nasal chamber against respiratory organisms, and in the digestive systems against enteric diseases." Preliminary research with mammals suggests that the Harderian gland may produce secretions of an antibacterial nature.

There is a positive correlation between the size of the olfactory bulbs and the importance of olfaction to the bird (Fig. 9). Bang and Cobb (1968) presented data on the sizes of the olfactory bulbs in relation to the cerebral hemispheres in 108 bird species. They concluded that "in kiwis, in the tube-nosed marine birds, and in at least one vulture, olfaction is of primary importance, and that most water birds, marsh dwellers, and waders, and possibly echo-locating species, have a useful olfactory sense. In other species it may be relatively unimportant." At the same time, they found that swifts and nightjars have a high index number, whereas cormorants have a low one.

Taste

Taste and smell are related in that both are types of *chemoreception,* and in man the sense of smell is important to overall taste sensations when eating. The *taste bud* is the highly specialized receptor for detecting the four primary taste sensations: sweet, salty, sour, and bitter (Fig. 10). Although taste has not been studied extensively in birds, it is probable that most birds have a poorly

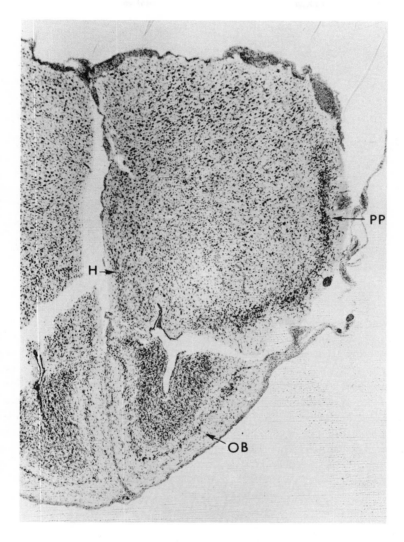

Figure 9. Frontal section through the rostral end of the hemispheres of the grebe *Podiceps au-*
ritus, showing the olfactory bulb (OB), praepyriform corticoid layer (PP), and the beginning of the
hippocampus (H); the bulbs are close together but are not fused. (By permission of Stanley Cobb
and the editor of *Epilepsia.*)

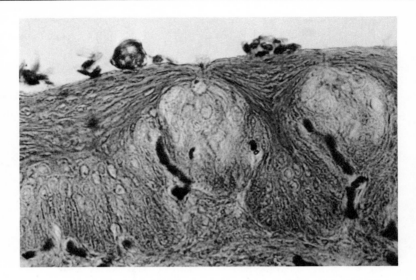

Figure 10. Two taste buds in the tongue of the Amakihi (*Loxops virens*, Drepanididae).

developed taste sense. There is a great reduction in numbers of taste nerve fibers in birds, and most are said to have only 30 to 70 taste buds, although some parrots apparently have about 400. (By comparison, a rabbit has some 17,000 taste buds.) Moreover, birds have very few taste buds on the tongue, and none near its tip; most are located in the palate, pharynx, and inferior surface of the epiglottis. In their study of Allen's Hummingbird (*Selasphorus sasin*), Weymouth, *et al.* (1964) found only a single taste bud on the tongue. Moore and Elliot (1946) reported from 27 to 59 taste buds in six pigeon tongues. Lindenmaier and Kare (1959) found an average of 8 (range: 5 to 12) taste buds in day-old chicks, and 24 taste buds in a 3-month-old cockerel. Taste buds have been found on the edges of the mandibles in some birds.

Tactile Sensations

Among the most sensitive areas in the skin of man are the tips of the fingers, where there are nerve endings for various types of external (*exteroceptive*) tactile sensations, such as pain, temperature, and texture (smooth, rough) discrimination. Birds, too, have several types of tactile end organs, but their precise functions are incompletely known. Named after the men who described them, the best known are the corpuscles of Herbst, Grandry (or Grandry-Merkel), and Key-Retzius. Most have been found in the bills of aquatic birds, but they also are widely distributed in the body. Herbst's corpuscles have been described for many unrelated birds: e.g., *Apteryx,* ducks, flamingos, parrots, woodpeckers, and passerines. They occur not only in the bill and oral cavity but also in the cloaca, crus, forearm, and skin, and especially in the pterylae (but not in the apteria) at the bases of the rectrices, remiges, and other contour feathers (Saxod, 1970). Vrabec (1961) found Herbst corpuscles in the anterior chamber of the eye of the goose, in which species the corpuscles are associated with a scleral venous plexus, and suggested that "their most probable function is to register pressure differences within those vessels." Schwartzkopff (1960) suggested that Herbst corpuscles may receive vibratory stimuli. Malinovský and Zemánek (1969) studied the sensory corpuscles in the coverings of the upper and lower mandibles and of the skin around the bill in the pigeon and found four types: from typical Herbst corpuscles to slender, greatly elongated, and small corpuscles with a simple structure. Less than half of more than 1400 corpuscles were of the typical Herbst structure (Figs. 11, 12).

Interoceptors and Proprioceptors

Interoceptors receive stimuli from internal organs that are controlled by the autonomic nervous system; they mediate such feelings as hunger, thirst, and diffuse pain. *Proprioceptors* are responsible for "position sense"; they are found in muscles, tendons, and ligaments. This unconscious position sense is essential for the integrated functioning of all voluntary muscle, whether the bird is asleep or flying. Malinovský and Zemánek (1970) studied the sensory nerve endings in the joint capsules of the wing and leg. The numbers of nerve endings in the various joints were correlated with the locomotor habits of the birds studied: e.g., ducks had more nerve endings in the joint capsules of the wings; chickens had more endings in the hip capsule; Rooks had about equal numbers in the capsules of the two limbs.

THE BASIS FOR BEHAVIOR

The special senses are discussed separately because it is convenient to do so. The entire nervous system, hormones, and muscles are integral parts of an animal's coordinating system. Behavior can be understood only in terms of the anatomy and physiology of the organism. Tinbergen (1951:5) commented that behavior is "always the outcome of a highly complex integration of muscle contractions." We believe that all behavior is derived from the neurophysiology of the organism, and that our inadequate knowledge of the mechanisms involved is not increased by postulating a "directive drive arising mainly outside the supposed system of channels in which it is 'flowing'" (Thorpe, 1963:35). When the ethologist writes of directive "drives," he is referring to the directiveness that is inherent in the physiology of the organism itself.

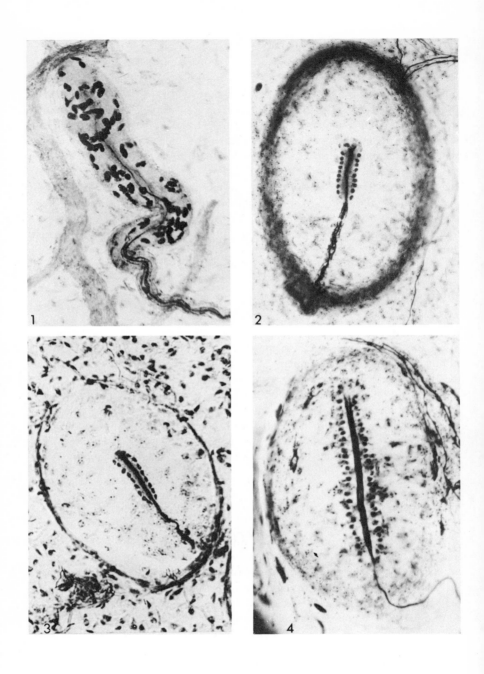

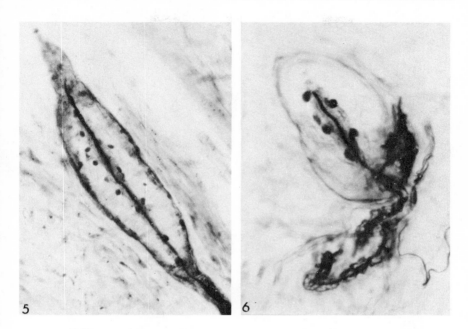

Figure 11. Sensory corpuscles: 1, from the joint capsule of the hip joint of the hen, approx. ×218.40; 2, typical Herbst corpuscle from the beak of the domestic duck, approx. ×142.80; 3, typical Herbst corpuscle from the beak of the domestic duck; 4, typical Herbst corpuscle as found in birds other than the Anatidae, approx. ×168; 5, an elongated, atypical corpuscle from the comb of the fowl, approx. ×201.60; 6, a simple, small corpuscle from the beak of the domestic hen, approx. ×310.80. All specimens were treated by Lavrentyev's modification of the Bielschowski-Gross impregnation method. (By permission of Lubomír Malinovský and the editors of *Folia Morphologica*, Prague, Vol. 18, 1970, pp. 206–212, and *Zeitschrift für mikroskopisch-anatomische Forschung*, Vol. 77, 1967, pp. 279–303.)

Johnsgard (1967), Klopfer and Hailman (1967), and Eibl-Eibesfeldt (1970) reviewed the historical developments of the study of animal behavior that led to the contemporary concepts of ethology, which is defined as the objective study of behavior or as the biology of behavior.

Instinct and Instinctive Behavior

Contemporary students of behavior recognize that the adult animal is the product of its genetic endowment and its environment. The living animal is a phenotype. The environment begins to exert its influence as soon as the sperm enters the egg. However, the environment can only modify—it cannot

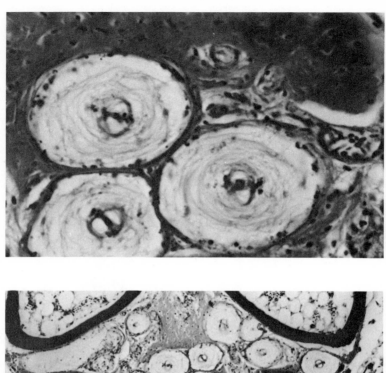

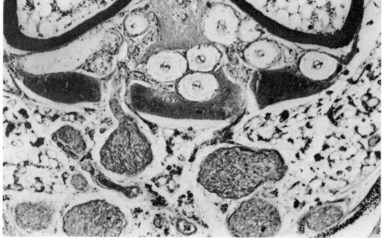

Figure 12. Sensory corpuscles in the tongue of the Apapane (*Himatione s. sanguinea*; Drepanididae).

negate—the genetic constitution. At the same time, the genic pattern unfolds through the medium of the total environment.

As interpreted originally by Lorenz, Tinbergen, Thorpe, and others, instinct is innate behavior, and innate behavior is that which has not been changed by learning processes.[1] Some characteristics of instinctive behavior are the following:

1.   Instincts or drives are innate, i.e., inherited.
2.   They evolve phylogenetically (rather than ontogenetically).
3.   They are expressed in complex and often highly rigid or stereotyped patterns (fixed-action patterns).
4.   They tend to be species-specific.
5.   They usually are evoked by relatively simple stimuli (sign stimuli or releasers).
6.   They should develop in young animals that are initially prevented from practicing them.

Thorpe (1963) listed six basic drives as the minimum number to be studied in birds and mammals: nutrition, fighting, reproduction, social relations, sleep, and care of the body surfaces.

Lorenz (1965) proposed that instinctive behavior is expressed by *fixed-action patterns*. Originally, these were thought to be as rigid and specific as anatomical structures, but later study showed this not to be true. The final behavior resulting from the sequential development of a fixed-action pattern is called the *consummatory act*. It is a simple, stereotyped activity; it is the last step in a chain of reactions related to a particular drive; and its completion "satisfies" an internal drive.

Tinbergen (1951) proposed a *tentative* hypothetical scheme (a nervous hierarchy) to account for these innate behavioral phenomena.[2] His hypothesis

---

[1] Students of bird behavior should compare the modern concepts with the conclusions reached by Morgan (1900:98).

[2] Lorenz presented a "psycho-hydraulic" model to explain the principles operating within the nervous system. Deutsch (1960) developed a model based on physiological changes in the internal medium which involves a receptor or analyzer discharge (a feedback system to the brain), a "central structure or link," and a motor discharge that results in a specific behavior pattern. Hayes *et al.* (1953) proposed a more detailed hierarchical "instinctive control system," which is based on information obtained by psychoanalysis but "applies directly to lower animals and has many implications for ethology." Deutsch wrote that "the doctrine that theories are premature is a convenient one for its exponents. It exonerates them from having to pay serious attention to existing attempts and saves them from the labor of rigorous thinking." In his comments on the values of models, Manning (1972) wrote that "if we find that one model consistently explains an animal's behaviour under a wide range of conditions, it may tell us a great deal about the *principles* upon which the nervous system is operating. It can tell us little or nothing about the *means* of operation." Similarly, Hinde (1970) remarked that models "serve merely to indicate the job which the physiological machinery must be doing and to provide the physiologist with an indication of what he must look for."

included a series of functional "centers" or neurological "motivational" processes within the CNS for the integration of the several fixed-action patterns of the major instincts. *Innate releasing mechanisms* (IRMs)[3] at different levels, however, block the discharge of the component steps in a behavior pattern until *sign stimuli* (*releasers*) are received, whereupon the blocking mechanism is removed. Thus the consummatory act cannot take place until appropriate releasers counteract the inhibitory impulses existing at different levels. In the absence of the releasers, the motivational drive finds expression in *appetitive* or *exploratory behavior*.

Later research revealed that fixed-action patterns are not inflexible and that the energy of the internal drive probably is specific to *classes* of behavior patterns rather than to individual reactions. Thorpe concluded that "the classic examples of appetitive behaviour and consummatory act can be regarded as the two ends of a series ranging from extreme variability and plasticity on the one hand to almost complete fixity on the other."

Lehrman (1953, 1970), Skinner (1966), and others were highly critical of the Lorenz-Tinbergen theory of behavior, in part because "the interaction out of which the organism develops is *not* one . . . between heredity and environment. It is between *organism* and environment! And the organism is different at each different stage of its development." Consequently, "to say a behavior pattern is 'inherited' throws no light on its development except for the purely negative implication that *certain types* of learning are not directly involved."

Derived Activities

Tinbergen proposed in 1951 that fixed-action patterns build up "tension" (*specific action potential*) in the CNS. This potential becomes 'dammed up" if the environmental situation is not appropriate for the normal release of the fixed-action pattern. This "results in a lowering of the threshold to stimuli effective in releasing the particular activity concerned," and it may accumulate to a point at which the instinct appears to "go off" without any external stimulation, resulting in *overflow* or *vacuum activity*. In such a case, the consummatory act is performed in the absence of the normal stimuli, so that the behavior does not fit into the environmental situation at the time. Overflow tension also may trigger the release of an irrelevant behavior pattern by *sparking over*.

A bird is subject to different drives. In general, only one drive can be dominant at a particular time; consequently, conflicts between drives arise. The behavior patterns that result from conflicts between drives are called *derived activities*. Three main types are recognized.

1. **Displacement** (or **allochthonous**; Tinbergen, 1952) **activities** result from high-intensity motivation, and the behavior appears irrelevant or "out of

---

[3] Some workers refer to this simply as the releasing mechanism (RM) in order to avoid debate over the word "innate."

context" to the immediately preceding activity. This often is demonstrated when two birds stop fighting and begin to preen or peck at the ground.

2.   **Intention movements** differ from displacement activities in that they are caused by the drives of their own behavior patterns; they also are called *autochthonous activities* because each is appropriate to the drive then dominant. They are incomplete or incipient movements that result from the low activation of an instinctive behavior pattern; they reveal to the observer what the bird "intends" to do. In preparing to fly, for example, a bird may bend its legs, lower its head, lift its wings away from the body, and raise its tail. The bird may engage in only one or two of these preparatory movements; but, if the sequence of events in a locomotor pattern is known, one can identify the intention movements.

When motivation is strong, only one drive can be dominant at a time; the drives are mutually exclusive. When, however, motivation is moderate, the animal may obey two drives simultaneously so that *ambivalent behavior* results. Depending on the intensity of the drive, the ambivalence involves intention movements or displacement activities from two drives, usually those of attack and escape. Tinbergen interprets the "upright threat posture" of the Herring Gull as "the simultaneous combination of the intention movements of attack and withdrawal." (See Baerends, 1958.)

3.   **Redirected activities** do not require an elaborate sparking-over process in the CNS, and they are similar to intention movements in that they are caused by the drives of their own behavior patterns. Moynihan (1955a) defined redirection movements as "autochthonous activities of a drive directed toward an object or animal other than the one releasing and usually directing them (although the releasing object or animal remains available, or partly available, as a potential goal at the time)." As an example, he cited the behavior of a Prairie Falcon when disturbed at its nest by a human being. "Both the attack and escape drives of this falcon were immediately activated; but they were largely incompatible, and the escape drive was strong enough to prevent the bird from venting its attack drive upon the real offending subject, the actual disturber. The falcon then found an outlet for its thwarted attack motivation by pouncing upon some other birds, a Barn Owl and a Raven, which happened to pass by at a convenient moment. This sort of 'unprovoked' attack upon an inoffensive scapegoat is the commonest type of redirection. It is also the type of redirection that has been most frequently confused with displacement." (Later interpretations are given by Hinde, 1970.)

Ritualization and Bird Display

"Whenever it is of advantage for an animal that some of its incidental behavior be understood by another, selection operates to transform the behavior pattern in question into a conspicuous signal. This modification of a behavior pattern

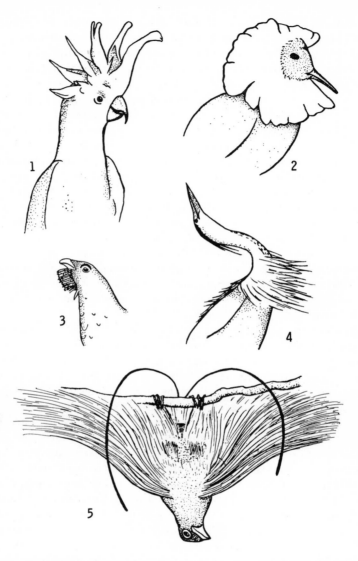

Figure 13. Specialized feathers used as signals: (1, crest; 2, ruff; 3, chin growth; 4, throat plumes; 5, flank plumes. (Courtesy of Desmond Morris and the editor of *Behaviour*.)

to serve communicative functions is called *ritualization*" (Eibl-Eibesfeldt, 1970). As a means of communication, displays may have either a psychological or a physiological effect on other birds (Fig. 13). Displays may stimulate the opposite sex (attract it or raise its sex drive), strengthen and maintain the pair-bond, "tie" the mate to the nest site, serve as a threat (intimidation) to another bird, serve to effect reproductive isolation, or divert a predator (distraction behavior).

Three evolutionary sources of displays are recognized (Hinde, 1970): intention movements, displacement activities, and autonomic responses. Intention movements of preening, bill-wiping, copulation, incubation, locomotion, and attack have evolved into display patterns. For example, the first intention movement (the bow) in walking or hopping in a secondarily adapted form becomes the crouching behavior of an alarmed Horned Lark. The "freezing" behavior (with bill and head pointed skyward) of a bittern represents the second

Figure 14. Young Least Bittern (*Ixobrychus exilis*) showing freezing behavior. (Courtesy of the late Walter P. Nickell.)

Figure 15. Freezing posture of a just-fledged Cedar Waxwing (*Bombycilla cedrorum*); Ann Arbor, Michigan, September 14, 1961.

phase of the walking intention movement (Figs. 14, 15). Such displays, often by a bird that is cryptically colored, aid in concealing the bird from predators. Other displays have been derived from the autonomic responses associated with movements of feathers and those accompanying urination and defecation. During the process of ritualization, the derived activity may be greatly modified (by exaggeration, by simplification, or by a shift in threshold of, or a loss of coordination among, its component elements). Hence the need for comparative studies was emphasized by Tinbergen (1952) when he wrote that "it may even be impossible to recognize a releaser as a displacement activity at all, until comparison with less specialized movements in closely related species reveals its origin."

Learning and Intelligence

Authorities disagree on the number of types of learning, and some insist that there is only one. Thorpe defined learning as "that process which manifests itself by adaptive changes in individual behaviour as a result of experience." It may be fairly evident where learning is involved, but the problem is rarely simple. We may cite, as an example, the many investigations of the pecking behavior of chicks. Some early results suggested that experience had little effect on efficiency in pecking; the ability was "innate." Later experiments indicated that peck aim became "virtually perfect" during the first 30 hours after hatching and, apparently, was due not to learning but to the physical maturation of the neuromuscular system.[4] When chicks were prevented from pecking for about 2 weeks, however, they had to be taught to peck. Thorpe (1963:353) pointed out that "we have always to bear in mind that birds which have been artificially prevented from performing a natural act may be suffering not merely from lack of practice but also from the setting up of contrary habits." (See also Hess, 1966.)

The following are the types of learning thought by some to apply to birds. Both Thorpe (1963) and Hinde (1970), however, emphasized two limitations to this classification as applied to birds: "First, the categories intergrade. Second, each contains examples of greatly differing complexity."

Habituation "is an activity of the central nervous system whereby innate responses to mild and relatively simple stimuli, especially those of potential value as warnings of danger, wane as the stimuli continue . . . without unfavourable results." In other words, the animal learns *not* to respond to stimuli that are without significance. Too great an efficiency in habituation would, as Thorpe pointed out, be disastrous for a species. Apparently, young birds do have to learn what not to be afraid of. A fledgling passerine bird, for example, may have to learn that a cabbage butterfly is something to eat and not a source of danger (see also Hogan, 1965).

Conditioning is the establishment of a substitute stimulus in an existing stimulus-response association. Nearly all of the experimental work with birds (primarily chickens and pigeons) has dealt with the special senses or with the pecking response. Pigeons, for example, readily learn to recognize which of several colored discs, when pecked, will reward them with food.

Trial-and-error learning is the procedure by which one response is selected from several possibilities in a problem situation. There is evidence, for example, that the young of some species have to learn which food is edible, which is inedible. Thorpe believes that trial-and-error learning probably plays an important part in the improvement of nest-building, especially in selecting the proper materials. (For a discussion of experiments with puzzle boxes and mazes, see Thorpe, 1963:358.)

---

[4] One is reminded of Morgan's "canon of interpretation" (1900:270–271): "We should not interpret animal behaviour as the outcome of higher mental processes, if it can be fairly explained as due to the operation of those which stand lower in the psychological scale of development."

Insight learning was defined by Thorpe (1963:110) as "the sudden production of a new adaptive response not arrived at by trial behaviour or as the solution of a problem by the sudden adaptive reorganization of experience." (For discussions on latent learning, social facilitation, local enhancement, and true imitation, see Thorpe, 1963, and Hinde, 1970.)

Thorpe discussed *imprinting* along with insight learning, although he stated that "there is no hard-and-fast line to be drawn between imprinting and other forms of learning"; he viewed imprinting as an "innate disposition to learn." One aspect (the following response) of imprinting was recognized by D. A.

Figure 16. Mallard ducklings showing the following response to Dr. Richard E. Phillips; Delta Waterfowl Research Station. (Courtesy of Frank McKinney.)

Figure 17. An imprinted Canada Goose gosling following a box containing an alarm clock. (Courtesy of A. Ogden Ramsay; photograph by Leland A. Graham.)

Spalding as early as 1873 and was described in greater detail by Oskar Heinroth in 1911 and by Konrad Lorenz in 1935. Lorenz emphasized that imprinting is a verb; it is the fixation of a drive or instinct on an object. In the classical example of imprinting, geese raised from artificially incubated eggs accepted their human keepers and followed them as they would have followed their own parents (Fig. 16). Imprinting behavior has since been demonstrated for many species of ducks, gallinaceous birds, coots, doves, ravens, and other birds. Incubator-hatched birds have become imprinted on such objects as a white box, a black box, a green box containing an alarm clock, a football, an orange ball, large and small canvas-covered frames of different sizes, and lights of different colors (Fig. 17). Most of these objects had one thing in common during the experiments: they were seen in motion by the young birds. Birds also have become imprinted on stationary models. Imprinted birds typically direct sexual displays toward the human figure on which they were first imprinted, although attachments to inanimate objects have also been reported.

Figure 18. Three newly hatched Dusky Crag Martins (*Hirundo c. concolor*); Baroda, India, March 8, 1965.

Lorenz believed at first that imprinting differed from learned behavior in that it could occur only during a very brief period immediately after the bird hatches, and that it is irreversible (i.e., the imprinted bird does not "forget"). Learning, on the other hand, is not restricted to a limited period, and learned behavior can be forgotten and relearned. Further study revealed, however, that imprinting is not always irreversible, nor is it invariably dependent on events occurring immediately after hatching. Klopfer and Gottlieb (1962) suggested that the critical period for imprinting "is best measured in terms of developmental age, i.e., time elapsed from the onset of incubation, rather than time since hatching." The many ramifications of imprinting have been summarized by Thorpe (1963), Eibl-Eibesfeldt (1970), Hinde (1970), and Manning (1972).

Ethology is a relatively new science, and current theories will be modified as more precise information becomes available on neurophysiological mechanisms (Barfield, 1969; Pearson, 1972). There is a need for neuroembryological studies in order to determine the rate of maturation of the nervous system for various integrative mechanisms (see Gottlieb, 1971). At the time of hatching, the brain of a precocial bird (e.g., *Megapodius*) may equal 40 percent of the adult weight, whereas in an altricial species (*Turdus merula*) it amounts to only 10 percent (Sutter, 1951). Furthermore, the relative proportions of the several parts of the brain approach the adult condition in *Megapodius,* whereas the proportions in *Turdus* are embryonic in that the cerebral hemispheres are very small and the brain stem is more highly developed. Myelinization of nerve fibers also progresses more rapidly in precocial birds, which can see and move about shortly after hatching. Myelinization of the optic nerve, for example, is completed several days before hatching in precocial species; it is not completed until several days after hatching in altricial species, birds in which the eyes remain closed for some time after hatching (Figs. 18, 19). Such differences in physical development are critical, of course, in interpreting the behavior of young birds (see Morgan, 1900:95–96).

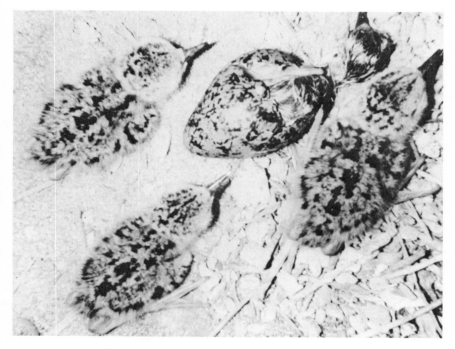

Figure 19. Three Hawaiian Black-necked Stilts (*Himantopus h. knudseni*) 23 hours old and a bird in the process of hatching.

SOME BIRD ACTIVITIES

All behavior is adaptive and has a genetic basis. Parts of eight other chapters in this book deal with major behavioral activities of birds; i.e., one may study behavior as it is related to migration, reproduction, social hierarchies, feeding habits, communication (visual and auditory), ecology, taxonomy, etc. The rest of this chapter deals with selected examples of named behavioral patterns.

Aggressive neglect is "the tendency of one species to neglect its nest or young owing to the release of excessive aggressive behavior in the presence of a second species" (Ripley, 1961). Ripley discussed this concept of competition between certain sunbirds and honeyeaters, between bulbuls and honeyeaters, and, in terms of territoriality, among several species of highly belligerent ducks, geese, and swans. With respect to sunbirds and honeyeaters in the Moluccan and New Guinea islands, he wrote that "it appears as if these species were in competition, not complete, in the absolute sense, but to a degree where the presence of one appears to affect the other. One evidence of this is the fact that on small islands throughout the area one species or the other may occur but not both."

Agonistic behavior has been defined as "any behavior appearing in conflict between animals, including fighting and escape behavior." "Aggressive behavior" and "hostile behavior" are essentially synonymous terms (Hinde, 1970:333–336; Manning, 1972). Moynihan (1955b) wrote: "Among the commonest social behavior patterns in most birds are a great variety of hostile activities, i.e., behavior patterns produced by attack and/or escape motivation." In his discussion of derived activities, Tinbergen (1952:11) commented that "there are, mainly, two types of situations in which displacement activities occur: a hostile situation, and a purely sexual one." During the breeding season, the first reaction of a male to a female is a mixture of sexual, aggressive, and escape responses. According to Lorenz, the escape drive suppresses sexuality in males, and drives of aggression suppress sexuality in females. Songs, call notes, displays, and postures have evolved that not only resolve the conflicting drives but also are species specific. Consequently, they usually ensure that male and female of the same species are brought together and that pair-bonding, copulation, and parental behavior follow.

Aggressive behavior exhibited between males of the same species or between one species and another is, in general, not destructive; i.e., few birds "fight to the death." The discharge of aggression is said to be "tension relieving," and one of the functions is the "spacing out" of individuals of a species during the breeding season. There appears to be, therefore, a selective pressure to save the loser in an encounter (Fig. 20).

One manifestation of aggressive behavior is "window fighting" or "shadow boxing," during which birds fight their own reflections from shiny surfaces:

Figure 20. Hostile postures of the Wood Thrush (*Hylocichla mustelina*). (*A, B*) Front and lateral views of high-intensity spread; (*C*) medium-intensity spread; (*D*) low-intensity spread (crest raising); (*E*) horizontal fluff. (Courtesy of William C. Dilger and the editor of *Auk*.)

windows, mirrors, automobile hubcaps, and chrome headlight rims. Gallup (1968) called this mirror-image stimulation (MIS). He viewed the reinforcing properties of mirrors "in terms of novel stimulation with social stimulus overtones. . . . An animal in front of a mirror is instrumental in producing changes in the behavior of the reflected image, yet most organisms respond to mirrors as if their image represented another animal." This behavior, which sometimes is repeated daily for many weeks, has been reported in a wide variety of passerine birds: e.g., the Crested Flycatcher, Mockingbird, Blackbird (*Turdus merula*), Cedar Waxwing, Chaffinch, Robin, Brown-headed Cowbird, Cardinal, House Finch, and Brown Towhee (Fig. 21).

Anting has been defined in two ways. McAtee (1938) said that "the term

Figure 21. Female Blackbird (*Turdus merula*) fighting its own reflection; Lower Hutt, New Zealand. (Courtesy of M. J. Daniel and the editor of *Notornis*.)

Figure 22. Female Orchard Oriole (*Icterus spurius*) anting. (Courtesy of the late Arthur A. Allen.)

should be restricted to phenomena involving ants, for the sake of logic, appropriateness, and due conservatism." During *active anting,* the bird holds one or more ants in its bill and rubs them among its feathers; during *passive anting,* the bird stands or sits among the ants, fluffs its feathers, and allows the ants to crawl into the plumage. A broader definition of the term is now in general use: "the application of foreign substances to the plumage and possibly to the skin" (Whitaker, 1957). More than 25 species of ants and 40 substitute substances have been used in anting by over 200 different species of birds, both passerine and nonpasserine. The substitute materials include such items as fruits, raw onion, grasshoppers, hair tonic, mustard, vinegar, burning matches, and moth balls (Figs. 22, 23). Whitaker discussed the several theories concerning the stimulus for anting and its function. She believed that her captive Orchard Oriole (*Icterus spurius*) was stimulated by the thermogenic property of certain ants, that is, the "burning or warming quality" of formic acid or other substances produced by these insects. V. B. Dubinin, a Russian parasitologist,

Figure 23. Common Grackle (*Quiscalus quiscula*) anting with a moth ball. (Courtesy of Leo H. Borgelt and the editor of *Wilson Bulletin*.)

presented evidence that anting results in the destruction of feather mites and perhaps also in cleaning "the oxidizing lipid film from feathers." In their review of Dubinin's research, Kelso and Nice (1963) concluded that "anting is an instinctive action present in many birds, perhaps aimed at defense against feather mites. It appears to be 'triggered' by the acid or burning taste of ants and other substances and apparently may be performed *in vacuo,* i.e., in the absence of mite infestation."

Goodman (1960) suggested that anting birds may be instrumental in the spread of fire. Several observers have seen birds ant with lighted cigarettes, matches, and burning straw, and there is evidence that birds sometimes carry such materials to their nests. Mayfield (1966) presented evidence that a House Sparrow and a European Starling had carried lighted materials to their nests in buildings.

Behavioral mimicry is especially characteristic of titmice and chickadees. When disturbed on the nest, these hole-nesting birds typically do not leave the nesting cavity but emit hissing sounds accompanied by a display. With open bill and partially spread wings, the bird may sway slowly back and forth for several seconds, after which it jumps upward very suddenly, hisses, and slaps the sides of the cavity with its wings. The explosive finale of this "snake display" is startling even to one acquainted with it.

Distraction behavior has been defined as any behavior pattern of an adult bird that tends to divert an intruder from the eggs or young (Fig. 24). Ethologists suggest that it results from the conflict between the drive to incubate (or brood) and the drive to flee from an enemy. E. A. Armstrong proposed that distraction display and displacement activities with a diversionary function (e.g., displacement brooding or feeding) be included under the term *diversionary display*. Distraction behavior has been reported in several hundred species of birds. Phylogenetically, it occurs from loons to finches; it is performed by ground nesters, tree nesters, and some hole-nesting species.

A common pattern is aptly named "injury-feigning" or "broken-wing" display. For example, a Killdeer frightened from its nest typically runs off a short distance, falls to the ground, leans to one side, spreads its tail (thus displaying its cinnamon-colored rump), and rapidly flaps one wing or both wings, while emitting alarm notes. The bird may then stand up, run a few feet, and repeat the display.

Another pattern, called the "rodent-run" display, has been reported for a

Figure 24. Distraction behavior of a Common Nighthawk. (Courtesy of Samuel A. Grimes.)

number of shorebirds (e.g., Golden, Ringed, and Kentish plovers, Sanderling, Spotted, Pectoral, and Purple sandpipers), for some tundra species, and for the Green-tailed Towhee (*Chlorura chlorura*) and the Red-billed Leiothrix (*Leiothrix lutea*). When a towhee is flushed from its nest, it may drop nearly straight to the ground without using its wings, and then run along with its tail elevated, greatly resembling a scampering chipmunk.

Distraction displays vary from species to species, and "there is good evidence of variation between different populations of the same species and between different individuals of the same population" (Williamson, 1956). An example showing individual differences is the Catbird (*Dumetella carolinenesis*), an abundant species that on only a few occasions has been reported to injury-feign (Berger, 1954).

Dominance involves the interrelationships among the members of flocks in social species; the word was first used in this sense by Schjelderup-Ebbe. Although, typically, one bird is dominant over all others, there also is a social hierarchy among the other members of the flock. Each bird is dominant over all birds below it in the hierarchy but subordinate to all birds above it. Among domestic chickens and some other species there is a relatively fixed *peck order* in each flock. The dominant bird (either cock or hen) has precedence in going to food or water supplies and can peck any other bird in the flock, usually without any retaliation from the subordinate members. As a result of his work with chickens, McBride proposed that a bird carries with it at all times a moving *personal field*. The dominant cock or hen has first choice of everything within that personal field (McBride *et al.*, 1969).

In writing of wild turkeys (*Meleagris gallapavo*) in Texas, Watts and Stokes (1971) said that young males of each brood leave the family unit when between 6 and 7 months old. These young males form a group which "continues to be an inseparable unit for life," but the birds flock with other groups of males for the winter. Each male in such a winter flock "is forced into two contests: one to establish his position within his own group of siblings, the other to determine the status of his sibling group with respect to other groups. Each sibling engages in physical combat with his brothers. The battle consists in wrestling, spurring in the fighting-cock style, striking with the wings and pecking at the head and neck. . . . The strongest fighter in the group becomes the dominant bird, and the order of rank established among the siblings is seldom challenged thereafter as long as the dominant bird lives." The social order tends to be less rigid in pigeons, canaries, and most wild birds, and the dominant and subdominant roles may vary seasonally. Birds subordinate in neutral or feeding grounds usually become dominant as soon as they reach their nesting territory. Interspecific dominance is obvious at feeding stations and among foraging winter flocks composed of several bird species.

Handedness or **footedness** in birds has been discussed by several authors. In 1938 Friedmann and Davis reported in *The Auk* that parrots tend to be left-footed in holding food. Fisher (1957) made mechanical recordings of more than 7000 landings by 11 pigeons and found that "seven pigeons used the right foot predominantly, three the left foot, and one bird showed no particular preference for either." He also noted that, although pigeons "may show an average preference for one foot over a long period of time, choice of foot seems in part to be a matter of daily preference." Thus handedness in birds appears to differ from the pattern in human beings. Rand (1958) summarized information available on this phenomenon in birds and nonhuman mammals.

Individual distance, a phenomenon demonstrated by man as well as by birds, is similar to McBride's concept of the personal field; i.e., except for its mate, a bird tries to maintain a certain minimum distance between itself and other members of the species. In highly territorial birds, the individual distance is great. Colonial nesting species, on the other hand, have a very high tolerance for other individuals: their nests often are built just far enough apart so that the birds cannot reach each other when incubating; this is a reflection of the low individual distances they can tolerate (Fig. 25). As originally defined by Heidigger, individual distance applied only when an animal was resting; many communal roosting birds demonstrate this sort of individual distance, but there are exceptions. Birds maintain their individual distances by aggressive or hostile acts.

Maintenance activities "are concerned with locomotion and general health and efficiency of the body" (Marler, 1956). We exclude flight, flight intention movements, and foraging behavior from this discussion. In his intensive study of the Anatidae, McKinney (1965) used the term *comfort movements,* under which he included shaking, stretching, and cleaning movements, oiling, preening, nibbling-preening, washing, and bathing. Many of these activities also have become modified (through ritualization) and are used as social signals.

1. Bathing. Sunbathing and water bathing are common bird activities, and a number of species also take dust baths.

While *sunbathing,* a passerine bird typically fluffs its feathers, leans to one side, opens its bill, spreads its tail feathers, and either droops or extends one wing or both wings (Fig. 26). The bird may face the sun but more commonly perches with one side toward the sun. Hand-raised Kirtland's Warblers (*Dendroica kirtlandii*) were first observed to sunbathe when between 17 and 23 days old (Berger, 1968). When 1 year old, a captive Yellow-bellied Flycatcher (*Empidonax flaviventris*) raised by Berger took a "sunbath" under a 150-watt floodlight in an outdoor cage at 9:45 P.M. The bird perched on the nearest branch to the light, 16 inches away; it leaned away from the light, fluffed its

Figure 25. A Gannet (*Morus bassanus*) nesting colony; Bonaventure Island, July 27, 1958. (Courtesy of Douglass H. Morse.)

feathers, fanned its tail and its left wing (the one closest to the light), and opened its bill. Japanese White-eyes (*Zosterops j. japonica*) and three species of Hawaiian honeycreepers (Amakihi, *Loxops virens*; Anianiau, *L. parva*; and Kauai Creeper, *L. maculata bairdi*) kept in indoor aviaries at the University of Hawaii also frequently took sunbaths close to floodlights. Sunbathing ducks and gulls typically sit with their backs to the sun. Some gallinaceous species, Bank Swallows and Cliff Swallows lie on one side and spread and raise one wing. Pigeons, doves, and passerine birds sometimes spread both wings.

The stimulus for sunbathing behavior remains obscure. It has been suggested that the bird's eyes absorb ultraviolet rays; that the secretions of the oil gland, when applied to feathers, provide a source of vitamin D; that the behavior is the result of direct exposure to the sun's rays; and that it is motivated by increased heat. From his observations of a hand-raised Western Meadowlark (*Sturnella neglecta*), Lanyon (1958) wrote that he was "inclined to regard a sudden warming of the bird's immediate environment as being extremely important in the motivation of sun-bathing behavior."

Water bathing is a common summer activity, and some birds bathe during the winter as well. With the possible exception of some desert species, all birds probably take water baths, although the methods vary widely. Simmons (1963) wrote that "there are five main types of true bathing in passerine birds: (1) bathing while standing in shallow water; (2) bathing by hopping in and out of the water; (3) bathing from the air, usually in deep water, in a series of dips and rises; (4) bathing in the rain; and (5) bathing in wet vegetation (rain or dew soaked)." Some swifts, swiftlets, swallows, kingbirds, and the Scissor-

Figure 26. Sun-bathing postures of some passerine birds, showing different levels of response to the sun. 1 and 2, White-throated Sparrow; 3, Cardinal; 4, Catbird; 5 and 6, Mockingbird; 7, Blue Jay. (Courtesy of the late Doris C. Hauser and the editor of *Wilson Bulletin*.)

tailed Flycatcher bathe during flight by swooping low over ponds and dipping into the water. Pigeons and doves may lie on one side and raise the opposite wing when it is raining. Larks, according to Simmons, "bathe only in the rain . . . by lying down with wings and tail spread." Some species congregate at shallow puddles or bird baths, there to bathe simultaneously or by turns. Although there are few published records of dew bathing, it probably is common among passerine species, at least. Berger (1967, 1968, 1972) described this behavior in several species of *Empidonax* flycatchers, Kirtland's Warbler, and the Japanese White-eye. Verbeek (1962) reported dew bathing by the Black-capped Chickadee, Golden-crowned Kinglet (*Regulus satrapa*), Red-eyed Vireo (*Vireo olivaceus*), Song Sparrow (*Zonotrichia melodia*), and three species of wood-warblers. There are published reports of birds taking baths in snow: e.g., the Hawk Owl (*Surnia ulula*), Downy Woodpecker, Horned Lark, Black-capped Chickadee, and Slate-colored Junco.

Nearly all bird species complete bathing behavior by alternately preening the plumage and rapidly fluttering the wings and tail until the feathers are dry again, but there are exceptions (e.g., cormorants, vultures, anhingas).

The ontogeny of bathing behavior has been little studied, although Morgan (1900:89–90) wrote about it many years ago: "Ten days after receiving two nestling jays I placed in their cage a shallow tin of water. They took no notice of it, having never seen water before; for they were fed chiefly on sopped food. Presently one of them hopped into it, whether attracted by the water or by accident it is difficult to say, squatted in it bending his legs, and at once fluttered his feathers, as such birds do when they bathe, though his breast scarcely touched the water. The other seized the tin in his bill, and then pecked at the water, thus wetting his beak. He, too, fluttered his feathers in a similar fashion, though he was outside the tin and not in the water at all."

Ficken (1962) first noted bathing in American Redstarts (*Setophaga ruticilla*) when the birds were 18 days old. Berger (1967, 1968, unpub.) first observed water bathing by Blue-winged Warblers (*Vermivora pinus*) when they were 17 days old; at that time, the birds got so wet that they could not fly, and they did not preen after the bath but perched quietly with wings drooped, infrequently shaking their wings. When older, the birds hopped from branch to branch after a bath, shaking their wings and fanning their tails vigorously; they preened primarily after the feathers were nearly dry. Kirtland's Warblers first took water baths when between 19 and 22 days old. One of two Yellow-breasted Chats (*Icteria virens*) took its first water bath when between 20 and 21 days old; the second bird took several baths the following day. Hand-raised flycatchers took their first water bath at the following ages: Acadian Flycatcher (*Empidonax virescens*), 16 days; Yellow-bellied Flycatcher, 37 days; Traill's (now Willow) Flycatcher, 38 or 39 days. Sutton (1943) reported that hand-raised Field Sparrows (*Spizella pusilla*) first took water baths when they were about 3 weeks old.

Dust bathing is characteristic of gallinaceous birds, bustards, cariamas, some hawks and owls, the Hoopoe (*Upupa epops*), and such passerine species as larks, Wrentits, House Sparrows, Vesper Sparrows, Lark Sparrows, and Field Sparrows. Bathing motions, especially in passerine birds, are similar to those used in taking a water bath; some ethologists believe that these movements are derived from feeding or nest-building movements. Gallinaceous birds have special dusting areas that they visit day after day. House Sparrows often engage in social dust bathing, in which a dozen or more sparrows bathe at the same time. The summer dust bathing of Vesper Sparrows is a notable feature of this species' behavior. Sutton observed dust bathing by a captive Vesper Sparrow when the bird was about 3 weeks old. Dust bathing is presumed to be effective in reducing the numbers of external parasites, particularly bird lice (Mallophaga).

2. Head-scratching.   In birds this is accomplished in one of two ways: *direct* and *indirect*. In direct head-scratching, the bird brings the leg and foot upward under the wing to reach the head. In the indirect method, the wing is lowered and the leg is then brought forward over the wing to the head. Oskar Heinroth believed that the pattern of head-scratching was a very rigid one, and Simmons wrote in 1957 that "one method is used by all members of the same family though not necessarily by all families in the same order." Nice and Schantz (1959), however, concluded that "the method of head-scratching proves to be less stereotyped than has been assumed. Diversity may exist within a family, a genus, or among individuals of the same species, and the same individual may use both methods." The truth appears to lie somewhere between the two extreme positions expressed by Simmons and by Nice and Schantz, and we believe that head-scratching behavior can be a useful taxonomic feature.

Many nonpasserine birds scratch directly. The Scolopacidae are said to scratch directly, whereas the Charadriidae scratch indirectly. Similarly, both methods are used among psittacine genera. Kilham (1959) reported direct head-scratching by six species of American woodpeckers. Although most passerine species scratch indirectly, there are some interesting exceptions. For example, the Jungle Babbler (*Turdoides striatus*), Cedar Waxwing (*Bombycilla cedrorum*), and Yellow-breasted Chat scratch directly (Berger, 1966c) (Fig. 27). Simmons (1961, 1963) summarized available information on indirect head-scratching among birds. He concluded that the two methods of head-scratching "may have evolved independently from a more primitive method or successively one from the other. If the two methods evolved successively, then it is argued, contrary to the classical view of Heinroth, that indirect-scratching evolved later than direct-scratching, probably after, or concurrently with, the development of the extended function of scratching. Support is provided for this view by the distribution of indirect-scratching in the bird-kingdom and by the occasional use, by species in which indirect-scratching is

Figure 27. Fully feathered Jungle Babblers shortly before leaving the nest—these babblers scratch the head directly; Baroda, India, August 30, 1964.

the rule, of a scratching method with some characters of the direct. The latter may be explained as either a reversion to the primitive method and/or the use of incomplete or wrongly co-ordinated indirect movements. Variation in scratching within the avian family may be similarly explained and/or is due to incorrect taxonomy (e.g. in Parulidae and Timaliidae)."

We believe that some confusion has resulted because of insufficient emphasis on the ontogenetic development of head-scratching ability in passerine birds. The first attempts at head-scratching by many, if not all, passerine birds are made under the wing, but only because the bird has not yet developed adequate muscular coordination to maintain its balance on one leg while scratching its head with the other leg. In hand-raising such species as Blue-winged Warblers, Kirtland's Warblers, Yellow-breasted Chats, Rose-breasted Grosbeaks, Red-vented Bulbuls (*Pycnonotus cafer*), Common Ioras (*Aegithina tiphia*), Tailor-birds (*Orthotomus sutorius*), and Common Bayas (*Ploceus philippinus*), Berger noted that these species typically developed neuromuscular coordination within a period of approximately 24 hours that permitted indirect scratching after the birds' first unsuccessful attempts. Thereafter the bird invariably scratched its head over the wing. In fact, nestling birds have the "drive" or "urge" to scratch their heads before they are physically capable of doing so. This can be observed easily as a bird draws one leg forward with the "intention" of scratching but abruptly retracts the leg to regain balance, simply because the bird is unable to support itself on one leg. To be sure, there are a few observations of birds that use the atypical method of scratching at times, but these occasions surely are rare. Only once during more than a decade of watching wild and captive Kirtland's Warblers did Berger see a bird scratch directly, and we agree with Simmons (1961) that the experiments performed by Nice and Schantz probably demonstrate "no more than indirect-scratching carried out abnormally in response to super-normal stimuli."

3. *Preening* the feathers with the bill.   Universal among birds, this is one of the first of the maintenance activities or comfort movements performed by nestling altricial birds. By "preening" one generally means the nibbling or other actions of the bill on the feathers. McKinney (1965) used the term "for activities which involve contact of the bird's bill or head with the feathers in the form of billing, rubbing, or combing movements." He recognized three types of preening: oiling, nibbling, and washing. By "oiling" he refers to the transfer of secretions from the oil gland to the feathers; "cleaning movements accompanying bathing are called 'washing'"; "all preening not associated with bathing or oiling is called 'nibbling.'" Nestling birds begin to preen at an early age as the contour feathers grow; these preening actions assist in removing the covering sheath of the "pin feathers" (see the discussion by Ficken, 1962). Nestling passerine birds often preen the top of the head with the under surface of the wing. Nightjars (such as the Poor-will) use the pectinate claw of the third toe in preening (Brauner, 1953).

Flight preening was described in detail for the Black Tern (*Chlidonias niger*) by Goodwin (1959). This tern has a characteristic undulatory flight during which it preens its lower neck and breast feathers while gliding downward; at times, a bird even preens "the inner wing coverts, scapulars and feathers of the back." Goodwin also summarized information on aerial scratching and preening in other species: e.g., albatrosses, the Mallard, some species of gulls, the Purple Martin (*Progne subis*), and the Bank Swallow (*Riparia riparia*).

Allopreening (mutual preening) has been observed in one or more species of 19 different orders of birds (Harrison, 1965). In some families, this behavior was observed in only one species (e.g., Phasianidae), but in others (e.g., Spheniscidae, Anhimidae, Zosteropidae) it occurred in all species. Captive Japanese White-eyes commonly practice group preening, during which one bird preens another while the latter is preening a third (Eddinger, 1967).

Harrison remarked that, "although performed by a variety of birds with

Table 1.   Parts of Plumage Preened during Allopreening[a,b]

Family	1	2	3	4	5	6	7	8	9	10
Spheniscidae	X	X	X	X						
Diomediidae	X	X	(X)							
Procellariidae	X	X		(X)	(X)		(X)		(X)	(X)
Pelecanoididae	X		(X)							
Hydrobatidae	X									
Sulidae	X	X								
Phalacrocoracidae	X	X								
Ardeidae	X	X					X			
Ciconiidae	X									
Threskiornithidae	X	X					(X)			
Anhimidae	X	X								
Anatidae	X	X	(X)							
Cathartidae	X	X								
Accipitridae	X									
Phasianidae	X									
Rallidae	X	X	(X)	(X)						
Laridae	X	X		X						
Alcidae	X	X								
Columbidae	X	X	X	(X)	(X)	(X)				
Psittacidae	X	X								
Cuculidae	X	X								
Steatornithidae	X									
Apodidae	X	X								

considerable differences in bill-shape, allopreening is remarkably stereotype in the form that it takes." Table 1 shows the parts of the plumage preened in different groups of birds. Harrison wrote that allopreening "was found to be closely linked with aggressive behaviour, and in many species attack appeared to give way to allopreening, which appeared to be a replacement, or outlet, for it. . . . Preening invitation postures are considered to be postures resulting from thwarted fleeing, appeasing, or withdrawing behaviour, combined with head positions which initially result from attempts to protect the eyes from, or evade, the bill of the preener without moving away. These head positions could function as cut-off postures in which the aggressor ceases to be visible and the tendency to flee is reduced. Allopreening can be regarded as a form of agonistic behaviour in which the normal tendencies of attacking or fleeing, when two individuals are in close proximity, are in conflict with sexual and opposing attacking and fleeing tendencies."

Table 1. (cont.)

Family	1	2	3	4	5	6	7	8	9	10
Bucerotidae	X	X								
Ramphastidae	X									
Pycnonotidae	X	X	(X)	(X)	(X)					
Troglodytidae	X	X			(X)					
Timaliinae	X	X	(X)	(X)		(X)	(X)			
Paradoxornithidae	X									
Malurinae	X									
Zosteropidae	X	X	X	X		X	X	X	(X)	(X)
Icteridae	X	X								
Estrildidae	X	X								
Sturnidae	X	X	(X)	(X)	(X)					
Corvidae	X	X								

[a] These data have been taken from the accumulated records and are relatively incomplete. Although the family names have been used, the entry relates only to species for which information is available. Records in parentheses refer to infrequent occurrences and may indicate atypical single records. Although no quantitative data are available, the information, as shown here, indicates the overwhelming importance of the preening of head and neck. The preening of chicks by adult birds, which tends to be random over much of the body, has been omitted. Reproduced by permission from Harrison (1965).

[b] Key to columns: 1. Head. 2. Neck. 3. Breast. 4. Mantle. 5. Wing. 6. Back and rump. 7. Flanks. 8. Belly. 9. Tail coverts. 10. Tail.

Selander (1964) described interspecific allopreening in which a captive male Bay-winged Cowbird (*Molothrus badius*) solicited preening from Chestnut-fronted Troupials (*Agelaius ruficapillus*) (Fig. 28). He also reported that both male and female Shiny Cowbirds (*Molothrus bonariensis*) solicited preening from a House Sparrow, and that he had many times observed free-living Brown-headed Cowbirds (*Molothrus ater*) solicit preening from House Sparrows, Red-winged Blackbirds, and Dickcissels (*Spiza americana*). This behavior among parasitic cowbirds "might function to forestall aggressive action by potential hosts"; the Bay-winged Cowbird cares for its young but appropriates the nests of other species for its use, so that "it is possible that an effective appeasing display would be advantageous" to this bird.

4. Stretching Movements.   Although little studied, these too are universal among birds. Four basic stretching patterns appear to be characteristic of

Figure 28. Upper: Male Bay-winged Cowbird (*Molothrus badius*) on the left solicits preening from male Chestnut-fronted Troupial (*Agelaius ruficapillus*). Lower: Male Chestnut-fronted Troupial gapes into the plumage of a displaying male Bay-winged Cowbird. (Courtesy of Robert K. Selander and the editor of *Auk*.)

passerine birds: *both-wings-up stretch,* in which both wings are raised simultaneously above the back; *both-wings-down stretch,* in which both wings are simultaneously stretched downward; *both-legs stretch,* which involves the simultaneous extension of both legs, the bird typically arching its back and lowering its head and neck; and the *wing and leg sideways stretch,* in which the wing and the leg of the same side are extended downward (see Ficken, 1962). These muscular movements are performed by adult birds and by nestlings that have matured enough to stand and preen.

Mobbing behavior describes the response of some species of birds to potential predators, such as dogs, cats, snakes, shrikes, and perched owls. The mobbing reaction may be performed by several individuals of the same species (such as crows or jays) or by individuals of several different species. Altmann (1956) experimented with several species of stuffed owls, and found that "the range of the number of species of birds actually attacking the owls was from 1 to 7, with a mean of 2.13. The number of individuals ranged from 1 to 300, with a mean of 6.1. The number of bird-minutes of reaction ranged from 1 to 300, with a mean of 63.7." Passerine birds most commonly take part in mobbing activities, but Altmann also recorded the behavior in two species of doves, four species of hummingbirds, and one woodpecker. In Hawaii, owls are mobbed by mixed groups of birds that sometimes include both introduced (Cardinal, Japanese White-eye) and endemic (e.g., Apapane, *Himatione sanguinea*) species. Mobbing appears to be more prevalent during the breeding season, and potential nest robbers (e.g., Blue Jay) also may be attacked. Red-winged Blackbirds and Kingbirds (*Tyrannus tyrannus*) are noted for flying considerable distances to dive-bomb a crow in flight, a behavior related to mobbing.

Andrew (1961) conducted intensive experimental studies on the causes and nature of mobbing calls of the Blackbird (*Turdus merula*). He found that this species used two calls, "one of which progressively replaced the other as mobbing increased in intensity. . . . The important variables affecting calling in flight were the distance of the Blackbird from the owl at take-off and the time that had elapsed since the beginning of mobbing, not direction of flight."

Among his conclusions after studying enemy recognition in birds, Curio (1963) wrote that "by learning, the bird either reduces the range of stimuli eliciting predator responses (habituation) or widens it. Recognition of learned and innate stimulus situations does not seem to differ basically." He found that a number of passerine species learned to recognize the calls of owls and were stimulated to mobbing behavior on hearing the calls alone.

Opening milk bottles by birds was first reported in England in 1921; more than 400 records of this interesting behavior had been reported by 1949, from widely scattered areas in Scotland, Ireland, Sweden, Denmark, Holland, and France. The behavior is more characteristic of several species of European titmice, but milk-bottle opening has been recorded also for such birds as the Great Spotted Woodpecker (*Dryobates major*), British Robin (*Erithacus rubecula*), Song Thrush (*Turdus ericetorum*), European Starling, House Sparrow, and Chaffinch (*Fringilla coelebs*). Eric Hosking wrote (letter, July 24, 1972) that the birds "do not appear to be doing it nearly so much now as they did during the 1950s."

Titmice are "curious" birds that investigate new and conspicuous objects they encounter. It was suggested, therefore, that the birds first found milk bottles in their appetitive behavior of food-seeking. Then (by trial-and-error learning) they learned to return to open the bottles. Small groups of titmice were observed to follow a milkman and fly to the bottles almost as soon as they were deposited. Titmice are sedentary in habit, rarely moving more than a few miles from their place of hatching. Consequently, some of the birds must have learned to open milk bottles *de novo* in different parts of their range, whereas others learned the behavior from other titmice while in flocks during the nonbreeding season. It was suggested that this "copying behavior" was due to the process called *local enhancement,* which modified the appetitive behavior and initially was independent of the reward provided by the milk.

Paper-tearing, reported repeatedly in the British Isles, is a peculiar behavior pattern exhibited especially by titmice. The birds enter houses and strip pieces of wallpaper from the walls; in some instances, titmice chased from a house persisted in flying about and pecking at a closed window in an effort to return to a particular room. The birds do not confine themselves to wallpaper but also attack covers of books, name cards on doors, lampshades, match boxes, clothing, leather, straw baskets, and oilcloth.

Play in animals has been called an experimental dialogue with the environment by Eibl-Eibesfeldt, and he suggests that the degree of learning is correlated with the amount of play engaged in by the animal. Sutton (1943) referred to playful behavior of hand-raised fringillids, and Thorpe (1963:362–364) discussed playful activities of both young and adult birds.

Roosting behavior, although engaged in by all birds, has not been studied intensively. After the breeding season, some species sleep in flocks containing from hundreds to a million birds (some mixed flocks of blackbirds and European Starlings in the Washington, D. C., area have been estimated to contain more than a million birds). Roosting flocks may consist of a single species (*homogeneous aggregations*) or of several species (*heterogeneous aggregations*). Red-winged Blackbirds, Brown-headed Cowbirds, and Common Grackles form tremendous postbreeding roosting flocks, often in marshy areas (Fig. 29). Studies of roosting Starlings have shown that roosting time is correlated with light intensity, although weather and other factors affect the time when birds return to the roost in the evening.

The Ruffed Grouse is well known for its habit of diving from the wing into a

Figure 29. Method used by Brooke Meanley to band blackbirds at night roosts during February and March in Arkansas. In order of relative abundance, species occupying the roost were Red-winged Blackbird, Common Grackle, Brown-headed Cowbird, and Rusty Blackbird (*Euphagus carolinus*); 4000 birds were banded in 10 nights. (Courtesy of Brooke Meanley and the editor of *Bird-Banding*.)

snowdrift and spending the night there; sometimes a freezing rain during the night forms a crust over the snow, so that the bird is unable to dig out the following morning. Coveys of Bobwhite roost together in a circle on the ground with each bird facing outward; when disturbed, the birds take off in different directions, an adaptive behavior for reducing predation. Woodpeckers excavate roosting cavities, especially for the winter months. Brown Creepers (*Certhia familiaris*) are said to sleep while clinging vertically in a hole or depression in a tree, and, at times, even to excavate a roosting shelter. Nuthatches, chickadees, and titmice also roost in natural cavities or old nesting cavities in trees, but Knut Borg found 10 Long-tailed Tits (*Aegithalos caudatus*) roosting together in a sandbank. Oniki (1970) described the roosting behavior of three species of woodcreepers (Dendrocolaptidae), either in tree cavities or on branches. Gray-crowned Rosy Finches (*Leucosticte tephrocotis*) have been observed to spend winter nights in the abandoned, retort-shaped, mud nests of Cliff Swallows (*Petrochelidon pyrrhonota*). Clancey (1962) reported that Carol's Penduline Tits (*Anthoscopus caroli*) roost in old nests of the Spectacled Weaver (*Ploceus ocularis*) in East Africa.

Communal roosts of Rough-legged Hawks (*Buteo lagopus*), Marsh Hawks (*Circus hudsonius*), and Short-eared Owls (*Asio flammeus*) have been reported, and Swinebroad (1964) described the roosting habits of migrating shorebirds. *Cock roosts* have been described for Common and Lesser nighthawks in the western United States (Selander and Preece, 1951). As many as 100 Common Nighthawks have been seen roosting on horizontal branches of cottonwood and black locust trees in Idaho; such daytime flocks are thought to be composed almost entirely of adult males. Edgar Hartowicz reported seasonal changes in the roosting habits of Wood Ducks in Missouri, where more than 1500 birds used a roost in September.

Frazier and Nolan (1959) found from 5 to 14 Eastern Bluebirds (*Sialia sialis*) roosting in a birdhouse during the winter in Indiana. "The birds slept heads together and bodies pointed downward, forming an inverted cone." Communal winter roosting in cedar trees has been reported for Purple Finches (*Carpodacus purpureus*), and Owen A. Knorr estimated that at least 150 Pygmy Nuthatches (*Sitta pygmaea*) roosted together in the hollow trunk of a dead yellow pine. Skutch (1961) found 16 Prong-billed Barbets (*Semnornis frantzii*) sleeping together in a small hole in a tree. Crested swifts (Hemiprocnidae) and wood-swallows (Artamidae) are said to sleep in "clusters" of a dozen or more birds, and colies (Coliidae) sleep in "bunches of six or more clinging onto the topmost shoots of the tree or bush which is their communal roost" (Van Someren, 1956:206). Skutch also described the roosting habits of Central American birds, including wrens that build "dormitories," nests used only for sleeping. Cactus Wrens (*Campylorhynchus brunneicapillus*) use nests for roosting throughout the year; male and female wrens sometimes build

Figure 30. Cactus Wren at its nest in cholla cactus. (Courtesy of Lewis Wayne Walker.)

Figure 31. A male Bananaquit in roosting nest; the brood nest of the pair was located in vines 15 feet from the roosting nest; flash photo at night. (Courtesy of the late Alfred O. Gross.)

roosting nests in the same cholla cactus (Fig. 30). Ricebirds (*Lonchura punctulata*) and Bananaquits (*Coereba flaveola*) also use old nests as dormitories (Berger, 1972; Fig. 31).

Sleeping postures vary among birds. The most common pattern for adult birds, especially passerines, involves turning the head backward and resting the bill on, or buried in, the scapular feathers. There is an ontogenetic change in sleeping postures (Ficken, 1962; Berger, 1967, 1968; Fig. 32). Very young birds simply close the eyes, the head and neck resting on the bottom of the nest or on a sibling. Traill's (Willow) Flycatchers first adopted the adult sleeping posture (with head turned backward) when they were estimated to be between 14 and 15 days old; an Acadian Flycatcher, when 15 days old. Before that time, the nestlings slept with the head and bill pointed straight ahead, but often with the bill tipped upward above the horizontal plane. At other times, the birds slept with the neck resting on the rim of the nest and with the bill and head hanging downward on the outer surface of the nest. An 8-day-old Kirtland's Warbler turned its head backward as though it were trying to put its bill in or under the scapular feathers, although these feathers were not yet long enough to conceal the bird's bill. Very soon, as the bird slept, the head slowly turned so that the bill rested at right angles to the body, and within a few minutes the head moved slowly until the bill pointed straight ahead. This suggests that the instinct for sleeping with the head turned backward was at work, but that the muscle development and coordination enabling the bird to maintain this posture were not yet fully developed. Another Kirtland's Warbler slept with its head turned backward over the shoulder when 10 days old.

Descriptions of the mechanism that enables a sleeping bird to hold its grasp on a perch usually suggest that the tendons of muscles that cross the back of the "heel" joint are stretched so that the toes are "locked" around the perch. This appears, however, not to be the case. Sleeping Kirtland's Warblers, Common Ioras, Tailorbirds, and other passerine species do not maintain a viselike grip on the branch. The three front toes may be curved around the perch, but only the basal portion of the hallux touches the perch, so that the claw of the hallux is not even in contact with it. The claw is extended and relaxed (but not noticeably flexed) at frequent intervals (every 15 seconds to as frequently as

Figure 32. One sleeping posture of a hand-raised Yellow-breasted Chat at 25 days of age.

every 2 to 5 seconds). Adult passerine birds often perch on one foot when sleep-ing, whereas nestlings perch on both feet until neuromuscular coordination develops. Berger (1968) reported that, at relatively short intervals (10 to 30 seconds), a sleeping Kirtland's Warbler lifted one foot off the perch, rapidly extended and flexed the toes, and then grasped the perch again. The bird may thus relax the muscles of the legs alternately or may raise one foot several times in succession. Occasionally a bird will raise first one foot and then the other within 2 or 3 seconds.

Tool-using in animals is considered evidence of insight learning. Thorpe (1963:121, 374) wrote that "we do not normally regard the use of a rock or other immovable object as a resistance against which to break open the protec-tive shells of prey as tool using, even though the prey is dropped from a considerable height." Hence the use of a stone by the Song Thrush to break snail shells is not considered tool using, nor is the habit of the Bearded Vulture (*Gypaetus barbatus*) of dropping bones from a great height onto rocks, a practice which frequently breaks the bones and exposes the marrow. The Aus-tralian Black-breasted Buzzard (*Haemirostris melanosterna*) is reported to frighten the emu from its eggs "and then fly aloft with a stone in its claws; this missile it drops on the eggs and then immediately swoops down to devour the contents. . . . This appears clearly as tool using for, to be a tool, an object must become a movable body extension.

Figure 33. New Caledonian Crow probing into the end of a hollow twig; the head was moved vertically (up and down) in this action. (Courtesy of Ronald L. Orenstein and the editor of *Auk*.)

Figure 34. A wing-flashing sequence of a 10-day-old Mockingbird, drawn from 16-mm motion picture film taken at 18 frames per second. Frame 1, the wings are close to the sides and the tail is parallel with the horizontal. Frame 3, the wings are being raised. Frames 6–11, the wings, having been partially raised, are now paused in the first hitch. Frames 14-17, the wings are now held at the second hitch and the tail is raised. (Courtesy of Robert H. Horwich and the editor of *Wilson Bulletin*.)

"On this basis also we can consider the Tailor Bird . . . and the Little Spider-hunter . . ., using spiders' lines as thread, as tool users. More certainly still the Satin Bower Bird, which uses fibrous material as a brush with which to 'paint' the sticks of its bower with a colouring material from berries or with charcoal . . ., can be classed as a tool user."

Other examples have been discovered since Thorpe wrote. The Egyptian Vulture (*Neophron percnopterus*) breaks Ostrich eggs by throwing stones at them, repeating the behavior until the shell cracks. The Woodpecker-finch (*Cactospiza pallida*) and the Mangrove-finch (*C. heliobates*) of the Galapagos Islands pick up cactus spines or twigs and probe into crevices or holes that they have pecked; when an insect or larva emerges, the bird drops the probe and grabs the prey (Curio and Kramer, 1964). Morse (1968) watched Brown-headed Nuthatches (*Sitta pusilla*) use bark scales to remove other bark scales in their search for food, and Orenstein (1972) reported a single observation of a New Caledonian Crow (*Corvus moneduloides*) employing a twig to probe under bark or into the end of a hollow branch (Fig. 33).

Figure 35. Mockingbird wing-flashing between attacks on a dummy Screech Owl. (Courtesy of Robert K. Selander and the editor of *Wilson Bulletin.*)

Wing-flashing refers to the display, first reported in Mockingbirds, in which the wings are extended upward in "archangel fashion" in a sequence so that successive movements of the wings are separated by momentary pauses (Figs. 34, 35). The wings often are extended in this manner after a short run along the ground. Several theories have been proposed to account for this behavior: e.g., that it is a means of flushing prey, or that it is performed as a displacement act when the bird is "frustrated," sexually excited, or alarmed. Hailman (1960) concluded that, in adults, wing-flashing "is used in foraging, possibly to flush insects; but in young birds it is often given irrelevantly, and seems to be motivated by hunger and curiosity." Selander and Hunter (1960), however, wrote: "All evidence considered, we are inclined to think that Sutton's suggestion (1946) that wing-flashing is primarily 'a gesture indicating wariness, suspicion, [and] distrust' is more nearly correct than any other. We suggest that wing-flashing in the Mockingbird represents a highly ritualized flight-intention movement of the wings, which, evolving originally as a social signal or wariness, has acquired a secondary function in food-getting. It is possible that it may also function to intimidate other birds in addition to indicating an apprehensive 'mood.' This theory would account for its frequent use by birds that are mobbing an owl or are reacting to the presence of a live or dummy Mockingbird in their territory. We further suggest as a working hypothesis that the Mockingbird's use of the display while foraging may be individually conditioned. Perhaps when first foraging on the ground, the young Mockingbird is apprehensive and gives the display; insects are flushed as a result, and in time the bird comes to associate wing-flashing with foraging reinforcement being provided by the capture and eating of insects."

Selander and Hunter remarked that "whatever the biological significance of wing-flashing in Mockingbirds may be, the behavior appears early in development and is almost undoubtedly innate, for we have seen it in nestlings and it has previously been observed in young fledglings." Horwich (1965) found that wing-flashing appeared when his Mockingbirds were between 9 and 13 days of age, and most commonly on the 10th or 11th day. Berger (1966a) agreed with Horwich that "in birds from 9 days to 10 months old, wing-flashing is definitely associated with some type of strange or uneasy situation."

Wing-flashing behavior has been reported in two species of Mockingbirds that do not have wing patches (Calandria Mockingbird, *Mimus saturninus,* and Graceful Mockingbird, *M. gilvus*), in the Brown Thrasher (*Toxostoma rufum*), and in the Catbird (*Dumetella*).

REFERENCES

Altmann, S. A. 1956. Avian mobbing behavior and predator recognition. *Condor,* **58:**241–253.

Andrew, R. J. 1961. The motivational organisation controlling the mobbing calls of the Blackbird (*Turdus merula*). I, II, III, IV. *Behaviour,* **17:**224–246, 288–321; **18:**25–43, 161–176.

Aronson, L. R., *et al.,* eds. 1970. *Development and Evolution of Behavior: Essays in Memory of T. C. Schneirla.* W. H. Freeman, San Francisco.

Baerends, G. P. 1958. The contribution of ethology to the study of the causation of behaviour. *Acta Physiol. Pharmacol. Neerl.,* **7:**466–499.

Bang, B. G. 1964. The nasal organs of the Black and Turkey vultures; a comparative study. . . . *J. Morphol.,* **115:**153–184.

———1971. Functional anatomy of the olfactory system in 23 orders of birds. *Acta Anat.,* Suppl., **58:**1–76.

———, and Stanley Cobb. 1968. The size of the olfactory bulb in 108 species of birds. *Auk,* **85:**55–61.

Barfield, R. J. 1969. Activation of copulatory behavior by androgen implanted into the preoptic area of the male fowl. *Horm. Behav.,* **1:**37–52.

Beecher, W. J. 1951. A possible navigation sense in the ear of birds. *Am. Midl. Nat.,* **46:**367–384.

Berger, A. J. 1954. Injury feigning by the Catbird. *Wilson Bull.,* **66:**61.

——— 1966a. Behavior of a captive Mockingbird. *Jack-Pine Warbler,* **44:**8–13.

——— 1966b. Experiences with insectivorous birds in captivity. *Ibid.,* **44:**65–73.

——— 1966c. Head-scratching behavior of some hand-raised birds. *Wilson Bull.,* **78:**469.

——— 1967. Behavior of hand-raised Empidonax flycatchers. *Jack-Pine Warbler,* **45:**131–138.

——— 1968. Behavior of hand-raised Kirtland's Warblers. *Living Bird,* **1968,** pp. 103–116.

——— 1972. *Hawaiian Birdlife.* University Press of Hawaii, Honolulu.

———, and D. V. Howard. 1968. Anophthalmia in the American Robin. *Condor,* **70:**386–387.

Brauner, J. 1953. Observations on the behavior of a captive Poor-will. *Condor,* **55:**68–74.

Clancey, P. A. 1962. Carol's Penduline Tits roosting in weaver nest. *Ostrich,* **33:**38.

Coulombre, A. J., and J. L. Coulombre. 1958. Corneal development. I. Corneal transparency. *J. Cell. Comp. Physiol.,* **51:**1–12.

———, and E. S. Crelin. 1958. The role of the developing eye in the morphogenesis of the avian skull. *Am. J. Phys. Anthropol.* (n.s.), **16:**25–37.

Curio, E. 1963. Probleme des Feinderkennens bei Vögeln. *Proc. XIII Int. Ornithol. Congr.,* **1:**206–239.

———, and P. Kramer, 1964. Vom Mangrovefinken (*Cactospiza heliobates* Snodgrass und Heller). *Z. Tierpsychol.,* **21:**223–234.

Deutsch, J. A. 1960. *The Structural Basis of Behavior.* University of Chicago Press, Chicago.

Duijm, M. 1959. On the position of a ribbon-like central area in the eyes of some birds. *Arch. Neerl. Zool.,* **13,** Suppl. **1:**128–145.

Dunlap, K., and O. H. Mowrer. 1930. Head movements and eye functions in birds. *J. Comp. Psychol.,* **11:**99–113.

Eddinger, C. R. 1967. Feeding helpers among immature White-eyes. *Condor,* **69:**530–531.

Eibl-Eibesfeldt, I. 1970. *Ethology: The Biology of Behavior.* Holt, Rinehart and Winston, New York.

Emlen, J. T., Jr. 1955. The study of behavior in birds. In *Recent Studies in Avian Biology*, University of Illinois Press, Urbana, pp. 105–153.

Ficken, M. S. 1962. Maintenance activities of the American Redstart. *Wilson Bull.,* **74:**153–165.

Fisher, H. I. 1957. Footedness in domestic Pigeons. *Wilson Bull.,* **69:**170–177.

Frazier, A., and V. Nolan, Jr. 1959. Communal roosting by the Eastern Bluebird in winter. *Bird-Banding,* 30:219–226.

Gallup, G. G. 1968. Mirror-image stimulation. *Psychol. Bull.,* **70:**782–793.

Goodge, W. R. 1960. Adaptations for amphibious vision in the Dipper (*Cinclus mexicanus*). *J. Morphol.,* 107:79–92.

Goodman, J. M. 1960. *Aves incendiaria. Wilson Bull.,* **72:**400–401.

Goodwin, R. E. 1959. Records of flight preening and related aerial activities in birds, particularly the Black Tern. *Auk,* **76:**521–523.

Gottlieb, G. 1971. *Development of Species Identification in Birds.* University of Chicago Press, Chicago.

Hailman, J. P. 1960. A field study of the Mockingbird's wing-flashing behavior and its association with foraging. *Wilson Bull.,* **72:**346–357.

Hardy, J. W. 1965. Flock social behavior of the Orange-fronted Parakeet. *Condor,* **67:**140–156.

Harrison, C. J. O. 1965. Allopreening as agonistic behaviour. *Behaviour,* **24:**161–209.

Hayes, J. S., W. M. S. Russell, C. Hayes, and A. Kohsen. 1953. The mechanism of an instinctive control system: a hypothesis. *Behaviour,* **6:**85–119.

Henton, W. W., J. C. Smith, and D. Tucker. 1966. Odor discrimination in pigeons. *Science,* **153:**1138–1139.

Hess, E. H. 1966. Origins of behavior. *Science,* **154:**1636–1637.

Hinde, R. A. 1970. *Animal Behaviour,* 2nd ed. McGraw-Hill Book Co., New York.

Hocking, B., and B. L. Mitchell. 1961. Owl vision. *Ibis,* **103a:**284–288.

Hogan, J. A. 1965. An experimental study of conflict and fear: an analysis of behavior of young chicks toward a mealworm. I. *Behaviour,* **25:**45–97.

Horwich, R. H. 1965. An ontogeny of wing-flashing in the Mockingbird with reference to other behaviors. *Wilson Bull.,* **77:**264–281.

Johnsgard, P. A. 1967. *Animal Behavior.* Wm. C. Brown, Dubuque, Iowa.

Kelso, L. and M. M. Nice. 1963. A Russian contribution to anting and feather mites. *Wilson Bull.,* **75:**23–26.

Kikkawa, J. 1961. Social behavior of the White-eye *Zosterops lateralis* in winter flocks. *Ibis,* **103a:**428–442.

Kilham, L. 1959. Head-scratching and wing-stretching of woodpeckers. *Auk,* **76:**527–528.

Klopfer, P. H., and G. Gottlieb. 1962. Imprinting and behavioral polymorphism. . . . *J. Comp. Physiol. Psychol.,* **55:**126–130.

———, and J. P. Hailman. 1967. *An Introduction to Animal Behavior.* Prentice-Hall, Englewood Cliffs, N.J.

Konishi, M. 1973a. How the owl tracks its prey. *Am. Sci.* **61:**414–424.

——— 1973b. Locatable and nonlocatable acoustic signals for Barn Owls. *Am. Nat.,* **107:**775–785.

Lanyon, W. E. 1958. The motivation of sun-bathing in birds. *Wilson Bull.,* **70:**280.

Lehrman, D. S. 1953. A critique of Konrad Lorenz's theory of instinctive behavior. *Quart. Rev. Biol.,* **28:**337–363.

——— 1970. Semantic and conceptual issues in the nature-nurture problem. In *Development and Evolution of Behavior*. W. H. Freeman & Co., San Francisco, pp. 17–52.

Lindenmaier, P., and M. R. Kare. 1959. The taste end-organs of the chicken. *Poult. Sci.,* **38**:545–550.

Lord, R. D., Jr., 1956a. An anomalous condition in the eye of some hawks. *Auk,* **73**:457.

——— 1956b. A comparative study of the eyes of some falconiform and passeriform birds. *Am. Midl. Nat.,* **56**:325–344.

Lorenz, K. Z. 1965. *Evolution and Modification of Behavior*. University of Chicago Press, Chicago.

——— 1973. The fashionable fallacy of dispensing with description. *Naturwissenschaften,* **60**:1–9.

McAtee, W. L. 1938. "Anting" by birds. *Auk,* **55**:98–105.

McBride, G., I. P. Parer, and F. Foenander. 1969. The social organization and behaviour of the feral domestic fowl. *Anim. Behav. Monogr.,* **2**, Pt. 3, 181 pp.

McKinney, F. 1965. The comfort movements of Anatidae. *Behaviour,* **25**:120–220.

MacNichol, E. F., Jr. 1964. Three-pigment color vision. *Sci. Am.,* December 1964:1–10.

Malinovský, L., and R. Zemánek. 1969. Sensory corpuscles in the beak skin of the domestic pigeon. *Folia Morphol.,* **17**:241–250.

———, and ——— 1970. Sensory nerve endings in the joint capsules of the large limb joints in the domestic hen (*Gallus domesticus*) and the Rook (*Corvus frugilegus*). *Ibid.,* **18**:206–212.

Manning, A. 1972. *An Introduction to Animal Behavior,* 2nd ed. Addison-Wesley Publishing Co., London.

Marler, P. 1956. Behaviour of the Chaffinch *Fringilla coelebs*. *Behaviour, Suppl.,* **5**:1–184.

———, and W. J. Hamilton III. 1966. *Mechanisms of Animal Behavior*. John Wiley & Sons, New York.

Mayfield, H. 1966. Fire in birds' nests. *Wilson Bull.,* **78**:234–235.

Medway, L. and D. R. Wells. 1969. Dark orientation by the Giant Swiftlet *Collocalia gigas*. *Ibis,* **111**:609–611.

Michelsen, W. J. 1959. Procedure for studying olfactory discrimination in pigeons. *Science,* **130**:630–631. (See also *Science,* **131**:1265.)

Moltz, H. 1961. An experimental analysis of the critical period for imprinting. *Trans. N.Y. Acad. Sci.,* **23**:452–463.

Moore, C. A., and R. Elliot. 1946. Numerical and regional distribution of taste buds on the tongue of the bird. *J. Comp. Neurol.,* **84**:119–131.

Morgan, C. L. 1900. *Animal Behaviour*. Edward Arnold, London.

Morse, D. H. 1968. The use of tools by Brown-headed Nuthatches. *Wilson Bull.,* **80**:220–224.

Moynihan, M. 1955a. Remarks on the original sources of displays. *Auk,* **72**:240–246.

——— 1955b. Types of hostile display. *Ibid.,* **72**:247–259.

Nice, M. M., and W. E. Schantz. 1959. Head-scratching movements in birds. *Auk,* **76**:339–342.

Oniki, Y. 1970. Roosting behavior of three species of woodcreepers (Dendrocolaptidae) in Brazil. *Condor,* **72**:233.

O'Rahilly, R. 1962. The development of the sclera and the choroid in staged chick embryos. *Acta Anat.,* **48**:335–346.

———, and D. B. Meyer. 1961. The development and histochemistry of the pecten oculi. In *The Structure of the Eye*, Academic Press, New York, pp. 207–219.

Orenstein, R. I. 1972. Tool-use by the New Caledonian Crow (*Corvus moneduloides*). *Auk,* **89:**674–676.

Pearson, R. 1972. *The Avian Brain.* Academic Press, London.

Pumphrey, R. J. 1961. Sensory organs: vision. In *Biology and Comparative Physiology of Birds,* Vol. 2, Academic Press, New York, pp. 55–68.

Rand, A. L. 1958. Patterns in the use of left and right limbs in vertebrates. *Wilson Bull.,* **70:**92–93.

Ripley, S. D. 1961. Aggressive neglect as a factor in interspecific competition in birds. *Auk,* **78:**366–371.

Sandoval, J. 1965. Sobre los órganos extrinsecos del globo ocular de las aves. *Anal. Anat.,* **14:**411–428.

Saxod, R. 1970. Etude au microscope électronique de l'histogenese du corpuscule sensoriel cutané de Herbst chez le Canard. *J. Ultrastruct. Res.,* **33:**463–482.

Schwartzkopff, J. 1963. Morphological and physiological properties of the auditory system in birds. *Proc. XIII Int. Ornithol. Congr.,* **2:**1059–1068.

——— 1968. Structure and function of the ear and the auditory brain areas in birds. In *Hearing Mechanisms in Vertebrates,* Little, Brown and Co., Boston.

Seaman, A. R., and T. M. Himelfarb. 1963. Correlated ultrafine structural changes of avian pecten oculi and ciliary body of *Gallus domesticus. Am. J. Ophthalmol.,* **56:**278–296.

Selander, R. K. 1964. Behavior of captive South American cowbirds. *Auk,* **81:**394–402.

———, and D. K. Hunter. 1960. On the functions of wing-flashing in Mockingbirds. *Wilson Bull.,* **72:**340–345.

———, and S. J. Preece, Jr. 1951. Cock roosts of nighthawks. *Condor,* **53:**302–303.

Shlaer, R. 1972. An eagle's eye: quality of the retinal image. *Science,* **176:**920–922.

Simmons, K. E. L. 1961. Problems of head-scratching in birds. *Ibis,* **103a:**37–49.

——— 1963. Some behaviour characters of the babblers (Timaliidae). *Avicult. Mag.,* **69:**183–193.

Skinner, F. 1966. The phylogeny and ontogeny of behavior. *Science,* **153:**1205–1213.

Skutch, A. F. 1961. The nest as a dormitory. *Ibis,* **103a:**50–70.

Stager, K. E. 1964. The role of olfaction in food location by the Turkey Vulture (*Cathartes aura*). *Contrib. Sci.,* **81:**1–63 (Los Angeles County Museum).

Stokes, A. W., ed. 1968. *Animal Behavior in Laboratory and Field.* W. H. Freeman & Co., San Francisco.

Stonor, C. R. 1939. Asymmetry of the external ear in Pesquet's Parrot (*Psittrichas pesqueti*). *Ibis,* **1939:**342–343.

Sutter, E. 1951. Growth and differentiation of the brain in nidifugous and nidicolous birds. *Proc. 10th Intn. Ornithol. Congr.,* **1950:**636–644.

Sutton, G. M. 1943. Notes on the behavior of certain captive young fringillids. *Univ. Mich. Mus. Zool. Occas. Pap.* No. 474, pp. 1–14.

——— 1947. Eye-color in the Green Jay. *Condor,* **49:**196–198.

Swinebroad, J. 1964. Nocturnal roosts of migrating shorebirds. *Wilson Bull.,* **76:**155–159.

Thompson, W. L. 1960. Agonistic behavior of the House Finch. I, II. *Condor,* **62:**245–271, 378–402.

Thorpe, W. H. 1963. *Learning and Instinct in Animals.* Methuen & Co., London.

———, and O. L. Zangwill, eds. 1961. *Current Problems in Animal Behaviour. Cambridge University Press,* England.

Tinbergen, N. 1951. *The Study of Instinct.* Oxford University Press, London.

——— 1952. "Derived" activities; their causation, biological significance, origin, and emancipation during evolution. *Quart. Rev. Biol.,* **27**:1–32.

——— 1965. *Animal Behavior.* Time-Life Books, New York.

——— 1971. Clever gulls and dumb ethologists—or: The trackers tracked. *Vogelwarte,* **26**:232–238.

Tucker, D. 1965. Electrophysiological evidence for olfactory function in birds. *Nature,* **207**:34–36.

Van Rossem, A. J. 1927. Eye shine in birds, with notes on the feeding habits of some goatsuckers. *Condor,* **29**:25–28.

Van Someren, V. G. L. 1956. Days with birds: studies of habits of some East African birds. *Fieldiana: Zool.* **38** (Chicago Natural History Museum).

Verbeek, N. A. M. 1962. On dew bathing and drought in passerines. *Auk,* **79**:719.

Vrabec, F. 1961. The topography of encapsulated terminal sensory corpuscles of the anterior chamber angle of the goose eye. In *The Structure of the Eye,* Academic Press, New York, pp. 325–333.

Walls, G. L. 1942. The vertebrate eye and its adaptive radiation. *Cranbrook Inst. Sci. Bull.* No. 19.

Watts, C. R. and A. W. Stokes. 1971. The social order of turkeys. *Sci. Am.,* **224**:112–118.

Weir, J. J. 1891. The colour of the iris in albino birds. *Zoologist,* **15**:358.

Weymouth, R. D., R. C. Lasiewski, and A. J. Berger. 1964. The tongue apparatus in hummingbirds. *Acta Anat.,* **58**:252–270.

Whitaker, L. M. 1957. A resume of anting, with particular reference to a captive Orchard Oriole. *Wilson Bull.,* **69**:195–262.

Williamson, K. 1956. Distraction displays. In *The Ornithologists' Guide,* Philosophical Library, New York, pp. 125–127.

Winter, P. 1963. Vergleichende Qualitative and Quantitative Untersuchungen an der Hörbahn von Vögeln. *Z. Morphol. Okol. Tiere,* **52**:365–400.

CHAPTER FIVE

Voice and Sound Production

Throughout the animal kingdom birds are outstanding for the versatility of their songs and for the variety of nonvocal or mechanical sounds that they produce.

MECHANICAL SOUNDS

The flightless and secretive *Apteryx* of New Zealand stamps its feet when annoyed. Boat-billed Herons (*Cochlearius*), Whale-billed Storks (*Balaeniceps*), and other storks rattle or clap their mandibles (Fig. 1). The Roadrunner (*Geococcyx*) also rattles or clatters its mandibles. Woodpeckers produce a drumming sound by pounding trees or metal roofs with their bills. "Drumming" is also accomplished with the wings, as is the case with the Ruffed Grouse (*Bonasa umbellus*). The wings of the Common Goldeneye (*Bucephala clangula*) make such a noticeable noise in flight that the bird has been called the "Whistler." Many species of doves announce their presence by the whistling sound made by the wings as the birds fly by. The members of one subfamily (Drepanidinae) of the Hawaiian honeycreepers have a characteristically noisy flight, which often draws attention to them before

Figure 1.   A nesting colony of Painted Storks (*Ibis leucocephalus*); Vittalgabh, India, November 22, 1964.

they are seen in the dense rain forests. The wings of the Woodcock are responsible for a melodious sound during springtime courtship flights. During the breeding season the male Nighthawk (*Chordeiles*) frequently interrupts its lazy and somewhat erratic flight high above ground to dive earthward. When close to the ground, the bird pulls out of its dive, and the air rushing through its wing feathers produces a characteristic booming noise.

Sutton (1945:241; *in litt.*) referred to the Wedge-tailed Sabre-wing or Singing Hummingbird (*Campylopterus curvipennis*) as a "species notable for its curiously curved, much stiffened wing feathers," which produce "an almost frightening roar" as the bird flies about the undergrowth in which it "sings." The tips of the two outer primaries of the male (but not the female) Broad-tailed Hummingbird (*Selasphorus platycercus*) are highly modified and produce a distinctive trilling or rattling sound in flight that is different from the buzzing sound characteristic of hummingbirds. The function of serrate outer primary feathers in the Rough-winged Swallow (*Stelgidopteryx*), the African swallow *Psalidoprocne*, and the Crested Sharpbill (*Oxyruncus cristatus*) apparently has not been determined.

The tail feathers, also, may produce mechanical sounds during flight. The Common Snipe (*Capella gallinago*) engages in an aerial display involving a "bleating" flight, produced by the "relatively slow quivering of the half open wings superimposed on the much more rapid vibration of the tail-feathers" (Witherby *et al.*, 1949:199).

Several unrelated birds make sounds by secondarily filling the esophagus with air from the lungs. During the breeding season the esophagus of the male Sage Grouse (*Centrocercus*) is capable of distention to a size 25 times that of the mature nonbreeding male. The hypertrophy of the esophagus is accomplished by an enlargement of fleshy pads at the sides of the glottis and of the skin and bare areas of the neck. "By sudden contraction of the muscles of the neck, the air trapped in the distended portion of the esophagus causes the membranes of the bare areas to vibrate," and results in a "soft, hollow 'plopping' sound of considerable intensity" (Honess and Allred, 1942). Chapin (1922) described anatomical specializations in the male American Bittern (*Botaurus lentiginosus*) and concluded that the "pumping" of the bittern is accomplished by distending the esophagus with air. He further pointed out that certain other birds produce sounds in a similar manner: e.g., a rail (*Sarothrura elegans*), a bustard (*Neotis cafra*), the Pectoral Sandpiper (*Calidris melanotus*), many pigeons, two African sylviids (*Camaroptera superciliaris* and *Bathmocercus rufus*), and the Mocking-thrush (*Donacobius atricapillus*).

The male Ruddy Duck has a specialized tracheal air pouch that opens as a "depression in the dorsal wall of the trachea immediately behind the larynx and lies between the trachea below and the esophagus above" (Wetmore,

1917). A tracheal bulla is absent in this duck, and the air pouch modifies sounds made during courtship displays.

VOCAL SOUNDS

The syrinx is the primary sound-producing organ in birds. Other anatomical structures may modify the voice, but usually are not responsible for it. That the mouth plays a minor role is suggested by the fact that some birds (e.g., Prothonotary and Kirtland's warblers) habitually sing complete songs while carrying a billfull of food, and birds often sing with the bill closed (Fig. 2). Miskimen (1951) concluded that sounds are produced by vibrations of the tympaniform (drumlike) membranes during expiration, and that no sound is produced during inspiration. Calder (1970) demonstrated by simultaneous recordings that there is "a 1 : 1 correspondence between song notes and respira-

Figure 2. A Prothonotary Warbler at its nesting cavity; Reelfoot Lake, Tennessee, July 8, 1940. (Courtesy of Lawrence H. Walkinshaw.)

tory movements [in the canary]. Even trilled notes as rapid as 25/sec occur in individual, shallow, 'mini-breaths.'" He noted that "all the chirps and notes of longer duration in the canaries, as well as the premature crowing of testosterone-treated chicks (*Gallus domesticus*), occur in expiration."

The function of the semilunar membrane in the syrinx remains obscure. After removing this membrane in a Starling, Miskimen could detect no modification of sounds when air was forced through the syrinx. Gaunt *et al.* (1973) found, during distress calls of the Starling, that "air sac pressures soar, often to more than 20 cm H_2O (×40 normal respiratory amplitudes), while tracheal pressures are the same or lower than during respiration. A distress call lasts twice as long as a normal exhalation, but flow is reduced and tidal volume increases only slightly. Seemingly, the syringeal valve can regulate flow across a large, sustained pressure differential. The inability of birds to vocalize when the interclavicular air sac is ruptured is more easily explained by an alteration of flow to bypass a closed, pressure independent valve than a necessity for a high pressure surrounding the syrinx during vocalization" (Fig. 3).

Elongation and coiling of the trachea result in a trombone effect, adding resonance to the voice. In the Chachalaca (*Ortalis vetula*), in which the tracheal loop is absent in females and young males, "the old males sing a full octave lower than the females and young males" (Sutton, 1951:127). Dilatations of the trachea (as in some male ducks and in the cotingas *Cephalopterus* and *Perissocephalus*) also modify the sounds made by the syrinx.

Adult males of some species of boobies have higher-pitched voices than females or immature birds. In writing of the Blue-footed Booby (*Sula nebouxii*), Murphy (1936:834) said that the male "is equipped with a double sounding-box at the junction of the trachea and bronchial tubes. The throats of the females, and the young of both sexes, show no such structure, and this observation led to the discovery that the young males, as well as their sisters, have the same resonant, ear-splitting call as the adult females. The change in the voice of the males comes with maturity, when the delicate vibrating membrane of the vocal organ grows out to form a hard egg-shaped chamber, thus converting a trumpet into a whistle!" Similarly, the voice of the male Masked or Blue-faced Booby (*Sula dactylatra personata*) is a shrill whistle, whereas that of females and immature birds consists of hoarse, gooselike honks (Fig. 4). Miller (1947) showed that the pitch of owl calls is related to the cross section of the bronchi, and that, in general, "the larger the species the larger is the syringeal segment of the air passages and the longer the vibratile membranes in the walls of the syrinx which produce the tone. The longer membranes of the larger species of course vibrate more slowly and yield lower-pitched tones." The call of the Flammulated Owl (*Otus flammeolus*), however, "is strikingly low in pitch in view of the small size of the bird," and Miller found that this owl has both an enlarged syrinx and thickened vibratory membranes.

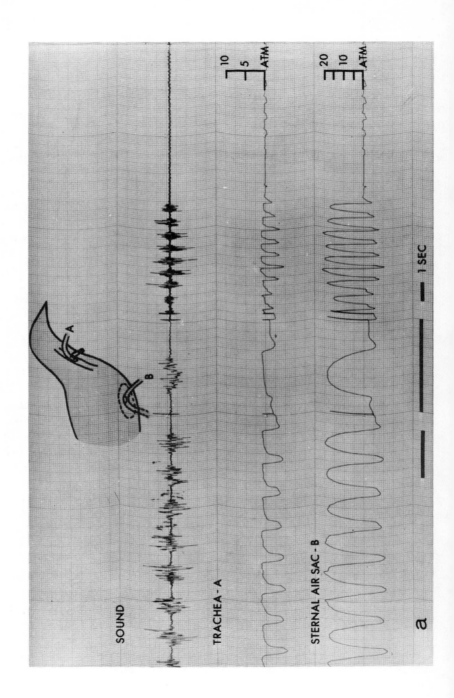

242

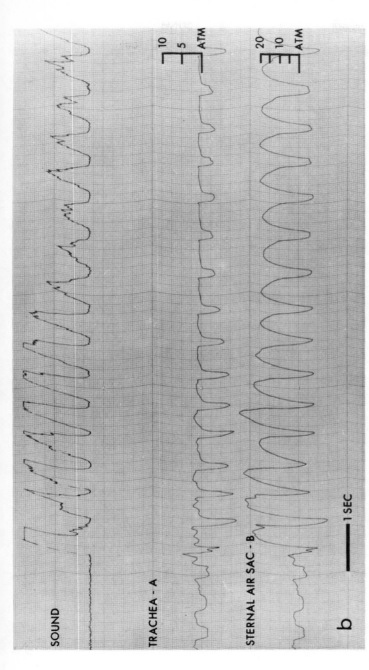

Figure 3. Simultaneous recordings of sound and tracheal and internal pressures. (*a*) Sound recorded in "direct" mode; three different time bases. Direct mode more precisely defines the onset and cessation of sound, but does *not* reproduce the wave form of distress calls. (*b*) Sound recorded in integrate mode. The very high level of sound associated with the first cry is probably due to a wing striking the microphone. The pressure record shows transition from excited but relatively regular breathing to vocalization. Note close correspondence of tracheal and air sac pressures during breathing. (*c*) A series of apparently related fluctuations in all three traces. The relationship is not simple and requires additional analysis. (Courtesy of Abbot S. Gaunt and the editor of *Journal of Experimental Zoology*.)

243

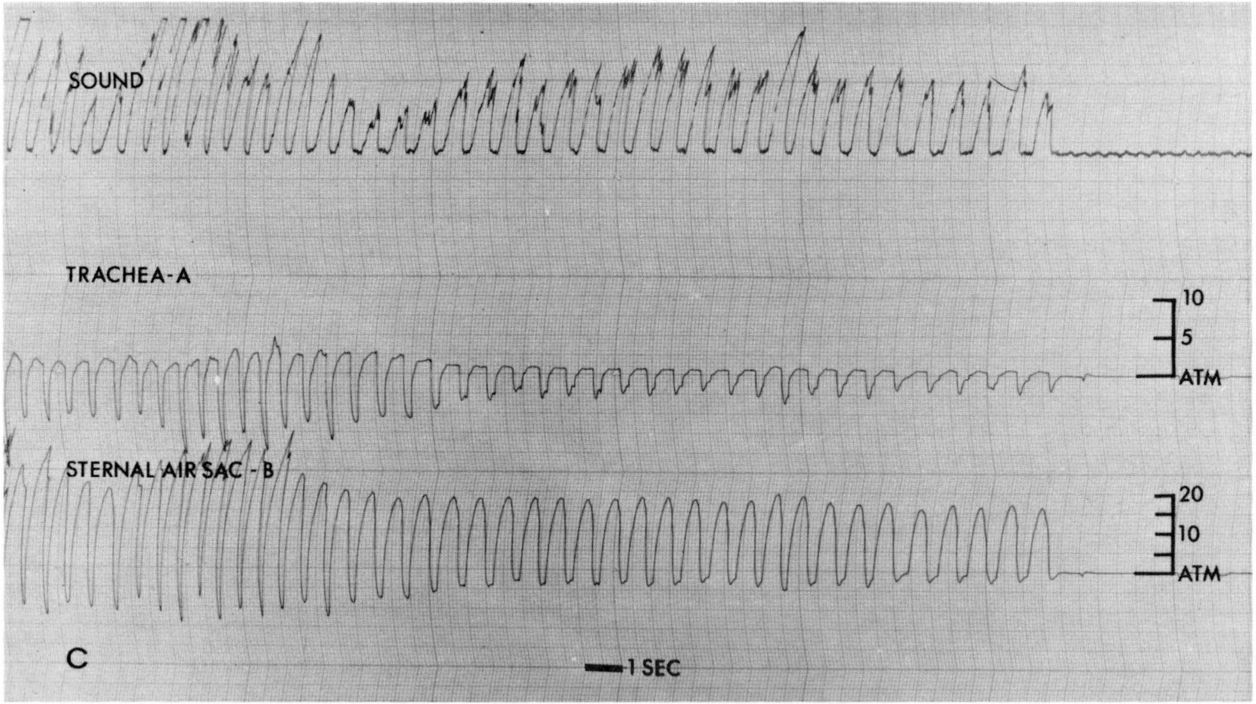

Figure 3 (*Continued*)

244

Figure 4. A Masked Booby incubating its egg; note the circular, cleared area around the bird which is the result primarily of defecation by the incubating bird; Christmas Island (Pacific Ocean), November 5, 1971.

An apparently unique diverticulum from the interclavicular air sac has been described for the Brown Jay (*Psilorhinus*) by Sutton and Gilbert (1942). This small, subcutaneous "furcular pouch" lies in the interclavicular space and is thought to be responsible for the *hiccup* that is part of the Brown Jay's vocabulary.

CLASSIFICATION OF VOCAL SOUNDS

The vocal sounds made by birds are classified as songs or call notes, depending on length and function. "Song is primarily under the control of the sex hormones and is in general concerned with the reproductive cycle, it is of great functional importance in the establishment and defence of territory, often serving as a substitute for physical combat. . . . Call-notes, on the other hand, are concerned with the co-ordination of the behaviour of other members of the species (the young, and the flock and family companions), mostly in situations which are not primarily sexual but rather concerned with maintenance activities—feeding, flocking, migration and responses to predators. In those birds which are not song birds and have no song in the ordinary meaning of the term, obviously the signals used in reproductive behaviour must—if vocal at all—be of the nature of call-notes: that is, acoustically simple, and not the complex affairs we normally call song" (Thorpe, 1961:15–16).

Call Notes

Call notes serve many purposes; Thorpe listed 10 general types: call notes of pleasure, distress, territorial defense, feeding, at the nest, flock, aggression, general alarm, and "specialised alarm calls, such as the ground predator call and the flying predator call." Chamberlain and Cornwell (1971) described 23 vocalizations of the Common Crow. The Song Sparrow has 21 "chief vocalizations." Female Song Sparrows announce their sex "by various notes, used only by females and typically in connection with a mate and nest" (Nice, 1943:172, 274). Other birds have special notes emitted primarily during the nest-building or incubation periods. Special notes often accompany copulation.

Gallinaceous species use a special gathering call when the members of a covey become separated. Both adults and young of such precocial species utilize similar calls to keep the brood together or to reunite it (Fig. 5). Fledglings of altricial species, as well, give location notes. Nestlings emit distinctive food calls when hungry, and other calls when being fed; they utter soft notes of well-being when they are warm and adequately fed; they may give distress calls when cold or when handled (after a certain age).

A species may utter several kinds of alarm notes, depending in part on the cause for alarm and the time of year. Colonial-nesting species and nonbreeding flocks have distinct warning notes. In describing the warning cry of Wagler's

Figure 5. A Rock Bush Quail (*Perdicula argoondah*) at its nest. (Courtesy of Shivrajkumar of Jasdan, India.)

Oropendola, Chapman remarked that at times not only did the colony "act as one bird and plunge from their nests or perches into the forest" but also birds of other species sometimes responded to the warning call. The alarm notes of titmice often serve to alert all of the species found in winter feeding flocks. Various other notes are associated with fear or pain. Some nocturnal migrants use special call notes which are rarely heard at other times.

Song

As more has been learned about bird vocalizations and behavior, most definitions of bird song have proved to be either too restrictive or too broad to be of more than historical interest (see Marler, 1969).

Tinbergen (1939:73–74), recognizing the "enormous diversity of bird calls that are commonly labeled as song," commented that the duration of the call and aesthetic appreciation are poor criteria for classifying bird sounds as call notes or as songs, and he emphasized the importance of function for understanding the vocal efforts of birds.

Thorpe (1961:15) wrote that a bird song "is a series of notes, generally of more than one type, uttered in succession and so related as to form a recognisable sequence or pattern in time. Thus the song as a whole displays the features of accent, increased duration, increased rhythmical complexity, . . . which are not discernible to anything like the same extent in call-notes." This *full song* also is called *true song* and *primary song*.

The following classification is of value insofar as it enables one to develop a concept of bird song. The discussion that follows each definition reveals the frequent overlap in functions of a particular song type among different birds.

Primary Song.   **Advertising (or territorial) song** is a "loud sound, given by a bird of one of the two sexes especially at the beginning of the reproductive period, that serves to attract a sex partner, to warn off a bird of the same sex, or both" (Tinbergen, 1939). This type of song is inseparably related to territoriality in birds, and it is one of the biological isolating mechanisms that "guarantees the mating of individuals belonging to the same species" (Nice, 1943:149). In the Song Sparrow and the Grasshopper Sparrow advertising song serves equally for attracting a mate and for warning other males (Fig. 6). In the Snow Bunting, on the other hand, attraction of a mate is the primary function. Mechanical sounds may fulfill the two functions of advertising song.

Signal song may be defined as any song that serves to coordinate the activities of birds, particularly of a mated pair. This definition arbitrarily excludes the song function of warning or threat to other males (or females) that is inherent in the concept of advertising song. Nice (1937:129) found that under normal conditions a female Song Sparrow "may come off the nest two-thirds of

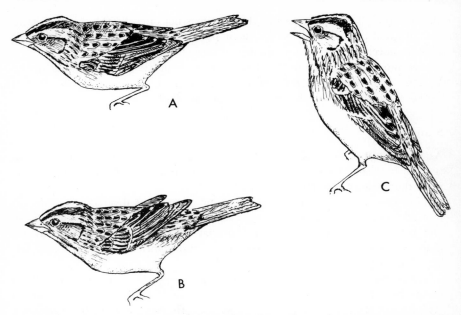

Figure 6. Attitudes of a male Grasshopper Sparrow during the grasshopper-song sequence. (*A*) Crouched position. (*B*) Wing-fluttering. (*C*) Delivering the song. (Courtesy of Robert L. Smith and the editor of *Wilson Bulletin.*)

the time in answer to 'signal songs' and one-third independently." Some signal songs may be an integral part of courtship behavior; still others may act as a stimulus to gaping by nestlings. Such a song may be the same as the territorial song, in which case only the function differs.

Emotional song encompasses a variety of songs that cannot be associated directly with securing a mate and defense of territory. Excluding early anthropomorphic interpretations of bird song, some ornithologists have suggested that song is a normal outlet of excess energy in many birds even when "there is no special function to be served, nor on the other hand, any inhibitory influence" (Nice, 1943:148). Twilight, postbreeding, and some winter songs may be included under this heading. Thorpe (1961:6–7) wrote that "it would, however, be dishonest to suggest that the biological theories at present available offer a complete explanation for all bird vocalisations. There are many instances of songs which seem to transcend biological requirements and suggest that the bird is actively seeking new auditory and vocal experience—'playing with sounds,' so to speak—and that this may represent the beginnings of a true artistic type. Thus the twilight song of the Wood Pewee appears to have no territorial function and is said to be independent of the breeding-cycle, and the day-time song also continues long after the end of the breeding-season."

Secondary Song. Secondary song or subsong is so soft or faint that it may be inaudible at a few yards; it has no territorial significance, it may be sung by either sex, and it may be more varied than the primary song. Lister (1953) proposed four major types.

1. **Whisper song** is the quiet, inward rendering of the primary song, with or without slight variations or additions and with "an audibility limit of no more than about 20 yards."

2. **Subsong** is the "very quiet, inward rendering of song which is intrinsically different from the primary song." (See Thorpe, 1961:64.)

3. **Rehearsed song** is the "random utterance of song-notes by young and sometimes old birds before they have attained perfection in the primary song."

4. **Female song,** according to Lister, should be considered a type of secondary song, but the latter certainly would be an inappropriate designation for female song in many species.

MALE AND FEMALE SONG

Although song is primarily a characteristic of passerine birds (the "songbirds"), not all songbirds sing and some nonpasserine birds do (e.g., Bobwhite, Mourning Dove, Upland Sandpiper). Song is typically the function of the male, but there are exceptions (see Nice, 1943:129). The songs of the female Mockingbird, Gray-cheeked Thrush (*Catharus minimus bicknelli*), Elepaio (*Chasiempis sandwichensis*), Cardinal, and Rose-breasted and Black-headed grosbeaks are nearly as elaborate as those of the male, and the females of some other species (e.g., Cactus Wren, Wrentit, European Robin, Wood Thrush, Bullock's Oriole, Summer Tanager, Violet-eared Waxbill, Indigo Bunting, Rufous-sided Towhee, and White-crowned, Grasshopper, and Song sparrows) sing occasionally or under special conditions (Fig. 7). Some females (e.g., Elepaio, Wood Thrush, Black-headed Grosbeak, and Orange-billed Sparrow, *Arremon aurantiirostris*) sing while incubating.

Both sexes may sing elaborate songs in courtship and as an aid in maintaining the pair-bond; this is true of some scrub-wrens (Sylviidae), several spinetails (Furnariidae), the European Dipper (*Cinclus cinclus*), the Gray-cheeked Thrush, and the Cardinal. In Australia, female song is very common "among resident territory-holders which appear to follow the pattern of the life cycle of the Wren-tits very closely" (Robinson, 1949). In a few species, in which the male incubates, the female has an elaborate song and the male may not sing at all: e.g., the Variegated Tinamou (*Crypturus variegatus*), hemipode-quails, Painted-snipe, Northern Phalarope (*Lobipes lobatus*).

Responsive or antiphonal singing and duetting have been reported in many species, in which both sexes sing either the same or different songs and either

Figure 7.   A Bullock's Oriole at the nest. (Courtesy of Samuel A. Grimes.)

simultaneously or alternately (Thorpe, 1972; Fig. 8). Unfortunately, many authors have used the terms "duetting" and "antiphonal" singing interchangeably, even though it seems more logical to refer to the simultaneous singing of two birds as duetting and the alternate singing as antiphonal or responsive singing. Haverschmidt (1947) wrote that one "can only speak of a duet when, in a mated pair of birds, both sexes utter the same sounds either simultaneously" or alternately; "community singing" cannot be called duetting. He considered true duetting to take place in *Certhiaxis, Donacobius,* and *Thamnophilus*, as well as in the White Stork (*Ciconia ciconia*), because "the well known bill-clattering is always uttered by the returning bird when landing on its nest."

Diamond and Terborgh (1968) described three types of duetting, of which their first is antiphonal singing:

1. "The leader (generally the male) sings one phrase, and at its conclusion the second bird (generally the female) sings another phrase. Often the follower begins so precisely upon cessation of the leader's phrase that the result may easily pass for a conventional single song, unless the observer can see the singers or is close enough to hear that different notes come from different places.

Figure 8. A Streaked Fantail Warbler (*Cisticola juncidis malaya*) at its nest; Singapore. This is one of several genera of Old-world Warblers that exhibit antiphonal singing. Photographed by the late Loke Wan Tho. (Courtesy of Salim Ali.)

For example, the Marbled Wood-Quail . . . calls a rapidly repeated 'corcoro-vado,' in which 'corcoro' is always sung by one member of a pair and 'vado' by the other.'' The audiospectrographic analysis of bird song has led to the recognition of a second type of antiphonal singing, in which there is a very brief interval between the calling of one bird and the response of the mate. Stokes and Williams (1968) described the closely synchronized calling of the Bob-white, California Quail, and Gambel's Quail. Thorpe (1963) reported a mean reaction time of only 144 msec between the onset of song by one bird and that of the mate of a pair of Black-headed Gonoleks (*Laniarius barbarus*). Hooker and Hooker (1969) described a similar elaborate pattern of antiphonal singing in the Slate-colored Bou-bou, *Laniarius aethiopicus* (Figs. 9, 10).

2.   "Members of a pair sing different phrases simultaneously. The synchro-nization between the two phrases may nevertheless be very exact. . . . In New Guinea the duets of *Megapodius freycinet, Campochaera sloetii,* and *Coracina montana* are of this type."

3.   "The male and female sing virtually identical phrases in unison." Examples are the cuckoo *Centropus toulou* and the Blue-throated Motmot (*Aspatha gularis*).

Figure 9. (*a*) *Laniarius aethiopicus*; sexes similar. A pair singing the "snarl/*bou-bou*" duet. (*b*) *L. erythrogaster*; sexes similar. Alert; note similarity of contrast pattern with that of *L. ae-thiopicus*; under parts scarlet except for buff under tail coverts; iris bright yellow. (*c*) *L. funebris*; sexes similar. Plumage—varying tones of glossy black and dull slate-gray with no marked contrast. (Courtesy of T. and Barbara I. Hooker.)

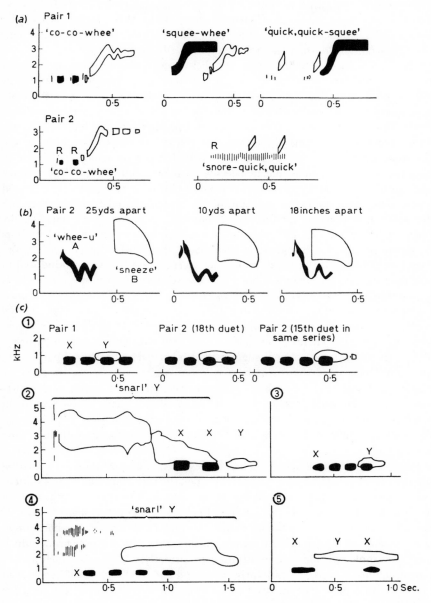

Figure 10. A selection of diagrammatic sound spectrograms of three species of the genus *Laniarius*: (*a*) *L. funebris*; (*b*) *L. erythrogaster*; (*c*) (1-5) *L. aethiopicus*. X and Y in (*c*) represent presumed female and male, respectively. (Courtesy of T. and Barbara I. Hooker.)

Dinsmore (1969) reported duetting by males of the Greater Bird-of-Paradise (*Paradisaea apoda*) on Little Tobago Island in the West Indies, where the species was introduced in 1909. Wickler (1972) reported the "perfect duetting" of a Slate-colored Bou-bou and a drongo (*Dicrurus adsimilis*).

The significance of female song remains obscure despite the many known examples. Nice and Ter Pelkwyk suggested that "song was originally present in both sexes, and that there were two lines of evolution, one where the male developed the more elaborate song, the other where the female did so" (Nice, 1943:129). But the causes, physiological and psychological, of female song are little understood. It seems likely that part of the answer is to be found in the fact that female birds are heterogametic and that certain endocrine glands of each sex secrete both male and female sex hormones, but data are fragmentary and confined to a small group of experimental birds. It has been demonstrated experimentally in ducks, by injection of sex hormones and by castration, that the ovary is responsible for the sexual dimorphism in the development of the osseus bulla and the genital tubercle; that is, the female sex hormone inhibits the development of the male-type syrinx and the penis (Witschi, 1961; van Tienhoven, 1961). By contrast, females of several species have been induced to sing after injections of male sex hormone: e.g., canary, *Turdus merula, Turdus migratorius, Fringilla coelebs, Junco phaeonotus, J. hyemalis oreganus, Zonotrichia leucophrys nuttalli,* and *Z. l. gambelii* (Kern and King, 1972). Such data, however, do not explain female song in wild birds. In discussing passerine birds in general, Ames (1971:135) wrote that "the uniformity of syringeal structure among oscines suggests that inherited and acquired patterns in the central nervous system usually play a far greater role in determining the characteristics of song [of each species] than do patterns of syringeal musculature." Although little is known about sexual dimorphism of the syrinx in passerine birds, it seems more likely that innate differences in the CNS between the sexes and different hormone titers are the important factors in female song in wild birds.

Possible functions of female song, however, can be postulated. In writing specifically about antiphonal singing, Thorpe (1961:50) said that "the explanation suggests itself that birds living in the dark and tangled undergrowth of tropical forests, conditions which are likely to hinder the development of mutual visual displays, are the more likely to replace these by the elaboration of vocal displays. It is perhaps also significant that in Europe the only group which seems to employ antiphonal vocalisation persistently is the owls which are, of course, mainly nocturnal." Similarly, Diamond and Terborgh (1968) concluded that "duetting may arise in response to two different situations: the need for birds living in dense vegetation where visual contact is difficult to evolve intricate vocal displays rather than visual displays, and the importance for birds whose nesting season depends upon unpredictable climatic changes, such as the start of the rainy season, to become and remain paired long in advance of breeding."

ECOLOGICAL CORRELATIONS

There is in general an ecological correlation with the development of song. Many ocean birds tend to be silent most of the year, although some are noisy at breeding colonies, where petrels, shearwaters, and terns are noted for their moaning or shrieking nocturnal calls. Forest-inhabiting birds, where communication by eye is limited, are pre-eminent as songsters. Several authors have suggested that "birds of bright plumage have poor songs," and Skutch (1954:177) wrote that "it is perhaps not wholly without significance that the two tanagers that dwell in the undergrowth of the heavy forests of the valley of El General are less brilliant in plumage than most of the local representatives of their family and are here its most notable songsters." Nevertheless, many exceptions can be cited to the generalization linking bright plumage to poor songs.

A flight song is featured by some birds that live in the open, especially arctic species that live beyond the tree limit; in many it serves as a territorial song. In the Song Sparrow, however, the flight song "has no direct relationship with any other bird" (Nice, 1943:118). The Skylark (*Alauda arvensis*), Horned Lark (*Eremophila alpestris*), and Sprague's Pipit (*Anthus spragueii*) often sing at considerable heights (at least 800 feet). Bobolinks (*Dolichonyx oryzivorus*), Cassin's Sparrows (*Aimophila cassinii*), and Lapland Longspurs (*Calcarius lapponicus*) habitually sing during flight, and many other birds occasionally do so. The Streaked Saltator (*Saltator albicollis*) has a flight song which is most often undertaken "in the dusk at the day's end, well after the saltator and most other birds have ceased their ordinary or sedentary singing" (Skutch, 1954:82). The flight song of the Willow Flycatcher (*Empidonax traillii*) is often given as a part of the evening song, less commonly as part of the morning song choruses. Bachman's Sparrow (*Aimophila aestivalis bachmanii*) has a special flight song that it gives at dusk. Among birds inhabiting wooded areas, the Ovenbird (*Seiurus aurocapillus*) and Blue-winged and Golden-winged warblers (*Vermivora pinus* and *V. chrysoptera*) have flight songs, but these seem to be exceptional among Temperate Zone species. The Hawaiian Thrush (*Phaeornis obscurus*) and the Small Kauai Thrush (*P. palmeri*), which inhabit dense rain forests, have flight songs, and several species of Hawaiian honeycreepers have flight songs or sing in flight.

PHYLOGENY OF SONG

Several writers have speculated about the origin in vertebrates of voice in general and of bird song in particular. S. J. Holmes suggested that voice in vertebrates developed primarily as a sex call, "as is now its exclusive function in the Amphibia." Charles Darwin, Eliot Howard, and others, however, imagined vocal sounds to be produced involuntarily under stress of fear and excitement. Hence birds first gave alarm notes. Lister (1953) thought it possible that "sub-song may be the basic utterance from which primary song has evolved."

Nottebohm (1972b) discussed the origins of vocal learning and concluded that "maximal female stimulation and the advantages of vocal dialects have been primary causes favoring vocal learning in birds. In either case females would have been the agents of selection in that their choice of mate, based upon vocalizations, determined their breeding success and fitness of their progeny. A very suggestive though imperfect correlation also exists between intense speciation and vocal learning. Presumably the interaction of these selective pressures, and others such as individual recognition, pair-bond maintenance, and recognition of close-blood relatives, has affected the manifestation of vocal learning as it exists today. In such a context, vocal learning in birds evolved not so much to convey new information, which dialects and familial tradition do in a way, but to restate in a more distinct and florid manner, as a readily changeable phenotypic expression, an old message: the presence of an individual male of a given species in breeding condition."

ONTOGENY OF SONG

The steps by which an individual bird learns its species's song have been outlined by several writers (Figs. 11, 12). The first song is generally an indefinite warble, nonspecific in character, and is sometimes given by nestlings (e.g., *Cinclus cinclus, Turdus ericetorum, Sylvia atricapilla*). Nice (1943:133) listed five stages in the development of the Song Sparrow's song. She stated that "young birds of 16 species are reported as starting to sing from the age of 13 to 24 days, and 15 species from the age of 4 to 8 weeks."

A problem that has intrigued ornithologists since the days of Daines Barrington in the eighteenth century is the method by which a species acquires its song: Is song inherited, or is it learned by imitation? Sauer (1954) hand-raised seven Whitethroats (*Sylvia communis*) in isolation from the egg (egg isolation), and found that all of the 25 call notes and three song types characteristic of the species were entirely innate: "a bird lacking all acoustical experience utters all these sounds in exactly the same manner, in the same phases of its life cycle and in the same specific moods as birds in the field." Similarly, Messmer and Messmer (1956) reported that the juvenile song of the European Blackbird (*Turdus merula*) "is fully innate" but that the adult song is partly learned from other birds. Konishi and Nottebohm (1969:31), however, wrote that the conclusions reached by Sauer and the Messmers "have been accepted rather uncritically. The musical notations, oscillographic display, and spectral band

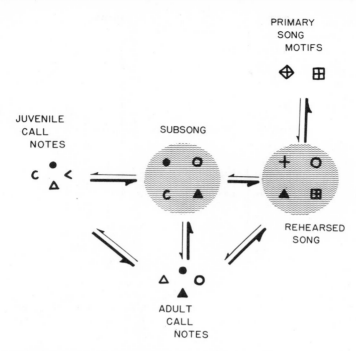

Figure 11. Diagram of sequential development of primary song in passerine birds. Double arrows suggest the reversible nature of the developmental process. (Courtesy of Wesley E. Lanyon and the American Institute of Biological Sciences.)

analysis used by these early workers were not adequate to reveal the subtler frequency or temporal characteristics of vocal signals. These are the very features which later turned out to be most susceptible to modification by isolation or deafening." At the same time, these authors pointed out that the great difficulty of hand-raising newly hatched passerine birds "forced most workers in this field to be content with isolating nestlings." They added: "While it may be generally true that the effects of isolation of nestlings approximate those of isolation of eggs, the early exposure of the young to the natural auditory environment might have some effects upon their vocal development. Judging from the audiospectrograms prepared by Thielcke-Poltz and Thielcke (1960) the songs of their nestling-isolates and those of the Messmers' egg-isolates are different. Since the song of the blackbird is very variable, it is difficult to say how many of the differences are due to individual variation and how many to the differences in their auditory experience." And, indeed, Konishi, Nottebohm, and others have shown conclusively that auditory feedback is of critical importance in song-learning.

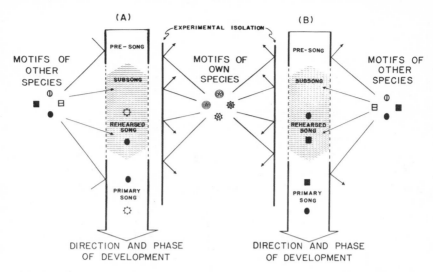

Figure 12. Development of primary song in some experimental birds, isolated from their own species but exposed to other species. (*A*) Some birds develop a mixed repertoire of original motifs and of motifs learned from other species. (*B*) Some birds develop a repertoire composed solely of motifs learned from other species. (Courtesy of Wesley E. Lanyon and the American Institute of Biological Sciences.)

Lanyon (1960) listed the different techniques used in studies of the ontogenetic development of song; all involve some degree of isolation of the experimental birds.

1.   The birds are deafened and therefore are isolated from all sound.

2.   The birds are isolated from all other birds but can hear their own sounds.

3.   The birds are isolated from all experienced (adult) birds but can hear the sounds made by siblings.

4.   The birds are isolated from experienced birds of their own species but can hear experienced individuals of other species.

All vocalizations are recorded on magnetic tape; they are analyzed by an oscillograph or sound spectrograph, which produces graphs that show amplitude and frequency modulations (see Borror, 1960; Greenewalt, 1968; Stein, 1968).

As of 1960, Lanyon wrote that "there are only four passerine species for which spectrographic evidence for the ontogeny of vocalizations has been published." These four were the Chaffinch, Eastern Meadowlark, Western Meadowlark, and canary. Data now are available on a number of other species: e.g., the Crested Lark (*Galerida c. cristata*), Hill Mynah (*Gracula re-*

ligiosa), European Blackbird, Red-backed Shrike (*Lanius collurio*), Blue-winged Warbler, Golden-winged Warbler, Oregon Junco, White-crowned Sparrow (see Marler *et al.*, 1962; Hooker, 1968; Bertram, 1970; Gill and Murray, 1972).

Lanyon (1960) wrote that the call notes of passerine birds "are as remarkably steryotyped and genetically fixed as the vocalizations of the non-passerines." However, "various workers have suggested that learning from experienced individuals may be essential for the refinement of certain calls with regard to exact pattern and frequency." For example, "a characteristic call of the male western meadowlark did not develop in hand-reared individuals until the latter were exposed to experienced males during their first spring." Call notes that are initially given at the onset of the bird's first breeding season may be caused by an increase in male sex hormone at that time.

The factors involved in the learning of song have proved to be much more complicated than originally believed, and they vary considerably among different species (Konishi and Nottebohm, 1969). Included are a "critical period" for song-learning, the establishment of memory of an auditory template, an auditory feedback, and, presumably, a proprioceptive feedback (Fig. 13). Young Song Sparrows, "deprived from the time of the egg stage of opportunity to hear adult Song Sparrow song, are nonetheless able to produce an approximately normal song. The isolated version is somewhat lacking in variety and complexity and there is less contrast in tempo between early and later parts of the song. The repertoires are smaller than in the wild, but nevertheless, number about ten basically normal songs" (Mulligan, 1966). The sensitive period for imitation in the Song Sparrow is between about 4 to 12 weeks of age, and improvisation "plays the most important role in the development of the repertoire." By contrast, the critical learning period in the White-crowned Sparrow occurs "before the actual incidence of full song. Motor learning in these cases cannot be invoked as the immediate cause for the observed end of the critical period. It is conceivable, however, that even in these species postponement of the motor learning of song would result in recurrence of a 'critical period' in successive years under normally recurring hormonal conditions" (Nottebohm, 1970). The critical period in the Chaffinch occurs during the first 10 months after hatching; "after song is established in its final stereotyped pattern it will not change in subsequent years, nor will new themes be added to the repertoire."

The general pattern for the development of song was described by Konishi and Nottebohm (1969:39–40): "Song development in the chaffinch is characterized by a series of progressive changes consisting of the early nestling stage, juvenile and adult subsong, 'plastic' song and final full song stages. The transition from one stage to the next is gradual: elements of nestling calls may be found in early subsongs, which gradually become more stereotyped, louder,

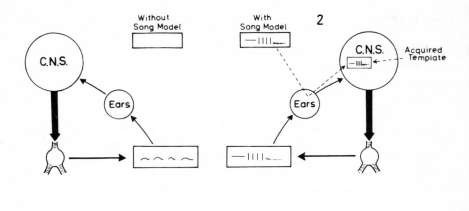

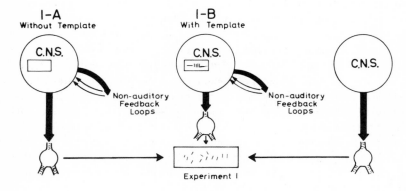

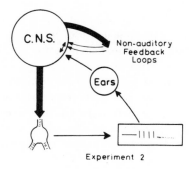

Figure 13. Analyses of the vocal control system. The thick arrows symbolize the motor output from the CNS to the vocal organ and other muscles involved in vocal control. 1, Intact birds develop atypical songs if they have not been exposed to a correct song model during the critical period. 2, The song model is reproduced by matching the vocal output to its template, which was established during the critical period. Experiment 1: Removal of auditory feedback before singing causes a drastic departure from the basic wild song pattern, regardless of whether a song template was formed (1-*B*) or not (1-*A*). The song template cannot be used without auditory feedback. The

and recurrent during transformation to 'plastic' songs. These are unstable but contain some of the basic features of chaffinch song. They finally develop into highly stereotyped and discrete full songs, which appear normally during the bird's first spring at about ten months of age" (Fig. 14). The defects in the songs given by deafened birds depend on the age at which the birds were deprived of hearing. Birds deafened when about 3 months old sing rudimentary and unstructured songs as adults, whereas experimental birds deafened in the spring, when in the plastic-song stage, give songs that show some resemblance

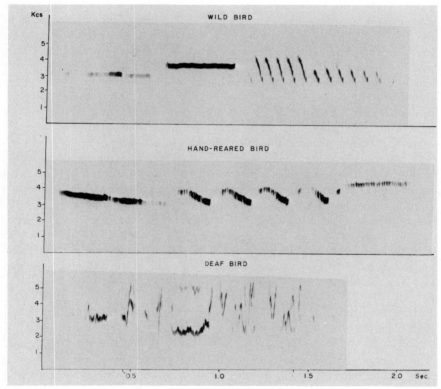

Figure 14. Songs of three White-crowned Sparrows raised in different auditory environments: a wild bird, a hand-reared bird, and a bird deafened before the onset of singing. (Courtesy of Masakazu Konishi, Fernando Nottebohm, and Cambridge University Press.)

resultant vocal motor output is patterned centrally and/or by nonauditory feedback. Experiment 2: A mechanism to evaluate nonauditory feedback information is established by matching this to auditory feedback information. Therefore nonauditory feedback alone can maintain the established vocal pattern in the absence of auditory feedback. (Courtesy of Masakazu Konishi and the editor of *Zeitschrift für Tierpsychologie*.)

to the wild-type song, and "adult male chaffinches deafened after at least one full season of singing experience retain their song pattern with great fidelity." The necessity for auditory feedback in the development of normal song also was demonstrated in Mulligan's (1966) study of the Song Sparrow. An isolated, hand-raised bird deafened at 12 weeks of age did not progress beyond the first stages of song development.

The primary nerve supply to the syrinx is carried by the hypoglossal nerve. Nottebohm (1970, 1972a) reported experiments in which this nerve was cut unilaterally in adult Chaffinches that had already established their stable song pattern. When the left nerve was cut, "either (i) the majority of song elements (for example, nine out of 12) disappears, so that their corresponding place in the song sequence is now vacant, while the remaining elements retain their structure and position within the song, or (ii) all components of song are highly modified so that their structure becomes that of short bursts of noise or very simple modulations." If, on the other hand, only the right nerve is cut, "either the structure of the song is not affected at all, or two or three of its simpler elements are lost." One Chaffinch in each experimental group also had been deafened as an adult, but these birds "performed no differently from the others." Nottebohm referred to this phenomenon as *lateralization of voice control* and remarked that "these observations indicate that song learning in the chaffinch is lateralized at the hypoglossus [*sic*; hypoglossal nerve]. This lateral commitment occurs at the time song is learned as a motor skill. Before this learning takes place, lateral control can be shifted from what is usually the dominant side to the other side. We do not know to what extent, if at all, this lateralization is represented at higher centers of control." Zigmond *et al.* (1973) found androgen-concentrating cells in the midbrain of the Chaffinch and suggested that "the nucleus intercollicularis is a site in the action of androgens on avian vocal behavior."

SONG DIALECTS

The songs of certain species vary in different parts of their range; the populations having different songs are called *song races*, and the songs, *song dialects*. Van Tyne noted that the calls of White-breasted Nuthatches in Michigan differed from those in the Chisos Mountains of Texas. Benson (1948) studied the songs of over 200 African species and found geographical variation in 33 species.

Nottebohm (1969) distinguished between *geographic variation* in song ("differences in song over long distances and between populations which normally do not mix") and *song dialects* ("differences between neighboring populations, or between populations of potentially interbreeding individuals"). He thought it best "to restrict the usage of the dialect concept to population differences within the same subspecies and to differences known to be phenotypic or suspected of being so." Nottebohm reported five song-dialect areas of the Chingolo or Rufous-collared or Andean Sparrow (*Zonotrichia capensis*) in Argentina (Figs. 15, 16). These song dialects were found in populations in areas where there were marked life-zone changes over short distances, resulting in some degree of isolation, whereas over vast areas of pampas country the dialect of the Chingolo remained unchanged. Nottebohm and Selander (1972) suggested, therefore, that "song dialects in this bird may have evolved to reduce gene flow between neighboring populations experiencing different selective pressures. Females born in a given dialect area would breed in that dialect area and choose as mates males with a song that matched that dialect; males, in turn, would learn to sing the dialect of their birth area."

One of the best examples of song dialects in the western United States is found among the races of the White-crowned Sparrow (*Zonotrichia leucophrys*), in which not only do the songs of the subspecies differ but also in the race *nuttalli* the songs differ from one local population to another (Milligan and Verner, 1971). Lanyon (1957) and Lanyon and Fish (1958) described geographical variation in the songs of the Western Meadowlark. Song dialects also have been reported for the European Blackbird, Snow Bunting, and Chaffinch. Geographical variation in song probably is more common than has been

Figure 15. The Rufous-collared Sparrow. (Courtesy of Philip Boyer.)

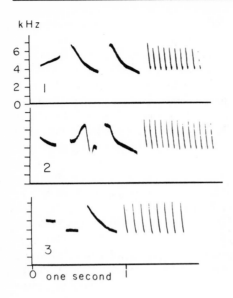

Figure 16. Songs of three different Rufous-collared Sparrows recorded, respectively, at Pinamar (1), La Brava (2), and La Azucena (3), Argentina, on October 26, November 2, and November 8, 1969. (Courtesy of Fernando Nottebohm, Robert K. Selander, and the editor of *Condor*.)

reported: some examples of species in which it occurs are the Nightingale-Wren (*Microcerculus philomela*), Carolina Wren (*Thryothorus ludovicianus*), Rufous-sided Towhee, (*Pipilo erythropthalmus*), Fox Sparrow (*Passerella iliaca*), and Song Sparrow.

Dialects in the call notes of Herring Gulls and several species of crows were described by Busnel *et al.* (1957) and by Frings *et al.* (1958). Four calls (food-finding, alarm, *mew*, and *trumpeting*) of Herring Gulls were recorded in the United States; when the recordings were broadcast in colonies of Herring Gulls and Black-headed Gulls (*Larus ridibundus*) in France, however, "the calls were totally lacking in observable effects on the French gulls." Similarly, four calls (assembly, alarm, *cawing*, and begging cries of young nestlings) of the Crow *Corvus brachyrhynchos* were recorded on tape. When the assembly call was played in the American habitat of the crow, "groups of crows numbering 2–30 came to the sound sources within 1–5 minutes. The alarm call proved to be repellant, even at low intensities. Tests with this could be made only on crows already attracted or where they could be seen." When, however, these calls were broadcast to flocks of three species of French corvids (*Corvus monedula, C. frugilegus, C. corone*), three of the calls "proved ineffective in eliciting any observable reactions," and the French birds reacted to the American assembly call essentially as they did to the distress call of the Jackdaw (*C. monedula*). Moreover, when the distress calls of the three French corvids were played to American crows, there were no observable reactions by the latter birds in Maine during the summer or Pennsylvania during the winter. However, when the distress cry of the Jackdaw was played in Pennsylvania during June, crows assembled. This difference in reaction of crow populations was explained as follows. The breeding crows of Maine winter in Pennsylvania and at no time associate with other species of crows. Therefore they respond only to their own call notes. The breeding crows of Pennsylvania, however, winter in the southern states where the Fish Crow (*Corvus ossifragus*) lives (Fig. 17). By hearing the calls of the Fish Crow, the Common Crows apparently learned to respond to the general features of crow calls. Consequently, they also responded to one of the calls of the Jackdaw. Similarly, Busnel found that the three species of French corvids, which feed and roost together, respond to each other's call notes.

INDIVIDUAL VARIATION IN SONG

Variation in song is the rule both among the members of a species and within individuals of the species (Fig. 18). Nevertheless, each species has a distinctive

Figure 17.   Fledgling Fish Crows calling for food. (Courtesy of Samuel A. Grimes.)

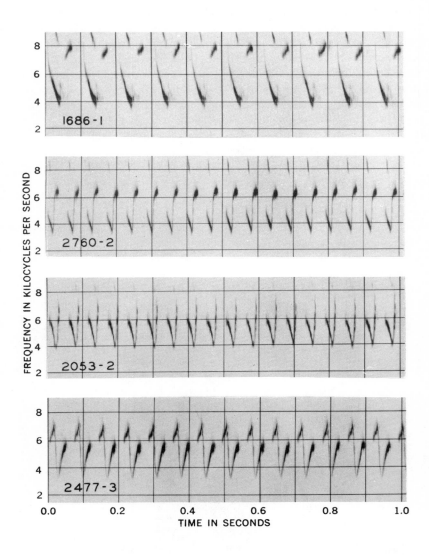

Figure 18. Audiospectrographs of songs of four male Chipping Sparrows. (Courtesy of Donald J. Borror.) 1686-1: Franklin County Ohio, April 13, 1956. 2760-2: Oscoda County, Michigan, May 30, 1957. 2053-2: Lincoln County, Maine, June 20, 1956. 2477-3: Franklin County, Ohio, April 27, 1957.

basic song pattern whereby it can be identified despite this variation; only rarely is a song encountered that cannot be identified. Surprises do occur, however: Short (1966) wrote of a Field Sparrow that sang the song of a Chipping Sparrow, and Thomas (1943) told of a wren singing the songs of both the House Wren and the Bewick's Wren.

The audiospectrographic analysis of bird song has revealed the basic pattern for each species's song, as well as the great individual variation among individuals of a species. Thus Mulligan (1966) wrote that each male Song Sparrow "possesses an average of sixteen song types, most of them unique, although one or two may be shared with one or several individuals nearby. A large number of syllable types, seventy-five or more, can be distinguished in these songs. . . . Individuals with large repertoires of fifteen or more song types show a graded preference in their use. Approximately five songs have recurrence numbers of less than ten, and ten songs in a repertoire of twenty account for 75 percent of the total number of songs." Similarly, Thompson (1970) found that some Indigo Buntings (*Passerina cyanea*) had virtually identical song patterns but that most of the birds sang different songs. Borror (1961), reporting on the variation in songs of a number of North American species, noted that "individuals of some species have advertising songs of only one pattern, while those of other species may have two to many patterns; 58 patterns were found in the songs of one Lark Sparrow. . . . The variation within a species may vary from only a few patterns to a situation in which different individuals seldom if ever sing songs of the same pattern" (see also Lanyon and Gill, 1964; Borror, 1965; Wildenthal, 1965; Gill and Murray, 1972). In writing of the variation in the song of the White-throated Sparrow (*Zonotrichia albicollis*), Borror and Gunn (1965) reported that "the relative incidence of different patterns in a given area may vary from year to year, and some patterns once common in certain areas are now rare or absent there. The pattern on which the widely quoted paraphrasing (*old Sam Peabody Peabody Peabody*) is based on is now apparently quite rare."

Little is known about the inheritance of song within a species. In her study of the Song Sparrow, Nice (1943:139) found that there was "no case of a male having the song of his father or grandfather," and, furthermore, that two males from the same nest "had no song in common." (See Nottebohm, 1972b.)

VOCAL MIMICRY

A large number of birds exhibit varying degrees of vocal mimicry and imitate call notes or songs of other species (Fig. 19). The Starling (*Sturnus vulgaris*) is a minor mimic in North America, frequently adopting the "pewee" call of the Wood Pewee and the "bobwhite" call; Starlings in Chicago, however, are said

Figure 19. A Rufous-backed Shrike (*Lanius schach erythronotus*) at its nest; this species is an accomplished and versatile mimic of other birds' calls. (Courtesy of Shivrajkumar of Jasdan, India.)

not to use these calls. Blue Jays sometimes imitate crows or the *meow* of a cat. Two of the better-known master mimics are the Mockingbird and the European Marsh Warbler (*Acrocephalus palustris*).

Borror and Reese (1956) presented audiospectrographs of Mockingbird imitations of Carolina Wren songs, and Borror (1964) said that mimicking by the Mockingbird is more noticeable in northern than in southern birds. Berger (1966) reported that a fledgling Mockingbird collected in California but hand-raised in Michigan included parts of the songs and call notes of 15 species of southern Michigan birds in its adult songs; among the species imitated were the Bobwhite, Great-crested Flycatcher (*Myiarchus crinitus*), Blue Jay, Tufted Titmouse (*Parus bicolor*), Wood Thrush (*Hylocichla mustelina*), Baltimore Oriole (*Icterus galbula*), Scarlet Tanager (*Piranga olivacea*), and Goldfinch

(*Spinus tristis*) (Fig. 20). Wildenthal (1965) compared the songs of two Kansas Mockingbirds with those of two Florida birds, and noted that "each bird uses a characteristic set of syllable-patterns. A number of syllable-patterns were shared between the two Kansas birds; few were shared between the Kansas and Florida birds. Both individual variation in syllable-patterns and sharing of syllable-patterns may function in individual recognition." She also stated that "there is no evidence that the specific syllable-patterns used convey discrete items of information or that syllable-patterns learned from other species play any part in interspecific communication."

Over 50 mimetic Australian species have been classified as "master," "minor," and "casual" mimics (Marshall, 1950). The best mimics inhabit wooded country and are strongly territorial. Marshall suggested that "lack of visibility places a premium on sound and that it is biologically advantageous for individuals to make more and more sound in order that territorial rivals and members of the opposite sex will be constantly aware of their presence." The same author showed a correlation between vocal mimicry and seasonal changes in the gonads. In *Ptilonorhynchus violaceus*, for example, "vocal mimicry is commonly heard only during the territorial and display season while spermatogenesis and interstitial modification are going on. Gonadectomy inhibits vocal mimicry in *P. violaceus*, and mimicry, along with other song and display, is re-established by injections of testosterone propionate."

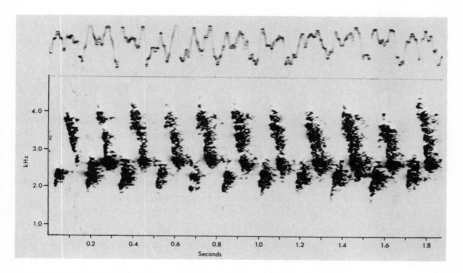

Figure 20. Sonagram and (above) amplitude display of a Mockingbird singing the song of a Cardinal (see Berger, 1966). Narrow band filter; calibration 1000 kHz.

Thorpe (1961:89) wrote that Marshall's argument "seems a rather doubtful one, however: it is not necessary for a bird to be imitative in order to make plenty of noise, and making a loud and persistent noise will not avail the species at all unless that noise is recognized as a specific signal by other birds. Only if—as is perhaps the case—the vocal quality is such that the species is recognisable *whatever* song-pattern it utters, would mimicry of this kind seem to be biologically permissible."

Parrots, mynahs, crows, and magpies apparently mimic only in captivity. There appears to be no neurological basis for Thorpe's (1955) assumption that "a considerable proportion of brain substance must be in some way ear-marked for voice control and so not available for other purposes" in parrots, and the bases for mimicking appear to be unknown. Thorpe (1961:113) and Greene-walt (1968:166) present opposing views on the acoustical processes involved in the production of human vowel sounds by talking birds.

SONG CYCLES

Few birds sing throughout the year, and those that do typically have periods of increased song activity during the nesting season. There is a daily cycle as well. Most species sing less during the middle of the day than they do in the early morning and late afternoon. A midday slackening of song is often correlated with high heat and wind and is especially noticeable in desert areas. Midday singing does take place on cloudy days, and light rain in itself has little effect on singing. Robinson (1949:303) noted that in Australia "whisper songs are generally sung by 'unemployed' birds from a shady position on a hot day or from a sheltered position on a very windy day."

A few species sing more or less continually throughout the day with little regard for weather. One of the best examples is the Red-eyed Vireo (*Vireo olivaceus*), a species that holds the world's record for number of songs given in one day: 22,197 songs (De Kiriline, 1954). Meanley (1968) reported that a male Swainson's Warbler (*Limnothlypis swainsonii*) in Virginia sang 1168 songs during one day; "it produced 280 songs the first hour and sang at a fairly constant rate from 5:00 to 8:00 A.M., 192, 194, 198 songs per hour." Harold Mayfield made observations from a blind at a Kirtland's Warbler (*Dendroica kirtlandii*) nest on June 21, 1956, the day before the first egg hatched (Fig. 21). The male sang 2212 songs between 4:57 A.M. (first song) and 7:56 P.M. (last song); sunset was about 8:25 P.M. E.S.T.

Some birds (e.g., Eastern Wood Pewee and Great Crested, Scissor-tailed, Willow, Acadian, and Least flycatchers) have special morning twilight (or dawn) and evening songs, which seem not to be related to territorial defense (Fig. 22). McCabe (1951) wrote that the flight songs of the Willow Flycatcher

Figure 21. A Kirtland's Warbler nest with a newly hatched nestling, two warbler eggs, and one Brown-headed Cowbird egg; Mack Lake, Oscoda County, Michigan, July 19, 1963.

are given "late in the day and continue until after dark. Song generally starts after sunset, and the time of starting and stopping varies in direct relation to the time of sunset. The onset of daily song is probably controlled by a light-intensity factor; cessation probably is not, inasmuch as it is totally dark (after civil twilight) when the singing stops." Leopold and Eynon (1961) studied the daybreak and evening singing of 20 bird species in southern Wisconsin. They wrote that "tabulation of the light intensities disclosed that daybreak song came at lower, more precise values than evening song. The lower range came close to minimum values reported for European species. Song was delayed on cloudy mornings until light reached an intensity not significantly different from that of

Figure 22.   An Acadian Flycatcher at its nest. (Courtesy of Samuel A. Grimes.)

clear mornings. Mean light intensities at evening song ranged from 29 to 94 times higher than for daybreak song. . . . It seems probable that some birds, at least, awaken endogenously, test their environment during an anticipatory period, and sing when the light reaches a certain intensity. Such entrained circadian rhythms have been suggested for other animals. Cessation at evening may be due to fatigue, waning drives, or a fading stimulus, but none of these concepts nor the endogenous components in activity rhythms can be verified in the field."

The Red Ant Tanager (*Habia rubica*) has a dawn song entirely different from its daytime song, and the Blue Honeycreeper (*Cyanerpes cyaneus*) has a "dawn-song which he rarely if ever utters after sunrise" (Skutch, 1954:390). The thrushes (Turdidae) are noted for evening singing, mainly after sunset, and many daytime-singing birds sing also at night, some of them often, others only occasionally: e.g., American cuckoos, the Bush-lark (*Mirafra javanica*), Magpie-lark (Grallinidae), Mockingbird, Nightingale, Willie Wagtail (*Rhipidura leucophrys*), Yellow-breasted Chat (*Icteria virens*), Field Sparrow (*Spizella pusilla*), and Henslow's Sparrow (*Passerherbulus henslowii*) (Fig.

23). Similarly, night-singing birds such as the nightjars (Caprimulgidae) occasionally call during the day.

The seasonal song cycle begins in the spring for most birds of the Northern Hemisphere. American wood-warblers and sparrows sing during migration, but in a few cases (e.g., Fox Sparrow) this may not be the full song. Some American thrushes, however, apparently rarely sing during migration. In Australia birds tend to sing year round: there are comparatively few migrants; a large number of the resident birds mate for life and hold a permanent territory throughout the year; the average territory occupies about 20 acres; and some species have a prolonged breeding season (Robinson, 1949).

The song behavior of birds during the nesting season varies widely among different species. In many nonpasserine birds, display and accompanying sounds are discontinued after pairing is accomplished. Tinbergen reported that the male Snow Bunting stops singing as soon as a mate is taken. There is a similar marked decrease in song by Grasshopper and Song sparrows after the males secure mates, but the males sing again when nest-building commences.

Figure 23. A Henslow's Sparrow at the nest; Kalamazoo County, Michigan, August 26, 1937. (Courtesy of Lawrence H. Walkinshaw.)

Davis (1958) wrote that unmated Rufous-sided Towhees in California "are the most persistent singers of all." Although Blue Jays are noisy during most of the year, they become very quiet and inconspicuous while nesting. The Hawaiian Elepaio sings most persistently throughout the day during the first week of nest-building; singing is infrequent during the incubation period, but there is an increase in singing after the eggs hatch, and a further rise in song frequency after the young leave the nest. By contrast, "the Swainson's Warbler sings vigorously from the time it arrives on its breeding ground in April, until the nesting season is over, which is usually in the latter part of June. Thereafter singing becomes more sporadic" (Meanley, 1968). Incubating male Warbling Vireos (*Vireo gilvus*) and Rose-breasted and Black-headed grosbeaks sing while incubating.

REFERENCES

Ames, P. L. 1971. The morphology of the syrinx in passerine birds. *Peabody Mus. Nat. Hist. Bull.* No. 37.

Benson, C. W. 1948. Geographical voice-variation in African birds. *Ibis*, **90**:48–71.

Berger, A. J. 1966. Behavior of a captive Mockingbird. *Jack-Pine Warbler*, **44**:8–13.

Bertram, B. 1970. The vocal behaviour of the Indian Hill Mynah, *Gracula religiosa*. *Anim. Behav. Monogr.*, **3**:79–192.

Borror, D. J. 1960. The analysis of animal sounds. In *Animal Sounds and Communication*, American Institute of Biological Sciences, Washington, D.C., pp. 26–37.

——— 1961. Intraspecific variation in passerine bird songs. *Wilson Bull.*, **73**:57–78.

——— 1964. Songs of the thrushes (Turdidae), wrens (Troglodytidae), and Mockingbirds (Mimidae) of eastern North America. *Ohio Jr. Sci.*, **64**:195–207.

——— 1965. Song variation in Maine Song Sparrows. *Wilson Bull.*, **77**:5–37.

———, and W. W. H. Gunn. 1965. Variation in White-throated Sparrow songs. *Auk*, **82**:26–47.

———, and C. R. Reese. 1956. Mockingbird imitations of Carolina Wren. *Bull. Mass. Audubon Soc.*, **40**:245–250, 309–318.

Brackbill, H. 1961. Duetting by paired Brown-headed Cowbirds. *Auk*, **78**:97–98.

Brown, J. L. 1969. The control of avian vocalization by the central nervous system. In *Bird Vocalizations*, Cambridge University Press, England, pp. 79–96.

Busnel, R., J. Giban, P. Gramet, H. and M. Frings, and J. Jumber. 1957. Interspécificité de signaux acoustiques ayant une valeur sémantique pour des Corvidés européens et nord-américains. *C. R. Acad. Sci.*, **245**:105–108.

Calder, W. A. 1970. Respiration during song in the canary (*Serinus canaria*). *Comp. Biochem. Physiol.*, **32**:251–258.

Chamberlain, D. R., and G. W. Cornwell. 1971. Selected vocalizations of the Common Crow. *Auk*, **88**:613–634.

———, W. B. Gross, G. W. Cornwell, and H. S. Mosby. 1968. Syringeal anatomy in the Common Crow. *Ibid.*, **85**:244–252.

Chapin, J. P. 1922. The function of the esophagus in the Bittern's booming. *Auk*, **39**:196–202.

Chapman, F. M. 1940. The post-glacial history of *Zonotrichia capensis*. *Bull. Am. Mus. Nat. Hist.*, **77**:381–438.

Davis, J. 1958. Singing behavior and the gonad cycle of the Rufous-sided Towhee. *Condor*, **60**:308–336.

De Kiriline (Lawrence), L. 1954. The voluble singer of the treetops. *Audubon Mag.*, **56**:109–111.

Diamond, J. M., and J. W. Terborgh. 1968. Dual singing by New Guinea birds. *Auk*, **85**:62–82.

Dinsmore, J. J. 1969. Dual calling by birds of paradise. *Auk*, **86**:139–140.

Frings, H. and M., J. Jumber, R. Busnel, J. Giban, and P. Gramet. 1958. Reactions of American and French species of *Corvus* and *Larus* to recorded communication signals tested reciprocally. *Ecology*, **39**:126–131.

Gaunt, A. S., R. C. Stein, and S. L. L. Gaunt. 1973. Pressure and air flow during distress calls of the Starling, *Sturnus vulgaris* (Aves; Passeriformes). *Jr. Exp. Zool.*, **183**:241–262.

Gill, F. B., and B. G. Murray, Jr. 1972. Song variation in sympatric Blue-winged and Golden-winged warblers. *Auk*, **89**:625–643.

Greenewalt, C. H. 1968. *Bird Song: Acoustics and Physiology*. Smithsonian Institution Press, Washington, D.C.

——— 1969. How birds sing. *Sci. Am.*, **221** (November):126–139.

Haverschmidt, F. 1947. Duetting in birds. *Ibis*, **89**:357–358.

Hinde, R. A., ed. 1969. *Bird Vocalizations: Their Relations to Current Problems in Biology and Psychology*. Cambridge University Press, England.

Honess, R. F., and W. J. Allred. 1942. Structure and function of the neck muscles in inflation and deflation of the esophagus in the Sage Grouse. *Wyoming Game Fish Dept. Bull.* No. 2, pp. 5–12.

Hooker, B. I. 1968. Birds. In *Animal Communication: Techniques of Study and Results of Research*, Indiana University Press, Bloomington, pp. 311–337.

Hooker, T., and B. I. Hooker. 1969. Duetting. In *Bird Vocalizations*, Cambridge University Press, England, pp. 185–205.

Kern, M. D., and J. R. King. 1972. Testosterone-induced singing in female White-crowned Sparrows. *Condor*, **74**:204–209.

Konishi, M. 1965. The role of auditory feedback in the control of vocalization in the White-crowned Sparrow. *Z. Tierpsychol.*, **22**:770–783.

———, and F. Nottebohm. 1969. Experimental studies in the ontogeny of avian vocalizations. In *Bird Vocalizations*, Cambridge University Press, England, pp. 29–48.

Lanyon, W. E. 1957. The comparative biology of the meadowlarks, *Sturnella*, in Wisconsin. *Publ. Nuttall Ornithol. Club*, **1**:1–67.

——— 1960. The ontogeny of vocalizations in birds. In *Animal Sounds and Communication*, American Institute of Biological Sciences, Washington, D.C., pp. 321–347.

———, and W. R. Fish. 1958. Geographical variation in the vocalizations of the Western Meadowlark. *Condor*, **60**:339–341.

———, and F. B. Gill. 1964. Spectrographic analysis of variation in the songs of a population of Blue-winged Warblers (*Vermivora pinus*). *Am. Mus. Novit.*, No. 2176, 18 pp.

Leopold, A., and A. E. Eynon. 1961. Avian daybreak and evening song in relation to time and light intensity. *Condor*, **63**:269–293.

Lister, M. D. 1953. Secondary song of some Indian birds. *J. Bombay Nat. Hist. Soc.*, **51**:699–706.

McCabe, R. A. 1951. The song and song-flight of the Alder Flycatcher. *Wilson Bull.*, **63**:89–98.

Marler, P. 1969. Tonal quality of bird sounds. In *Bird Vocalizations*, Cambridge University Press, England, pp. 5–18.

———, M. Kreith, and M. Tamura. 1962. Song development in hand-raised Oregon Juncos. *Auk*, 79:12–30.

Marshall, A. J. 1950. The function of vocal mimicry in birds. *Emu*, 50:5–16.

Marshall, J. T., Jr. 1964. Voice in communication and relationships among brown towhees. *Condor*, 66:345–356.

Meanley, B. 1968. Singing behavior of the Swainson's Warbler. *Wilson Bull.*, 80:72–77.

Messmer, E., and I. Messmer. 1956. Die Entwicklung der Lautäusserungen und einiger Verhaltensweisen der Amsel (*Turdus merula merula* L.) unter natürlichen Bedingungen und nach Einzelaufzucht in schalldichten Räumen. *Z. Tierpsychol.*, 13:341–441.

Miller, A. H. 1947. The structural basis of the voice of the Flammulated Owl. *Auk*, 64:133–135.

Milligan, M. M., and J. Verner. 1971. Inter-population song dialect discrimination in the White-crowned Sparrow. *Condor*, 73:208–213.

Miskimen, M. 1951. Sound production in passerine birds. *Auk*, 68:493–504.

Mulligan, J. A. 1966. Singing behavior and its development in the Song Sparrow *Melospiza melodia*. *Univ. Calif. Publ. Zool.*, 81:1–76.

Murphy, R. C. 1936. *Oceanic Birds of South America*. Macmillan Co., New York, 2 vols.

Nice, M. M. 1937–1943. Studies in the life history of the Song Sparrow. I, II. *Trans. Linn. Soc. N. Y.*, 4, 1937: vi + 247 pp.; 6, 1943: viii + 329 pp.

Nottebohm, F. 1969. The song of the Chingolo, *Zonotrichia capensis*, in Argentina: description and evaluation of a system of dialects. *Condor*, 71:299–315.

——— 1970. Ontogeny of bird song. *Science*, 167:950–956.

——— 1972a. Neural lateralization of vocal control in a passerine bird. II. Subsong, calls, and a theory of vocal learning. *J. Exp. Zool.*, 179:35–49.

——— 1972b. The origins of vocal learning. *Am. Nat.*, 106:116–140.

———, and R. K. Selander. 1972. Vocal dialects and gene frequencies in the Chingolo Sparrow (*Zonotrichia capensis*). *Condor*, 74:137–143.

Robinson, A. 1949. The biological significance of bird song in Australia. *Emu*, 48:291–315.

Sauer, F. 1954. Die Entwicklung der Lautäusserungen vom Ei ab schalldicht gehaltener Dorngrasmücken (*Sylvia c. communis*, Latham) im Vergleich mit später isolierten und mit wildlebenden Artgenossen. *Z. Tierpsychol.*, 11:10–93.

Short, L. L. 1966. Field Sparrow sings Chipping Sparrow song. *Auk*, 83:665.

Skutch, A. F. 1954. Life histories of Central American birds. *Pacific Coast Avifauna*, No. 31.

Stein, R. C. 1968. Modulation in bird songs. *Auk*, 85:229–243.

Stokes, A. W., and H. W. Williams. 1968. Antiphonal calling in quail. *Auk*, 85:83–89.

Sutton, G. M. 1945. At a bend in a Mexican river. *Audubon Mag.*, 47:239–242.

——— 1951. *Mexican Birds: First Impressions*. University of Oklahoma Press, Norman.

——— and P. W. Gilbert. 1942. The Brown Jay's furcular pouch. *Condor*, 44:160–165.

Thielcke-Poltz, H., and G. Thielcke. 1960. Akustisches Lernen verschieden alter Schallisolierter Amseln (*Turdus merula* L.) und die Entwicklung erlernter Motive ohne und mit künstlichen Einfluss von Testosteron. *Z. Tierpsychol.*, 17:211–244.

Thomas, E. S. 1943. A wren singing the songs of both Bewick's and the House Wren. *Wilson Bull.*, 55:192–193.

Thompson, W. L. 1970. Song variation in a population of Indigo Buntings. *Auk,* **87**:58–71.

Thorpe, W. H. 1955. The analysis of bird song with special reference to the song of the Chaffinch (*Fringilla coelebs*). *Acta XI Congr. Internat. Ornithol.,* **1954**:209–217.

——— 1961. *Bird-Song.* Cambridge University Press, England.

——— 1963. Antiphonal singing in birds as evidence for avian auditory reaction time. *Nature,* **197**:774–776.

——— 1972. Duetting and antiphonal song in birds: its extent and significance. *Behaviour,* Suppl., **18,** 197 pp. (See *Auk,* **90,** 1973:451–453.)

Tinbergen, N. 1939. The behavior of the Snow Bunting in spring. *Trans. Linn. Soc. N. Y.,* **5**:1–95.

Van Someren, V. G. L. 1956. Days with birds: Studies of habits of some East African species. *Fieldiana: Zool.,* **38** (Chicago Natural History Museum).

Van Tienhoven, A. 1961. Endocrinology of reproduction in birds. In *Sex and Internal Secretions,* Vol. 2, 3rd ed., Williams & Wilkins Co., Baltimore, pp. 1088–1169.

Wetmore, A. 1917. On certain secondary sexual characters in the male Ruddy Duck, *Erismatura jamaicensis* (Gmelin). *Proc. U. S. Natl. Mus.,* **52**:479–482. (See also *Condor,* **20,** 1918:19–20).

——— 1926. Observations on the birds of Argentina, Paraguay, Uruguay, and Chile. *U.S. Natl. Mus. Bull.,* **133.**

Wickler, W. 1972. Duettieren zwischen artverschiedenen Vögeln im Freiland. *Z. Tierpsychol.,* **31**:98–103.

Wildenthal, J. L. 1965. Structure in primary song of the Mockingbird (*Mimus polyglottos*). *Auk,* **82**:161–189.

Witherby, H. F., F. C. R. Jourdain, N. F. Ticehurst, and B. W. Tucker. 1949. *The Handbook of British Birds,* Vol. 4. H. F. & G. Witherby, London.

Witschi, E. 1961. Sex and secondary sexual characters. In *Biology and Comparative Physiology of Birds,* Vol. 2, Academic Press, New York, pp. 115–168.

Zigmond, R. E., F. Nottebohm, and D. W. Pfaff. 1973. Androgen-concentrating cells in the midbrain of a songbird. *Science,* **179**:1005–1007.

CHAPTER SIX

Bird Distribution

For more than 100 years men have been systematically studying the distribution of birds and, for nearly as long, the factors that control it. The ranges of many species are still incompletely known, but for a large number the geographical distribution can be adequately described. Therefore ornithologists are perhaps better qualified than most zoologists to discuss this subject and to draw sound conclusions.

ZOOGEOGRAPHY AND TAXONOMIC CONCEPTS

Correct interpretation of bird distribution is, of course, completely dependent on correct classification. Our zoogeography can be only as good as our taxonomic system. For example, when Chapman and Griscom discussed the distribution of wrens (Troglodytidae) in 1924, they recognized about 48 species and subspecies in Asia and Europe in addition to the numerous forms in the Western Hemisphere, but 22 years later all but 1 of the 48 were classified as timaliids rather than as wrens. Therefore, when Mayr (1946) discussed the North American avifauna, he could treat the wrens as one great American group, of which a single species (*Troglodytes troglodytes*) has spread westward across the Bering Straits, ranging over all Asia and over Europe as far as Iceland.

ZOOLOGICAL REGIONS

Of the more than 8000 species of birds in the world, no 2 have exactly the same distribution (if we exclude a few species on small islands). Yet many species have distributions that overlap to a considerable extent, and this suggested to early ornithologists a way of organizing the facts of distribution.

Gadow (1913) summarized the history of zoogeography, beginning with Buffon's *Histoire naturelle* (1770). Sclater first proposed dividing the world into six major regions (Fig. 1) as follows: Palaearctic, Aethiopian, Indian (later called Oriental), Australian (called Australasian), Nearctic, and Neotropical (see Udvardy, 1969). Later it was proposed that the northern regions of the Old and New worlds (Palaearctic and Nearctic) be united and called the "Holarctic."

LIFE ZONES

Merriam's (1894) system of transcontinental "life zones" quickly received widespread approval by naturalists generally. Merriam decided that tempera-

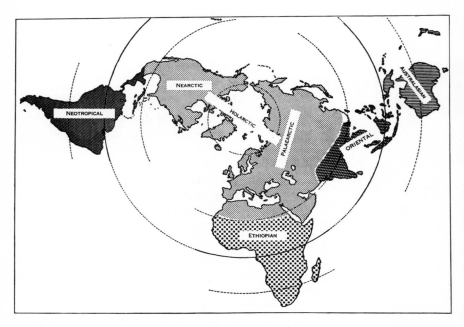

Figure 1. The six zoogeographical regions of the world, approximately as conceived by Sclater. (By permission of *Encyclopaedia Britannica,* Vol. 23, 1949, p. 964.)

ture was the most important limiting factor and proposed two temperature "laws."[1] He tried to place the life zones on an exact basis by having the U.S. Weather Bureau calculate temperature "sums" for a large number of localities. Although these figures were later discovered to be erroneous (Merriam, 1899), the promised corrected tables were never published. Merriam claimed, however, that because of temperature factors the distribution of animals and plants could be expressed in terms of a few great transcontinental zones (Fig. 2).

Subsequently Daubenmire (1938) and others showed that these zones do not fit the known distribution of animals and plants except in limited mountain areas. Moreover, temperature is not the only factor controlling animal distribution.

BIOMES

Another method of analyzing the distribution of animals and plants is by "biomes" (Fig. 3). According to this concept, the world is divided into natural biomes in which, under conditions undisturbed by man, the vegetation eventually reaches a stable "climax" that theoretically will maintain itself indefinitely. A familiarity with the major biomes is essential to an understanding of bird distribution in North America, and therefore a brief description of them is given here. Pettingill (1970) lists characteristic birds for each biome. (Space limitations prevent a discussion of savannas, tropical rain forests, etc.; see Udvardy, 1969.)

1. The **tundra biome** is characterized by low temperatures and a short growing season (about 2 months). Only the surface of the ground thaws during the short summer, and vegetation is typified by lichens, sedges, grasses, and dwarfed woody plants. Similar conditions are found at high latitudes and elevations (above the timberline), so that two types of tundra are recognized.

   a. Arctic tundra: northern Alaska and Canada, south along the coasts of Hudson Bay, and throughout most of Labrador and Newfoundland (Fig. 4).
   b. Alpine tundra: discontinuous areas above 10,000 feet[2] in the Rocky Mountains and the Sierra Nevada-Cascade system.

---

[1] *a.* Animals and plants are restricted in northward distribution by the total quantity of heat during the season of growth and reproduction. *b.* Animals and plants are restricted in southward distribution by the mean temperature of a brief period covering the hottest part of the year (Merriam, 1894:233–234).

[2] The elevation at which the timberline occurs depends on the latitude, climate, and direction of slopes in the mountain range. The timberline in the western United States varies in elevation from about 6000 to 11,500 feet. Alpine tundra also occurs on high mountains in the tropics: e.g., Africa, Hawaii.

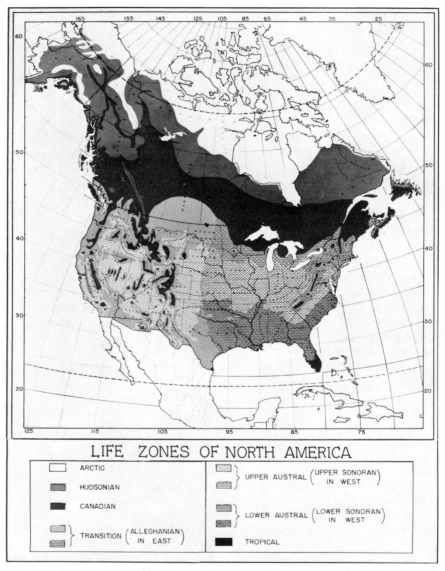

Figure 2. Life zones of North America. (Courtesy of C. F. W. Muesebeck and Arthur D. Cushman, U.S. Bureau of Entomology and Plant Quarantine.)

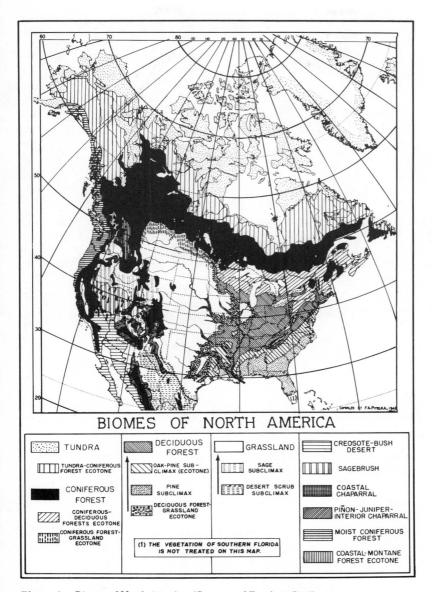

Figure 3. Biomes of North America. (Courtesy of Frank A. Pitelka.)

283

Figure 4. A female Smith's Longspur, an Arctic tundra species, ready to settle on its ground nest at Churchill, Manitoba. (Courtesy of Samuel A. Grimes.)

2. **Coniferous forests** cover vast areas of Canada and the United States. Three main subdivisions are described. The species of conifers vary from region to region.

a. The *transcontinental* coniferous forest of southern Canada and Alaska is bounded by the arctic tundra to the north (Fig. 5).

b. The *eastern montane* coniferous forest extends from northeastern United States southward at high elevations in the Appalachian Mountains to Tennessee and North Carolina.

c. The *western montane* coniferous forest occupies the area just below the alpine tundra in the Rocky Mountains, extending southward to New Mexico and Mexico, and in the Sierra Nevada-Cascade system into California (Fig. 6).

3. The **deciduous forest biome** is found, at elevations below those of the coniferous forests, in the eastern and southern United States, from southern New England and the Appalachian Mountains west to the bottom lands and bluffs along the Mississippi River and its tributaries. Although the deciduous forest biome extends far south along the Mississippi River system, in a large

segment of the southeastern United States historical and climatic factors have produced several types of mixed or subclimax forests. One of these is the southeastern pine subclimax forest, "characterized by the presence of open pine forests which, because of the poor soils and many fires, have never been succeeded by deciduous stands" (Pettingill, 1970:215).

4. The **grassland biome** (the prairie) extends from the forested bottom lands along the Mississippi River west to the Rocky Mountains, and from south-central Texas north into Canada (the prairie provinces). The vegetation changes from east to west with increasing elevation and decreasing rainfall, so that three general east-west zones are recognized: tall grass prairie, mixed, and short grass prairie.

Figure 5. A Lincoln's Sparrow standing in its nest; Schoolcraft County, Michigan, July 1, 1956. This is a characteristic bird in tamarack bogs in eastern coniferous forests. (Courtesy of L. H. Walkinshaw.)

Figure 6. A Goshawk at its nest; this species is found especially in coniferous forests in Canada, the United States, and Eurasia. (Courtesy of Heinz Meng.)

5. The **southwestern pine-oak woodland** occurs mainly on hills and mountain slopes in New Mexico, Arizona, Nevada, Utah, California, and parts of Colorado and Oregon (Fig. 7).

6. The **piñon-juniper woodland** is found on hills and mountain slopes above the deserts or grasslands and below the coniferous forest in Nevada, Arizona, and New Mexico, and on the eastern side of the Sierra Nevada-Cascade system in California.

7. The **chaparral biome** occurs in canyons and on mountain slopes from southern Oregon to Baja California and Mexico. Vegetation consists of a variety of broad-leaved evergreen shrubs, bushes, and dwarf trees.

8. The **desert biome** occurs, in general, in regions having about 5 inches of annual rainfall (Kendeigh, 1961). Two main types in the United States have been classified as "hot" or "low" and "cool" or "high" deserts. Both have typical xeric plants with a characteristic spacing or pattern of vegetation, but many intermediate types result from regional differences in temperature, elevation, rainfall, and geological nature of the country (Fig. 8). Dixon (1959), in a study of desert birds in Brewster County, Texas, found 13 species "that might be considered to comprise a 'standard' desert avifauna" of the hot deserts; some of the species occur also in cool deserts and in the chaparral biome. Dixon concluded that, to a considerable extent, the "occurrence of desert species in adjacent formations indicates an attraction to the shrub form and not the desert climate *per se.*"

Pettingill characterized, in part, two chief types of desert communities in the western United States:

a. *Sagebrush* communities are found at elevations above the scrub desert between the Rocky Mountains and the Sierra Nevada-Cascade system. Dominant plants include sagebrush (*Artemisia*), shadscale (*Atriplex*), greasewood (*Sarcobatus*), rabbit brush (*Chrysothamnus*), winterfat (*Eurotia*), and some grasses.

b. *Scrub desert* occurs in the lowlands and valley floors from western Texas west to southwestern California. Here the plants, 3 to 6 feet high, are widely spaced. Creosote bush (*Larrea*) is often the dominant plant; other common plants include mesquite (*Prosopis*), paloverde (*Cercidium*), catclaw (*Acacia*), ironwood (*Olneya*), ocotillo (*Fouquieria*), agaves, cacti, and yuccas (Figs. 9, 10).

On a world basis, Logan (1968) discusses five basic types of desert areas: subtropical, cool coastal, rain shadow, continental interior, and polar deserts. "The drier areas of the earth are commonly divided into two groups: arid and semiarid (or desert and semidesert, or desert and steppe, each pair of terms being essentially synonymous with the other pairs), based upon the severity of their aridity." The boundary between humid and semiarid areas is drawn at

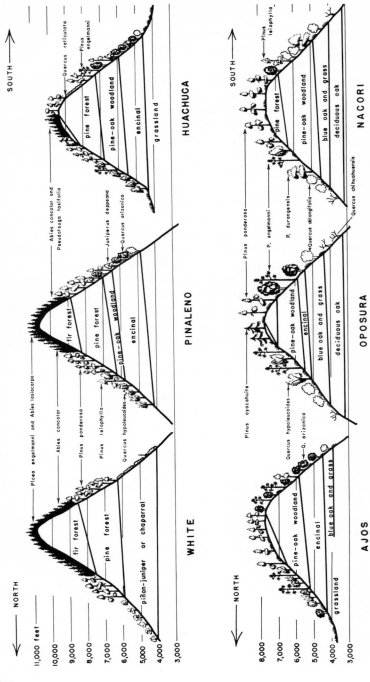

Figure 7. Sequence of montane vegetation types with differences in altitude and latitude in the pine-oak woodland regions of Arizona, New Mexico, and northern Mexico. (Courtesy of Joe T. Marshall, Jr., and the Cooper Ornithological Society.)

288

Figure 8. Typical desert east of Tucson, Arizona. A Curve-billed Thrasher's nest is located in the center cholla cactus. Dry grama grass of the previous fall occupies the foreground; the low tree in the background is mesquite, and in the distance is a single saguaro cactus, an uncommon plant on the flats in this area; the Tanque Verde Mountains can be seen in the distance. (Courtesy of George Olin.)

the point where the potential evaporation equals the precipitation. Because of the interrelations among several factors (e.g., evaporation, temperature, seasonality of precipitation), it is impractical to use a fixed precipitation figure for the boundary between humid and semiarid regions.

The biome concept was first applied seriously to American bird distribution by Pitelka (1941). Kendeigh (1954) published an excellent illustrated history and evaluation of the various concepts of plant and animal communities in North America; he concluded by recommending the biome system. Since that time, however, the trend has been to study birds in relation to the ecosystems of which they are a part.

THE BREEDING-SEASON DISTRIBUTION OF SOME BIRD FAMILIES

Although the modern ornithologist no longer attaches to the several zoological regions the importance that Sclater and his followers did, some important

Figure 9.   An Elf Owl at the entrance to its nesting cavity in a saguaro cactus. (Courtesy of Lewis W. Walker.)

general facts about the distribution of bird families can be concisely expressed in terms of these regions. The ornithologist may, for example, list as follows the regions and the families of birds that, in breeding season, are restricted to them.

NEOTROPICAL

TINAMIDAE	Tinamou family	FURNARIIDAE	Ovenbird family
RHEIDAE	Rhea family	FORMICARIIDAE	Antbird family
COCHLEARIIDAE	Boat-billed Heron family	RHINOCRYPTIDAE	Tapaculo family
		COTINGIDAE	Cotinga family
ANHIMIDAE	Screamer family	PIPRIDAE	Manakin family
CRACIDAE	Curassow family	PHYTOTOMIDAE	Plantcutter family
OPISTHOCOMIDAE	Hoatzin family		
PSOPHIIDAE	Trumpeter family	OXYRUNCIDAE	Sharpbill family
		DULIDAE	Palmchat family
EURYPYGIDAE	Sunbittern family	CYCLARHIDAE	Pepper-shrike family

CARIAMIDAE	Cariama family	VIREOLANIIDAE	Shrike-vireo family
THINOCORIDAE	Seedsnipe family		
STEATORNITHIDAE	Oilbird family	ZELEDONIIDAE	Wren-thrush family
NYCTIBIIDAE	Potoo family		
TODIDAE	Tody family	TERSINIDAE	Swallow-tanager family
MOMOTIDAE	Motmot family		
GALBULIDAE	Jacamar family	CATAMBLYRHYNCHIDAE	Plush-capped Finch family
BUCCONIDAE	Puffbird family		
RAMPHASTIDAE	Toucan family		
DENDROCOLAPTIDAE	Woodcreeper family		

NEARCTIC plus NEOTROPICAL

No bird family is restricted to the Nearctic region. The following families are *restricted to the New World* but occur in both the Neotropical and the Nearctic regions.

| CATHARTIDAE | American Vulture family | MIMIDAE | Mockingbird family |
| | | PTILOGONATIDAE | Silky-flycatcher family |

Figure 10. A Curve-billed Thrasher at its nest in cholla cactus. (Courtesy of George Olin.)

MELEAGRIDIDAE	Turkey family	VIREONIDAE	Vireo family
ARAMIDAE	Limpkin family	PARULIDAE	American wood-
TROCHILIDAE	Hummingbird family		warbler family
TYRANNIDAE	Tyrant-flycatcher	ICTERIDAE	Troupial family
	family	THRAUPIDAE	Tanager family

PALAEARCTIC

No bird family is restricted to the Palaearctic region.

HOLARCTIC (Palaearctic plus Nearctic)

GAVIIDAE	Loon family	ALCIDAE	Auk family
TETRAONIDAE	Grouse family	BOMBYCILLIDAE	Waxwing family
PHALAROPODIDAE	Phalarope family		

ETHIOPIAN (an asterisk indicates restriction to Madagascar)

BALAENICIPITIDAE	Whale-billed Stork family	LEPTOSOMATIDAE*	Cuckoo-roller family
		PHOENICULIDAE	Wood-hoopoe family
SCOPIDAE	Hammerhead family	PHILEPITTIDAE*	Asity family
SAGITTARIIDAE	Secretarybird family	HYPOSITTIDAE*	Coral-billed Nuthatch
NUMIDIDAE	Guineafowl family		family
MESOENATIDAE*	Mesite family	VANGIDAE*	Vanga-shrike family
MUSOPHAGIDAE	Touraco family	PRIONOPIDAE	Wood-shrike family
COLIIDAE	Coly family		

ORIENTAL

IRENIDAE   Leafbird family

AUSTRALIAN (an asterisk indicates restriction to New Zealand)

CASUARIIDAE	Cassowary family	PTILONORHYNCHIDAE	Bowerbird family
DROMICEIIDAE	Emu family	PARADISAEIDAE	Bird-of-paradise
APTERYGIDAE*	Kiwi family		family

RHYNOCHETIDAE	Kagu family	GRALLINIDAE	Mudnest-builder family
AEGOTHELIDAE	Owlet-frogmouth family	NEOSITTIDAE	Australian nuthatch family
XENICIDAE*	New Zealand wren family	CALLAEIDAE*	Wattlebird family
ATRICHORNITHIDAE	Scrubbird family	CLIMACTERIDAE	Australian treecreeper family
CRACTICIDAE	Bellmagpie family family		

THE ORIGINS OF THE NORTH AMERICAN AVIFAUNA

Mayr (1946) studied the distribution of bird families and postulated the areas of origin of many of them. He found that some families defy such analysis. For example, most families of oceanic birds are widely distributed around the world, since they are independent of continental boundaries. Mayr found 29 families, including oceanic birds, shorebirds, freshwater birds, and even landbirds (such as hawks, falcons, nightjars, swifts, woodpeckers, and swallows), in the unanalyzable category. We may group the families that Mayr classified as follows.

OF OLD WORLD ORIGIN

PHASIANIDAE	Pheasant family	PARIDAE	Titmouse family
GRUIDAE	Crane family	SITTIDAE	Nuthatch family
COLUMBIDAE	Pigeon family	CERTHIIDAE	Creeper family
CUCULIDAE	Cuckoo family	TIMALIIDAE	Babbler family
TYTONIDAE	Barn-owl family	TURDIDAE	Thrush family
STRIGIDAE	Typical-owl family	SYLVIIDAE	Old-world warbler family
ALCEDINIDAE	Kingfisher family	MOTACILLIDAE	Pipit family
ALAUDIDAE	Lark family	LANIIDAE	Shrike family
CORVIDAE	Crow family		

OF PAN-AMERICAN ORIGIN

CRACIDAE	Curassow family	THRAUPIDAE	Tanager family
TROCHILIDAE	Hummingbird family	ICTERIDAE	Troupial family
TYRANNIDAE	Tyrant-flycatcher family		

OF SOUTH AMERICAN ORIGIN

TINAMIDAE	Tinamou family	FURNARIIDAE	Ovenbird family
EURYPYGIDAE	Sunbittern family	FORMICARIIDAE	Antbird family
NYCTIBIIDAE	Potoo family	RHINOCRYPTIDAE	Tapaculo family
GALBULIDAE	Jacamar family	COTINGIDAE	Cotinga family
BUCCONIDAE	Puffbird family	PIPRIDAE	Manakin family
RAMPHASTIDAE	Toucan family	OXYRUNCIDAE	Sharpbill family
DENDROCOLAPTIDAE	Woodcreeper family		

OF NORTH AMERICAN ORIGIN

CATHARTIDAE	American Vulture family	BOMBYCILLIDAE	Waxwing family
		PTILOGONATIDAE	Silky-flycatcher family
TETRAONIDAE	Grouse family	DULIDAE	Palmchat family
MELEAGRIDIDAE	Turkey family	CYCLARHIDAE	Pepper-shrike family
ARAMIDAE	Limpkin family	VIREOLANIIDAE	Shrike-vireo family
TODIDAE	Tody family	VIREONIDAE	Vireo family
MOMOTIDAE	Motmot family	PARULIDAE	American wood-warbler family
TROGLODYTIDAE	Wren family		
MIMIDAE	Mockingbird family		

FACTORS CONTROLLING DISTRIBUTION

Several factors control the present distribution of species of birds: past history, physical barriers, the ecological conditions that the birds can tolerate, and mobility (see Udvardy, 1969; Lack, 1971).

Time and Geology

The importance of the historical factor is most strikingly demonstrated in certain isolated islands, such as the Galapagos and Hawaiian islands. Similarly, on a continental scale, the absence of woodpeckers (Picidae) and finches (Fringillidae) from Australia, where physical conditions are perfectly suitable for them, must be explained on purely historical grounds.

Physical Barriers

Any considerable area of unfavorable environment obviously constitutes a barrier to the expanding range of any bird species. Seabirds are limited by bodies of land, and landbirds by large bodies of water. But such zones must be very wide indeed before they will serve as complete barriers to birds that are both strong-flying and tolerant of short periods of change. For example, the 50-mile-wide Isthmus of Panana is no barrier to certain strictly marine birds: Brown Pelicans (*Pelecanus occidentalis*) and Frigatebirds (*Fregata magnificens*) fly daily across the Isthmus (Fig. 11). On the other hand, certain birds of equally powerful flight, e.g., the Common Booby (*Sula leucogaster*), apparently never venture across, and distinct subspecies are found on the two sides of the Isthmus.

Different races of a sedentary tropical bird may live on adjacent islands (as certain woodpeckers, warblers, and tanagers in the West Indies) without mingling, or broad rivers may provide impassable barriers. Describing the segregation of subspecies of the barbet *Capito auratus* (a tree-top bird with well-developed powers of flight), Chapman (1928) said, "In Amazonia it is evident that this segregation is supplied by a river system whose broad streams cut this vast area into districts" where races may live completely separated though within sight of each other. The smaller rivers nearer the Andes, however, prove to be no barriers to the barbets, and the same race lives on either side.

Figure 11.   A Brown Pelican at its nest. (Courtesy of Samuel A. Grimes.)

Mobility

If we exclude the few birds that are flightless, most species seem perfectly capable of transporting themselves far beyond the boundaries of their present ranges. Indeed, there are some examples of birds which in historic time have suddenly jumped a wide gap and occupied new areas. Thus in 1937 the Fieldfare (*Turdus pilaris*) invaded North America from Europe. A flock of these large, strong-flying thrushes, apparently stimulated by a severe cold spell, started on January 19 to migrate from Norway southwest toward England. Caught that night in a strong southeast wind, they apparently were swept far off their course to the northwest: to Jan Mayen and the northeast coast of Greenland, at both of which places individuals were seen and collected on January 20. They then moved rapidly southward and by January 27 and 28 were reported on the coast of southwestern and southern Greenland. They became established as nonmigratory colonies in several southern Greenland localities, where the milder climate of recent years has permitted them to persist (Salomonsen, 1951). According to Bond (1971), the Glossy Cowbird (*Molothrus bonariensis*) of South America was unknown in the Lesser Antilles until 1899, but has since become established as far north as Martinique, as well as in Puerto Rico.

More spectacular has been the presumed natural colonization of the New World by the Cattle Egret (*Bubulcus ibis*) and its subsequent range expansion (Fig. 12). This species "colonized northern South America across the Atlantic around 1930" (Mayr, 1965:564); it reached southern Florida by 1941 or 1942. Davis (1960) summarized the distribution of the Cattle Egret in the eastern United States as of 1959 and noted that "the lack of recent extension suggests that the spread of the Cattle Egret has stopped." However, Cattle Egrets were found nesting in Canada in 1962 (Buerkle and Mansell, 1963), and 2 years later the species was seen in California (McCaskie, 1965).

Eisenmann (1971) described a great population increase of the White-tailed Kite (*Elanus leucurus*), particularly during the 1950s and 1960s. "Even more startling was the extension of range (including the breeding range) through Central American countries and Panama, where the species historically had been unknown."

Environmental Requirements

One of the first facts that strikes the student of bird distribution is that most birds, in spite of the very great mobility resulting from the power of flight, have sharply demarcated distribution.

The physical factors that determine some of the geographical boundaries of many species are well known. In many instances these are clearly coasts, mountains, deserts, or other obvious geographical features, but more often we

Figure 12. Nestling Cattle Egrets; St. Croix, Virgin Islands, May 7, 1957. The first Cattle Egret was collected in the Virgin Islands on February 21, 1955. (Courtesy of G. A. Seaman.)

can say only that the bird's range ceases where certain necessary factors in the physical environment are no longer present.

Although food has often been singled out as an important element in determining the limits of the range of bird species, few cases can be documented as regards any one kind of food. However, a few striking examples have been recorded. For instance, the Everglade Kite (*Rostrhamus sociabilis*; Fig. 13) and the Limpkin (*Aramus guarauna*) live almost exclusively on the large freshwater snail *Pomacea*, and the ranges of *Pomacea* and these birds coincide wherever they have been studied. The Common Jay (*Garrulus glandarius*) of Eurasia feeds chiefly on acorns, and its range is closely correlated with that of the oaks *Quercus* (Turček, 1950); Fig. 14.

Ecological Tolerance

It is important, however, not to overemphasize the fact that many bird species have strongly marked ecological requirements and, as a result, restricted distribution. Equally striking, though less understood, are the examples of birds of

Figure 13.   A male Everglade Kite on nest. (Courtesy of Samuel A. Grimes.)

very broad tolerance and very wide distribution. For example, Murphy (1936:109–110) says of the Olivaceous Cormorant: "Still more remarkable . . . is the case of an American cormorant (*Phalacrocorax oliva- ceus*). The specific range is both inland and coastal from Cape Horn to the southern United States, but the birds occurring north of Nicaragua have been separated as subspecies *mexicanus*, while a third race seems to inhabit Tierra del Fuego. The typical South American form disregards not only latitude and climate, but also altitude, water temperature and salinity, precipitation, the na- ture or presence of vegetational ground-cover, and almost every other obvious environmental factor."

Other wide-ranging species of the New World avifauna that largely ignore zonal restrictions include such well-known birds as the Turkey Vulture (*Cathartes aura*), Barn Owl (*Tyto alba*), Great Horned Owl (*Bubo virginianus*), Black Phoebe (*Sayornis nigricans*), and Rough-winged Swallow (*Stelgidopteryx ruficollis*).

Psychological Factors

The psychological (or behavioral) factor must also be considered when we study bird distribution.

Thus the Harlequin Duck (*Histrionicus histrionicus*) in Iceland is confined to swift, rocky streams, where it feeds by swimming against the current and poking its bill under the stones to secure insect larvae. It is absent from lakes, where insect larvae are equally abundant and where many other ducks of more generalized habits are common.

The Olive-sided Flycatcher (*Nuttallornis borealis*) in the coastal belt of California originally nested only in fairly open or interrupted stands of tall coniferous trees. But this flycatcher now nests in introduced eucalyptus trees. The chief feature that the eucalyptus and the conifers have in common seems to be height; the insect fauna, the nest sites, and the perching places are all very different (Miller, 1942).

RANGES OF BIRDS

Size of Range

The geographic ranges of birds vary greatly in size. The smallest ranges are those of species that occur on only one small island. For example, *Calandrella razae*, a small species of lark found only on Razo Island (Cape Verde Islands),

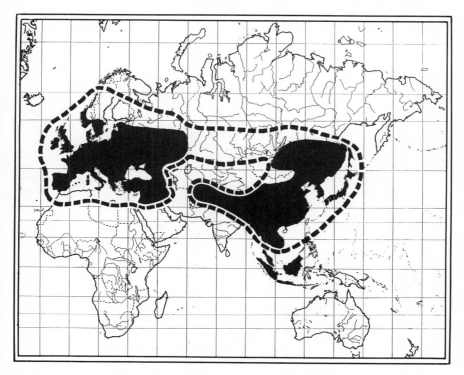

Figure 14. The range of the Common Jay (*Garrulus glandarius*) (dotted line) and the oaks *Quercus* (in black). (After Turček, 1950.)

has a range of about 3 square miles. The total range of the Nihoa Millerbird (*Acrocephalus familiaris kingi*) and the Nihoa Finch (*Psittirostra cantans ultima*) is limited to Nihoa Island, a volcanic remnant of 156 acres in the Hawaiian chain (Fig. 15). Another small landbird whose breeding range is restricted to an island is the Ipswich Sparrow (*Passerculus princeps*) of Sable Island (about 20 sq mi), Nova Scotia.

A number of small oceanic islands have (or had) species of rails peculiar to them, e.g., Wake (total of 3 sq mi), Laysan (2+ sq mi), and Henderson islands (about 5 sq mi) in the Pacific; and Gough (about 30 sq mi), Inaccessible (less than 2 sq mi), and Tristan da Cunha (about 40 sq mi) in the Atlantic.

A few well-marked species of water birds are restricted to a single lake, e.g., a grebe (*Podiceps rufolavatus*) and a duck (*Aythya innotata*), which are restricted to Lake Alaotra (40 mi long) on Madagascar; the flightless grebe (*Rollandia micropterum*), found only on Lake Titicaca (138 mi long) in the Andes; and the Giant Pied-billed Grebe (*Podilymbus gigas*), restricted to Lake Atitlán (24 mi long) in Guatamala.

Small ranges are rare among continental species, but the Kirtland's Warbler (*Dendroica kirtlandii*) is one well-studied exception. The total known breeding range in historic time is about 100 by 60 miles (Van Tyne, 1951); but, because

Figure 15. Nihoa Island, a 156-acre volcanic remnant. This is the southeasternmost island in the Hawaiian Islands National Wildlife Refuge.

of this warbler's peculiar ecological requirements, it has a shifting local distribution and occupies only part of that total area in any one year. An intensive survey in 1951 (Mayfield, 1953) revealed breeding pairs in 91-square-mile sections, but singing males were found in only 27 sections during June 1971 (Mayfield, 1972).

Most species have much larger ranges than those mentioned thus far. If examples are taken from among the landbirds of eastern North America, we find that the average breeding range is of the order of 1000 miles from east to west and the same, or a little less, from north to south. The breeding ranges of other North American species extend completely across the continent (Fig. 16).

A few species extend over a large part of the world. Those with the largest ranges are a small, curiously assorted group, and no one seems to have suggested what they have in common that puts them in this exclusive category. Omitting a few species whose taxonomic boundaries are still being debated, we may mention the following striking cases.

Golden Eagle (*Aquila chrysaetos*): Mountainous areas of much of the northern Temperate Zone.

Osprey (*Pandion haliaetus*): Australia, Celebes, and most of the northern Temperate Zone, extending into the tropics at some points (e.g., northern Africa).

Peregrine Falcon (*Falco peregrinus*): Most of the Northern Hemisphere, Chile and the Falkland Islands, southern Africa and Madagascar, much of the East Indies, Australia, and some of the Pacific Islands.

Barn Owl (*Tyto alba*): South America and the West Indies, North America and Eurasia except the northernmost areas, Africa, East Indies, Australia, and many oceanic islands.

Common Gallinule, or "Moorhen" (*Gallinula chloropus*): The Americas (except the extreme north), including the West Indies, the Galapagos, and the Hawaiian Islands; Africa, including Madagascar; Eurasia, except the extreme north.

Horned Lark (*Eremophila alpestris*): Africa north of the Sahara; much of Eurasia (wherever aridity or severe climate produces the necessary bare or nearly bare ground); North America, south through Mexico; and the Bogotá area of Colombia.

House Wren (*Troglodytes aedon* and *musculus*) (considering *aedon* and *musculus* to be members of a single species): A range exceeded by no other American passerine bird. It is found through South and North America (except the far north) and on many of the adjacent islands.

It is notable that another small wren of the same genus, *Troglodytes troglodytes*, has a similarly vast range. It occurs from Newfoundland across the northern states (and down the Appalachians to Georgia); across most of Eu-

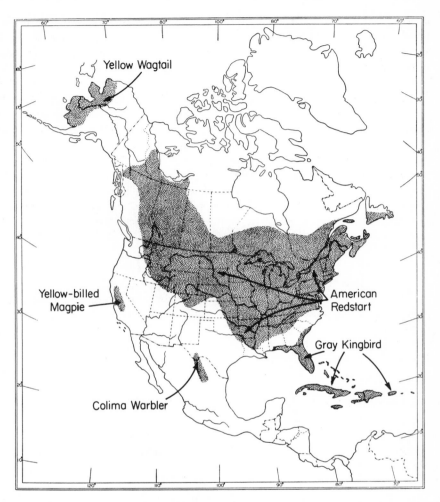

Figure 16. Approximate breeding ranges of five selected species. The large breeding range of the American Redstart (*Setophaga ruticilla*) is shown in comparison with the limited breeding ranges of three other American species. The Yellow Wagtail (*Motacilla flava*) has a vast distribution in the Old World (from Scotland to northern Siberia). The Alaskan birds apparently migrate across the Bering Sea and southward along the Pacific coast of Asia.

rasia and the northern mountains of Africa to Great Britain, the Faroes, and Iceland (this is the Wren of Britain and Europe and the Winter Wren of the New World).

Discontinuity of Range

Many striking cases of discontinuous distribution are known among birds. They may be considered to be of three main types.

1.   Species that are presumed to have had a far wider distribution in ancient times, but that have now decreased in numbers and disappeared from much of their old ranges, surviving in widely separated "islands" of relict populations. (This statement, of course, expresses the extreme case; the populations of certain other species have shrunk only a little, leaving a moderate gap in the range, as in the case of the Hudsonian Curlew, Fig. 17). Examples are the following:

Caspian Tern (*Hydroprogne caspia*): Breeds very locally in widely separated parts of North America, Eurasia (Denmark to northern China), Africa, Australia, and New Zealand.

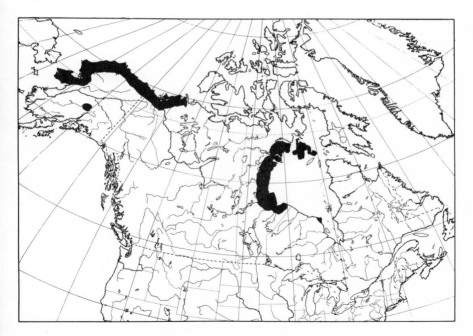

Figure 17.   The discontinuous breeding range of the Hudsonian Curlew (*Numenius phaeopus hudsonicus*). (After Taverner.)

Southern Pochard (*Netta erythrophthalma*): Found in western South America from Venezuela to Peru, also eastern and south Africa (Abyssinia and Angola to the Cape of Good Hope).

Fulvous Tree-duck (*Dendrocygna bicolor*): Occurs in Mexico and southern United States, northern and southeastern South America, southeastern Africa, and India.

Azure-winged Magpie (*Cynopica cyanus*): Occurs in southern and central Spain and Portugal, also eastern Asia (China, Japan, Manchuria, and a little of adjacent Siberia).

2. Species with much smaller ranges, the continuity of which has been broken by some such factor as glaciation, vulcanism, forest loss, forest growth, or changed humidity. Examples are the following:

Black Tanager (*Diglossa carbonaria brunneiventris*): Occurs along the Andes from Bolivia to northern Peru. The bird has never been recorded in Ecuador or southern Colombia, but is found in the eastern and western Andes of northwestern Colombia.

Red-shouldered Hawk (*Buteo lineatus*): Occurs throughout the eastern half of North America (south of about 48°) and on the west coast (California west of the Sierra, and northern Baja California). Undetermined factors (perhaps altitude and aridity) have made a 1000-mile gap in the range of this species.

3. Species of strong flight which have probably wandered, colonizing certain remote outposts where conditions are favorable. Examples are the following:

Hispaniolan White-winged Crossbill (*Loxia leucoptera megaplaga*): This outpost subspecies in the mountains of Hispaniola in the West Indies is doubtless derived from the Common White-winged Crossbill (*Loxia leucoptera leucoptera*) of Canada and the northern edge of the United States. The narrowest gap is now between Hispaniola and the Adirondacks, New York; during the greatest extension of the Pleistocene glaciers the gap was probably much less.

Bogotá Horned Lark (*Eremophila alpestris peregrina*): This representative of the great complex of Horned Lark forms which inhabit North America south into southern Mexico is isolated in the Highlands of Colombia. The forms nearest to its range occur in Oaxaca and Veracruz.

Changes in Range

Seasonal Changes. Some species of birds are completely sedentary, but the greater number (at least outside of the tropics) are in some degree migratory. In many species, the breeding range and the wintering range are far—even many thousands of miles—apart. For example, some shorebirds such as the Pectoral, Baird's, and White-rumped sandpipers (*Calidris melantos, C. bairdii,* and *C. fuscicollis*) nest in arctic North America (Fig. 18) but winter on the pampas of southern South America, more than 8000 miles away.

Figure 18. A White-rumped Sandpiper on its nest; Jenny Lind Island, June 23, 1962. (Courtesy of David F. Parmelee.)

Sporadic "Flights." Some birds undergo occasional sporadic movements that differ from the annual migratory patterns of many bird species.

The cyclic flights of the Snowy Owl (*Nyctea scandiaca*) into northern United States and adjacent Canadian provinces are perhaps the best recorded (Fig. 19). These great flights occur, on the average, every 4 years.

Other northern species make great southward invasions at less regular intervals. In America such flights have been reported, for example, for the Arctic or Black-backed Three-toed Woodpecker (*Picoides arcticus*), the Great Horned Owl (*Bubo virginianus*), the Dovekie (*Alle alle*), and the Brown-capped Chickadee (*Parus hudsonicus*). Davis and Williams (1964) described irruptions of Clark's Nutcracker (*Nucifraga columbiana*) in which the birds moved from their usual winter range in the mountains into desert and coastal areas. In this instance, the winter movement was attributed to low pine-cone crops in the mountain areas.

ECOLOGICAL DISTRIBUTION

MacArthur (1971:194) wrote that "it is not clear where community ecology ends and biogeography begins. . . ." Great expanses of climax vegetational

Figure 19. A female Snowy Owl at her nest on Victoria Island, Canadian Arctic Archipelago, July 21, 1960. (Courtesy of David F. Parmelee.)

types existed before man began to exploit the land. Even in undisturbed areas, however, local variations in climatic and topographical features (such as streams, rivers, lakes, cliffs, mountains) resulted in discontinuities in the climax vegetation so that "islands" of different plant forms occurred, each of which had its characteristic bird species. Hence, to say that a particular species is found throughout the eastern United States means only that it is to be found merely where suitable habitat exists for it.

Ecotones

Only rarely is there an abrupt change in the vegetation at the boundary between adjacent biomes. Usually there are areas of transition, called *ecotones*, from one type of plant community to another. The ecotones have a wider variety of plants than are found in the relatively uniform composition of a biome. Such areas of transition in vegetational types are found, for example, between coniferous and deciduous forests, between forests and shrub-grown fields, and between marshes or bogs and upland communities. Ecotones are of special interest to ornithologists because they usually support a larger variety of bird species than either of the communities they separate. This tendency for increased variety or diversity of species in ecotones is called the *edge effect*. Some species are almost entirely limited to such areas of transition. Johnston and Odum (1956) reported that from 40 to 50 percent of the breeding bird species in the Piedmont region of Georgia occupied an ecotone type of vegetation.

Middleton (1957) studied an unusual area of about 300 acres in southeastern Michigan, where he found bird species typical of northern coniferous forests and of both northern and southern deciduous forests. Conifers along a spring-fed stream included arbor vitae, tamarack, white pine, hemlock, and black spruce. This distinctly northern-type habitat, with its associated ground cover of sphagnum moss, wintergreen, pipsissewa, goldthread, and clintonia, supported such typically northern species as the Magnolia Warbler, Northern Waterthrush (*Seiurus noveboracensis*), Mourning Warbler (*Oporornis philadelphia*), and Canada Warbler (Fig. 20). There also was a swamp forest

Figure 20. A male Mourning Warbler at its nest; Luce County, Michigan; June 24, 1957. (Courtesy of Lawrence H. Walkinshaw.)

in which the principal woody plants were red maple, black ash, slippery elm, red-osier dogwood, and several species of hawthorns. Here the nesting species were those characteristic of deciduous forests in southern Michigan (Wood Thrush, Veery, Red-eyed Vireo, Ovenbird, *Seiurus aurocapillus,* Scarlet Tanager, Rose-breasted Grosbeak). The third type of habitat was an upland beech-maple forest with a scattering of oaks, hickories, basswood, cottonwood, black cherry, and tulip poplar. Southern species that have moved into Michigan within the past 75 years and/or that approach the northern limit of their breeding range in Michigan were found in this upland forest (Acadian Flycatcher, Tufted Titmouse, Yellow-throated Vireo, Cerulean Warbler, and Cardinal) or in the wetter areas (Blue-winged Warbler, Yellow-breasted Chat).

In his thorough study of the birds of the pine-oak woodlands of southern Arizona, Marshall (1957:64) wrote that "life-zone categories have obscured an understanding of the important environments for birds in this area; it is my impression that the biotic areas have been concocted by taxonomic slight-of-hand. They seem meaningless unless defined so generously as to consist of an entire biome. Under the community concept, the pine-oak association of this study area might be regarded as a sort of self-perpetuating ecotone. It consists of a permanent mixture of tree forms and birds from two biomes, represented locally by ponderosa pine forest and encinal, enabling it to support a greater variety of birds than do the biomes separately." He noted (p. 40) that the varied avifauna (93 species) "was apt to reflect the peculiar juxtaposition of vegetation in the canyon chosen rather than to typify the mountain, its latitude, and climate." Consequently, the breeding species found in his census stations were similar to those present in other foothill areas in the southwestern United States.

Soikkeli (1965) compared the bird populations in coastal meadows in Finland, and noted that "coastal meadows form narrow transition zones between two major ecological communities: dry land and water. This ecotonal zone is inhabited by bird species belonging originally to both of these communities, viz. species of the orders of *Passeriformes* and *Anseriformes,* but also by a group of species adapted to circumstances predominating just in this transition zone, namely members of the order of *Charadriiformes.*" In fact, charadriiform species constituted from 50 to 60 percent of the birds found in two primary study sites; each of the other orders contributed approximately 20 percent of the bird population.

It is still possible to study virgin, or near-virgin, areas in a few parts of the world, but man has created many ecotonelike habitats, and therefore has changed the distribution of certain bird species. For example, the effect that cutting the deciduous and coniferous forests in the Great Lakes area had on the relative abundance and distribution of the Brown-headed Cowbird (*Molothrus ater*) and other grassland species was described by Mayfield (1965).

Ecosystems, Communities, and the Ecological Niche

Kendeigh (1961:18) defined a community as "an aggregate of organisms which form a distinct ecological unit. Such a unit may be defined in terms of flora, of fauna, or both. Community units may be very large, like the continent-wide coniferous forest, or very small, like the community of invertebrates and fungi in a decaying log." Plant and animal communities together "make up the *biotic community*, and the biotic community along with its habitat is termed an *ecosystem*."

An ecosystem can be studied by examining either the abiotic components (climatic factors and inorganic materials) or the biotic components (the producer and consumer organisms); and one may study nutrient cycles and energy exchange, food chains, diversity of plants or animals, stability and fragility, evolution, or control of ecosystems. Moreover, a number of different ecosystems may exist in a relatively small area. So varied are the climatic areas in Hawaii, for example, that plant ecologists have defined 30 different ecosystems just on the eastern flank of Mauna Loa, a mountain that rises 13,680 feet above sea level.

Each species of bird lives in a particular habitat in an ecosystem. Within that habitat it occupies an *ecological niche*; i.e., "the ecological niche of an organism depends not only on where it lives but also on what it does (how it transforms energy, behaves, responds to and modifies its physical and biotic environment), and how it is constrained by other species" (Odum, 1971:234). Milstead (1972) proposed that there is a "finite number of broad ecological niches" in any given type of biome and that the niches can be defined by comparing the biology of convergent animal forms on different continents. He concluded, however, that "past natural history studies do not yield the detailed type of information needed" in order to make such comparisons.

These concepts of the ecosystem, community, and ecological niche emphasize the fact that the dynamics of a bird's life cycle can be understood only after one determines the interrelationships of the bird with other members of its own species, with other species that occupy the same habitat, and with the vegetation that forms the growth form of the habitat. Because of the complexities of ecosystems, therefore, bird ecologists usually study some aspect of bird communities, and they may do so in one of many ways. Only selected examples can be discussed here (see MacArthur, 1971, for other examples and references).

Species Diversity.   There are many more species of birds in tropical regions than in temperate regions, but among passerine species the number of individuals of each species per unit area tends to decrease as one moves from temperate to tropical regions. Nonpasserine species, however, although they form a higher proportion of the species in tropical areas than in North Temperate regions, do not show a reduction in numbers per unit area in the tropics. Klopfer and MacArthur (1960) interpret these phenomena "as support for the notion that the phylogenetically older non-passerines species are insufficiently plastic in their niche requirements to colonize temperate areas, tropical niches being smaller and less subject to change."

Mixed forests (particularly of hardwoods) also provide suitable habitats for more species of birds than do forests of a homogeneous composition. MacArthur and MacArthur (1961) attempted to determine what factors account for species diversity in selected forest types from Maine to Florida and in a savanna in Panama (Fig. 21). They concluded that, in deciduous forests, "bird species diversity can be predicted in terms of the height profile of foliage density. . . . The layers 0-2´, 2´-25´, > 25´ seem equally important in determining bird species diversity; these layers presumably correspond to different configurations of foliage." They acknowledged that there is "nothing biological about the number of layers chosen. Four or 5 layers in a roughly similar subdivision would be more cumbersome to analyse but would presumably be more accurate."

After studying deciduous forests in the eastern United States, MacArthur and his co-workers reached four conclusions regarding bird species diversity.

1.   The amount of vegetation (as measured in area of leaves per unit volume of space) in each of three horizontal layers (herbs, shrubs, and trees over 25 feet tall) determines the diversity of bird species that will breed in 5 acres of habitat.

2.   The number of breeding bird species is greatest when each of the three layers has the same amount of foliage.

3.   A knowledge of the number (or volume) of plant species does not improve the chances of predicting the number of bird species actually present.

4.   In fields, bushy fields, or seral stages in ecological succession, the bird species present and their abundance can be predicted from measurements of the variety of patches of vegetation "by showing that each bird species has a preference for a certain characteristic proportion of foliage in each layer."

When, however, MacArthur (1964) extended his studies to include the more complex habitats on the slopes of the Chiricahua Mountains of southeastern Arizona, he found that, although "the prediction of bird species diversity from the foliage profile needed only slight modification, . . . the prediction based on foliage profile of just what bird species these would be, failed."

Marshall (1957:64) had reported earlier that, in his pine-oak study area in Arizona, "only a small proportion of bird species seems to depend upon the *combined* life forms of pine and oak, as does the Painted Redstart, for instance. Therefore an interdependent community of vertebrates and plants was hardly evident. As the dominant plants were inconsistent in their groupings, so were the birds. Aside from a few instances of predation, competition, parasitism, and

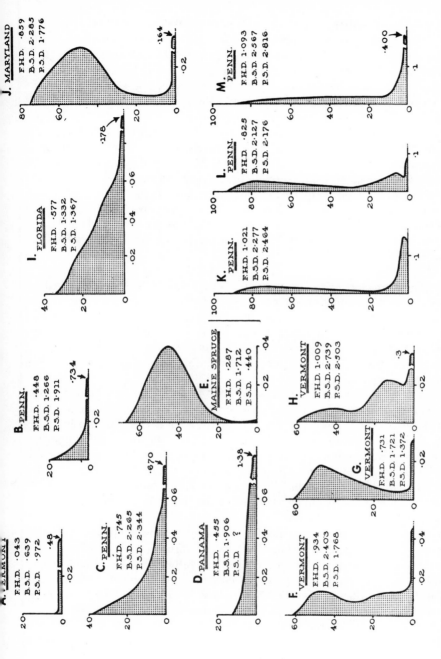

Figure 21. The densities of foliage (measured in square feet of leaf silhouette per cubic foot of space) are plotted along the abscissae. The height in feet above the ground is the ordinate. F.H.D. is foliage height diversity; B.S.D., bird species diversity; and P.S.D., plant species diversity. (Courtesy of Robert H. and John W. MacArthur and the editor of *Ecology*.)

the dependence of several species upon holes of Acorn Woodpeckers for nest sites, each species seemed to obey its unique laws of distribution, oblivious to its fellows. Whole niches were thus left vacant in some pine-oak areas. . . ."

Similarly, Balda (1969) failed to find a neat correlation between bird species diversity and foliage height diversity in ponderosa pine and oak-juniper forests in Arizona, but rather observed three different patterns of tree use by birds. Juniper (*Juniperus deppeana*) was the most abundant tree in the oak-juniper habitat; although this species showed a "good fit for bird use according to the height distribution of the foliage," it was sparsely used by the birds in the area. Two species of oaks were heavily used by the birds and all parts of a tree were exploited by the several bird species. In the pine forest, however, "total bird use of Douglas fir foliage by height class is not close, as there is a large section of low foliage which is underused while the upper heights show a heavy concentration of bird activity"; at the same time, "the very few tall Douglas firs in the area . . . were used greatly out of proportion to their availability" (Fig. 22).

Foraging Behavior and Competition. MacArthur (1964) suggested three ways in which a population of birds could be distributed within a forest so as to reduce serious competition: *vertical,* in which different species occupy more or less separate strata in the forest; *horizontal,* in which different species occupy "different patches in the environment"; and *temporal,* in which breeding seasons are staggered, thereby ostensibly reducing competition for food when young are being fed.

In a study of vertical stratification of birds in a dry forest in Peru, Pearson (1971) found that 8 of 37 species foraged in four of five strata that he had arbitrarily selected and, moreover, that edge birds, characteristic of brushy clearings, "did not remain constant relative to fixed heights above the ground, but instead seemed to follow any area of direct exposure to light; thus they apparently treated the upper canopy and emergent trees [75 to more than 100 feet tall] as a 'scrub' area as well." Pearson also found a daily vertical shift in the foraging activities of populations, birds of the upper strata moving to lower strata between 9 A.M. and 2 P.M.

Willis (1966) studied interspecific competition between two species that follow swarms of army ants. The Ocellated Antthrush (*Phaenostictus mcleannani*) is behaviorally dominant over the Plain-brown Woodcreeper (*Dendrocincla fuliginosa*), and the woodcreepers forage closer to the ground when the antthrushes are absent. On Trinidad, for example, where low-foraging antbirds usually are absent, the woodcreeper forages close to the ground, whereas on Barro Colorado Island, Panama Canal Zone, the woodcreeper is forced to forage at higher levels in the vegetation. Willis cites this as an example of competitive exclusion by dominance.

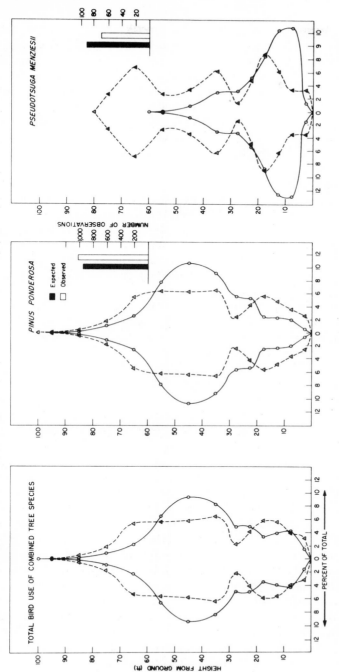

Figure 22. Total bird use of tree foliage in a ponderosa pine forest. The shaded bar is the expected bird use in each species of tree, based on available foliage volume of that tree species. The open bar is the actual number of observations. The solid line is the percentage of total foliage volume in each height class. The broken line is the percentage of total observations of birds in each height class. The middle and right figures show total bird use and available foliage volume in the two most abundant tree species. (Courtesy of Russell P. Balda and the editor of *Condor*.)

313

Cody (1966, 1968) presented a detailed comparison of the species composition and diversity of birds in grassland communities in widely scattered areas in North, Central, and South America. He proposed that grassland communities tend to be "saturated," i.e., the habitat contains the maximum number of bird species that it can support. He believed this assumption to be justified because, in 14 different areas, the number of species and the bird species diversity "fall between 3–4 and 2.44–3.83, respectively, exhibiting remarkable constancy." Cody approached the problem in a variety of ways, and the student is referred to the original papers for details, including an elaborate model purported to enable one to make predictions of community variables. It will suffice here to quote Cody's chief conclusions: "The members of avian communities resident in simple grassland habitats coexist by virtue of differences in habitat preferences and feeding behavior and in very tall vegetation by differences in feeding height. The sum of these ecological differences is constant for all communities. Using only two habitat indexes, vegetation height and its standard deviation, it is possible to predict (1) the number of species, (2) the differences in their feeding ecology, and (3) their relative habitat separation in the community which occupies this habitat. As these predictions are made for South American communities on the basis of the communities studied in North America, and as the predictions hold regardless of grazing or irrigation programs, it is suggested that these communities have a full quota of species which are optimally adapted to their current environment."

After completing an intensive study of grassland birds in Wisconsin and comparing his results with those of Cody and of Ricklefs (1966), Wiens (1969), however, remarked that "foraging behavior is sometimes regarded as the only behavioral feature to be considered in analyses of coexistence mechanisms. . . . Certainly it does not seem proper to restrict consideration, *a priori*, to a few readily-measurable habitat features which may or may not have any direct relevance to the activity of the birds. Cody (1968), for example, in his analysis of the habitats of grassland birds, restricted measurements to vertical and horizontal vegetation density, vegetation height, and 'profile area' (the area under the graph of height vs. average horizontal density). Despite the degree of separation between species that Cody was able to demonstrate with these few measures and the elegant model which they enabled him to build, it seems to me that many other features undoubtedly may influence the choice of habitats by grassland birds and that restriction of habitat analysis and description may preclude any detailed insight into the dynamics of the organization of grassland avian communities."

Wiens studied such features as density and height of vegetation, light intensity at various levels in the vegetation, depth of ground litter, soil characteristics, and territorial and other behavioral activities of the bird species (Figs. 23, 24). He found both a higher species diversity and a higher density of indi-

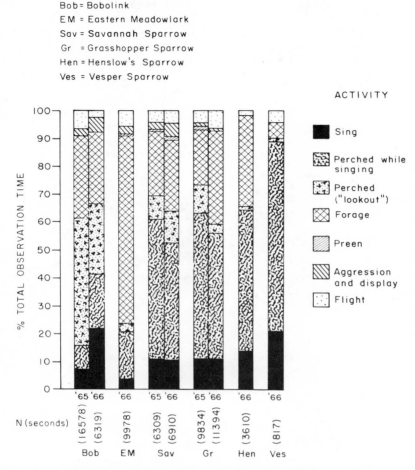

Bob = Bobolink
EM = Eastern Meadowlark
Sav = Savannah Sparrow
Gr = Grasshopper Sparrow
Hen = Henslow's Sparrow
Ves = Vesper Sparrow

ACTIVITY

■ Sing

Perched while singing

Perched ("lookout")

Forage

Preen

Aggression and display

Flight

Figure 23. Activity budgets of six breeding grassland species at Fitchburg, Wisconsin, during 1965 and 1966, from continuous tape-recorded observations. (Courtesy of John A. Wiens and the American Ornithologists' Union.)

viduals in Wisconsin than had been reported by Cody in his comparative study. The density of breeding birds (from 85 to 126 pairs per 100 acres) in Wisconsin was similar to that reported by Graber and Graber (1963) for pastures and ungrazed grasslands in Illinois (59 to 168 pairs per 100 acres).

Of special significance are Wiens's comparisons of observational data obtained from "opportunistic spot-observations" and from tape-recorded continuous observations of individuals. The results shown in Table 1 clearly

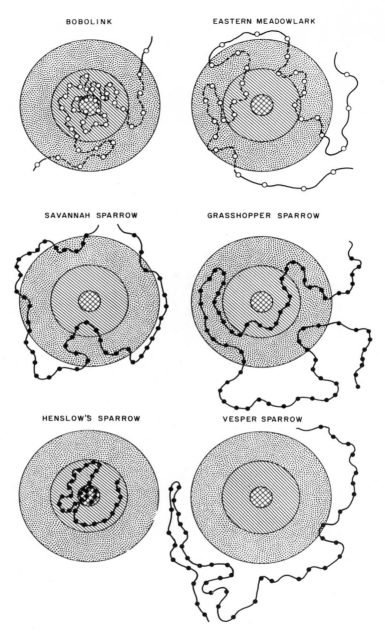

Figure 24. General characteristics of foraging patterns of six grassland species at Fitchburg, Wisconsin, 1966. The concentric circles indicate contours of grass depth, increasing toward the center of a hypothetical grass clump surrounded by low, lawnlike vegetation. The length of the foraging path of each species in each of the grass depth categories indicates the relative degree of utilization of the category. Key: solid line, individual walks rapidly (more than 1 meter per

demonstrate the inadequacy of many field census techniques if one is, indeed, interested in how birds actually spend their time.

Temporal Separation.   The proposition that drawn-out or staggered breeding seasons effectively reduce competition for food or nest sites appears to be a logical one, and Berger (1954) reported that, in southern Michigan, three species of birds built their nests in the same type of vegetation but at different times during the summer: Yellow Warblers began nesting during the first or second week of May; Willow Flycatchers, during the first week of June; and American Goldfinches, usually not until the first or second week of July (and often later) (Fig. 25). Although competition for food does not occur among these species, competition for nest sites is possible in some habitats. During one summer, all three species built nests in the same clump of panicled dogwood, but construction of a nest by the Willow Flycatcher was not begun until a Yellow Warbler's nest held nestlings, and the Goldfinch nest was not started until the Willow Flycatcher nest contained young. Other examples could be cited, but the doctrine of temporal separation was thought originally to apply primarily to tropical areas where the overall breeding season often extends throughout much of the year. However, MacArthur (1971:197) wrote that "the remarkable thing is that the season of breeding of all species combined is about 1.35 times the average season of the component species, whether we study the tropic, the temperate, or the Arctic zones! Thus, in Lapland, the season for nesting was 1.32 months while the separate species had average seasons of 0.96 months (1.32/0.96 = 1.37); and West Java, with the longest nesting seasons, had a combined nesting season of 9.89 months with the component species nesting over an average of 7.54 months (9.89/7.54 = 1.31). Thus, there are about one-and-a-third nesting shifts the world over." Indeed, Ricklefs (1966) concluded that "the average length of the breeding season of individual species of birds occupies very nearly the same proportion of the total breeding season in all localities. This essentially eliminates a temporal component to increased tropical species diversity. Further analysis gives no evidence that closely related sympatric species stagger their nesting seasons to avoid competition, but in fact, the seasons of such pairs consistently overlap more than 90 per cent. This and other evidence indicates that the temporal diversity found within the total breeding season is the result of specific feeding differences together with temporal diversity in the availabilities of the different food sources."

Whether or not these conclusions will prove to be valid after more detailed

minute); dashed line, walks slowly (less than 1 meter per minute); open circles, foraging pecks which are slow and deliberate; solid circles, foraging pecks rapid; cross hatching, grass depth of 16 to 20 cm; single-line hatching, grass depth 11 to 15 cm; stippling, grass depth 6 to 10 cm; white, grass depth 0 to 5 cm. (Courtesy of John A. Wiens and the American Ornithologists' Union.)

Table 1. Comparison of Values Obtained with Continuous Tape Recordings of Activity and those Obtained by Spot Observations, 1965[a]

Values are percentages of total recorded time or total observations. The differences reflect strong biases introduced in the spot-observation procedure.

Species and Activity	Continuous Observations	Spot Observations
Savannah Sparrow	N = 6309 seconds	N = 307 observations
Activity		
Singing	61.3	90.9
Foraging	23.6	3.6
Perching	7.7	3.3
Aggression and display	3.1	2.3
Flying	3.8	—
Preening	0.4	—
Substrate		
Grass	30.4	13.0
Heavy-stemmed forb	32.5	31.9
Medium-stemmed forb	17.6	18.2
Wire	3.7	22.1
Post	9.6	13.0
Tree	—	1.6
Flight	6.2	—
Grasshopper Sparrow	N = 9834 seconds	N = 161 observations
Activity		
Singing	63.8	96.9
Foraging	19.3	3.1
Perching	10.1	—
Aggression and display	0.9	—
Flying	4.4	—
Preening	1.4	—
Substrate		
Grass	24.3	9.3
Heavy-stemmed forb	37.1	43.5
Medium-stemmed forb	16.6	13.0
Wire	10.3	20.5
Post	6.7	13.7
Tree	—	—
Flight	5.0	—

[a] Reproduced by permission from Wiens (1969).

Figure 25. An American Goldfinch nest with six eggs; Ann Arbor, Michigan, July 27, 1962.

studies have been conducted remains to be seen. Ricklefs, Klopfer, and others, of necessity, used previously published data on breeding seasons, and few were as detailed as the authors would have liked. Moreover, data that demonstrate that food is a limiting factor are available for very few bird species (see Lack, 1966). Wiens (1969:85) remarked that "the existence of competition between species, however, demands more than overlap in occupancy of a habitat or utilization of a resource; the resource in question must also be in 'short supply' . . . , and it is exceedingly difficult to discern the fuzzy line which separates 'short supply' from 'sufficient for both.' This would seem especially true with food resources. On the other hand, if food is superabundant, species with identical foraging patterns and food habits may continue to coexist. Evans (1964) concluded that food was superabundant in an old field in Michigan in which Field, Chipping, and Vesper sparrows nested."

Altitudinal Distribution.   Terborgh (1971) studied a different aspect of ecological distribution by examining species composition (diversity) between elevations of 585 to 3510 meters on the Cordillera Vilcabamba in Peru. He collected
his data (4961 birds) by capturing birds in mist nets at 15 stations along the
environmental gradient that extended from lowland rain forest upslope through
cloud forest, elfin or stunted forest, to alpine grassland. Terborgh designed
three models in an attempt to explain the distributional limits of 261 bird
species along the transect. He stressed certain limitations in the models and
noted that, "as evaluated by this preliminary analysis, the three mechanisms of
distributional limitation differ appreciably in their importance in the Vilcabamba avifauna. Ecotones account for less than 20% of the distributional
limits, competitive exclusion for about one-third of the limits and gradually
changing conditions along the gradient for about one-half of the limits."
Terborgh's important study emphasizes the difficulties involved in the analysis
of extensive bird faunas found over environmental gradients that span significant changes in elevation. The problems are compounded in tropical regions
because of the numbers of bird species and the difficulties of working in rain
forests and cloud forests.

Habitat Selection.   The question involved here is, Why does a bird live where
it does? Thorpe suggested the possibility of habitat imprinting in young birds,
and James (1971) developed the concept of an ecological *niche-gestalt,* by which
she meant that each bird species has a set of characteristic vegetational requirements. "Inherent in the term *gestalt* are the concepts that each species has a
characteristic perceptual world . . . , that it responds to its perceptual field as
an organized whole . . . , and that it has a predetermined set of specific search
images. This is assumed to be at least partially genetically determined, but is
surely also modifiable by experience and subject to ecological shift under varying circumstances." James measured 15 vegetational features of the habitats for
46 species of Arkansas breeding birds and then subjected the data to two
methods of multivariate analysis. Her stimulating paper deserves careful reading, but space permits only a quotation from her conclusions: "Although all 15
vegetational variables contributed significantly to the ordinations, the most
powerful variables for describing habitat differences were per cent canopy
cover, canopy height, and the number of species of trees per unit area. If one
considers the vegetation of a geographic area to be a set of continuously-varying
phenomena, and if one assumes that bird distribution is at least partly based on
species-specific adaptiveness to the resources offered by this heterogeneous
structure, then ordination procedures are appropriate methods for its
expression" (Fig. 26).

Actually, we know very little about habitat selection in birds, or the reasons

Figure 26. Outline drawings of the niche-gestalt for five species of vireos, representing the visual configuration of those elements of the structure of the vegetation that were consistently present in the habitat of each. Numbers give the vertical scale in feet. (Courtesy of Frances C. James and the editor of *Wilson Bulletin*.)

why some species are ecologically tolerant and others are ecologically intolerant (Hildén, 1965; Klopfer and Hailman, 1965).

Islands.   The peculiar features of islands were discussed by Charles Darwin and A. R. Wallace during the last century and, more recently, by Carlquist (1965), MacArthur and Wilson (1967), Nelson (1968), and MacArthur (1972). There are *continental islands* and *oceanic islands.* A continental island once was connected with a continental land mass. At the time such an island was cut off, it had the same general groups of plants and animals as the adjacent mainland; its biota is said to be *harmonic.* "As the age of a continental island increases, some groups originally present may become extinct, and those that persist will follow different evolutionary paths than those taken by their mainland relatives, but traces of the mainland origin of the biota are never completely eliminated" (Hubbell, 1968). An oceanic island has never been connected with a major land mass; this type has emerged above the surface of the ocean, primarily as a result of volcanic action; its plants and animals are the descendants of ancestors that reached the island across a saltwater barrier. Oceanic islands are colonized through long-distance dispersal, and the capacity for such dispersal varies tremendously among plants and animals. Consequently, only a limited number of kinds of continental forms ever reach oceanic islands and become established on them; many forms are absent, and the biota is said to be *disharmonic.*

Oceanic islands have been referred to appropriately as "natural laboratories for the study of evolution," because the results of adaptive radiation from a presumed single ancestral species frequently are most striking on such islands: e.g., among the Hawaiian honeycreepers (Drepanididae) and the Galapagos finches (Geospizinae). Consequently, *endemism,* the evolution of unique forms found nowhere else, typically is high on oceanic islands. The degree of endemism tends to be directly related to the distance between an island and major land masses. The Hawaiian Islands, for example, are separated by more than 2000 miles of ocean from both North America and Asia. Approximately 95 percent of the flowering plants and more than 98 percent of the birds, land molluscs, and insects are endemic to those islands. Since the discovery of the Hawaiian Islands by Captain James Cook in 1778, ornithologists have described 17 genera, 39 species, and 38 subspecies of endemic land and freshwater birds that are presumed to have evolved from 15 ancestral species. The age of the islands is such that only one successful colonization by an ancestral species every 300,000 years would have been necessary to account for this assemblage of birds. Indeed, during the past century, more than 100 species of migrants, stragglers, and chance arrivals have been recorded in the Hawaiian Islands, but none of these succeeded in becoming established (Berger, 1972b). By contrast, 8 species of birds have been successful in colonizing New

Zealand (about 1230 miles from Australia) since about 1850 (Wodzicki, 1965).

The size of an island, as well as its distance from other land, is significant for colonization. Thus the islands in the Hawaiian chain contain a total land area of approximately 6400 square miles, and they extend for a distance of some 1660 nautical miles from southeast to northwest. By contrast, Christmas Island in the Central Pacific (2° N., 157° W.), the largest coral atoll in the Pacific Ocean, is about 35 miles long and 14 miles in maximum width; the total land area is about 94 square miles (Fig. 27). It has but a single species of endemic land bird, *Conopoderas a. aequinoctialis* (Sylviidae).

Populations and Their Study. A *population* may be defined as the total number of individuals found in a given area or as the total population of a species throughout its range. A population possesses characteristics that are peculiar to the population rather than to the individuals themselves: density, natality (birth rate), mortality (death rate), biotic potential, growth form, dispersion, and distribution. The many ramifications of population analysis have been discussed by Lack (1966), MacArthur and Connell (1966), and von Haartman (1971).

One fundamental approach to populations is determining the density of a

Figure 27. Cook Islet, Christmas Island (Pacific Ocean); the birds are White Terns (*Gygis alba*); November 15, 1971. The islet also serves as nesting habitat for Crested Terns (*Thalasseus bergii*) and many other seabirds.

particular population. The number of birds that actually occupy a unit of habitat constitutes the *ecological density* of the species; this number can be learned only by making a direct count. More often, however, students of bird populations present their results in terms of the number of individuals or pairs per 100 acres. Usually data obtained from a study of a small area (sample-plot census) are projected mathematically to the 100-acre unit (census by sampling); the total density obtained in this manner is referred to as *crude density*.

Emlen (1971) discussed seven census methods and compared their efficiency ratings (acres per man-hours) for comparable accuracy. He pointed out that all methods have their complications and limitations (Fig. 28); several rely on a count of singing males. He then described a new method (count × detectability, Table 2) that involves the use of a *coefficient of detectability* (C.D.). The C.D. value has to be determined for each species through preliminary (but thorough) field analysis in order to determine a lateral distance conversion figure and a basal detectability adjustment. For most species, Emlen selected a band 412 feet (126 meters) wide on either side of a mile-long transect because this incorporates 100 acres, a frequently used standard in population studies. Emlen remarked that his method "appears to be applicable to most nonflocking, temperate zone doves, cuckoos, hummingbirds, woodpeckers, and passerines. It is poorly suited for wide-ranging water birds, shorebirds, and hawks, for nocturnal birds, for treetop birds in tall dense forests, and for swifts and swallows that cruise about above the vegetation." Moreover, "low values indicating less than 20 percent coverage were obtained for nonsinging ground feeders that characteristically remained undetected until flushed at close range (wintering Bobwhite Quail, Savannah Sparrows, and Ovenbirds), quiet species frequenting dense brush (nonsinging Song Sparrows and Yellowthroats), and small quiet arboreal and subarboreal birds (Kinglets and Brown Creeper in the fall and most wintering warblers)."

We emphasize the problems involved in censusing birds because so many kinds of field studies depend in large part on population size (e.g., number of breeding pairs per unit area), and because invalid conclusions result from inaccurate data. Davis (1965) demonstrated clearly the potential significant error in singing-male censuses if one is not thoroughly aware of the daily, as well as seasonal, differences in a species's song cycle: "a difference of only thirty minutes can influence daily census results in major fashion, even when censuses are made very early in the morning and are timed to a schedule which has been set by the birds themselves."

INTRODUCED BIRDS

For more than 2000 years, man has been an agent for the dispersal of birds. The Jungle Fowl may have been the first species to be transported from its na-

Table 2. Characteristics and Evaluations of Census Methods for Nonflocking Terrestrial Birds[a]

	A	B	C	D	E	F	G
Characteristics	Mark portion and tally ratio	Map nests	Map territories	Count in fixed strip	Count to flushing distance	Count × effectivity	Count × detectability
Areal unit	Plot	Plot	Plot	Strip (fixed)	Strip = 2 flushing distances	Strip (indef.)	Strip (wide)
Tally unit	Marked observations: total observation	Nests	Singing males	Max. count for each species	Total count	Singing ♂♂	Total count
Conversion	Marked population: tally	×2 (for ♀♀)	×2 (for ♀♀)	(None)	(None)	Divided by effectivity ×2 for ♀♀	Divided by C.D.
Subjective evaluations							
Seasonal applicability	Summer-winter	Summer	Summer	All	All	Summer	All
Acres/3 man-hours[b]	10	10	50	100	50	300	300
Replications needed[b]	10	5	10	7	3	3	3
Total man-hours[b]	40	15	35	21	9	9	9
Efficiency (acres/hour)	0.25	0.7	1.4	4.9	5.6	33.0	33.0

[a] Reproduced by permission from Emlen (1971).
[b] Estimated.

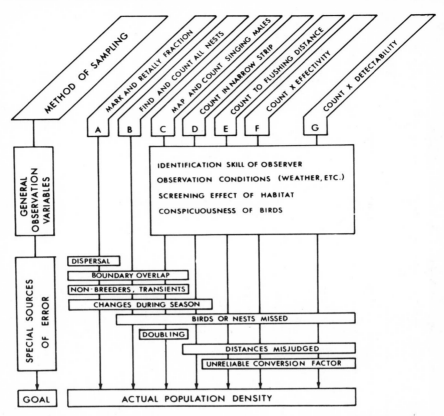

Figure 28. Factors limiting the accuracy and efficiency of census methods for nonflocking terrestrial birds. (Courtesy of John T. Emlen and the editor of *Auk.*)

tive range (Ceylon, India, and eastward to Java) to Europe and elsewhere. The Ring-necked Pheasant is so well known in many parts of North America that one is likely to forget that it is an exotic species. The House Sparrow and European Starling are well-known examples of adaptable exotic species that have spread across the United States, northward into Canada and Alaska and southward into Mexico; Yocum (1963) reported three Starlings near Fort Yukon, Alaska, north of the Arctic Circle. By contrast, three species that have been established for many years have not undergone extensive range expansion: Crested Mynah (*Acridotheres cristatellus*), established on Vancouver Island and casual in parts of Washington and northwestern Oregon; European Tree Sparrow (*Passer montanus*), released in St. Louis and now found there and in adjacent areas in Illinois; and European Goldfinch (*Carduelis carduelis*), resident in southwestern Long Island, New York. Representatives of seven bird families have become established in southern Florida since about 1950 (Rohwer and Woolfenden, 1969); the Monk Parakeet (*Myiopsitta monachus*) has spread rapidly from the New York-New Jersey area since its escape from John F. Kennedy Airport in 1967 and has become established in other widely separated states.

The largest number of exotics, however, have been introduced on certain islands. As of 1962, at least 130 species of foreign birds had been introduced to New Zealand (Wodzicki, 1965); of these, 36 species had become established. A significant point is that only one-third is to be found in both the introduced and the native habitat, the others living in the man-made habitat only. None of the colonists is found only in the native habitat. A minimum of 160 bird species had been intentionally or accidentally released on the Hawaiian Islands as of 1972; about 50 species had become established. Several of these exotics have invaded the near-virgin Hawaiian rain forests, where most species of endemic

Figure 29. A Japanese Crane (*Grus japonensis*), an endangered species; Hokkaido, Japan, March 19, 1969. (Courtesy of Lawrence H. Walkinshaw.)

forest birds live. Exotics also have been introduced on many other islands in the Pacific and elsewhere; in most instances, they have proved to be a nuisance.

Except for some gamebirds, little study has been made of introduced birds, particularly of their relationships to, and effects on, native birds. Some of the problems have been discussed by Wodzicki (1965), Gullion (1965), and Berger (1972a).

RARE AND ENDANGERED SPECIES

At least brief mention of this subject is mandatory at this time in history. James Fisher and Roger Tory Peterson reported that 76 species of birds have become extinct since 1681, the last year that a dodo was seen on the island of Mauritius. Approximately 13 species were exterminated by excess hunting for food; 11 species are presumed to have become extinct as a result of the introduction of cats, rats, mongooses, and other predators into a foreign habitat; and about 14 species succumbed because of destruction of their habitats by draining wetland areas, cutting forests, and clearing land for agriculture. The

Figure 30. A female Great Indian Bustard (*Choriotis nigriceps*) settling on her nest; another endangered species. (Courtesy of Shivrajkumar of Jasdan, India.)

Wake Island Rail undoubtedly became extinct about 1945, a casualty of World War II, presumably because the birds were needed for food by a starving Japanese garrison on the island. The Laysan Island Rail population on the islands of Midway Atoll also was a victim of the war—rats got onto the islands from naval ships. Of the 76 species reported by Fisher and Peterson, 14 were Hawaiian birds, and these have become extinct during the past century. The International Union for Conservation of Nature and Natural Resources tallied 337 species and subspecies of rare and endangered birds in its *Red Data Book* as of 1968 (Figs. 29, 30).

REFERENCES

Balda, R. P. 1969. Foliage use by birds of the oak-juniper woodland and ponderosa pine forest in southeastern Arizona. *Condor,* **71:**399–412.

Beals, E. 1960. Forest bird communities in the Apostle Islands of Wisconsin. *Wilson Bull.,* **72:**156–181.

Berger, A. J. 1954. Association and seasonal succession in the use of nest sites. *Condor,* **56:**164–165.

——— 1972a. Hawaiian birds 1972. *Wilson Bull.,* **84:**212–222.

——— 1972b. *Hawaiian Birdlife.* University Press of Hawaii, Honolulu.

Bond, J. 1971. *Birds of the West Indies.* Collins, London.

Buerkle, U., and W. D. Mansell. 1963. First nesting of the Cattle Egret (*Bulbulcus ibis*) in Canada. *Auk,* **80:**378–379.

Carlquist, S. 1965. *Island Life: A Natural History of the Islands of the World.* Natural History Press, Garden City, N. Y.

Chapman, F. M. 1928. Mutation in *Capito auratus. Am. Mus. Novit.,* No. 335.

Cody, M. L. 1966. The consistency of intra- and inter-continental grassland bird species counts. *Am. Nat.,* **100:**371–376.

——— 1968. On the methods of resource division in grassland bird communities. *Ibid.,* **102:**107–147.

——— 1974. *Competition and the Structure of Bird Communities.* Princeton University Press, Princeton.

Daubenmire, R. F. 1938. Merriam's life zones of North America. *Quart. Rev. Biol.,* **13:**327–332.

Davis, D. E. 1960. The spread of the Cattle Egret in the United States. *Auk,* **77:**421–424.

Davis, J. 1965. The "singing male" method of censusing birds: a warning. *Condor,* **67:**86–87.

———, and L. Williams. 1964. The 1961 irruption of the Clark's Nutcracker in California. *Wilson Bull.,* **76:**10–18.

Dice, L. R. 1943. *Biotic Provinces of North America.* University of Michigan Press, Ann Arbor.

Dixon, K. L. 1959. Ecological and distributional relations of desert scrub birds of western Texas. *Condor,* **61:**397–409.

Eisenmann, E. 1971. Range expansion and population increase in North and Middle America of the White-tailed Kite (*Elanus leucurus*). *Am. Birds,* **25:**529–536.

Emlen, J. T. 1971. Population densities of birds derived from transect counts. *Auk,* **88:**323–342.

Evans, F. C. 1964. The food of Vesper, Field and Chipping sparrows nesting in an abandoned field in southeastern Michigan. *Am. Midl. Nat.,* **72:**57–75.

Fisher, H. I. 1973. Pollutants in North Pacific albatrosses. *Pacific Sci.,* **27:**220–225.

Fisher, J., N. Simon, and J. Vincent. 1969. *Wildlife in Danger.* Viking Press, New York.

Gadow, H. 1913. *The Wanderings of Animals.* Cambridge University Press, London.

Goodman, I. J., and M. W. Schein. 1974. *Birds Brain and Behavior.* Academic Press, N.Y.

Graber, R. R., and J. W. Graber. 1963. A comparative study of bird populations in Illinois, 1906–1909 and 1956–1958. *Bull. Ill. Nat. Hist. Surv.,* **28:**378–528.

Greenway, J. C., Jr. 1958. *Extinct and Vanishing Birds of the World.* American Committee for International Wild Life Protection, New York.

Gullion, G. W. 1965. A critique concerning foreign game bird introductions. *Wilson Bull.,* **77:**409–414.

Hildén, O. 1965. Habitat selection in birds. *Ann. Zool. Fenn.,* **2:**53–75.

Hubbell, T. H. 1968. The biology of islands. *Proc. Natl. Acad. Sci.,* **60:**22–32.

James, F. C. 1971. Ordinations of habitat relationships among breeding birds. *Wilson Bull.,* **83:**215–236.

Johnston, D. W., and E. P. Odum. 1956. Breeding bird populations in relation to plant succession on the Piedmont of Georgia. *Ecology,* **37:**50–62.

Keast, A. 1959. Australian birds: their zoogeography and adaptations to an arid continent. In *Biogeography and Ecology in Australia, Monogr. Biol.* **8:**89–114.

Kendeigh, S. C. 1954. History and evaluation of various concepts of plant and animal communities in North America. *Ecology,* **35:**152–171.

———— 1961. *Animal Ecology.* Prentice-Hall, Englewood Cliffs, N. J.

Klopfer, P. H., and J. P. Hailman. 1965. Habitat selection in birds. *Advan. Study Behav.,* **1:**279–303.

————, and R. H. MacArthur. 1960. Niche size and faunal diversity. *Am. Nat.,* **94:**293–300.

————, and ———— 1961. On the causes of tropical species diversity: niche overlap. *Ibid.,* **95:**223–226.

Lack, D. 1966. *Population Studies of Birds.* Oxford University Press, England.

———— 1968. *Ecological Adaptations for Breeding in Birds.* Methuen, London.

———— 1971. *Ecological Isolation in Birds.* Harvard University Press, Cambridge, Mass.

Logan, R. F. 1968. Causes, climates, and distribution of deserts. In *Desert Biology,* Vol. 1, Academic Press, New York, pp. 21–50.

MacArthur, R. H. 1964. Environmental factors affecting bird species diversity. *Am. Nat.,* **98:**387–397.

———— 1971. Patterns of terrestrial bird communities. In *Avian Biology,* Vol. 1, Academic Press, New York, pp. 189–221.

———— 1972. *Geographical Ecology: Patterns in the Distribution of Species,* Harper and Row, New York. (See also *Science,* **178,** 1972: 389–393.)

————, and J. H. Connell. 1966. *The Biology of Populations.* John Wiley & Sons, New York.

————, and J. W. MacArthur. 1961. On bird species diversity. *Ecology,* **42:**594–598.

————, and E. O. Wilson. 1967. *The Theory of Island Biogeography.* Princeton University Press, Princeton, N.J.

McCaskie, R. G. 1965. The Cattle Egret reaches the west coast of the United States. *Condor,* **67:**89.

Marshall, J. T., Jr. 1957. Birds of pine-oak woodland in southern Arizona and adjacent Mexico. *Pacific Coast Avifauna,* No. 32.

Mayfield, H. 1953. A census of the Kirtland's Warbler. *Auk,* **70:**17–20.

——— 1965. The Brown-headed Cowbird, with old and new hosts. *Living Bird,* **1965:**13–28.

——— 1972. Third decennial census of Kirtland's Warbler. *Auk,* **89:**263–268.

Mayr, E. 1946. History of the North American bird fauna. *Wilson Bull.,* **58:**3–41.

——— 1965. *Animal Species and Evolution.* Harvard University Press, Cambridge, Mass.

Merriam, C. H. 1894. Laws of temperature control of the geographic distribution of terrestrial animals and plants. *Natl. Geogr. Mag.,* **6:**229–238.

——— 1899. Zone temperatures. *Science,* **9:**116.

Middleton, D. S. 1957. Notes on the summering warblers of Bruce Township, Macomb County, Michigan. *Jack-Pine Warbler,* **35:**71–77.

Miller, A. H. 1942. Habitat selection among higher vertebrates and its relation to intraspecific variation. *Am. Nat.,* **76:**25–35.

Milstead, W. W. 1972. Toward a quantification of the ecological niche. *Am. Midl. Nat.,* **87:**346–354.

Morse, D. H. 1967. Competitive relationships between Parula Warblers and other species during the breeding season. *Auk,* **84:**490–502.

Murphy, R. C. 1936. *Oceanic Birds of South America.* American Museum of Natural History, New York, 2 vols.

——— 1955. Feeding habits of the Everglade Kite (*Rostrhamus sociabilis*). *Auk,* **72:**204–205.

Nelson, B. 1968. *Galapagos: Islands of Birds.* William Morrow & Co., New York.

Odum, E. P. 1971. *Fundamentals of Ecology,* 3rd ed. W. B. Saunders Co., Philadelphia.

Pearson, D. L. 1971. Vertical stratification of birds in a tropical dry forest. *Condor,* **73:**46–55.

Pettingill, O. S., Jr. 1970. *Ornithology in Laboratory and Field,* 4th ed. Burgess Publishing Co., Minneapolis.

Pitelka, F. A. 1941. Distribution of birds in relation to major biotic communities. *Am. Midl. Nat.,* **25:**113–137.

Ricklefs, R. E. 1966. The temporal component of diversity among species of birds. *Evolution,* **20:**235–242.

Rohwer, S. A., and G. E. Woolfenden. 1969. Breeding birds of two Florida woodlands: comparisons with areas north of Florida. *Condor,* **71:**38–48.

Salomonsen, F. 1951. The immigration and breeding of the Fieldfare (*Turdus pilaris* L.) in Greenland. *Proc. 10th Int. Ornithol. Congr.,* **1950:**515–526.

Sauer, J. D. 1969. Oceanic islands and biogeographical theory: a review. *Geogr. Rev.,* **59:**582–593.

Skead, C. J. 1965. The ecology of the ploceid weavers, widows and bishop-birds in the southeastern Cape-Province, South Africa. In *Ecological Studies in Southern Africa, Monogr. Biol.,* **14:**219–232.

Snyder, L. L. 1947. The Snowy Owl migration of 1945–1946: second report of Snowy Owl committee. *Wilson Bull.,* **59:**74–78.

Soikkeli, M. 1965. On the structure of the bird fauna on some coastal meadows in western Finland. *Ornis Fenn.* **42:**101–111.

Terborgh, J. 1971. Distribution on environmental gradients: theory and a preliminary interpretation of distributional patterns in the avifauna of the Cordillera Vilcabamba, Peru. *Ecology,* **52:**23–40.

Turček, F. J. 1950. (The Continental Jay in relation to the oak and its distribution.) *Lesni. Práce,* **29**:385–396. (English summary.)

Udvardy, M. D. F. 1969. *Dynamic Zoogeography, with Special Reference to Land Animals.* Van Nostrand Reinhold Co., New York.

Van Tyne, J. 1951. The distribution of the Kirtland Warbler (*Dendroica kirtlandii*). *Proc. 10th Int. Ornithol. Congr.,* **1950**:537–544.

Von Haartman, L. 1971. Population dynamics. In *Avian Biology,* Vol. 1, Academic Press, New York, pp. 391–459.

Wiens, J. A. 1969. An approach to the study of ecological relationships among grassland birds. *Ornithol. Monogr.* No. 8.

Willis, E. O. 1966. Interspecific competition and the foraging behavior of Plain-brown Wood-creepers. *Ecology,* **47**:667–672.

Wodzicki, K. 1965. The status of some exotic vertebrates in the ecology of New Zealand. In *The Genetics of Colonizing Species,* Academic Press, New York, pp. 425–460.

Yocum, C. F. 1963. Starlings above the Arctic Circle in Alaska, 1962. *Auk,* **80**:544.

CHAPTER SEVEN

Migration

Every environment has unfavorable aspects for any given animal; often these unfavorable factors are seasonal. An effective solution to the problem posed by a seasonally changing environment is migration. The animal simply moves to an environment that *is* favorable.

WORLDWIDE SCOPE OF BIRD MIGRATION

Bird migration is a worldwide phenomenon. Wetmore (1926) was one of the first to give a general account of bird migration in the Southern Hemisphere (southern South America). Chapin (1932) documented many instances of bird migration in Africa, a number of them entirely within tropical latitudes; e.g., two species of bee-eater (*Merops nubicus* and *M. nubicoides*) nest, respectively, north and south of the equator, and both migrate toward the equator after breeding (Fig. 1). (See also Moreau, 1961, 1972.)

Remarkable migrations have been mapped for two species of New Zealand cuckoos: the Bronze Cuckoo (*Chalcites l. lucidus*) migrates northward toward the equator, more than 2000 miles across the ocean, to winter in the region of the Solomon Islands (Fig. 2); and the Long-tailed Cuckoo (*Eudynamis taitensis*) makes an even longer transocean migration to Pacific islands as much as 4000 miles to the northwest, north, and northeast.

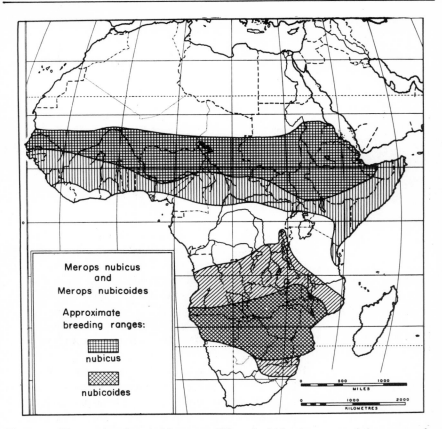

Figure 1. Two species of tropical bee-eaters (*Merops*) which migrate toward the equator after their respective breeding seasons. (After Chapin, 1932.)

Hitchcock and Carrick (1958) published the first summary report of the migratory movements of banded birds between Australia and other parts of the world (Fig. 3). McClure (1974) analyzed the migrations of birds in southeast Asia on the basis of a banding program involving over a million birds.

POSSIBLE CAUSES OF MIGRATION

Failure of Food Supply on the Breeding Ground

It is often assumed that the winter scarcity of food in the North is the simple explanation for most bird migration. But many species leave the North long before there is the slightest reduction in the food supply, and we must therefore assume that other factors are involved.

Cold Weather of Winter

However plentiful food and shelter might be, cold would certainly prevent many birds from wintering in the North. But we observe species after species leaving the northern United States in July or early August before the heat of summer is over; also, there are many cases of extended migration within the tropics.

Fluctuations of the Pleistocene Ice Front

To bird students in the North Temperate Zone the complex of many major and minor advances and recessions of the Pleistocene ice front has always loomed large as a probable cause of bird migration, but this is now generally conceded to have been greatly overemphasized. Moreau (1951) observed that the whole

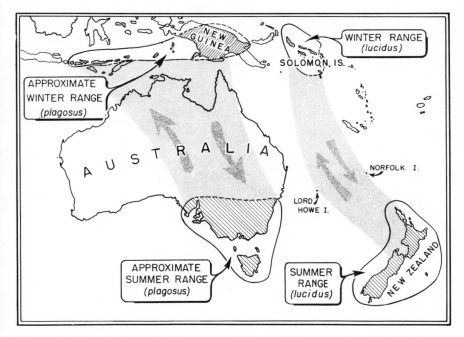

Figure 2. The migration paths of two geographical races (*plagosus* and *lucidus*) of the Bronze Cuckoo, *Chalcites lucidus*. The New Zealand race traverses more than 2000 miles of ocean. (After Fell, 1947.

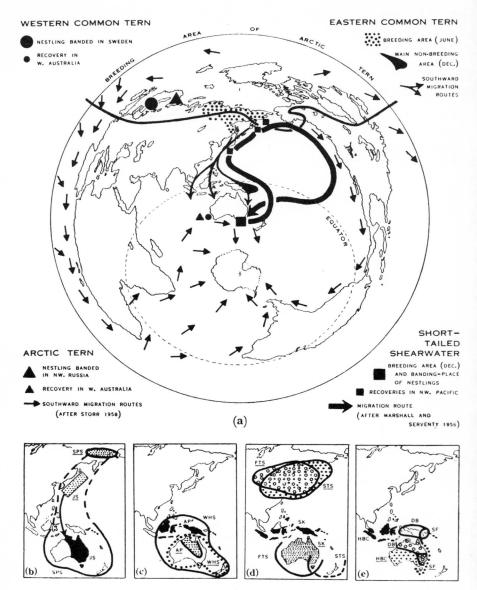

Figure 3. Map showing equatorial and transequatorial migrations of the following Australian birds: (*a*) Western Common Tern, *Sterna hirundo hirundo;* Eastern Common Tern, *S. h. longipennis;* Arctic Tern, *Sterna paradisaea;* Short-tailed Shearwater, *Puffinus tenuirostris.* (*b*) Siberian Pectoral Sandpiper (=Sharp-tailed Sandpiper) (SPS), *Calidris* (=*Erolia*) *acuminata;* Japanese Snipe (JS), *Gallinago hardwickii.* (*c*) Australian Pratincole (AP), *Glareola* (=*Stiltia*) *isabella;* White-headed Stilt (WHS), *Himantopus himantopus leucocephalus.* (*d*) Fork-tailed Swift (FTS), *Apus pacificus pacificus;* Spine-tailed Swift (STS), *Hirundapus caudacutus caudacutus;* Sacred Kingfisher

Pleistocene glaciation occupied "a period less than one hundredth part of the age of the class Aves."

Return to "Ancestral Home" in the South

Some have postulated that the Southern Hemisphere is the original home of many species of birds that now breed only in the Far North. This belief is based largely on the Wegener hypothesis of two great original land masses which split into the present continents and then gradually drifted apart. However, Wegener placed the time of splitting of the original land masses before the Jurassic—the earliest period in which we find the fossils of even the most primitive birds.

ANNUAL STIMULUS FOR MIGRATION

In searching for the causes of migration, ornithologists have separated environmental factors into two groups. *Ultimate factors* are those which have exerted a long-term positive selective evolutionary influence on the members of a species that developed migratory behavior, and a negative selective influence on those that failed to develop this behavior. Ultimate factors, therefore, are those which have made migratory behavior advantageous to a species; they include an abundant food supply, suitable nesting conditions, and optimum daylight hours for raising young. *Proximate factors* are those which stimulate a migratory condition in the individual bird and which lead to actual migration. The "migratory state" of the individual bird consists of physiological changes that include the deposition of subcutaneous and peritoneal fat and a change in hormonal secretions. In the spring it also includes a great increase in the size and activity of the gonads. The proximate factor which has received the greatest amount of experimental attention is increasing day length (photoperiod) in the spring.

Photoperiod

The first experimental approach to this problem was made by Rowan (1946), who kept Dark-eyed Juncos (*Junco hyemalis*) and Crows (*Corvus brachyrhynchos*) in outdoor aviaries at Edmonton, Alberta, where he simulated increasing day length by electric lights. As a result of his experiments, Rowan suggested that increased exercise resulting from the increasing day length caused the

---

(SK), *Halcyon sancta sancta.* (*e*) Dollar-bird (DB), *Eurystomus orientalis pacificus;* Horsfield Bronze Cuckoo (HBC), *Chalcites basalis;* Satin Flycatcher (SF), *Myiagra cyanoleuca.* (Courtesy of W. B. Hitchcock, R. Carrick, and CSIRO.)

gonads to develop and that the developing gonads, in turn, actuated northward migration; he thought that the secretions of the interstitial cells may provide the primary stimulus for migration.

As long ago as 1939, Hann attempted to determine the relation of the gonads to migration. He castrated 37 males of three species (Slate-colored Junco, Rufous-sided Towhee, White-throated Sparrow); most of the birds left the area after their incisions healed, and one towhee was recaptured 2 years later, presumably having migrated twice during that period. Morton and Mewaldt (1962) similarly castrated about 40 Golden-crowned Sparrows (*Zonotrichia atricapilla*) and maintained them in outdoor aviaries along with an equal number of controls. They concluded that "the fact that the castrates used in this investigation deposited premigratory fat and exhibited night restlessness suggests that gonadal recrudescence and spring migration do not represent a cause and effect relationship in male birds."

Wolfson (1959) also concluded that "the relation between physiological changes induced by increasing day length and the actual release of migratory behavior are not known. . . . gradually increasing day lengths (or an increase in day length) are not necessary to induce spring migration. The role of day length, once the birds are ready to respond, is the *regulation of the rate* at which the response proceeds." He added: "After the breeding season, the gonads regress, the birds molt, and subsequently, there is a physiological change, which precedes the onset of fall migration. Nothing is known about the factors which regulate this state. When the fall migration gets underway in September or October the day lengths have reached a value which is effective for the beginning of the preparatory phase of the next spring migration. And thus a new cycle begins."

Kendeigh *et al.* (1960) proposed the following sequence of events leading to the migration of birds that winter north of the tropics: "(1) Increasing photoperiods are important, not because they give longer daily periods for activity and feeding but perhaps because they bring (2) the recrudescence of gonadal activity. (3) Increasing temperatures not only diminish the energy requirements for existence but perhaps more significantly they along with increasing photoperiods induce (4) nightly unrest on the part of the birds. This nightly unrest results in increased rate of feeding beyond the needs of existence so that (5) fat deposition occurs. . . . All this puts the bird into proper physiological and psychological readiness for migration. Nightly unrest continues to augment with increasing favourable conditions until its intensity reaches a requisite threshold. Probably then the actual stimulus that releases the migratory behaviour is (6) the passage of warm fronts with clear weather and favourable winds."

Premigratory Fat Deposition

The fuel for the muscular energy required in migration is primarily stored fat (George and Berger, 1966:204). McGreal and Farner (1956) described 15 different fat bodies in White-crowned Sparrows. Hussell (1969) estimated the mean rates of weight loss of nocturnal migrants killed on the same night at a lighthouse in Ontario. A sample of 80 Veeries (*Catharus fuscescens*) had a mean rate of weight loss of 0.41 gram per hour; 96 Ovenbirds (*Seiurus aurocapillus*), a rate of 0.20 gram per hour. As a percentage of mean body weight, the rate of weight loss was 1.3 percent per hour for the Veeries and 1.0 percent per hour for the Ovenbirds.

Nisbet and Drury (1967) studied Blackpoll Warblers in Massachusetts. The birds weighed between 11 and 12 grams upon arrival in early September. They gained weight slowly, but "then, suddenly, at the end of September, in the course of a four-day rainstorm, they quickly rose to 21 grams and departed at once, carrying almost their own weight in fat! Their departure from our netting area on the evening of 1 October coincided with a massive south-southeast flight seen on radar." Then James Baird, working with Nisbet and Drury, caught 14 Blackpoll Warblers at a lighthouse in Bermuda on the night of October 2–3; these birds weighed about 17 grams. Nisbet and Drury exclaimed: "Our question was answered: after flying for 32 hours or 750 miles, these birds had used little more than a third of the fuel which they had started with."

Johnston and McFarlane (1967) studied Pacific Golden Plovers (*Pluvialis dominica fulva*) that winter on Wake Island, located about 1200 miles west of Midway Atoll in the Hawaiian chain. They found that, although the birds weighed more in April (an average of 153 grams) than in August (133 grams), the average lipid or fat storage was similar: 26.5 grams in April and 22.8 grams in August. Golden Plovers have been reported to fly between 60 and 70 miles per hour, and Johnston and McFarlane postulate that a plover can fly 2400 miles nonstop in 37 hours, utilizing about 18 grams of fat for the required energy. They also suggest that "only a plover containing at least 18 g of lipid and weighing about 150 g will attempt the 2400-mile flight from Wake Island to the Aleutian Islands or the Kamchatka Peninsula, or vice versa."

INFLUENCE OF THE WEATHER ON MIGRATION

The problem of the relation between weather (as contrasted with seasonal climatic change) and bird migration is exceedingly difficult. Lack (1960) concluded that warm temperatures in the spring and cold temperatures in the fall were the primary weather factors that influence migration. According to Lack, "migration is unaffected by the general weather situation as such or by barometric pressure, while the available evidence suggests that it is also unaffected by stable air conditions or by wind direction as such."

An apparently valid general distinction has been made between "weather migrants" of early spring and "instinct migrants" of late spring. The former are greatly influenced by weather; the latter, only a little. Of all Swedish migratory birds the Common Swift (*Apus apus*) is the most sensitive to weather changes; it may make a southward migration during an unseasonable cold spell even in the very midst of the breeding season. Such behavior has been called *reverse migration*.

Bagg (1955) analyzed the influence of weather on the migration of the Indigo Bunting; he concluded that numbers of this species flew nonstop from Yucatan to points as far north as Maine and Nova Scotia on April 17–18, 1954. He proposed that these long flights were possible because of the unusually favorable strong tropical airflow (Figs. 4, 5). It was estimated that fat deposits would make it possible for these birds to remain airborne for about 36 hours, long enough to make the flight from Yucatan to Maine. Many of the birds were deflected from the course they would have followed normally by the direction of the winds upon which their flight was based.

Williamson (1955) referred to such deflection as *migrational drift*. He believed that wind "is the migrant bird's greatest enemy." Such vagaries of weather offer a partial explanation, at least, for the occurrence of birds far out of their normal range. For example, at least 100 different species of birds have been reported as stragglers or chance arrivals in the Hawaiian Islands, among them such species as the Belted Kingfisher (*Megaceryle alcyon*), Barn Swallow (*Hirundo rustica gutteralis*), and Snow Bunting (*Plectrophenax nivalis townsendi*).

LATITUDINAL MIGRATION

It is now well known that individual birds migrate regularly back and forth between a definite breeding spot and an equally definite wintering spot. Van Tyne (1932) was the first ornithologist to show that birds return to the same wintering areas in the tropics. He banded 99 Indigo Buntings at Uaxactum, Guatemala, during March and April 1931; several of these birds were trapped in the same jungle clearing the following year.

Nickell (1968) summarized the data accumulated between 1932 and 1966 on birds returning to the same tropical areas in successive winters (15 species: 12 in Central and South America, 2 in Africa, and 1 in Asia), and he reported

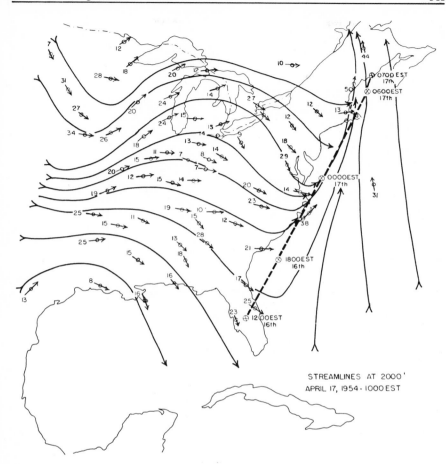

Figure 4. Streamlines for 10:00 A.M. E.S.T., April 17, 1954. Hypothetical track at 2000 feet for an Indigo Bunting arriving in Washington County, Maine, is indicated by the dashed line. Positions of the bird at various times are indicated. (Courtesy of the late Aaron M. Bagg and the Massachusetts Audubon Society.)

data obtained during six expeditions to Stann Creek Valley, British Honduras. He banded 7178 individuals of 73 migrant species during the 6-year period; 3 species formed 80.1 percent of all birds banded: Orchard Oriole, 62.4 percent, Indigo Bunting, 12.0 percent, Catbird, 5.7 percent. Recoveries of banded birds returning to the same habitat (often to a mist net set in exactly the same spot where the bird was first captured) during the 5-year period amounted to 108 (2.5%) Orchard Orioles, 27 (3.1%) Indigo Buntings, and 5 (1.5%) Catbirds.

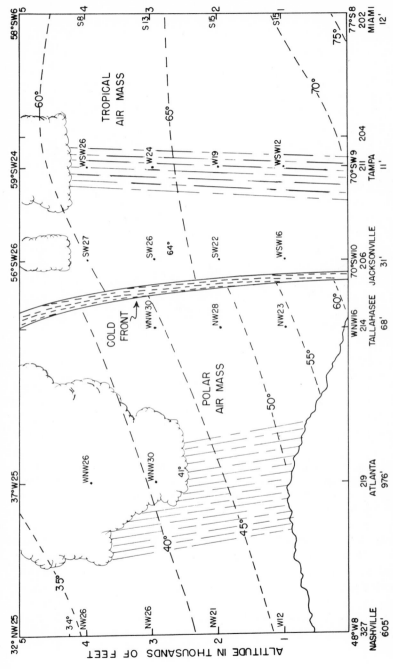

Figure 5. Vertical cross section along line from Nashville to Miami at 10:00 P.M. E.S.T., April 16, 1954. Observed wind directions, speeds in knots, and temperatures in degrees Fahrenheit are indicated above each reporting station. Lines of equal temperature (isotherms) for every 5°F are shown as dashed lines, and relative humidities aloft are reported by radiosone. The cold front at this time lies a short distance northwest of Jacksonville and is moving from left to right across the section. Scattered showers are falling in the tropical air ahead of the front, while rain is falling within the cold air some distance to the rear of the front. Data are given for levels from the surface to 5000 feet. (Courtesy of the late Aaron M. Bagg and the Massachusetts

Sixteen Orchard Orioles returned to the same habitat 2 years later, 18 returned 3 years later, and 2 were netted 4 years after being banded.

In view of this widespread pattern of returning to the same general location, both for nesting and for wintering, it is reasonable to suppose that the birds follow the same route year after year. A few species of North American water-fowl are known to have restricted summer and winter ranges connected by rather narrow migration routes (though the routes may shift laterally to some degree from year to year). For example, the little Ross's Goose (*Chen rossii*) nests in the Perry River region on the central arctic coast of Canada and mi-grates south and then southwest, along a very narrow route, to winter in the Great Valley of California (Fig. 6). The Blue Goose and Lesser Snow Goose breed north of Hudson Bay and migrate by a narrow route to their wintering grounds on the Gulf coast of Louisiana and Texas, returning by a somewhat different, but also narrow, route in spring.

Some species, restricted by their ecological requirements, migrate along extremely narrow routes determined by the seacoast. The Ipswich Sparrow (*Passerculus sandwichensis princeps*), for example, migrates from Sable Island, Nova Scotia, down the Atlantic coast to its wintering ground (Massachusetts to Georgia). Apparently the route is rarely so much as a mile wide and along many stretches may be only a few hundred yards across. Other coastal species, such as the Sharp-tailed and Seaside sparrows (*Ammospiza caudacuta* and *A. maritima*), have similar narrow migration routes along the Atlantic coast.

In some cases the routes followed by migrating birds involve quite clearly the retracing of paths by which the species historically reached given breeding areas. For example, the Yellow Wagtail (*Motacilla flava*) of Eurasia has spread across Bering Strait into Alaska. The New World population has been breeding there so long that a well-marked subspecies (*tschutschensis*) has developed, but after the breeding season the whole population flies back across Bering Strait and migrates down the Pacific coast of Asia to winter in the Philippines and New Guinea (Grant and Mackworth-Praed, 1952). Similarly, the Wheatear (*Oenanthe oenanthe*) of Europe has spread into Greenland and Labrador and developed a geographical race (*leucorhoa*) there. At the end of the nesting season the Labrador individuals start their migration to their tropical winter home by flying northward to Greenland (along the route by which their ancestors doubtless reached Labrador); there the Labrador Wheatears turn eastward to Europe and finally southward along the European coast to Africa, where they winter (Fig. 7). (The Wheatear has also spread into Alaska from northeastern Asia and, similarly, retreats in winter to its ancestral home.)

Some young birds use a migration route in autumn that is very different from that of the adults. W. W. Cooke discovered the two-route migration of the

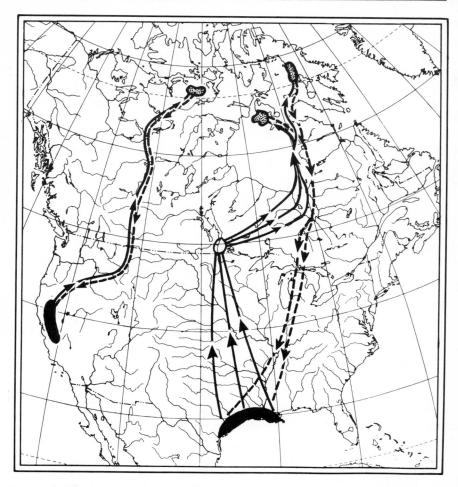

Figure 6. The narrow migration paths of Ross's Snow Goose (between Mackenzie and California) and of the Blue and Snow Geese (between the Southampton-Baffin Island region and the Gulf coast). The route of the latter species is now somewhat broader. (After Soper, 1942.)

Eastern Golden Plover (*Pluvialis d. dominica*) northward through the Mississippi Valley, but southward from Labrador over the Atlantic to the Lesser Antilles—presumably the ancient, ancestral spring migration route of all the Eastern Golden Plover. Clench (1969) presented evidence suggesting that immature Least Flycatchers (*Empidonax minimus*) "migrate through eastern North America at approximately the same time. Migration of the

WORLD, MERCATOR

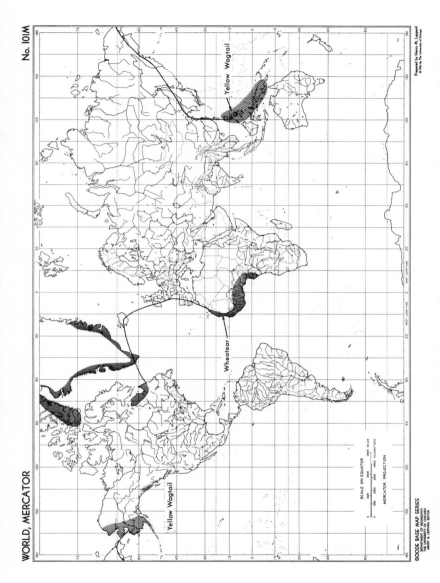

GOODE BASE MAP SERIES
DEPARTMENT OF GEOGRAPHY
THE UNIVERSITY OF CHICAGO
HENRY M. LEPPARD, EDITOR

Prepared by Henry M. Leppard
© 1944 by The University of Chicago

Figure 7. The approximate migration route of Yellow Wagtails and Wheatears that nest in North America but winter in southeast Asia and Africa, respectively. Relatively few data are available for either migration route, and the Wheatear has been seen repeatedly far south of Iceland over the North Atlantic during fall migration (see *Ibis*, Vol. 95, p. 376). The route shown here is presumably the one the birds follow in spring.

adults is concentrated inland, and may be differently timed from that of the young birds." By contrast, Johnson (1970) reported that adults and immature birds of both sexes of Hammond's Flycatcher (*Empidonax hammondii*) of western North America migrate essentially in synchrony through each region in the fall.

Mention must be made of the "flyway" theory, according to which populations of North American waterfowl "adhere with more or less fidelity" to one or another of four "flyways" (the Atlantic, Mississippi, Central, and Pacific) (Fig. 8.) This doubtless has value as a legal concept for classifying game administrative areas; however, when Lincoln (1950) admits that "during the nesting season extensive areas may be occupied by birds of the same species but which belong to different flyways," and when Aldrich *et al.* (1949) show by actual banding records (cf. Mallard, Baldpate, and Lesser Scaup) that ducks banded on their breeding grounds in Alberta, for example, go south via all four flyways, we cannot give the flyway theory much weight as a biological concept.

DETERMINATION OF "HOMING" POINT

Some attention has been given to the age at which birds fix the locality to which they will return after their absence on the wintering ground. The classic experiment, reported by Välikangas (1933), demonstrated that in the Mallard (*Anas platyrhynchos*) the "homing" point is not determined by heredity. Mallard eggs from England (where the species is nonmigratory) were hatched in Finland, and the young raised there and banded for identification. In the fall these Mallards migrated southwest to the wintering ground of Finnish Mallards in western Europe, and a large number returned to Finland to breed the next spring. McCabe (1947) reported that Wood Ducks (*Aix sponsa*) from central Illinois were moved 200 miles north, to Madison, Wisconsin, at the age of 3 to 5 weeks and raised there. Released at about 7 weeks, they left with the fall migration a couple of months later. A number returned to Madison and nested there in subsequent seasons.

LONGITUDINAL MIGRATION

Even in northern latitudes, not all migration routes are north and south. Many years ago W. W. Cooke showed that the White-winged Scoter (*Melanitta deglandi*) migrates from its central Canada breeding ground almost due east and due west to the Atlantic and Pacific coasts. Magee (1934) demonstrated by banding that the Evening Grosbeak (*Hesperiphona vespertina*) of northern

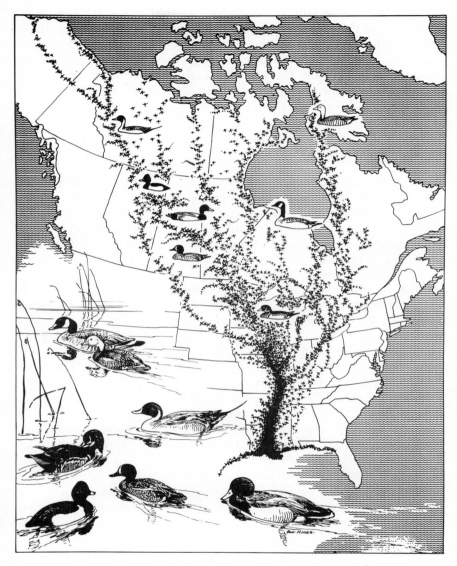

Figure 8. The Mississippi flyway. (Courtesy of the late Frederick C. Lincoln and the U.S. Fish and Wildlife Service; from a drawing by Robert Hines.)

Michigan migrates directly east, to winter in New England. Later, Shaub (1964) showed that Evening Grosbeaks from Quebec moved to the southwest (Fig. 9). McClure (1974) showed that the Hawfinch (*Coccothraustes coccothraustes*) moves eastward from breeding grounds in Russia to Korea and Japan for the winter (Fig. 10). Most Asian species, however, exhibit a north-south migration pattern, as shown in Fig. 11 for the House or Common Swallow (*Hirundo rustica*).

ALTITUDINAL MIGRATION

Many species that live in mountainous regions adjust to the changing seasons with a minimum of migratory effort by making an altitudinal migration. These altitudinal migrations are, of course, usually downward to lower altitudes and milder climates in winter, with a return to a higher breeding ground in spring. Many examples could be given: the junco *Junco hyemalis* of the Great Smoky Mountains of Tennessee makes only a shift in altitude, though its relatives in nonmountainous parts of the eastern United States regularly migrate many hundreds of miles southward in the fall. Mountain Quail (*Oreortyx pictus*) nest at altitudes up to 9500 feet in the central California mountains, but in September they leave the region of deep snow, little parties of 10 to 30 individuals migrating on foot, single file, down to areas below 5000 feet. In spring they return, again on foot, to the higher altitudes.

In a few cases altitudinal migration may be in the reverse direction: a few birds migrate from lower to higher altitudes to spend the winter. One example is the Blue Grouse (*Dendragapus obscurus*), which winters in the fir-pine forest of the northern Rocky Mountains well above its nesting grounds. This movement is apparently correlated with food supply and absence of predators.

Another example is furnished by the Ibisbill (*Ibidorhyncha struthersii*), a long-billed "shorebird" of the high plateaus of central Asia. According to La Touche, the Ibisbill breeds almost at sea level in Chihli, northeastern China, but retires in winter to the Shanhaikuan Mountains, where warm springs and the swiftness of the streams ensure open water all winter, when the lowland streams are frozen hard.

PARTIAL MIGRATION AND VARIATION IN THE URGE TO MIGRATE

Among many birds we find what Thomson (1926) called "individual migration"; Lack (1943–1944), thinking in terms of species, used the better term

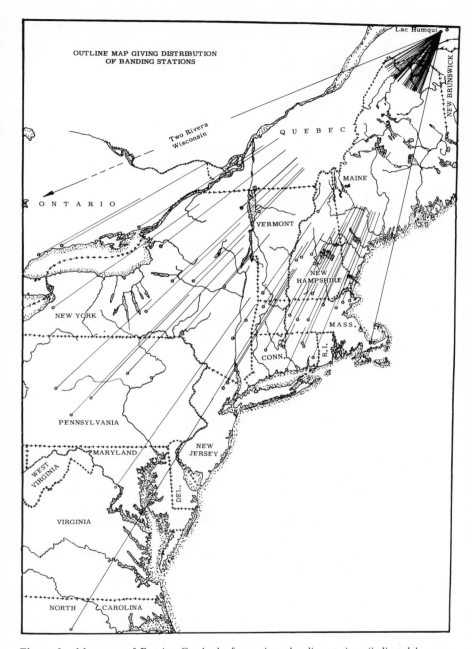

Figure 9. Movement of Evening Grosbeaks from winter banding stations (indicated by open circles) to the breeding ground in Quebec; 94 birds were shot near Lac Humqui under the mistaken notion that the Bureau of Sport Fisheries and Wildlife would pay $1.00 for each banded bird returned to the bureau. (Courtesy of B. M. Shaub and the editor of *Wilson Bulletin*.)

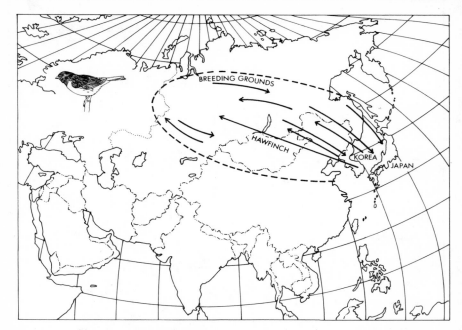

Figure 10. The east-west migration pattern of the Hawfinch in Asia. (Courtesy of H. Elliott McClure.)

"partial migration." Briefly, it appears that some individuals of a species stay through the winter on their breeding grounds, whereas others migrate —in some instances to distant countries.

Nice (1937:32–42) found that some Song Sparrows of a southern Ohio population migrated whereas others did not. A few individuals migrated some years, remained resident other years. Thomson suggested that there might prove to be migratory and nonmigratory genetic strains in such mixed populations, but Nice found that the offspring of two regularly migratory parents might be either migratory or resident.

In some species variation in the urge to migrate is apparently correlated with age, with sex, or with both age and sex. Dixon and Gilbert (1964) reported that first-year Mountain Chickadees (*Parus gambeli*) migrated altitudinally, whereas the adults were sedentary. Lack (1943–1944) found that some individuals of several passerine and shorebird species remained in the region where they had been banded as breeding adults or nestlings, whereas others migrated south to France and the Iberian Peninsula or west to Ireland. Females and (except in westward migration) immature males showed a greater tendency to migrate than adult males.

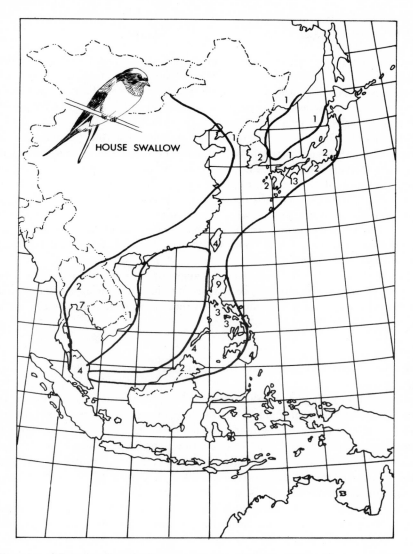

Figure 11. Migration route of the House Swallow from Japan and Russia to southeast Asia. (Courtesy of H. Elliott McClure.)

PHYSICAL PROPORTIONS OF BIRDS IN RELATION TO MIGRATION

There is a close correlation between the physical proportions of birds and their migratory habits. Averill (1920) showed that among related species and subspecies of North American birds the migratory forms have smaller bills, feet, and tails; the wings of the migrants are significantly longer and frequently are also characterized by a reduced outermost primary.

Chapman's classic study (1940) of the Andean Sparrow (*Zonotrichia capensis*) of South America strikingly illustrates the correlation of wing form with migratory habits (Fig. 12). All but two of the many geographical races are nonmigratory birds with short, rounded wings. The race (*Z. capensis australis*) that inhabits the cold southern tip of South America makes an annual migration of over 1800 miles, and its wing is long and pointed. The remaining form, *Z. capensis sanborni* of northern Chile, has a wing almost as long, but it is a bird of high mountains where wind conditions may necessitate strong powers of flight.

Meinertzhagen (1951) reported that migratory species of larks have long, pointed wings—the primaries are longer and the secondaries shorter than in resident species, though the wing area in relation to weight is actually less in the migrant larks (Fig. 13).

DIURNAL AND NOCTURNAL MIGRATION

Brewster (1886) divided North American birds into groups according to the time of day of their migrations. Most small birds (and larger birds that are shy and secretive) migrate by night: e.g., rails, cuckoos, woodpeckers, wood-warblers, wrens, and vireos. Some birds migrate "chiefly, or exclusively" by day: e.g., hawks, hummingbirds, swallows, crows and jays, pipits, and shrikes. Others migrate more or less equally by day and night: e.g., loons, ducks, and geese, auks and murres, and most shorebirds.

Diurnal migration is most readily seen and studied at certain promontories which have the effect of concentrating the migrants, often to a spectacular degree (Hofslund, 1966). Day migrants are strongly affected by topography and often seem reluctant to start out across large expanses of water. Famous points of concentration of diurnal migrants are Point Pelee, Ontario; Cape May, New Jersey; and Falsterbo, Sweden. Miller (1957) described eight local flyways through the mountains of California, classifying them as "fault lines" and "erosion gaps" in the mountains.

Lowery (1951) used the long-neglected method of telescopic observation of night migrants passing across the face of the moon and developed it into a

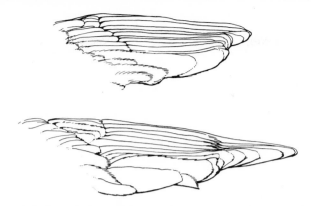

Figure 12. The folded wings of two races of Andean Sparrows (*Zonotrichia capensis*): a relatively sedentary form (*costaricensis,* Columbia, upper figure) and a highly migratory form (*australis,* Chile, lower figure). (After Chapman, 1940.)

procedure of considerable precision, employing data gathered by hundreds of cooperating observers scattered across the continent (Figs. 14, 15).

Lack and Varley (1945) first revealed the potential use of radar in the study of bird migration, and many papers have been published on this subject since 1958. Radar observations suggest that nocturnal migration involves many more birds than anyone had suspected, and Swinebroad (1964) remarked that "compared to what we see by daylight, even on a big day in the spring, we must be looking at only a tiny fraction of the nocturnal movement." He also noted that "perhaps the most spectacular findings yet reported have to do with the direction of migration. Radar has shown directions of migration not previously known." Lack (1963), writing about migration over the North Sea, said: "Hence of the six main passerine movements, two were previously unsuspected, two others in part were wrongly interpreted, and the largest was thought to occur in a different direction from that actually taken." Similarly, Nisbet and Drury (1967) found that "there were not just two but six prime directions of movement through our area in autumn."

The radar method has the advantage that the radarscope can be used continuously and photographs taken for long periods, thus providing photographic film that later can be correlated with information obtained by moon-watching and daytime observations (Fig. 16). To be most effective, studies of nocturnal migration must include all three techniques. Gauthreaux (1971, 1972) conducted such studies in Louisiana. Among his findings was that "the total number of birds entering Louisiana from over the Gulf [of Mexico] in early April was on the order of 20,000 to 25,000 birds per mile of front per day and increased to as high as 50,000 birds per mile of front per day near the

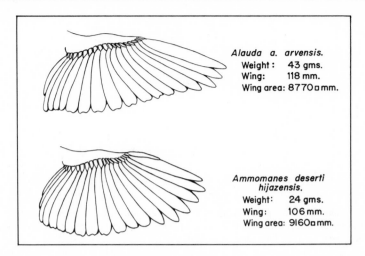

Figure 13. Spread wing of the highly migratory Skylark (*Alauda*) of Europe, compared with that of the nonmigratory Sandlark (*Ammomanes*) of Arabia. The migratory lark has a long, pointed wing, but actually *less* wing area per weight. (After Meinertzhagen, 1951.)

end of April and the first week of May." He added that the passerine birds usually began their nocturnal migrations from southern Louisiana 30 to 45 minutes after sunset, and that the departures were essentially finished an hour later.

Gauthreaux (1972) also concluded that "most nocturnal passerine migrants fly individually in the night sky," rather than in flocks as others had suggested. He believed that many reports of radar studies were inaccurate because of poor resolution of the radar equipment used (see also Eastwood, 1967; Bellrose, 1971).

Radio telemetry has become a widely used tool in biological research. In a small aircraft, Graber (1965) followed a Gray-cheeked Thrush which he had fitted with a transmitter. This bird flew its nocturnal migratory flight from Urbana, Illinois, northward the length of Lake Michigan at a speed of about 50 miles per hour. Its course over the lake suggested that the thrush would have flown nearly 400 miles during an 8-hour flight beginning at 7:55 P.M. (see also Graber and Wunderle, 1966; Cochran *et al.*, 1967; Cochran, 1972).

POSTBREEDING NORTHWARD MIGRATION

A few species of North American birds, notably herons and Bald Eagles, migrate many hundreds of miles northward immediately after the nesting season.

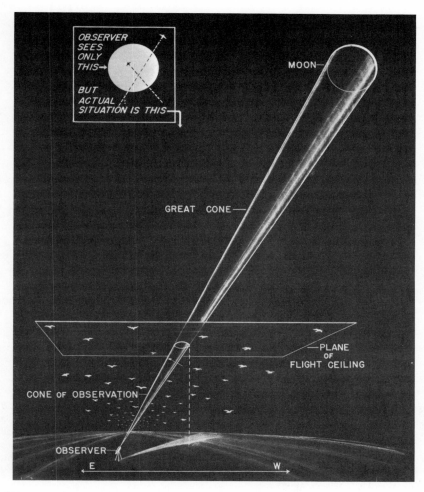

Figure 14. The field of observation in moon-watching, showing its two-dimensional aspect as it appears to the observer and its three-dimensional actuality. The *flight ceiling* is the highest level at which birds are flying; the *great cone* is the entire observation space between the telescope and the moon; the *cone of observation* is the part of the great cone lying beneath the flight ceiling. The breadth of the cone is greatly exaggerated in the illustration. (Courtesy of George H. Lowery, Jr., and the University of Kansas Museum of Natural History.)

355

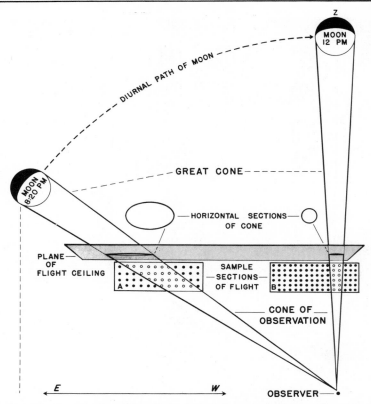

Figure 15. The changing size of the effective field of observation. The sample sections *A* and *B* represent the densities of flight at 8:20 and 12:00 P.M., respectively. With twice as many birds in the air at midnight, when the moon is at its zenith (Z), as there were at the earlier hour, only half as many are visible because of the decrease in the size of the cone of observation. Note that when the moon is overhead a horizontal section of the cone is circular, but when the cone is inclined this circle becomes elongated into an ellipse. (Courtesy of George H. Lowery, Jr., and the University of Kansas Museum of Natural History.)

In July and August of some years, great numbers of American (Common) Egrets (*Casmerodius albus egretta*) and Little Blue Herons (*Florida caerulea*), as well as a few Snowy Egrets (*Leucophoyx thula*), appear in northern United States and southern Canada. The flights consist mainly of young of the year. Broley (1947) showed by banding that young Bald Eagles fledged in Florida move northward immediately after leaving the nest; by May many of them are 1000 miles north of the nesting ground.

The term "abmigration" was proposed by Thomson (1931) for "a spring

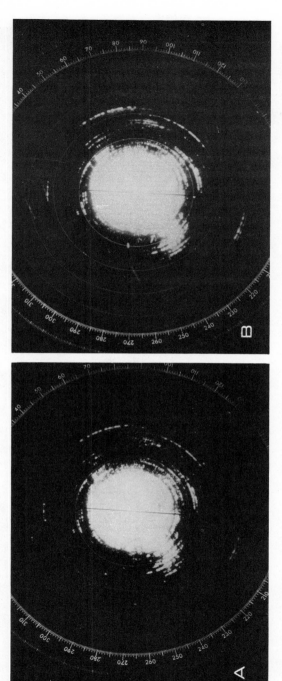

Figure 16. Photographs of the radar screen at Lake Charles, Louisiana, during the spring of 1965. (*A, C, E*) Exposures for a single revolution of the antenna. (*B, D, F*) Five-minute time exposures. *A* and *B*—March, 19, 23:28 to 23:33 C.S.T., 4° antenna elevation, no migration. *C* and *D*—May, 15, 20:11 to 20:16 C.S.T., 3° ant. elev., nocturnal migration. *E* and *F*—May 15, 18:46 to 18:51 C.S.T., 3° ant. elev., daytime migration. (Courtesy of Sidney A. Gauthreaux, Jr., and the editor of *Wilson Bulletin*.)

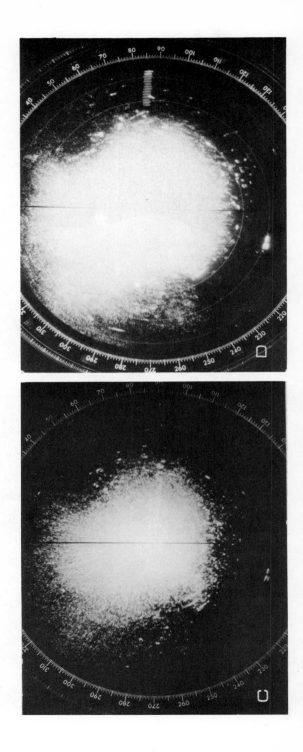

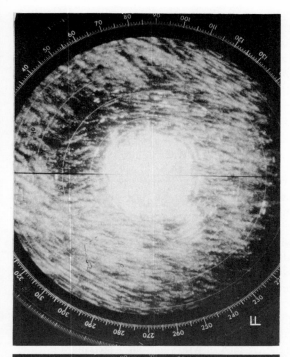

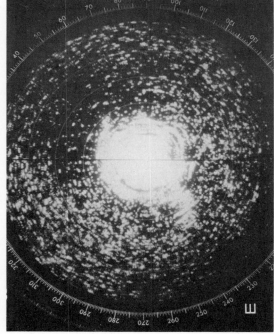

Fig. 16. *(Continued)*

movement by a bird which has performed no autumn movement at all but has passed the winter in its native area." Thus a bird native to one area may be found in a subsequent summer in quite a different area. Abmigration apparently occurs with some frequency among ducks but perhaps only very rarely among other birds.

ALTITUDE

Early writers believed that birds usually migrate at very great heights (20,000 or even 40,000 feet). Bellrose (1971) used a light aircraft equipped with auxiliary landing lights to study nocturnal migration and reported many significant findings. About 50 percent of the birds between ground level and 5000 feet were seen at the 500- and 1000-foot levels, under both clear and overcast skies. He noted that "from the 1,000-foot level upwards, the number of migrants decreased at a fairly constant rate: for each 500-foot span of altitude, density declined about one-third. . . . Our evidence suggests that shortly after taking off, small birds quickly ascend to their migrating altitude. In the next few hours they climbed slightly, if at all. Shortly after midnight, they began to descend, and, by dawn, were all below the 2,000-foot level" (Figs. 17, 18).

In 1962 a commercial airplane crashed in Maryland after striking a flock of Whistling Swans (*Cygnus columbianus*) at an altitude of 6000 feet, and a Mallard is thought to have struck another plane while flying at 21,000 feet.

In Europe, the Lapwing (*Vanellus*) is the species most frequently found at greater altitudes. Many of these strong-flying shorebirds have been recorded at 5000 and 6000 feet, an extreme record being 8500 feet. Ducks have been recorded at 7500 feet, geese at 9000 feet, and Rooks (*Corvus frugilegus*) at 11,000 feet. Two large birds (probably cranes) at 15,000 feet are among the highest-flying birds so far recorded, except over mountainous country.

All these altitude records refer to regions where the level of the land is not significantly above sea level. It is well known that considerable numbers of birds go up to 18,000 feet to migrate through passes in the Himalayan Mountains. Such birds as the Snow Partridge (*Lerwa lerwa*), the Snow Pigeon (*Columba leuconota*), and the Yellow-billed Chough (*Pyrrhocorax graculus*) even nest up to 15,000 feet in these ranges. The Mount Everest expedition of 1924 recorded these choughs up to 27,000 feet. Phelps (1961) reported that two Yellow-billed Cuckoos (*Coccyzus americanus*) and one Connecticut Warbler (*Oporornis agilis*) were killed by flying into cables at an elevation of more than 13,000 feet in Venezuela during the night of October 30, 1959. On the other hand, migrants may fly very low, especially when flying against the wind or over water.

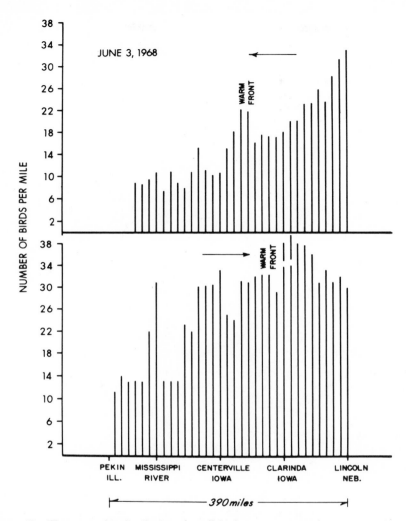

Figure 17. The geographic distribution of small birds east to west and west to east between Pekin, Illinois, and Lincoln, Nebraska, at the 1000-foot level, the night of June 3-4, 1968. Time of departure from Pekin: 20:40; arrival at Lincoln: 23:50; departure from Lincoln: 00:05; arrival at Pekin: 03:35. (Courtesy of Frank Bellrose and the editor of *Auk*.)

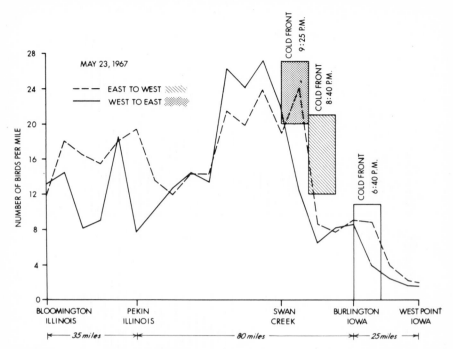

Figure 18. The geographic distribution of nocturnal bird migrants between Bloomington, Illinois, and West Point, Iowa, in relation to passage of a cold front on May 23, 1967. (Courtesy of Frank Bellrose and the editor of *Auk*.)

FLIGHT VELOCITY DURING MIGRATION

In regard to flight velocity, two quite distinct phenomena are involved: the migrating speed of individual birds and the rate of advance of the "species front."

Although in general very little is known about the rate at which individual birds migrate, there are some notable examples. A Lesser Yellowlegs (*Totanus flavipes*) was banded at North Eastham, Cape Cod, Massachusetts, on August 28, 1935, and was killed 6 days later 1900 miles away at Martinique in the West Indies; the average distance traveled was about 316 miles per day. Cooke (1940) gave additional records: two Mallards (*Anas platyrhynchos*) traveled over 500 miles (510 and 550) in 2 days; Chimney Swifts (*Chaetura pelagica*) covered 80 miles in 1 day, and 600 miles in 4 days. A Wheatear (*Oenanthe oenanthe*) banded at Skokholm, Wales, on August 16, 1949, was recovered at Capbreton, southwestern France, 43 hours later, having traveled about 600 miles. Amerson (1971) told of a Ruddy Turnstone (*Arenaria interpres*) banded on St. George Island, Alaska, that was recovered 4 days later at French Frigate Shoal (Hawaiian Islands) some 2200 miles to the south. (See Thompson, 1973.)

The species front, on the other hand, advances northward in the spring much more slowly. W. W. Cooke, the pioneer student of bird migration in North America, found that the earlier migrants advance into the North more slowly than do species that start later in the spring. The Robin (*Turdus migratorius*) takes 78 days, he reported, to advance from the edge of its winter range in Iowa to its northwestern outpost in Alaska, 3000 miles away (a rate of about 38 miles per day). The Black-and-White Warbler (*Mniotilta varia*) averages 20 to 25 miles per day in its migration from southern Florida to Lake Superior. The Blackpoll Warbler (*Dendroica striata*) Cooke mapped as starting late in spring and advancing at an ever-increasing rate (from 30 to more than 200 miles per day)—a habit characteristic, in varying degree, of most migrants that have been studied (Figs. 19, 20, 21).

PATTERN OF MIGRATION IN THE AMERICAS

The migrations of American birds are extremely varied. The principal types of postbreeding migrations to winter quarters (Fig. 22) may be summarized as follows.

1. Many North American species migrate southward. The distance covered varies from only a few miles to 7000 miles or more (Arctic American shorebirds that winter in Patagonia, and the Arctic Tern, Fig. 23).

2. A few species leave the tropical West Indies, Mexico, and even Central America and migrate to South America at the end of their breeding seasons. The Gray Kingbird (*Tyrannus dominicensis*; Fig. 24) and the Black-whiskered Vireo (*Vireo altiloquus*), for example, leave Cuba to winter in South America; the Sulphur-bellied Flycatcher (*Myiodynastes luteiventris*) nests in Mexico and throughout Central America to Costa Rica, but after breeding leaves all the area north and northwest of Costa Rica, moving into Panama and South America (south to Bolivia).

3. A number of species that nest in the South Temperate Zone of South America migrate northward after breeding, "wintering" as far north as the Caribbean coast of that continent (Eisenmann and Haverschmidt, 1970); and several species have been recorded from Central America and Mexico (Eisenmann, 1959); e.g., Ashy-tailed Swift (*Chaetura andrei meridionalis*), Blue-and-White Swallow (*Atticora cyanoleuca patagonica*), Brown-chested Martin (*Phaeoprogne tapera fusca*), Southern Martin (*Progne modesta elegans*).

4. Certain petrels and shearwaters (e.g., the Sooty Shearwater, *Puffinus*

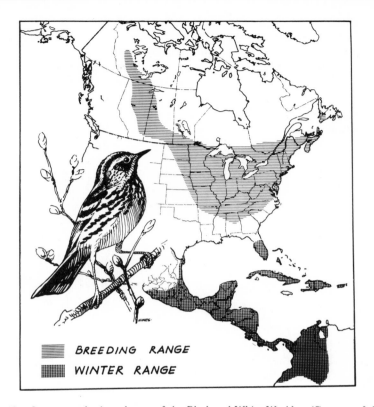

Figure 19. Summer and winter homes of the Black-and-White Warbler. (Courtesy of the late Frederick C. Lincoln and the U.S. Fish and Wildlife Service; drawing by Robert Hines.)

griseus), which nest in southernmost South America (and the subantarctic islands), migrate in their winter to the north Atlantic (Greenland) and Pacific (Alaska) oceans.

5. At least one species, *Vireo olivaceus*, breeds from northern North America to southern South America, and the migrants from Canada mingle in tropical South America with individuals of another subspecies (*Vireo olivaceus chivi*), which migrate north from Argentina. Furthermore, the wintering ground also supports other, nonmigratory, forms of this same species.

BIRD NAVIGATION

One of the most remarkable things about bird migration is the ability shown by birds to navigate a route of thousands of miles to an exact point. At the same time, there are great differences among species in the ability to navigate. Hinde (1952) surveyed many tests of homing ability in nonmigratory titmice (of the genus *Parus*) and concluded that these birds find their way home over distances of only a few miles (maximum: about 7 miles in the studies listed), the returns seeming to "depend on the bird coming by chance into known country."

By contrast, some seabirds have extraordinary powers of navigation. Mazzeo (1953) took an adult Manx Shearwater (*Puffinus puffinus*) by airplane from the breeding colony in Wales to Boston, Massachusetts, and released it

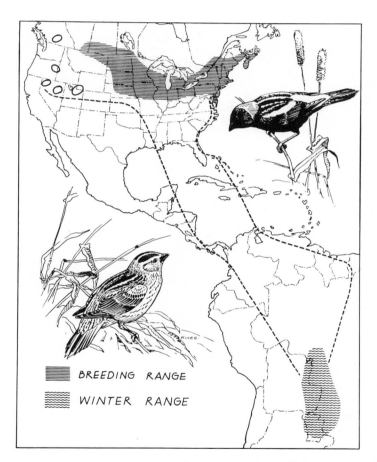

BREEDING RANGE

WINTER RANGE

Figure 20. Distribution and migration of the Bobolink. Although colonies have become established in western areas, these birds show no tendency to take a short cut across Arizona, New Mexico, and Texas, but instead adhere to the ancestral route of migration. (Courtesy of the late Frederick C. Lincoln and the U.S. Fish and Wildlife Service; drawing by Robert Hines.)

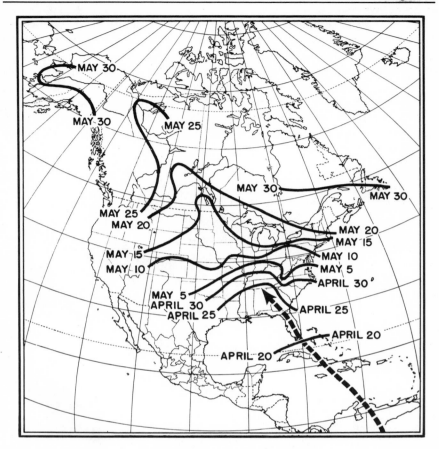

Figure 21. Migration of the Blackpoll Warbler, a species which starts late in spring and then migrates more and more rapidly as the season progresses. (After Cooke.)

there—3200 land miles airline from the nest; the bird was recaptured at the nest 12.5 days later! This shearwater returned from outside the range of the species, along an east-west route (at right angles to its normal migration), and at a rate of more than 250 miles per day, a speed that leaves little time for any deviation from the shortest possible route. Kenyon and Rice (1958) tested the homing ability of 18 nesting Laysan Albatrosses from Midway Atoll, Hawaiian Islands (Fig. 25). The birds were transported by air to various points in the Pacific, 5 of them to Kwajalein Atoll outside the known range of the species; all 5 of these birds returned to their nests, averaging 9.6 days for the 1665-mile flight. The fastest return (from Whidby Island, Washington) was made by a bird that flew the 3200 statute miles in 10.1 days, or at an average speed of 317

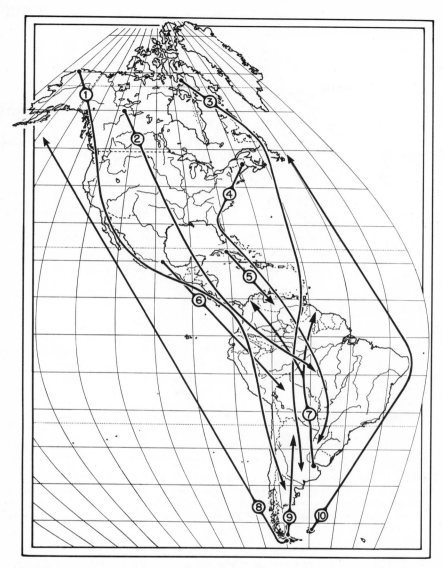

Figure 22. The postbreeding migrations of some American birds. 1, Sanderling (*Calidris alba*): Northern Alaska to southern Argentina. 2, Red-eyed Vireo (*Vireo o. olivaceus*): Mackenzie to Mato Grosso. 3, Golden Plover (*Pluvialis d. dominicus*): Melville Peninsula to Argentina. 4, Bobolink (*Dolichonyx oryzivorus*): Maine to southern Brazil. 5, Gray Kingbird (*Tyrannus dominicensis sequax*): Cuba to Venezuela. 6, Sulphur-bellied Flycatcher (*Myiodynastes l. luteiventris*): Southern Mexico to Bolivia. 7, Brown-chested Martin (*Phaeoprogne tapera fusca*): Argentina to British Guiana. 8, Sooty Shearwater (*Puffinus griseus*): Magellanic Islands to Alaska coast. 9, Rufous-backed Ground-tyrant (*Lessonia rufa*): Tierra del Fuego to northern Argentina. 10, Wilson's Petrel (*Oceanites oceanicus*): Falkland Islands to Newfoundland.

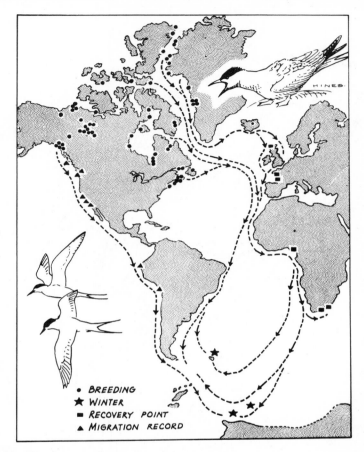

Figure 23. Distribution and migration of the Arctic Terns of eastern North America. The extreme summer and winter homes of the Arctic Tern are about 11,000 miles apart, and, as the route taken is circuitous, the terns probably fly at least 25,000 miles during their annual migration. (Courtesy of the late Frederick C. Lincoln and the U.S. Fish and Wildlife Service; drawing by Robert Hines.)

miles per day. The longest return was made by a bird released at Sangley Point, Philippine Islands, 4120 miles from Midway; during its 32.1-day return flight, this bird averaged 128 miles per day. (See also Fisher, 1971.)

Perhaps more surprising, however, is the report by Mewaldt (1964), who sent 411 wintering Golden-crowned and White-crowned sparrows from California to Baton Rouge, Louisiana, during the winter of 1961–1962; 26 of these birds returned to the California wintering grounds during the following winter, presumably after having spent the summer in their usual nesting area

Figure 24. A Gray Kingbird at its nest. (Courtesy of Samuel A. Grimes.)

Figure 25. Laysan and Black-footed albatrosses; Laysan Island, Hawaiian Islands.

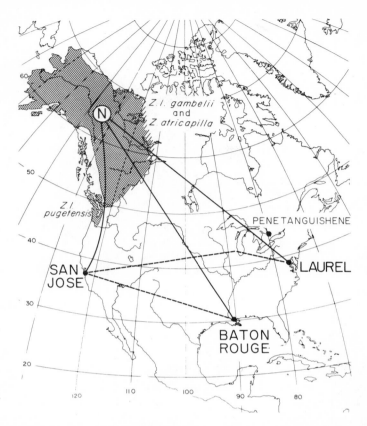

Figure 26. Probable nesting limits of *Zonotrichia leucophrys gambelii, Z. atricapilla,* and *Z. l. pugetensis,* which winter at San Jose, California, are represented by the diagonally ruled areas. Broken lines represent approximate aircraft displacement routes from San Jose to Baton Rouge, Louisiana (1961–62), and to Laurel, Maryland (1962–63). Solid lines between Baton Rouge and Laurel and the nesting grounds (*N* in the circle) represent the probable approximate routes taken from the displacement stations to the nesting area for *Z. l. gambelii* and *Z. atricapilla*. The solid line between nesting grounds and San Jose traces the probable migratory route for far northern birds. A *Z. atricapilla* released at Laurel on March 21, 1963, was found injured May 13, 1963, at Penetanguishene, Ontario. (Courtesy of L. Richard Mewaldt and the editor of *Science*; Vol. 146, November 13, 1964, pp. 941–942; copyright 1964 by the American Association for the Advancement of Science.)

in Washington and northward to Alaska. A year later, he sent 660 birds from California to Laurel, Maryland, where they were released; 15 of them were re-captured at the California banding site during the following winter, including 6 birds that previously had returned after being displaced to Louisiana (Fig. 26).

Types of Homing

Griffin (1952) outlined three types of homing in birds.

Type I relies on visual landmarks within familiar territory and the use of wandering "exploration" when the bird is released in unfamiliar country.

Type II depends on the ability to maintain a certain direction even when crossing unfamiliar territory.

Type III depends on the additional ability which some birds show, when released in unfamiliar territory, to choose approximately the correct direction to their destination. This is the most remarkable type, and it is necessarily the type used by all birds that migrate long distances between limited areas. For example, the Kirtland's Warbler (*Dendroica kirtlandii*) winters exclusively in the small Bahama Island group and breeds only in a small area of central Michigan, migrating annually between the two localities, a distance of about 1375 miles (Fig. 27).

Bird Navigation Related to Infrared Light

Wojtusiak (1949) discussed the theory that birds may be sensitive to infrared rays. If the theory is correct, birds are able to see not only through fog but also at night. Wojtusiak suggested that "during autumn migrations the birds would be guided towards warmer regions having stronger radiation and thus appear-ing brighter." Griffin (1952), however, dismissed this theory of bird navigation as scarcely deserving serious consideration.

Bird Navigation and the Coriolis Force

Ising (1946) proposed that bird navigation may depend on the mechanical ef-fects which result from the rotation of the earth (the "Coriolis force"). As a result of the earth's rotation, the weight of a flying bird would change, depend-ing on the direction of the flight (which would either add to or subtract from the velocity of its rotation about the axis of the earth) and its latitude. However, the force is so small and the confusing factors, such as variation in the speed of the bird, and the effects of wind and air turbulence, are so great that measurement of the changes which the theory requires seems well outside the sensitivity of the bird's sense organs. For example, according to Wynne-Edwards (1948), a change of speed from 40 to 39 miles per hour in a bird's

Figure 27. Lawrence H. Walkinshaw banded the first Kirtland's Warbler (a female) on June 25,1932. He had photographed this very tame bird perched on his hand on June 23 (see *Bird-Banding*, Vol. 39, p. 161). The first banded Kirtland's Warbler to be found in a subsequent year was a male banded by J. Van Tyne on June 30, 1932; he captured this bird at the same location on May 21, 1933. (Photograph courtesy of Lawrence H. Walkinshaw.)

flight speed could alter the Coriolis effect by 2.5 percent, the same change that would be registered by a geographical displacement of about 150 miles.

Bird Navigation Related to Terrestrial Magnetism and Coriolis Force

Yeagley (1951) published accounts of experiments with homing pigeons that he believed provide evidence of the use of both terrestrial magnetism and the Coriolis force in bird navigation. He suggested that birds sense their "latitude" by the effects of the Coriolis force and their "longitude" by the strength of the vertical component of the earth's magnetic field. Wynne-Edwards (1948) and Griffin (1952) presented fundamental criticisms of Yeagley's work.

Tanner and his colleagues investigated the nonthermal effects of microwave radiations on chickens in an attempt to find a means of reducing the bird hazard to aircraft (Romero-Sierra et al., 1969, 1970; Tanner et al., 1967). This research was initiated because it has long been known that birds become disoriented when they intercept a radar beam. These authors report several reactions within a few seconds after the onset of radiation: fluffing of the feathers; agitation, initiating flight; or collapse of the bird. In the last instance, the "wing outside the field of radiation becomes collapsed and the opposite wing is extended." They also studied the possible role of feathers as microwave sensors by subjecting birds to X-band and infrared frequencies. Preliminary experiments showed that plucked birds did not respond to microwave radiation for 10 days but exhibited the escape reaction after the eleventh day, by which time the new feathers had grown about 0.25 inch. Plucked birds responded immediately, however, to infrared stimulation. Although the full implication of this preliminary research is unknown, it does demonstrate that a great deal remains to be learned about the sensory system of birds.

Bird Navigation Related to Vision

There can be no question about the importance of vision in bird navigation. All published accounts of diurnal migration include many details that point unmistakably to the use of vision in modifying the migration route in relation to local physiography.

Experiments with captive Starlings (*Sturnus vulgaris*), pigeons (*Columba livia*), and other birds, made by Kramer (1952) and his associates in Germany, demonstrated that birds have a strong tendency to orient in relation to the sun. By the use of elaborate experimental equipment that included a series of windows opening on the sky, artificial light, mirrors, and an "environment" which could be easily revolved about the cage, it was possible to show that the birds oriented their migratory efforts ("migratory restlessness" or *Zugenruhe*) in relation to the source of light, natural or artificial. A heavy cloud cover prevented sun orientation, though a medium cover did not, the birds' ability to estimate the sun's position behind the clouds being apparently no better than man's. Kramer believed his study to demonstrate beyond question not only that birds orient strongly in relation to light sources but also that the daily movement of the sun "is mastered by the bird superlatively well."

Working with Manx Shearwaters, Matthews (1953b) and his collaborators took from their nesting burrows 338 marked shearwaters and released them at distances of 200 to 400 miles, largely at inland points that would be completely strange to these pelagic birds. A significantly large number of the shearwaters not only flew home with a promptness that demonstrated great navigational ability, but also were observed to turn in the correct compass direction within 3 minutes of their release. However, these shearwaters were able to navigate thus only when the sun was visible; cloudy weather greatly reduced their homing ability. Matthews proposed, therefore, that birds use the sun arc (the path followed by the sun across the sky) for direction-finding. The theory requires that a bird have an accurate memory of the sun-arc characteristics at its home locality, an accurate internal clocklike mechanism, and an eye capable of measuring small angles. This seemed reasonable to Matthews because the stimulus for the onset of migration among North Temperate Zone species appears to be governed by the rising and sinking of the sun arc (see Kramer, 1961).

Many similar studies have been conducted during the intervening years, and still others were designed to test the theory that birds migrate with reference to the stars. Most of the latter experiments have involved the recording of a bird's activity or migratory restlessness in a small cage at night under a variety of conditions (e.g., Mewaldt et al., 1964; Emlen, 1967; Brown and Mewaldt, 1968; Smith et al., 1969). Determination of a bird's navigational tendencies is based on the percentage of perches or movements in the direction in which wild birds would be migrating at that time of year (e.g., north in the spring). More sophisticated experiments have used an artificial planetarium sky like that at a particular latitude. Emlen (1969b) tested two groups of Indigo Buntings simultaneously under an artificial, spring planetarium sky. He found that differences exist between the physiological states of migratory readiness in the spring and autumn, and that there is the possibility that the same celestial clues are used during both migration seasons. "Birds in spring migration oriented northward; those in autumnal condition, southward. . . . It thus seems plausible that differences in hormonal states may determine the manner in which an orientational clue will be used."

Matthews (1968) reviewed the status of the various theories of how birds manage to navigate so precisely, and the student is referred to his book for detailed discussions. Several of his conclusions are pertinent as a summary as of 1968. "While it thus appears likely that bi-coordinate navigation by birds may have something to do with the interpretation of the sun's position, there is much less evidence to suggest that the interpretation of the stars' positions at night has a similar role. Indeed we have little evidence that birds can orientate towards home from an unknown release point at night. . . . Until definite evidence of night navigation is forthcoming we should rather consider astronavigation as the possible basis of the undoubted faculty for position finding by day. By this is not meant literally navigation by a plurality of stars, though the possibility of birds seeing stars during the day should not be discounted out of hand" (pp. 117, 121). Matthews pointed out that many birds have cone-rich retinas that are poorly adapted for night vision. At the same time, he noted (p. 144) that "it is a strong point in favour of the hypotheses of astronavigation that they rely on a sense organ that is particularly well developed in birds, and do not have to invoke unknown sense organs. . . . it is interesting to note that a tradition has developed among pigeon fanciers that good homers can be detected by their 'eye sign'. . . . However, the 'sign' is provided by shadowy marks on the iris, adjacent to the pupil, whose nature or relation to function are not explained."

In further discussing star navigation, Matthews wrote (p. 118): "A single Bobolink, shipped from North Dakota, on its first night of exposure to the natural San Francisco skies, showed a north-easterly tendency (i.e., in the direction of home); on the third night the tendency was SSE (the normal migration direction from North Dakota). The scatter of vectors was wide and the case is of interest only because the bird then escaped and was retrapped the following summer in the place from which it was originally taken"—i.e., after having first spent the winter in South America. Hence a caged, experimental bird may be confused, but it knows how to navigate when freed.

Matthews ended his book with this statement: "The ways in which the various forms of navigational ability, from fixed angle orientation through to bi-coordinate navigation, can have become fitted together in the course of phylogeny (or, for that matter, ontogeny) remain matters of fascinating speculation." In brief, therefore, it must be concluded that we still do not understand precisely what factors stimulate migratory behavior or how birds navigate.

Bellrose (1972) developed an hypothesis for establishing "an evolutionary hierarchy in navigation." He proposed that "navigational sophistication evolved as adaptive behavior to expansion of ranges of permanent-resident birds. Other traits being equal, the farther a bird expanded its range, the greater its navigational attributes.

"Because birds are basically diurnal animals, the adaptations by many species for nocturnal migration are considered progressive evolutionary development in navigational sophistication. . . . The landscape is assumed to be the single most important cue for orientation among nonmigratory birds. Therefore, landscape features are considered the most fundamental cues to migrants. The less reliance migratory birds place on landscape for orientation, the more advanced their navigational mechanisms. Therefore, long-ranging pelagic birds are considered to be among those with the most highly developed navigational attributes. . . . The ability of many birds to remain oriented when visual cues are minimal is probably one of the more significant findings derived from radar studies. It points to wind structure and/or the Earth's magnetic field as potential sources of directional information."

REFERENCES

Adler, H. E. 1963. Psychophysical limits of celestial navigation hypotheses. *Ergeb. Biol.* **26**:235–252.

Aldrich, J. W., *et al.* 1949. Migration of some North American waterfowl. *U.S. Fish Wildl. Serv. Spec. Sci. Rept. (Wildlife)* No. 1.

Amerson, A. B., Jr. 1971. The natural history of French Frigate Shoals, Northwestern Hawaiian Islands. *Atoll Res. Bull.,* **150**:1–383.

Averill, C. K. 1920. Migration and physical proportions: A preliminary study. *Auk,* **37**:572–579.

Bagg, A. M. 1955. Airborne from Gulf to Gulf. *Bull. Mass. Audubon Soc.,* **39**:106–110, 159–168.

Baird, J., C. S. Robbins, A. M. Bagg, and J. V. Dennis. 1958. "Operation recovery"—the Atlantic Coastal netting project. *Bird-Banding,* **29**:137–168.

Bellrose, F. C. 1971. The distribution of nocturnal migrants in the air space. *Auk,* **88**:397–424.

——— 1972. Possible steps in the evolutionary development of bird navigation. In *Animal Orientation and Navigation,* NASA SP-262, pp. 223–257.

Brewster, Wm. 1886. Bird migration. *Mem. Nuttall Ornithol. Club,* No. 1.

Broley, C. L. 1947. Migration and nesting of Florida Bald Eagles. *Wilson Bull.,* **59**:3–20.

Brown, I. L., and L. R. Mewaldt. 1968. Behavior of sparrows of the genus *Zonotrichia,* in orientation cages during the lunar cycle. *Z. Tierspychol.,* **25**:668–700.

Chapin, J. P. 1932. The birds of the Belgian Congo. I. *Bull. Am. Mus. Nat. Hist.,* **65**:x + 756 pp.

Chapman, F. M. 1940. The post-glacial history of *Zonotrichia capensis*. *Bull. Am. Mus. Nat. Hist.,* **77**:381–438.

Clench, M. H. 1969. Additional observations on the fall migration of adult and immature Least Flycatchers, *Bird-Banding,* **40**:238–243.

Cochran, W. W. 1972. Long-distance tracking of birds. In *Animal Orientation and Navigation,* NASA SP-262, pp. 39–59.

———, G. G. Montgomery, and R. R. Graber. 1967. Migratory flights of *Hylocichla* thrushes in spring: a radiotelemetry study. *Living Bird,* **6**:213–225.

Cooke, M. T. 1940. Notes on speed of migration. *Bird-Banding,* **11**:21.

D'Arms, E., and D. R. Griffin. 1972. Balloonists' reports of sounds audible to migrating birds. *Auk,* **89**:269–279.

Dixon, K. L., and J. D. Gilbert. 1964. Altitudinal migration in the Mountain Chickadee. *Condor,* **66:**61–64.

Eastwood, E. 1967. *Radar Ornithology.* Methuen & Co., London.

Eisenmann, E. 1959. South American migrant swallows of the genus *Progne* in Panama and northern South America; with comments on their identification and molt. *Auk,* **76:**529–532.

———, and F. Haverschmidt. 1970. Northward migration to Surinam of South American martins (*Progne*). *Condor,* **72:**368–369.

Emlen, S. T. 1967. Migratory orientation in the Indigo Bunting, *Passerina cyanea.* I, II. *Auk,* **84:**309–342, 463–489.

——— 1969a. Bird migration: influence of physiological state upon celestial orientation. *Science,* **165:**716–718.

——— 1969b. The development of migratory orientation in young Indigo Buntings. *Living Bird,* **1969:**113–126.

Fell, H. B. 1947. The migration of the New Zealand Bronze Cuckoo, *Chalcites lucidus lucidus* (Gmelin). *Trans. Roy. Soc. N.Z.,* **76:**504–515.

Fisher, H. I. 1971. Experiments on homing in Laysan Albatrosses, *Diomedea immutabilis. Condor,* **73:**389–400.

Gauthreaux, S. A., Jr. 1971. A radar and direct visual study of passerine spring migration in southern Louisiana. *Auk,* **88:**343–365.

——— 1972. Behavioral responses of migrating birds to daylight and darkness: a radar and direct visual study. *Wilson Bull.,* **84:**136–148.

George, J. C., and A. J. Berger. 1966. *Avian Myology.* Academic Press, New York.

Graber, R. R. 1965. Night flight with a thrush. *Audubon Mag.,* **67:**368–374.

——— 1968. Nocturnal migration in Illinois—different points of view. *Wilson Bull.,* **80:**36–71.

———, and J. W. Graber. 1962. Weight characteristics of birds killed in nocturnal migration. *Ibid.,* **74:**74–88.

———, and S. L. Wunderle. 1966. Telemetric observations of a Robin (*Turdus migratorius*). *Auk,* **83:**674–677.

Grant, D. H. B., and C. W. Mackworth-Praed. 1952. On the species and races of the Yellow Wagtails from western Europe to western North America. *Bull. Brit. Mus. (Nat. Hist.) Zool.,* **1:**255–268.

Griffin, D. R. 1952. Bird navigation. *Biol. Rev.,* **27:**359–400.

Hamilton, W. J. III. 1962. Evidence concerning the function of nocturnal call notes of migratory birds. *Condor,* **64:**390–401.

Hann, H. W. 1939. The relation of castration to migration in birds. *Bird-Banding,* **10:**122–124.

Hinde, R. A. 1952. The behavior of the Great Tit (*Parus major*) and some other related species. *Behaviour,* Suppl., **2.**

Hitchcock, W. B., and R. Carrick. 1958. First report of banded birds migrating between Australia and other parts of the world. *CSIRO Wildl. Res.,* **3:**54–70.

Hofslund, P. B. 1966. Hawk migration over the western tip of Lake Superior. *Wilson Bull.,* **78:**79–87.

Howell, T. R. 1953. Racial and sexual differences in migration in *Sphyrapicus varius. Auk,* **70:**118–126.

Hussell, D. J. T. 1969. Weight loss of birds during nocturnal migration. *Auk,* **86:**75–83.

Ising, G. 1946. Die physikalische Möglichkeit eines tierischen Orientierungssinnes auf Basis der Erdrotation. *Ark. Mat., Astron., Fys.,* **32A,** No. 18.

Johnson, N. K. 1970. Fall migration and winter distribution of the Hammond Flycatcher. *Bird-Banding,* **41:**169–190.

Johnston, D. W. 1966. A review of the vernal fat deposition picture in overland migrant birds. *Bird-Banding,* **37:**172–183.

——— 1970. Caloric density of avian adipose tissue. *Comp. Biochem. Physiol.,* **34:**827–832.

———, and R. W. McFarlane. 1967. Migration and bioenergetics of flight in the Pacific Golden Plover. *Condor,* **69:**156–168.

Kendeigh, S. C., G. C. West, and G. W. Cox. 1960. Annual stimulus for spring migration in birds. *Anim. Behav.,* **8:**180–185.

Kenyon, K. W., and D. W. Rice. 1958. Homing of Laysan Albatrosses. *Condor,* **60:**3–6.

Kramer, G. 1952. Experiments on bird orientation. *Ibis,* **94:**265–285.

——— 1961. Long-distance orientation. In *Biology and Comparative Physiology of Birds,* Vol. 2. Academic Press, New York, pp. 341–371.

Lack, D. 1943–44. The problem of partial migration. *Brit. Birds,* **37:**122–130, 143–150.

——— 1960. The influence of weather on passerine migration: A review. *Auk,***77:**171–209.

——— 1963. Migration across the southern North Sea studied by radar. 4: Autumn. *Ibis,* **105:** 1–54.

———, and G. C. Varley. 1945. Detection of birds by radar. *Nature,* **156:**446.

Lasiewski, R. C. 1962. The energetics of migrating hummingbirds. *Condor,* **64:**324.

Lawrence, L. de Kiriline. 1962. A noteworthy reverse migration of Snow Geese in central Ontario. *Auk,* **79:**718.

Lincoln, F. C. 1950. Migration of birds. *U.S. Fish Wildl. Serv. Circ.* No. 16.

Lowery, G. H., Jr. 1951. A quantitative study of the nocturnal migration of birds. *Univ. Kans. Publ. Mus. Nat. Hist.,* **3:**361–472.

———, and Robert J. Newman. 1966. A continentwide view of bird migration on four nights in October. *Auk,* **83:**547–586.

McCabe, R. 1947. The homing of transplanted young Wood Ducks. *Wilson Bull.,* **59:**104–109.

McClure, H. E. 1974. *Migration and Survival of the Birds of Asia.* Bangkok, Thailand.

McGreal, R. D., and D. S. Farner. 1956. Premigratory fat deposition in the Gambel White-crowned Sparrow: some morphologic and chemical observations. *Northwest Sci.,* **30:**12–23.

Magee, M. J. 1934. The distribution of Michigan recovered Eastern Evening Grosbeaks near the Atlantic Seaboard. *Bird-Banding,* **5:**175–181.

Matthews, G. V. T. 1952. An investigation of homing ability in two species of gulls. *Ibis,* **94:**243–264.

——— 1953a. Sun navigation in homing Pigeons. *J. Exp. Biol.,* **30:**243–267.

——— 1953b. Navigation in the Manx Shearwater. *Ibid.,* **30:**370–396.

——— 1968. *Bird Navigation,* 2nd ed. Cambridge University Press, Cambridge, England.

Mazzeo, R. 1953. Homing of the Manx Shearwater. *Auk,* **70:**200–201.

Meinertzhagen, R. 1951. Review of the Alaudidae. *Proc. Zool. Soc. London,* **121:**81–132.

Mewaldt, L. R. 1964. California sparrows return from displacement to Maryland. *Science,* **146:**941–942.

————, S. S. Kibby, and M. L. Morton. 1968. Comparative biology of Pacific coastal White-crowned Sparrows. *Condor,* **70:**14–30.

————, M. L. Morton, and I. L. Brown. 1964. Orientation of migratory restlessness in *Zonotrichia. Ibid.,* **66:**377–417.

Miller, L. 1957. Some avian flyways of western America. *Wilson Bull.,* **69:**164–169.

Moreau, R. E. 1951. The migration system in perspective. *Proc. 10th Int. Ornithol. Congr.,* **1950:**245–248.

———— 1961. Problems of Mediterranean-Saharan migration. *Ibis.,* **103:**373–427, 580–623.

———— 1972. *The Palaearctic-African Bird Migration Systems.* Academic Press, New York.

Mortensen, H. C. C. 1950. Studies in bird migration, being the collected papers of H. Chr. C. Mortensen, 1856–1921. Dansk Ornithologisk Forening, Copenhagen.

Morton, M. L., and L. R. Mewaldt. 1962. Some effects of castration on a migratory sparrow (*Zonotrichia atricapilla*). *Physiol. Zool.,* **35:**237–247.

Murray, M. D., and R. Carrick. 1964. Seasonal movements and habitats of the Silver Gull, *Larus novaehollandiae* Stephens, in south-eastern Australia. *CSIRO Wildl. Res.,* **9:**160–188.

Nice, M. M. 1937. Studies in the life history of the Song Sparrow. I. *Trans. Linn. Soc. N.Y.,* **4:**vi + 247 pp.

Nickell, W. P. 1968. Return of northern migrants to tropical winter quarters and banded birds recovered in the United States. *Bird-Banding,* **39:**107–116.

Nisbet, I. C., and W. H. Drury, Jr. 1967. Scanning the sky birds on radar. *Bull. Mass. Audubon Soc.,* **51:**166–174.

Odum, E. P., C. E. Connell, and H. L. Stoddard. 1961. Flight energy and estimated flight ranges of some migratory birds. *Auk,* **78:**515–527.

Olrog, C. C. 1969. Birds of South America. *Monogr. Biol.,* **19:**849–878.

Phelps, W. H. 1961. Night migration at 4,200 meters in Venezuela. *Auk,* **78:**93–94.

Romero-Sierra, C., A. O. Quanbury, and J. A. Tanner. 1970. Feathers as microwave and infrared filters and detectors—preliminary experiments. *Natl. Res. Counc., Control Syst. Lab. Tech. Rept.* LTR-CS-40, Ottawa, Canada.

————, J. A. Tanner, and F. Villa. 1969. EMG changes in the limb muscles of chickens subjected to microwave radiation. *Ibid.,* LTR-CS-16.

Rowan, W. 1946. Experiments in bird migration. *Trans. Roy. Soc. Can.,* **40** (Ser. 3, Sect. 5):123–135.

Shaub, B. M. 1964. Notes on the destruction of banded Evening Grosbeaks in Quebec in 1960. *Wilson Bull.,* **76:**179–185.

Smith, R. W., I. L. Brown, and L. R. Mewaldt. 1969. Annual activity patterns of caged non-migratory White-crowned Sparrows. *Wilson Bull.,* **81:**419–440.

Soper, J. D. 1942. Life history of the Blue Goose, *Chen caerulescens* (Linnaeus). *Proc. Boston Soc. Nat. Hist.,* **42:**121–222.

Stoddard, H. L., Sr., and R. A. Norris. 1967. Bird casualties at a Leon County, Florida, TV tower: an eleven-year study. *Bull. Tall Timbers Res. Sta,* **8:**1–104.

Swinebroad, J. 1964. The radar view of bird migration. *Living Bird,* **1964:**65–74.

Tanner, J. A., C. Romero-Sierra, and S. J. Davie. 1967. Non-thermal effects of microwave radiation on birds. *Nature,* **216:**1139.

Thompson, M. C. 1973. Migratory patterns of Ruddy Turnstones in the central Pacific Region. *Living Bird,* **1973:**5–23.

Thomson, A. 1926. *Problems of Bird-Migration*. H. F. & G. Witherby, London.

————1931. On "abmigration" among the ducks: An anomaly shown by the results of bird-marking. *Proc. 7th Int. Ornithol. Congr.*, **1930**:382–388.

Välikangas, I. 1933. Finnische Zugvögel aus englischen vogeleiern. *Vogelzug*, **4**:159–166. (Summarized in English by M. M. Nice in *Bird-Banding*, **5**, 1934:95).

Van Tyne, J. 1928. A diurnal local migration of the Black-capped Chickadee. *Wilson Bull.*, **40**:252.

———— 1932. Winter returns of the Indigo Bunting in Guatemala. *Bird-Banding*, **3**:110.

Wetmore, A. 1926. Observations on the birds of Argentina, Paraguay, Uruguay, and Chile. *U.S. Natl. Mus. Bull.*, **133**.

Williamson, K. 1955. Migrational drift. *Acta. XI Congr. Int. Ornithol.*, **1954**:179–186.

Winterbottom, J. M. 1965. The migrations and local movements of some South African birds. *Monogr. Biol.*, **14**:233–243.

Wojtusiak, R. J. 1949. Polish investigations on homing in birds and their orientation in space. *Proc. Linn. Soc. London*, **160**:99–108.

Wolfson, A. 1940. A preliminary report on some experiments on bird migration. *Condor*, **42**:93–99.

———— 1945. The role of the pituitary, fat deposition, and body weight in bird migration. *Condor*, **47**:95–127.

———— 1952. Day length, migration, and breeding cycles in birds. *Sci. Mon.*, **74**:191–200.

———— 1959. The role of light and darkness in the regulation of spring migration and reproductive cycles in birds. In *Photoperiodism* and *Related Phenomena in Plants and Animals*, American Association for the Advancement of Science, Washington, D.C., pp. 679–716.

————, and D. P. Winchester. 1960. Role of darkness in the photoperiodic responses of migratory birds. *Physiol. Zool.*, **33**:179–189.

Wynne-Edwards, V. C. 1948. Yeagley's theory of bird navigation. *Ibis*, **90**:606–611.

Yeagley, H. L. 1947, 1951. A preliminary study of a physical basis of bird navigation. I, II. *J. Appl. Phys.*, **18**:1035–1063; **22**:746–760.

CHAPTER EIGHT

Flight and

Flightlessness

The aerodynamic principles of bird flight have been discussed by Brown (1961), Raspet (1960), Houghton (1964), and Herzog (1968). Greenewalt (1960:217) wrote that "the resulting complexity of motion [of a bird's wing in flight] simply does not lend itself to detailed analysis. Perhaps with modern electronic computers something might be done, but until that very great effort is made, we shall have to rest content with our imperfect and rudimentary knowledge."

THE FEATHERS

Storer (1948) likened the take-off of the Wood Stork (Fig. 1) to that of a helicopter: "The wings sweep forward and back nearly horizontally, producing a climb that takes more power than forward flight requires." The American Egret uses its legs to catapult itself into the air, and its wings operate like a variable propeller, in which they concentrate on forward speed.

The primary feathers serve as the propellers in bird flight. The outermost primary is attached to the distal phalanx of digit II and can function independently. The remaining primaries act as a group because of their

Figure 1. A group of Wood Storks (*Mycteria americana*). Although long called an "ibis," this is the only North American representative of the stork family. (Courtesy of Samuel A. Grimes.)

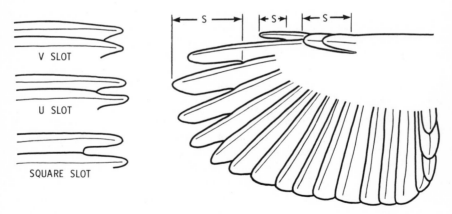

Figure 2. Left: Types of slot formed by emargination of the primaries. The V slot is primitive; the more efficient U slot is found in many woodland birds; the highly efficient square slot is found in birds of prey. Right: Diagrammatic plan view of the Catbird's wing, in which the slotted portions (S) make up 55 percent of the total wing length. (Modified by courtesy of D. B. O. Savile and the editor of *Evolution*.)

382

Figure 3.   Yellow-breasted Sunbird (*Leptocoma jugularis microleuca*) returning to its nest; note slotting in the wing and the position of the feet. (Courtesy of the late Loke Wan Tho.)

mode of attachment in the elastic tissue along the posterior margin of the hand. The tips of all the primaries, however, respond individually to air pressures.

The presence of wing slots increases lift, which is needed especially at take-offs. The alula functions as a wing slot when it is drawn forward away from the rest of the hand. According to the emargination of the primaries, an additional series of V-shaped, U-shaped, or square slots may be present (Figs. 2, 3). Savile (1957) reported that the slotted length of a Catbird's (*Dumetella*) wing may be as much as 55 percent of the total length of the wing.

Savile described four basic wing types in birds.

The **elliptical wing** is characterized by a low aspect ratio and only a slight amount of wing-tip vortex. This type is found in birds which "must move easily through restricted openings in vegetation": gallinaceous birds, doves, Woodcocks, woodpeckers, and most passerine birds. Slots are well developed in gallinaceous birds and in corvids, and are better developed in small, active birds (e.g., wood-warblers) than in more sedentary species (House Sparrow).

The **high-speed wing** is characterized by having a lower camber (flatter section) than is general in bird wings, a moderately high aspect ratio, pronounced sweepback of the leading edge, lack of wing-tip slots, and the development of a pronounced "fairing" at the root of the wing that blends the trailing edge of the wing into the body. Examples of birds having such wings are plovers, sandpipers, swifts, hummingbirds, swallows, and falcons.

The **high-aspect-ratio wing** is, by definition, one in which the length greatly exceeds the width. Savile gave approximate aspect ratios of 15 to 18 for two species of albatross; by comparison, the aspect ratio of the Catbird's wing is 4.7. The high-aspect-ratio wing is characteristic of oceanic soaring birds: albatrosses, frigatebirds, tropicbirds.

The **high-lift or slotted soaring wing** results from the combination of a moderate aspect ratio with pronounced slotting and camber (Fig. 4). This produces an efficient soaring wing for large birds that, in general, "inhabit forested areas or other impeded terrain. . . . Eagles, ospreys and harriers, of relatively open habitats, seem usually to have appreciably higher aspect ratios than the woodland hawks and owls." Savile includes the owls in this group because "their wings have evolved similarly to develop exceptionally high lift."

Storer (1948) pointed out that birds, such as pelicans, that fly in formation take advantage of the rising currents produced by the wing-tip vortex phenomenon: "The inner wing of each bird in a V formation gains support from the upward rising side of the whirl left by the outer wing of the bird ahead. Thus the energy wasted by each bird in making the vortex is passed on and used by the bird behind it."

Graham (1934) described the stiff, comblike fringe on the anterior margin of wing feathers that form the leading edge of the wing in most owls. It is this fringe that is responsible for their noiseless flight. The fringe is lacking, however, in some owls: e.g., Little Owl (*Athene noctua*) and Asian Fish Owls (*Ketupa*).

THE BONY FRAMEWORK

Theories of the origin of flight were discussed in Chapter 1, and selected osteological characters were described in Chapter 2. Little mention was made, however, of the following specializations for flight in the bird skeleton.

1.   The bones are light and many are pneumatic in flying birds[1] (Fig. 5).

[1] Pneumatization of bones reaches an extreme in hornbills and South American screamers, in which even the pygostyle and the phalanges of the fingers and toes are pneumatic; neither group of

Figure 4. A Sharp-shinned Hawk (*Accipiter striatus*) that has just left its nest. (Courtesy of Heinz Meng.)

2.   Uncinate processes on the ribs strengthen the thoracic cage.

3.   The pectoral girdle is firmly attached to the sternum.[2]

4.   The coracoid is robust, opposing the pull of the pectoral muscles.

5.   The clavicles are fused (forming the furcula) in most flying species.

6.   The sternum is strong and keeled, the keel being more highly developed in birds that use flapping (as contrasted to soaring) flight.

---

birds contains especially strong fliers. Pneumatization of the skull is decreased in some fast-flying birds (swifts), in some birds that dive from air into water (some diving-petrels, terns, kingfishers), and in birds such as woodpeckers that "hammer" into wood with their bills. Similarly, a notarium is not found consistently among strong-flying birds, and we believe that the importance of "rigidity" in the bird skeleton in relation to flight has been overemphasized by many writers.

[2] The wing, however, has no direct connection with the trunk of the bird; the only connection is an indirect one by means of the glenoid fossa of the pectoral girdle. Moreover, the girdle bones themselves have no direct connection with the vertebral column or any other part of the axial skeleton except where the coracoids articulate with the sternum. The scapula is connected to the vertebral column and ribs only by muscles that extend between them and the scapula.

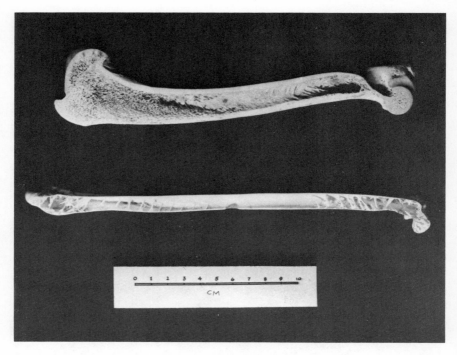

Figure 5. Longitudinal sections of the humerus of a Brown Pelican (below) and of a Gray Wolf (*Canis lupus*). The cavity in the dry humerus of the wolf is filled with bone marrow in life.

7.   The tail is shortened and there is fusion of some of the coccygeal vertebrae, forming a pygostyle in flying birds.

8.   The supracoracoideus muscle (which raises the wing) lies on the ventral side of the sternum and by running over a pulley in the shoulder accomplishes the same results as though it were on the dorsal side of the wing. Thus the center of gravity remains low (Fig. 6).

9.   The bones of the wing are reduced in number: two free carpals and three fused metacarpals instead of ten or more independent bones.

THE WING MUSCLES, BONES, AND JOINTS

About 50 different muscles and muscle slips affect the movements of the wing. As long ago as 1886 Edwards tabulated the ratio of body weight to the weight of the pectoral musculature (Mm. pectoralis and supracoracoideus) of over 50 species of North American birds; e.g., the pectoral muscles of the Broad-winged Hawk equal about 6 percent of the body weight; of the Tree Sparrow, 10 percent; and of the Mourning Dove, about 16 percent.

Fisher's (1946) data demonstrate how great may be the differences in structure among birds that are closely related and have a similar locomotor pattern. The ratio of length of bony wing to body weight in *Gymnogyps* is 31.9; in *Coragyps*, 93.3; and in *Cathartes*, 110.1. In this same series, the ratio of the volume of the wing musculature to body weight is 24.7, 33.1, and 52.3. A com-

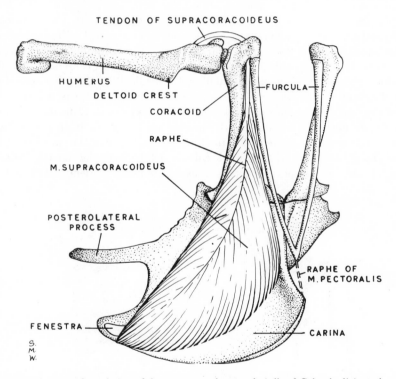

Figure 6. An anterolateral view of the sternum and pectoral girdle of *Columba livia* to show the general relationships of the belly and tendon of insertion of M. supracoracoideus. M. pectoralis has been removed, and the humerus has been rotated cephalad to show the site of insertion of M. supracoracoideus on the dorsal surface of the humerus. M. pectoralis (which lowers the wing) arises from all portions of the ventral surface of the sternum not preoccupied by M. supracoracoideus, as well as from the anterolateral surface of most of the clavicle and the adjacent area on the coracoclavicular membrane. In addition, the right and left pectoralis muscles share a common raphe extending between the inferior margin of the furcula and the anterior edge of the carina of the sternum. (By permission of J. C. George and A. J. Berger from *Avian Myology*, Academic Press, New York, copyright 1966.)

parison of body weight and the total supporting surface of the wings, tail, and body revealed that the "California condor has a supporting surface of 497 square centimeters per pound; in the turkey vulture the ratio is 1125 square centimeters per pound." Fisher correlated these differences with amount of wing flapping, relative stability in flight, and manner of take-off.

Extreme differences in relative proportions of wing segments are seen when the bony wing of a hummingbird is compared with that of a pelican (Fig. 7). The manus of the hummingbird *is longer than the arm and forearm combined*; the manus of the pelican is short, and both the arm and forearm are very long. Certain peculiarities in the hummingbird wing demonstrate that the actions of individual muscles or of muscle groups cannot be understood without reference to the wing bones and their mode of articulation. The hummingbird humerus is extremely short and robust and is much modified in configuration. The proximal articular surface is very small and is displaced to the inner side of the articular head of the humerus, so that rotation of the humerus is facilitated. Moreover, the relationship of the tendon of the supracoracoideus muscle to the head of the humerus is such that the muscle appears to function primarily in rotating the humerus, rather than in elevating it. The articular surfaces at the distal end of the humerus, the weak collateral ligaments, the lateral displacement of the tendon of the triceps muscle, and the presence of a greatly

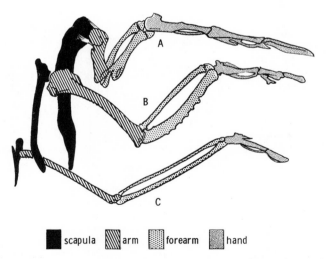

scapula arm forearm hand

Figure 7. Comparison of the wing skeletons of a hummingbird (*A*), the Roadrunner *Geococcyx* (*B*), and a pelican (*C*), to show the relative proportions of the wing segments to total wing length. (Based on Stolpe and Zimmer, 1939.)

hypertrophied sesamoid in the humeroulnar pulley—all these favor sliding and rotatory movements rather than simple flexion and extension at the elbow, as in most birds.

These osteological characters are correlated with several peculiar myological features. The deltoideus major muscle (of importance in wing elevation in other birds) and the biceps brachii muscle (the main flexor of the forearm) are either greatly reduced or absent in hummingbirds. The absence of the biceps is compensated for in part, at least, by the very high origin of M. extensor metacarpi radialis and by the hypertrophy and peculiar insertion of M. tensor patagii brevis. Musculus pectoralis is only 1.7 to 2.0 times larger than M. supracoracoideus; the ratio between these muscles in the Black-backed Gull is 15 : 1 (Stolpe and Zimmer, 1939). Hartman (1961) said that M. supracoracoideus "ranged from about 0.40 percent of the body in species of the genus *Buteo* to 11.5 percent of the body in the trochilids, or 1.8 to 30 percent respectively of the flight muscles," and Greenewalt (1962) wrote that the two pectoral muscles in hummingbirds "account for 25 to 30 percent of the total weight as compared with an average of 17 percent for ordinary birds."

Both Stolpe and Zimmer (1939) and Greenewalt (1960) studied the hovering flight of hummingbirds, in which the wing tip traverses a figure-eight pattern in the course of a single complete cycle of wing beat. During the backward stroke, the under (ventral) surface of the wing is turned *upward*. Then, at the rearmost reversal point in the stroke, the wing moves upward and rotates so that the under surface is turned downward for the forward stroke. The primaries form most of the functional wing, the secondaries being few in number and telescoped together. While hovering at a flower, the body of a hummingbird may "hang" in a vertical position, and the bird can even fly backwards. Greenewalt reported that "in hovering flight the wings move backward and forward in a horizontal plane. On the down (or forward) stroke the wing moves with the long leading edge forward, the feathers trailing upward to produce a small positive angle of attack. On the back stroke the leading edge rotates nearly a hundred and eighty degrees and moves backward, the underside of the feathers now uppermost and trailing the leading edge in such a way that the angle of attack varies from wing tip to shoulder, producing a substantial twist in the profile of the wing." He also stated that female Ruby-throated Hummingbirds have an average wing-beat rate of 52 times per second, and that, according to his experiments, "the top speed of which these birds are capable is something just under 30 miles per hour," which is considerably less than has been stated in the literature. Moreover, the wing-beat rate is essentially constant for an individual bird: 52 $\pm$ 3 beats per second. Greenewalt also noted that "there is no difference between the duration of the downbeat and that of the upbeat. Each occupies just one half of the full cycle. This result also is in accordance with oscillator theory. The wing, essentially a flat plane, changes neither its weight nor length, hence the same wing beat rate would obtain throughout the cycle."

Huey (1962) reported that a House Finch "weighing 20.85 g carried a 4.88 g load, which is 23 per cent of her weight, and the Golden Eagle weighing 4,169.4 g carried a 907 g load, which is 21 per cent of his weight."

Of interest physiologically is the report by Austin and Bradley (1969) that the Poor-will is capable of flight with a body temperature as low as 27.4°C. They suggested that "by using a daytime roost situated so that it is exposed at least to the afternoon sun, a torpid bird could be sufficiently warmed passively by sunset to be able to forage actively with a minimum of thermogenic expenditure."

FLIGHT SPEED

Flight speed is dependent on several factors: velocity and direction of the wind, air currents, thermal updrafts, presence of moving automobiles or airplanes, and motivation, such as pursuit of prey or escape from predator. The flight speed of small passerine birds has been reported to vary from about 15 to 55 miles per hour. Several ducks have been timed at speeds of 50 to 60 miles per hour when followed by airplanes. Under similar conditions, a Peregrine Falcon (*Falco peregrinus*) is reported to have flown at a rate of about 175 miles per hour, but Broun and Goodwin (1943) gave the average speed for migrating birds as 30 miles per hour. They reported an average speed of 41.5 miles per hour for several Ospreys but noted that one bird flew 80 miles per hour with a wind of only 4 miles per hour. Lokemoen (1967) said that nine flocks of Wood Ducks flew at an average speed of 47.1 miles per hour. Tucker and Schmidt-Koenig (1971) attempted to correlate flight speed with wind direction for 21 species of free-flying birds. They reported flight speeds ranging from about 25 miles per hour (American Kestrel, *Falco sparverius*) to 46 miles per hour (Gadwall, *Anas strepera*). Canada Geese, several species of gulls and terns, and the Red-winged Blackbird flew at speeds between 29 and 31 miles per hour.

Wing-beat frequencies of 200 strokes per second have been attributed to hummingbirds, but studies using stroboscopic light and slow-motion pictures have revealed much lower frequencies: 36 to 39 wing beats per second for *Chlorostilbon*; 50 to 51, for *Phaethornis*; and 52 for *Archilochus* (Greenewalt, 1960). Kahl (1971) tabulated wing-flapping rates for the 17 living species of storks; average rates varied from 139 (*Leptoptilos dubius*) to 208 (*Anastomus oscitans*) per minute.

BIRDS UNDER WATER

Also of interest are birds that use their wings both in the air and under water: the diving-petrels; some shearwaters (*Puffinus*); cormorants; murres, puffins, and other alcids (Fig. 8). The diving-petrels, especially, are noted for their habit of diving into the water from the air and then literally flying under water in search of their food (see Storer, 1971). Subcutaneous air cells are well developed in tropicbirds, boobies, and pelicans. Many other birds (e.g., loons, grebes, coots, ducks) that secure their food under water propel themselves with their feet alone.

Raikow (1970) reported on the anatomical adaptations for diving in three species of stifftail ducks (tribe Oxyurini) that exhibit specializations for an aquatic life. The Black-headed Duck (*Heteronetta atricapilla*) of South America is the least specialized of the three species; it tends to inhabit small ponds, dives skillfully for food, walks poorly, and takes off directly from the water rather than running along the surface. The North American Ruddy Duck (*Oxyura jamaicensis*) "cannot walk on land for more than a few steps

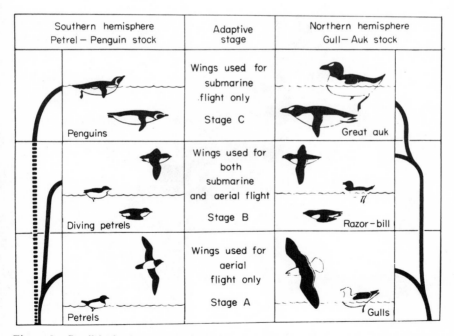

Figure 8. Parallel adaptive stages in the evolution of two stocks of wing-propelled diving birds. (Courtesy of Robert W. Storer.)

without falling forward onto its breast." The Musk Duck (*Biziura lobata*) and Blue-billed Duck (*Oxyura australis*) of Australia are completely aquatic and never come on land, "though they do sometimes crawl onto clumps of cumbungi to rest" (Frith, 1967; Fig. 9). Anatomical specializations for the more aquatic species include a lengthening of the tail and an enlargement of the

Figure 9. Female Blue-billed Duck swimming under water. (Courtesy of J. H. Frith; photo by Ederic Slater.)

levator muscles, a narrowing of the pelvis, and a reduction in the size of the thigh muscles, coupled with an increase in the shank muscles correlated with the change from walking to swimming.

Several observers have discussed the length of time that diving birds spend under water in search of food. Jenni (1969) reported an average diving time of 12.5 seconds for the Least Grebe (*Podiceps dominicus*) and of 21 seconds for the Masked Duck (*Oxyura dominica*).

The dippers (family Cinclidae) are unique among passerine birds in that they not only swim on the surface but also fly under water, using their wings for propulsion. They "have been known to fly down through 20 feet of water to feed on the bottom. The length of their average dive is about 10 seconds, with a

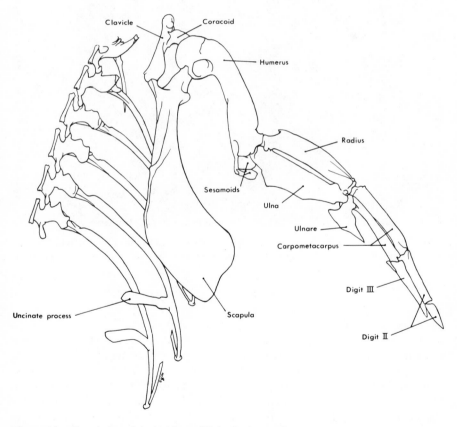

Figure 10. Dorsal view of the highly modified wing bones of a penguin.

Figure 11. Three Emperor Penguins swimming under water. Note the tucked-in position of the head, attitude of the feet with soles directed upward, and nearness of birds to each other. (Courtesy of G. L. Kooyman and the editor of *Auk*.)

probable maximum of perhaps 30 seconds" (Austin, 1961). The birds have a thick undercoat of down and a highly developed oil gland, but other specializations for their underwater life are poorly understood. Some authors refer to a "movable scale" (an operculum) over the nostrils (probably based on a statement made by Joseph Grinnell in 1924), but William R. Goodge wrote to us (1972) that he found "no muscles or vascular plexus which would open or close the operculum," adding that the American Dipper has a large nasal gland, whose function remains unknown.

FLIGHTLESSNESS

The most completely water-adapted birds are the penguins, which have lost the power of flight in air. The scapula is highly modified (very broad and thin), and all of the wing bones are much flattened (Fig. 10). The humeroradial and humeroulnar joints are modified so that flexion and extention are limited at the elbow. Musculus biceps brachii, a powerful flexor of the forearm in other birds, is absent, but some of its function is accomplished by specializations of Mm. pectoralis and tensor patagii longus.

Kooyman *et al.* (1971) published a fine study of the diving behavior of Emperor Penguins (*Aptenodytes forsteri*; Fig. 11). They reported a maximum swimming speed of 9.6 km per hour (5.2 mph), which was accomplished with an average wing beat of 29 per minute. The majority of dives lasted for less than 1 minute, but dives lasting as long as 12 minutes were recorded. The maximum measured rate for ascent or descent during a dive was approximately 120 meters per minute. These authors estimated that one bird dove to a depth

Figure 12. The Flightless Cormorant of the Galapagos Islands. (Courtesy of John Henry Dick.)

Figure 13. The Giant Pied-billed Grebe (*Podilymbus gigas*) of Lake Atitlan, Guatamala. Although this species is nearly twice the size of the Pied-billed Grebe, its wings are said to be so small that the giant grebe "can fly scarcely, if at all." (Courtesy of Philip Boyer.)

of 265 meters; the previous depth record was about 60 meters for the Common Loon and the Old-squaw (*Clangula hyemalis*).

A number of other birds have become flightless; although there are exceptions, most of them (both living and extinct) are island forms.

Among the best-known flightless carinate birds are these: the grebe *Rollandia* (*Centropelma*) *micropterum* of Lake Titicaca in the Andes, the flightless cormorant *Phalacrocorax* (*Nannopterum*) *harrisi* of the Galapagos

(Fig. 12), two Steamer Ducks (*Tachyeres brachypterus, T. pteneres*), the Brown Teal (*Anas aucklandica*) of the Auckland Island archipelago, several species of island-inhabiting rails (e.g., Laysan Island Rail, *Porzanula palmeri*; Wake Island Rail, *Rallus wakensis*; Tristan Island Gallinule, *Gallinula n. nesiotis*; Notornis or Takahe, *Porphyrio mantelli*, of New Zealand), the extinct Great Auk (*Pinguinus impennis*), and *Hesperornis* (Fig. 13). In addition, members of several other families (e.g., the mesite *Monias*, the Kagu, one species of New Zealand wren) are either flightless or nearly so. The Zapata Wren (*Ferminia cerverai*), an inhabitant of the Zapata swamp in Cuba, has a very weak flight and is said rarely to take to wing.

REFERENCES

Austin, G. T., and W. G. Bradley. 1969. Additional responses of the Poor-will to low temperatures. *Auk,* **86:**717–725.

Austin, O. L., Jr. 1961. *Birds of the World.* Golden Press, New York.

Broun, M., and B. V. Goodwin. 1943. Flight-speeds of hawks and crows. *Auk,* **60:**487–492.

Brown, R. H. J. 1961. Flight. In *Biology and Comparative Physiology of Birds,* Vol. 2. Academic Press, New York, pp. 289–305.

Clark, R. J. 1971. Wing-loading—a plea for consistency in usage. *Auk,* **88:**927–928.

Edgerton, H. R., R. J. Niedrach, and W. Van Riper. 1951. Freezing the flight of hummingbirds. *Nat. Geogr. Mag.,* August 1951:245–261.

Fisher, H. I. 1946. Adaptations and comparative anatomy of the locomotor apparatus of New World vultures. *Am. Midl. Nat.,* **35:**545–727.

——— 1956a. The landing forces of domestic Pigeons. *Auk,* **73:**85–105.

——— 1956b. Apparatus to measure forces involved in the landing and taking off of birds. *Am. Midl. Nat.,* **55:**334–342.

——— 1957a. Bony mechanism of automatic flexion and extension in the Pigeon's wing. *Science,* **126:**446.

——— 1957b. The function of M. depressor caudae and M. caudofemoralis in Pigeons. *Auk,* **74:**479–486.

——— 1959. Some functions of the rectrices and their coverts in the landing of pigeons. *Wilson Bull.,* **71:**267–273.

Frisch, O. v. 1963. Zur Anatomie des distalen Humerusendes der Limicolen und Lariden. *Zool. Jahrb. Anat.,* **80:**459–468.

Frith, H. J. 1967. *Waterfowl in Australia.* East-West Center Press, Honolulu, Hawaii.

George, J. C., and A. J. Berger. 1966. *Avian Myology.* Academic Press, New York.

Goodge, W. R. 1959. Locomotion and other behavior of the dipper. *Condor,* **61:**4–17.

Graham, R. R. 1934. The silent flight of owls. *Jr. Roy. Aeronaut. Soc.,* **38:**837–843.

Greenewalt, C. H. 1960. *Hummingbirds.* Doubleday & Co., Garden City, N.Y.

——— 1962. Dimensional relationships for flying animals. *Smithson. Misc. Collect.,* **144:**1–46.

Hamilton, T. H. 1961. The adaptive significance of intraspecific trends of variation in wing length and body size among bird species. *Evolution,* **15:**180–195.

Hartman, F. A. 1961. Locomotor mechanisms of birds. *Smithson. Misc. Collect.,* **143**:1–91.

Herzog, K. 1968. *Anatomie und Flugbiologie der Vögel.* Gustav Fischer, Stuttgart.

Houghton, G. 1964. Fluttering flight mechanisms in insects and birds. *Nature,* **4957**:447–449.

Huey, L. M. 1962. Comparison of the weight-lifting capacities of a House Finch and a Golden Eagle. *Auk,* **79**:485.

Jenni, D. A. 1969. Diving times of the Least Grebe and Masked Duck. *Auk,* **86**:355–356.

Kahl, M. P. 1971. Flapping rates of storks in level flight. *Auk,* **88**:428.

Kooyman, G. L., C. M. Drabek, R. Elsner, and W. B. Campbell. 1971. Diving behavior of the Emperor Penguin, *Aptenodytes forsteri. Auk,* **88**:775–795.

Kurotchkin, E. N., and V. G. Vasiliev. 1966. (Some functional fundamentals of the birds' swimming and diving; in Russian; English summary.) *Zool. Zh.,* **14**:1411–1420.

Lokemoen, J. T. 1967. Flight speed of the Wood Duck. *Wilson Bull.,* **79**:238–239.

Penney, R. L., and J. T. Emlen. 1967. Further experiments on distance navigation in the Adelie Penguin, *Pygoscelis adeliae. Ibis,* **109**:99–109.

Raikow, R. J. 1970. Evolution of diving adaptations in the stifftail ducks. *Univ. Calif. Publ. Zool.,* **94**:1–52.

Raspet, A. 1960. Biophysics of bird flight. *Science,* **132**:191–200. (Also, *Smithson. Rept.* No. 4447, 1961:405–424.)

Salt, W. R. 1967. Loads lifted by homogeneous muscle in flapping flight. *Can. J. Zool.,* **45**:73–79.

Savile, D. B. O. 1957. Adaptive evolution in the avian wing. *Evolution,* **11**:212–224.

Stolpe, M., and K. Zimmer. 1939. Der Schwirrflug des Kolibri im Zeitlupenfilm. *J. Ornithol.* **87**:136–155.

Storer, J. H. 1948. The flight of birds analyzed through slow-motion photography. *Cranbrook Inst. Sci. Bull.* No. 28.

Storer, R. W. 1955. Weight, wing area, and skeletal proportion in three Accipiters. *Acta XI Congr. Int. Ornithol.,* **1954**:287–290.

——— 1971. Adaptive radiation of birds. In *Avian Biology,* Vol. 1, Academic Press, New York, pp. 149–188.

Tucker, V. A. 1971. Flight energetics in birds. *Am. Zool.,* **11**:115–124.

———, and K. Schmidt-Koenig. 1971. Flight speeds of birds in relation to energetics and wind directions. *Auk,* **88**:97–107.

Van Tets, G. F. 1966. A photographic method of estimating densities of bird flocks in flight. CSIRO Wildl. Res., **11**:103–110.

CHAPTER NINE

Food and
Feeding Habits

It is a familiar fact that all animals are directly or in-directly dependent on plants for nourishment. As a group, however, birds feed mainly on animals, and only a rather small proportion of the birds of the world have become sufficiently specialized to utilize plant food directly. One indication that plant-eating is a secondary specialization is the fact that most plant-eating birds start their newly hatched young on an animal diet. Only a few birds do not start their young on animal food: e.g., pigeons, cardueline finches.

ANATOMICAL SPECIALIZATIONS

Some of the most obvious differences among birds are seen in the numerous bill adaptations for feeding. The bill shape is very similar within the genera of some families, but in other groups of closely related genera the bill exhibits wide adaptive specialization for particular feeding habits; examples are the Hawaiian honey-creepers, larks, and hummingbirds (Fig. 1).

Most species of Galapagos finches have relatively large, seed-crushing bills, but several have smaller bills and feed predominantly on insects. The Woodpecker-

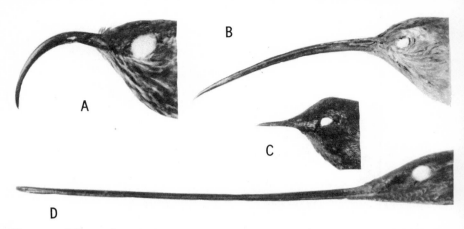

Figure 1. Bill specialization for different types of flowers in four genera of hummingbirds: (*A*) *Eutoxeres aquila*; (*B*) *Phaethornis superciliosus*; (*C*) *Ramphomicron microrhynchum*; (*D*) *Ensifera ensifera*.

finch (*Cactospiza pallida*) and the Mangrove-finch (*C. heliobates*) are noted for using cactus spines, leaf petioles, etc., for probing into holes or crevices to dislodge insects and their larvae. Another member of the family has evolved the unusual habit of eating blood. *Geospiza difficilis septentrionalis* inhabits Wenman Island, on which Red-footed and Masked boobies nest. The boobies tolerate the finches that land on their backs. By biting the skin of a booby at the base of a secondary flight feather in the elbow region, a wound is opened and the finch feeds on the blood that oozes from the skin (Bowman and Billeb, 1965).

The highly modified bill of the Spoon-billed Sandpiper (*Eurynorhynchus pygmeum*) is used for feeding on minute life in mud (Fig. 2); Burton (1974) gives a detailed discussion of the anatomy and feeding habits of a number of shorebirds. The Pink-eared Duck (*Malacorhynchus membranaceus*) of Australia has a spoon-shaped bill with membranous flaps near the tip, which is used in filtering water (Fig. 3). Other ducks have *lamellae* on the inner margin of the mandibles that aid in straining food from water or mud.

In some caprimulgids the palate is enlarged and reinforced, presumably against the force of insects caught in flight; some species (e.g., Red-necked Nightjar, *Caprimulgus ruficollis*) have a highly vascular and membranous palate, which Cowles (1967) suggested might be an adaptation for nocturnal aerial feeding. Rictal bristles at the margins of the gape are adaptations in many insectivorous birds (e.g., caprimulgids, swifts, flycatchers); rictal bristles may be well developed in "flycatching" species but greatly reduced or absent in

Figure 2. Ventral view of the highly specialized bill of the Spoon-billed Sandpiper (*Eurynorhynchus pygmeum*).

Figure 3. Upper: Male (front) and female Pink-eared Duck (photo by Ederic Slater). Lower: Pink-eared Ducks filter-feeding (photo by H. J. Frith). (Courtesy of H. J. Frith.)

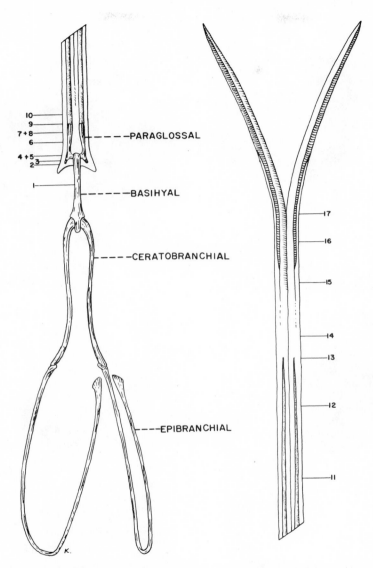

Figure 4. Reconstruction of the tongue and hyoid apparatus of Allen's Hummingbird (*Selasphorus sasin*). (By permission from Weymouth, Lasiewski, and Berger, 1964, and the editor of *Acta Anatomica*.)

Figure 5. (*A*) Natural pecan (*Carya pecan*). (*B*) Pecan after remaining in the gizzard of a domestic turkey for 1 hour. (*C*) A hickory nut (*Carya ovata*) showing incipient fracture. (*D*) Hickory nut after remaining in the gizzard for 31 hours. (Courtesy of the late A. W. Schorger and the editor of *Auk*.)

nonflycatching relatives (e.g., the wood-warbler genera *Setophaga* and *Dendroica*).

The great diversity in tongue structure among birds was described in Chapter 2. The hyoid apparatus is the support for the base of the tongue, and muscles attached to the hyoid are responsible for protruding and retracting the tongue. In some wrynecks, woodpeckers, and hummingbirds, which have especially long tongues, the posterior horns (epibranchials) of the hyoid wind upward around the back of the skull for varying distances. In hummingbirds (Fig. 4) and some woodpeckers, the epibranchials extend forward as far as the nostrils; in some woodpeckers, they enter the premaxilla and extend forward to end about midway between the anterior margin of the narial opening and the tip of the premaxilla. The two epibranchials may be either symmetrical or asymmetrical in pattern; when asymmetrical, the right epibranchial usually is much longer than the left; nestling and fledgling birds are said to have symmetrical epibranchials. In wrynecks and the Green Woodpecker (*Picus viridis*) the epibranchials form a loop at the sides of the esophagus in the neck.

Although most birds carry food in their bills, food may also be carried in a gular pouch (pelicans) or in a crop. The Rosy Finch (*Leucosticte tephrocotis*) has a pair of buccal food pouches, which apparently are used only during the breeding season. Bock *et al.* (1973) described a sublingual pouch in Clark's Nutcracker (*Nucifraga columbiana*) and in the Thick-billed Nutcracker (*N. caryocatactes*) in which the birds carry pine seeds and hazelnuts from harvesting areas to caching sites. The stored seeds provide food for the adults during the winter and constitute the primary source of food for the nestlings in the following breeding season. Fisher and Dater (1961) described a single esophageal diverticulum (a "false crop," also used for seed storage) in the Redpoll (*Acanthis flammea*).

In some birds the efficiency of the gizzard in grinding up food is truly remarkable (Fig. 5). Wild Turkeys and some ducks (e.g., Mallard, Wood Duck) swallow acorns, and even harder nuts, whole; eiders and other diving ducks swallow mussels and other shellfish. Many seed-eating and herbivorous birds eat *grit* (quartz, pebbles, small shells, very hard seeds), which facilitates the grinding process; Simpson (1965) wrote that the Southern Skua (*Catharacta skua lonnbergi*) uses pumice as grit or gizzard stones. Hummingbirds sometimes eat sand. Many caprimulgids also eat grit, presumably to aid in breaking down the chitinous bodies of beetles and other insects (Jenkinson and Mengel, 1970). Grit also may be an important source of calcium and phosphorus.

TYPES OF FOOD

It is probable that all classes of living things furnish food for birds. Not many birds are known to eat the more primitive plants such as algae and fungi, but a Pygmy Parrot (*Micropsitta*) of New Guinea apparently subsists largely on fungus in rotten wood, the Glaucous Gull (*Larus hyperboreus*) eats marine algae, and ptarmigan feed on lichens. Some tanagers (*Tanagra*), flowerpeckers (*Dicaeum*), silky-flycatchers (Phainopepla, *Ptilogonys cinereus*), the Cedar Waxwing, and a South American tyrant-flycatcher (*Phaeomyias*) feed on mistletoe berries. Yellow-bellied and Williamson sapsuckers (*Sphyrapicus varius* and *S. thyroideus*) have a diet consisting largely of the bark, cambium, and sap of trees. The Ruffed Grouse (*Bonasa umbellus*) in New York State is known to eat parts of 414 species of plants and 580 species of animals!

Figure 6.   A Broad-winged Hawk at its nest with a mole. (Courtesy of Samuel A. Grimes.)

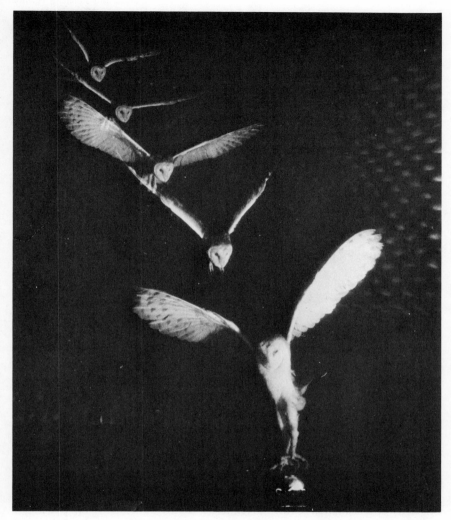

Figure 7. An infrared photograph of a Barn Owl flying toward a mouse. The photograph was taken in total darkness by using infrared film and electronic flashes (200 msec between exposures) in the laboratory of Masakazu Konishi. (Courtesy of Masakazu Konishi.)

Figure 8. A hen Rock Ptarmigan (*Lagopus mutus*); Ellesmere Island, May 31, 1955. (Courtesy of David F. Parmelee.)

Rodents, bats, and other mammals provide a large share of the diet of many hawks and owls (Figs. 6, 7). Sloths are eaten by the Harpy Eagle (*Harpia harpyja*) and monkeys by the Philippine Monkey-eating Eagle (*Pithecophaga jefferyi*).

American Kestrels are known to prey on such birds as the Bobwhite, Starling, Robin, Meadowlark, Brewer's and Red-winged blackbirds, and Ruby-throated Hummingbird; the Goshawk, on Sharp-tailed Grouse, Greater Prairie Chicken, and Common Crow; Gyrfalcons (*Falco rusticolus*), on ptarmigan (Fig. 8); Rough-legged Hawk on juncos (*Junco hyemalis*); and the New Zealand Falcon (*Falco novaeseelandiae*) on the Skylark (*Alauda arvensis*), Song Thrush (*Turdus ericetorum*), and other birds. Carothers *et al.* (1972) told of Steller's Jays (*Cyanocitta stelleri*) killing Gray-headed Juncos (*Junco caniceps*) and Pygmy Nuthatches (*Sitta pygmaea*) in Arizona during periods of heavy snow, and there is a report of Mexican Jays (*Aphelocoma ultramarina*) killing small birds, also in Arizona under similar conditions.

Although reptiles are eaten mainly by raptors and herons, representatives of other families sometimes eat them: cariamas, Roadrunner, trogons, kingfishers, motmots, toucans, ovenbirds, tyrant-flycatchers, birds-of-paradise, vanga-shrikes, and shrikes. Groups that feed to some extent on amphibians are cormorants, anhingas, herons, storks, hawks, rails, avocets, thick-knees, cuckoos, woodcreepers, and certain corvids and icterids (Fig. 9).

Fish are eaten by many birds from penguins to kingfishers and by a few passerine species (e.g., dippers, grackles, some tyrant-flycatchers, Robins). The Everglade (*Rostrhamus sociabilis*) and Slender-billed (*Helicolestes hamatus*) kites are noted for subsisting primarily on snails of the genus *Pomacea* (Fig. 10). The Bristle-thighed Curlew (*Numenius tahitiensis*) is known to have eaten scorpions (*Hormurus australasiae*). Whelan (1953) reported that Tree Swallows in Michigan fed their nestlings freshwater snails and clams. Finally, vultures and some gulls regularly feed on carrion and/or garbage, and many species of birds have been observed feeding from road-killed animals (Fig. 11).

Blue Jays and House Wrens (*Troglodytes aedon*) are noted for destroying the eggs of other birds, and a number of other unrelated species also demonstrate a predilection for eggs. Skuas and jaegers feed themselves and their nestlings largely on the eggs (and chicks) of seabirds and penguins; Herring Gulls, Clapper Rails (*Rallus longirostris*), and Fish Crows eat eggs; the

Figure 9. A Lesser Crow-pheasant (*Centropus bengalensis javanicus*) at its nest with a frog. (Courtesy of the late Loke Wan Tho.)

Figure 10.   An Everglade Kite carrying a snail. (Courtesy of Noel Snyder.)

Ruddy Turnstone has been observed to consume the eggs of the Sooty Tern; and there is a record of a Red-winged Blackbird destroying eggs of the Common Tern (*Sterna hirundo*) and the Roseate Tern (*Sterna dougallii*). Of special interest is egg-eating among several species of island passerines. At least two species of Galapagos Mockingbirds (*Nesomimus parvulus bauri, N. macdonaldi*) eat the eggs of seabirds and of the Galapagos Dove (*Zenaida galapagoensis*). Harris (1968) reported that the Darwin's Finch *Geospiza difficilis* on Wenman Island eats the eggs of gulls and seabirds but is not able to crack the shells. The Laysan Finch (*Psittirostra c. cantans*) and the Nihoa Finch (*P. c. ultima*) of the Hawaiian Islands eat the eggs of seabirds. The race on Laysan Island is able to crack the shells with its bill, whereas the smaller Nihoa race can puncture only the eggs of smaller terns. By pecking at the under surface of a large egg, however, the Nihoa Finch manages to roll the egg, which soon falls to a lava rock a foot or so below and breaks (Fig. 12).

In some carnivorous birds, undigestible materials such as teeth, bones, hair, feathers, and invertebrate exoskeletons are formed into a neat pellet in the gizzard and then are regurgitated (see Grimm and Whitehouse, 1963). This habit is characteristic of grebes, owls, hawks, herons, gulls, grouse, some rails (e.g., King and Clapper rails), caprimulgids, swifts, shrikes, the Black Wheatear (*Oenanthe leucura*), and some thrushes. Grebes eat feathers and feed them to their young. The strong acid in the grebe's stomach dissolves most of the bones of the fish eaten by the bird, and it is thought that the feathers act as retainers in the stomach to hold the bones there long enough for digestion to take place.

FOOD-FINDING AND CAPTURE

Some birds watch for their prey to appear and then descend upon it; other birds actively seek food through obstacles. Mud- and ground-feeders that pat and probe the earth as they move across their feeding grounds are presumed to be "flushing" to the surface the small animals on which they feed. Simmons (1961) and Meyerriecks (1966) summarized information on *foot-trembling* and *foot-paddling* by a variety of herons, shorebirds, and gulls, as well as by the Hermit Thrush (*Catharus guttatus*). The Black Heron (*Melanophoyx ar-*

Figure 11. White-backed Vultures (*Gyps bengalensis*) at the remains of a cow; Baroda, India, January 16, 1965.

Figure 12. Nihoa Finches at the egg of a Common Noddy; Nihoa Island, August 1, 1966.

desiaca) of Africa and Madagascar is noted for a peculiar behavior pattern used in catching fish: the bird spreads its wings and extends them forward and downward until the tips of the feathers touch the water, thus forming a canopy over the bird's head and presumably bewildering fish by the darkness created.

Food preparation in birds is very often left entirely to the digestive system. A Secretarybird (*Sagittarius*), for example, may swallow as much of a snake as it can, wait for the digestive processes to act on that part, and then swallow the rest (Fig. 13).

Food Storage

Although birds generally require an almost continuously available supply of food, a few, notably the crow family (Corvidae), store food in time of plenty and use it when supplies are scarce. For example, Swanberg (1951) reported that nutcrackers (*Nucifraga*) in Sweden fly considerable distances (up to 3.75 miles) from their haunts in the evergreen-forested hills to gather hazelnuts in the lowland. Filling their sublingual pouches with nuts, the nutcrackers fly back to the evergreen forest and carefully bury the load of nuts in a single hole which they dig in the ground. This process is carried on with hardly a break from sunrise to sunset during two or three autumn months. Then all through the winter and early spring they feed on these nuts, digging down through as much as 18 inches of snow to recover them. Swanberg believes that in their very successful (86 percent, or better) search for these caches they depend entirely on memory. The jay *Garrulus glandarius* also stores food in the ground.

Figure 13. A nestling American Kestrel with a garter snake hanging from its mouth; the nestling was taken from the nest (in a nesting box for flickers) to be banded; Detroit, Michigan. (Courtesy of the late Walter P. Nickell.)

In higher latitudes of Eurasia and North America certain shrikes (*Lanius*) store some of their prey (large insects and small mammals and birds) by hanging it on thorns or in forked branches. The Gray Butcherbird (*Cracticus torquatus*) of Australia stores food in a similar manner, and American Kestrels, owls, and some nuthatches and titmice store food more or less regularly.

Amounts of Food Required

Because of their high metabolic rates most birds require large amounts of food at relatively frequent intervals. This is especially true of small birds, which, in proportion to their weight, eat more food per day than do large birds. Lack (1954:131) gave the following approximate ratios between body weight and daily food consumption. Landbirds weighing between 100 and 1000 grams eat from 5 to 9 percent of their body weight daily; songbirds weighing 10 to 90 grams eat 10 to 30 percent of their own weight daily; and hummingbirds may eat twice their weight in syrup in a day. Kendeigh (1934:383) reported that adult seed-eating birds consume about 10 percent of their weight and that insectivorous birds eat an amount equal to 40 percent of their weight daily, though the high water content of larvae largely accounts for this high percentage. Calculations based on the dry weight of grains and meal worms indicated that granivorous and insectivorous birds eat about the same relative amounts of food. (See also Kendeigh and West, 1965.)

Experiments with captive passerine birds have revealed that the average time for the passage of artificially stained food through the digestive tract is about 1.5 hours, whether the food be grain, fruit, or insects. Grandy (1972) reported that Black Ducks (*Anas rubripes*) "digested and passed a blue mussel in 30–40 minutes." He added, therefore, that "Black Ducks killed after more than 30 minutes of feeding may have eaten, digested, and passed certain animal foods."

It is well known that nestlings are fed large amounts of food, but few quantitative data are available. On the basis of known amounts of food given to nestling Starlings at individual feedings, Kluijver estimated that the bird received about six-sevenths of its own weight in food per day during the last part of the nestling period (see Lack, 1954).

VARIATIONS IN THE FOOD OF BIRDS

Seasonal Variation

Seasonal variation is well illustrated by the Great Spotted Woodpecker (*Dendrocopus major*) of Finland, which subsists primarily on pine and spruce seeds during the winter but on ants and other insects during the summer (Lack, 1954:127). The widely distributed Savannah Sparrow (*Passerculus sandwichensis*) maintains a diet largely of plant foods (90%) in the fall and winter months while in the southern United States. After it migrates north to the breeding grounds, however, the diet becomes increasingly insectivorous, until animal food comprises as much as 75 percent of this sparrow's diet in midsummer.

Changes in diet during periods of bad weather have been reported for the European White Stork (*Ciconia ciconia*), which shifts from fish and frogs to mice during periods of drought.

Annual Variation

Banfield (1947) analyzed nearly 3000 Short-eared Owl (*Asio flammeus*) pellets over a period of 6 years in an area where two genera of rodents constituted about 99 percent of the food. The percentage of *Peromyscus* remains in the pellets varied from about 5 to 30 percent; of *Microtus*, from 68 to 95 percent. During four summers in Germany, the vole *Microtus arvalis* formed from 21 to 77 percent, and small birds from 2 to 34 percent, of the food of Long-eared Owls, *Asio otus* (Zimmerman, 1950).

Geographical Variation

In Finland the winter food of the Great Spotted Woodpecker is mainly conifer seeds; in England, mainly insects. Adult insects form the main diet for nestling titmice (*Parus*) in Lapland, whereas defoliating caterpillars constitute the bulk of their food in England (Lack, 1954).

In one 6-year study on the Arctic tundra, 88 percent by weight of the Gyrfalcon's food consisted of ptarmigans, and less than 1 percent of the Arctic ground squirrel. In other reports from Alaska, however, ptarmigans formed only 18 percent of the diet, the ground squirrel comprising about 79 percent. Similarly, Leopold and Wolfe (1970) found that rabbits and hares formed the predominant prey species of the Wedge-tailed Eagle (*Aquila audax*) near Canberra (Australia) but that kangaroos and lizards were more important in the dry interior of the continent. Kuroda (1962) reported that Gray Starlings (*Sturnus cineraceus*) in rural areas in Japan fed their nestlings primarily on mole-crickets, whereas nestlings in urban areas were fed both animal and vegetable matter, although cherries formed the bulk of the diet.

Variation with Age

Reference was made above to the frequent difference in the food taken by adults and that given to nestlings. Van Tyne (1951) reported observations of a male Cardinal that came to a feeding shelf with a billful of larvae. The bird put the larvae on the shelf and began to eat sunflower seeds, but later picked up the

Figure 14. Two Brown-headed Cowbirds (about 48 hours old) and a Kirtland's Warbler that has just hatched; Oscoda County, Michigan, July 21, 1963.

larvae and flew toward its nest, which contained young. White (1938, 1939) reported both a size difference and a species difference in the fish eaten by adult kingfishers and those given to the nestlings. The food of Ruffed Grouse chicks consists of 70 percent insects during their first 2 weeks; by the age of 1 month their intake of insect food drops to 30 percent, and late in the summer to 5 percent. An interesting problem, about which little is known, is the relationship between food and growth in parasitic cowbirds and cuckoos, whose diet depends on the food preferences of the host species (Fig. 14).

FOOD AS A LIMITING FACTOR

Many writers have assumed that the numbers of birds are limited by the available food supply and that territoriality developed from the special need to provide adequate food during the nesting season. There is, however, little direct evidence that this is so. It has been estimated for a number of bird species that during outbreaks of a prey species (insect or mammal) the birds eat only from 0.05 to less than 5 percent of the available supply. When the population of the prey species is not abnormally high, however, the birds may consume as much as 25 percent of the available food (Lack, 1954). Lack discussed four types of indirect evidence that "suggest that many species are limited in number by their food supply."

SOCIAL FEEDING HABITS

Evolved to exploit the food sources of the environment, social feeding is highly developed in many birds. This behavior can be discussed under several headings.

Cooperation

1. **Mixed flocks** of seabirds (petrels, shearwaters, terns, boobies) prey on schools of fish and on other organisms frightened to the surface by the fish. This flocking behavior of different species of seabirds is so pronounced (and effective) that feeding bird flocks still provide fishermen with the best available clue as to the location of schools of fish.

Davis (1946) in Brazil and Winterbottom (1949) in South Africa described mixed flocks of insectivorous forest birds made up of four to nine species which travel through the forest together, feeding as they go. A considerable number of species are members of these flocks on occasion, but certain "nucleus species" seem to be essential to the formation of such flocks.

2. **Feeding flocks** may consist **of a single species** of bird, such as the Double-crested Cormorant (*Phalacrocorax auritus*). Concerted fishing activities develop when a dozen or more cormorants congregate, but there are often many hundred in a fishing flock. As the line advances, cormorants throughout the flock keep making short dives, a quarter to a half of the total number being under water at any given moment (Fig. 15).

Parasitic or Symbiotic Feeding Relationships

Preying on Organisms Flushed (or Exposed) by Another Animal.   In tropical America the Groove-billed Ani (*Crotophaga sulcirostris*) and Smooth-billed

Figure 15. Cooperative autumn feeding habits of the Double-crested Cormorant. (Courtesy of George A. Bartholomew.)

Ani (*C. ani*) associate with cattle, horses, and mules, feeding on the insects flushed by the grazing animals. The European Starling, Common Mynah (*Acridotheres tristis*), Cattle Egret, and some other herons associate with cattle or wild grazing animals. Certain hornbills, drongos (Fig. 16), and leafbirds (Fig. 17) accompany troops of monkeys, apparently obtaining insects disturbed by them. The Fork-tailed Drongo (*Dicrurus adsimilis*) follows cattle, horses, and elephants; the Piapiac (*Ptilostomus afer*) perches on the backs of elephants; and the Common Sandpiper (*Tringa hypoleucos*) alights on the backs of hippopotamuses (Rice, 1963).

North (1944) described the behavior of the Carmine Bee-eater (*Merops nubicus*) of Africa, which habitually perches on the backs of bustards (*Choriotis*) as they walk through the grass, flushing many of the cryptically colored grasshoppers in their path. The bee-eater flies down to seize the grasshopper and then returns to the back of the bustard. Willis (1966) told of birds that follow ant swarms in Central and South America (see also Wiley, 1971).

The honeyguides (Indicatoridae) provide the most remarkable example of birds that benefit from the activities of other animals (Friedmann, 1955; Friedmann and Kern, 1956). Two African species of honeyguides are known to have the habit of leading a mammal (usually a badger or man) to a bees' nest.

The bird chatters noisily in the presence of the mammal, displays its tail feathers, and makes short flights, waiting or returning if the mammal does not follow promptly. The bird may travel a half mile or more but stops when it comes to the beehive, and after the mammal has broken into the beehive the honeyguide eats the wax, for which no other bird is known to have a comparable appetite.

Preying "Helpfully" on the Parasites of Another Animal.   A few birds prey on the external parasites of certain large mammals and reptiles. The Rook (*Corvus frugilegus*) and Fish Crow sometimes pick parasites from hoofed mammals. The finch *Geospiza fuliginosa* eats ticks from marine iguanas (*Amblyrhynchus cristatus*) in the Galapagos.

The Red-billed (*Buphagus erythrorhynchus*) and Yellow-billed (*B. africanus*) oxpeckers of Africa seem to represent the extreme of this line of specialization. Living only in regions where there are large numbers of big game or cattle, they spend most of their time on the animals, climbing actively about

Figure 16.   Malayan Racket-tailed Drongo (*Dicrurus paradiseus platurus*) at its nest. (Courtesy of the late Loke Wan Tho.)

Figure 17. A male Marshall's Iora (*Aegithina nigrolutea*) at its nest. (Courtesy of Shivrajkumar of Jasdan, India.)

by means of their strongly curved claws and stiff-pointed tails and clipping off the bloodsucking parasites with their very strong, sharp-edged bills. Since the mammals benefit by the removal of these parasites and also by the alarms which the keen-eyed birds sound at the approach of danger, this can clearly be called a symbiotic relationship.

Stealing from Other Animals. Some species of birds have developed great ability in securing certain difficult prey. An excellent example, familiar to everyone, is the American Robin (*Turdus migratorius*) pulling worm after worm from the lawns of our towns. Starlings (*Sturnus vulgaris*), Brown Thrashers (*Toxostoma rufum*), House Sparrows, and other birds sometimes also utilize this easy source of food (Van Tyne, 1946).

Robbing Other Animals by Force. Frigatebirds (*Fregata*) are swift and aggressive robbers. They often resort to robbing boobies and gulls, attacking them savagely and injuring or even killing birds that strongly resist. As a rule, the booby or gull quickly drops its food and the frigatebird dexterously catches it, usually before it strikes the water. American Bald Eagles (*Haliaeetus*) habitually rob Ospreys (*Pandion*) of fish they have captured by plunging from the air.

One whole family of gull-like birds, the jaegers and skuas (Stercorariidae), has specialized as robbers. All hunt for their own food in part, but frequently subsist largely on what they take from their less agile and less aggressive relatives, the gulls and terns. Attacking a successfully fishing gull or tern in the air, the jaeger makes the weaker bird disgorge its fish and then catches the food in midair.

WATER AND SALT

Water Requirements

Many birds apparently need no other water than that contained in a diet that includes some animal and some green-plant food. California Quail (*Lophortyx*

Figure 18. Adult male sandgrouse (*Pterocles namaqua*) soaking his belly feathers at a water hole in the Kalahari Desert; the bird is resting after a period of rocking. (Courtesy of Tom J. Cade, Gordon L. Maclean, and the editor of *Condor*.)

californicus) kept experimentally without water for a year in an outdoor enclosure nested and raised young (see Bartholomew and MacMillen, 1961). Similarly, the Budgerigar (*Melopsittacus undulatus*) can survive with little or no water (Uemura, 1964). Bartholomew (1972) discussed the physiological adaptations of seed-eating birds that survive without drinking water.

On the other hand, certain birds that live on dry seeds must have a regular supply of fresh water. In Australia, some parrots and pigeons come in flocks from points as much as 10 miles away to drink at tanks and water holes. Cade and Maclean (1967) confirmed the remarkable behavior of certain male sandgrouse that bring water to their young: "The adult male flies to a water hole in the morning, soaks his ventral feathers in a special way, and then flies back to the nesting grounds, where he alights, walks to the hiding brood, and presents his abdomen by standing in an upright posture with his feathers fluffed out. The young run from their hiding places, cluster around the male's belly, take the wet feathers in their beaks, and remove the absorbed water by a 'stripping' motion" (Fig. 18). White Storks regurgitate water from their crops to nestlings during July in Germany.

Cormorants and many gulls that live along the edge of the sea not only

Figure 19. Evening Grosbeaks (*Hesperiphona vespertina*) eating grains of table salt from the ground; St. Leon le Grande, Rimouski County, Quebec. (Courtesy of B. M. Shaub and the editor of *Wilson Bulletin*.)

prefer to drink fresh water but even seem to require it. Albatrosses, petrels, shearwaters, alcids, and some other seabirds are obligatory saltwater drinkers and will die of thirst even in the presence of fresh water. The Adelie Penguin (*Pygoscelis adeliae*) drinks salt water most of its life but abruptly changes to the purest of fresh water (antarctic snow) during its month of courtship on land; after the eggs are laid, it resumes drinking sea water.

Salt-eating

Little is known about the physiological basis of salt-eating in birds. Relatively small quantities of salt may be lethal to chickens and pigeons, however, and it is clear that tolerance to salt varies among birds.

The salt-eating habit appears to be most highly developed in birds that subsist largely on seeds or fruit during certain periods of the year. The best-known examples are found among the gallinaceous birds, doves, colies, and carduelines (Ploceidae). The propensity for salt by North American carduelines is well known (Fig. 19). A few other birds have also been observed to eat salt: e.g., woodpeckers (*Asyndesmus, Dendrocopos*), crows, jays (*Cyanocitta*), Black-billed Magpie, White-breasted Nuthatch (*Sitta carolinensis*), Rock Wren (*Salpinctes obsoletus*), and House Sparrow. Cade (1964) discussed water and salt balance in a wide variety of granivorous birds.

REFERENCES

Baird, J., and A. J. Meyerriecks. 1965. Birds feeding on an ant mating swarm. *Wilson Bull.,* 77:89–91.

Baker, J. K. 1962. The manner and efficiency of raptor depredations on bats. *Condor,* 64:500–504.

Banfield, A. W. F. 1947. A study of the winter feeding habits of the Short-eared Owl (*Asio flammeus*) in the Toronto region. *Can. J. Res., D,* 25:45–65.

Barnett, L. B. 1970. Seasonal changes in temperature acclimatization of the House Sparrow, *Passer domesticus. Comp. Biochem. Physiol.,* 33:559–578.

Bartholomew, G. A., Jr. 1972. The water economy of seed-eating birds that survive without drinking. *Proc. XVth Int. Ornithol. Congr.,* pp. 237–254.

———, and R. E. MacMillen. 1961. Water economy of the California Quail and its use of sea water. *Auk,* 78:505–514.

Bartonek, J. C., and J. J. Hickey. 1969. Food habits of Canvasbacks, Redheads, and Lesser Scaup in Manitoba. *Condor,* 71:280–290.

Berger, A. J. 1953. Remains of banded birds found in Screech Owl pellets. *Bird-Banding,* 24:19.

Bock, W. J., R. P. Balda, and S. B. Vander Wall. 1973. Morphology of the sublingual pouch and tongue musculature in the Clark's Nutcracker. *Auk,* 90:491–519.

Bowman, R. I., and S. L. Billeb. 1965. Blood-eating in a Galápagos Finch. *Living Bird,* 1965:29–44.

Bühler, P. 1970. Schädelmorphologie und Kiefermechanik der Caprimulgidae (Aves). *Z. Morphol. Tiere,* **66:**337–339.

Burton, P. J. K. 1974. *Feeding and the Feeding Apparatus in Waders.* British Museum (Natural History), London.

Buskirk, W. H., *et al.* 1972. Interspecific bird flocks in tropical highland Panama. *Auk,* **89:**612–624.

Cade, T. J. 1962. Wing movements, hunting, and displays of the Northern Shrike. *Wilson Bull.,* **74:**386–408.

———— 1964. Water and salt balance in granivorous birds. In *Thirst,* Pergamon Press, Oxford, pp. 237–343.

————, and L. I. Greenwald. 1966. Drinking behavior of mousebirds in the Namib Desert, southern Africa. *Auk,* **83:**126–128.

————, and G. L. Maclean. 1967. Transport of water by adult sandgrouse to their young. *Condor,* **69:**323–343.

Carothers, S. W., N. J. Sharber, and R. P. Balda. 1972. Steller's Jays prey on Gray-headed Juncos and a Pygmy Nuthatch during periods of heavy snow. *Wilson Bull.,* **84:**204–205.

Cowles, G. S. 1967. The palate of the Red-necked Nightjar *Caprimulgus ruficollis* with a description of a new feature. *Ibis,* **109:**260–265.

Davis, D. E. 1946. A seasonal analysis of mixed flocks of birds in Brazil. *Ecology,* **27:**168–181.

Dow, D. D. 1965. The role of saliva in food storage by the Gray Jay. *Auk,* **82:**139–154.

Eisenmann, E. 1961. Favorite foods of Neotropical birds: flying termites and *Cecropia* catkins. *Auk,* **78:**636–638.

Fisher, C. D., E. Lindgren, and W. R. Dawson. 1972. Drinking patterns and behavior of Australian desert birds in relation to their ecology and abundance. *Condor,* **74:**111–136.

Fisher, H. I., and E. E. Dater. 1961. Esophageal diverticula in the Redpoll, *Acanthis flammea. Auk,* **78:**528–531.

Friedmann, H. 1955. The Honey-guides. *U.S. Natl. Mus. Bull.,* **208.**

————, and J. Kern. 1956. *Micrococcus cerolyticus,* Nov. Sp., an aerobic lipolytic organism isolated from the African honey-guide. *Can. J. Microbiol.,* **2:**515–517.

Galushin, V. M. 1959. Data on the nesting of the European Short-toed Eagle in the Ryazan region. *Ornitologiya,* **2:**153–156. (In Russian; available in English, PST Cat. No. 627, Office of Technical Service, Dept. of Commerce, Washington, D.C.)

Grandy, J. W. 1972. Digestion and passage of blue mussels eaten by Black Ducks. *Auk,* **89:**189–190.

Grimm, R. J., and W. M. Whitehouse. 1963. Pellet formation in a Great Horned Owl: a roentgenographic study. *Auk,* **80:**301–306.

Gullion, G. W. 1966. A viewpoint concerning the significance of studies of game bird food habits. *Condor,* **68:**372–376.

Harris, M. P. 1968. Egg-eating by Galápagos Mockingbirds. *Condor,* **70:**269–270.

Haverschmidt, F. 1962. Notes on the feeding habits and food of some hawks in Surinam. *Condor,* **64:**154–158.

Heppner, F. 1965. Sensory mechanisms and environmental clues used by the American Robin in locating earthworms. *Condor,* **67:**247–256.

Horak, G. J. 1970. A comparative study of the foods of the Sora and Virginia rail. *Wilson Bull.,* **82:**206–213.

Jackson, J. A. 1970. A quantitative study of the foraging ecology of Downy Woodpeckers. *Ecology,* **51**:318–323.

Jenkinson, M. A., and R. M. Mengel. 1970. Ingestion of stones by goatsuckers (Caprimulgidae). *Condor,* **72**:236–237.

Kahl, M. P., Jr., and L. J. Peacock. 1963. The bill-snap reflex: a feeding mechanism in the American Wood Stork. *Nature,* **4892**:505–506.

Kendeigh, S. C. 1934. The rôle of environment in the life of birds. *Ecol. Monogr.,* **4**:299–417.

———, and G. C. West. 1965. Caloric values of plant seeds eaten by birds. *Ecology,* **46**:553–555.

Kuroda, N. H. 1962. Comparative growth rate in two Grey Starling chicks, artificially raised with animal and plant foods. *Misc. Rept. Yamashina's Inst. Ornithol. Zool.,* **3**:174–184. (In Japanese; English summary.)

Lack, D. 1954. *The Natural Regulation of Animal Numbers.* Oxford University Press, London.

Leck, C. F. 1972. Seasonal changes in feeding pressures of fruit- and nectar-eating birds in Panama. *Condor,* **74**:54–60.

Leopold, A. S., and T. O. Wolfe. 1970. Food habits of nesting Wedge-tailed Eagles, *Aquila audax,* in south-eastern Australia. *CSIRO Wildl. Res.,* **15**:1–17.

Maher, W. J. 1970. The Pomarine Jaeger as a brown lemming predator in northern Alaska. *Wilson Bull.,* **82**:130–157.

Meyerriecks, A. J. 1966. Additional observations on "foot-stirring" feeding behavior in herons. *Auk,* **83**:471–472.

Moody, D. T. 1970. A method for obtaining food samples from insectivorous birds. *Auk,* **87**:579.

North, M. E. W. 1944. The use of animate perches by the Carmine Bee-eater and other African species. *Ibis,* **86**:171–176. (See also Jackson, *Ibis.* **87**:284–286.)

Petrides, G. A. 1959. Competition for food between five species of East African vultures. *Auk,* **76**:104–106.

Rice, D. W. 1963. Birds associating with elephants and hippopotamuses. *Auk,* **80**:196–197.

Simmons, K. E. L. 1961. Foot-movements in plovers and other birds. *Brit. Birds,* **54**:34–39, 418–422.

Simpson, K. G. 1965. The dispersal of regurgitated pumice gizzard-stones by the Southern Skua at Macquarie Island. *Emu,* **65**:119–124.

Smith, S. M. 1971. The relationship of grazing cattle to foraging rates in anis. *Auk,* **88**:876–880.

Snow, B. K., and D. W. Snow. 1971. The feeding ecology of tanagers and honeycreepers in Trinidad. *Auk,* **88**:291–322.

Storer, R. W. 1966. Sexual dimorphism and food habits in three North American accipiters. *Auk,* **83**:423–436.

Swanberg, O. 1951. Food storage, territory and song in the Thick-billed Nutcracker. *Proc. 10th Int. Ornithol. Congr.,* **1950**:545–554.

Turcek, F. J., and L. Kelso. 1968. Ecological aspects of food transportation and storage in the Corvidae. *Commun. Behav. Biol.,* **1**:277–297.

Twente, J. W., Jr. 1954. Predation on bats by hawks and owls. *Wilson Bull.,* **66**:135–136.

Uemura, J. H. 1964. Effects of water deprivation on the hypothalamo-hypophysial neurosecretory system of the Grass Parakeet, *Melopsittacus undulatus. Gen. Comp. Endocrinol.,* **4**:193–198.

Van Tyne, J. 1946. Starling and Brown Thrasher stealing food from Robins. *Wilson Bull.,* **58**:185.

―――― 1951. A Cardinal's, *Richmondena cardinalis,* choice of food for adult and for young. *Auk,* **68:**110.

Weedin, R. B. 1967. Seasonal and geographic variation in the foods of adult White-tailed Ptarmigan. *Condor,* **69:**303–309.

West, G. C., and M. S. Meng. 1966. Nutrition of Willow Ptarmigan in northern Alaska. *Auk,* **83:**603–615.

Weymouth, R. D., R. C. Lasiewski, and A. J. Berger. 1964. The tongue apparatus in hummingbirds. *Acta Anat.,* **58:**252–270.

Whelan, M. 1953. Fresh-water mollusks fed to young Tree Swallows. *Wilson Bull.,* **65:**196.

White, H. C. 1938. The feeding of Kingfishers: Food of nestlings and effect of water height. *J. Fish. Res. Bd. Can.,* **4:**48–52.

―――― 1939. Change in gastric digestion of Kingfishers with development. *Am. Nat.,* **73:**188–190.

Wible, M., and K. C. Parkes. 1955. Barn Owls feeding on box turtles. *Fla. Nat.,* **28:**74–75.

Wiley, R. H. 1971. Cooperative roles in mixed flocks of antwrens (Formicariidae). *Auk,* **88:**881–892.

Willis, E. O. 1966. The role of migrant birds at swarms of army ants. *Living Bird,* **1966:**187–231.

Winterbottom, J. M. 1949. Mixed bird parties in the tropics, with special reference to northern Rhodesia. *Auk,* **66:**258–263.

Wolf, L. L. 1970. The impact of seasonal flowering on the biology of some tropical hummingbirds. *Condor,* **72:**1–14.

Zimmerman, K. 1950. Jährliche Schwankungen in der Ernährung eines Waldohreulen-Paares zur Brutzeit. *Vogelwelt,* **71:**152–155.

CHAPTER TEN

Courtship and
Nest Building

The breeding season is the focal point of a bird's life, and an annual nesting cycle is characteristic for most species. Although no other class of animals has attracted the interest of so many students, there are great gaps in our knowledge of the breeding activities of the majority of the world's birds.

INITIATION OF THE BREEDING SEASON

One of the intriguing physiological phenomena in birds is the tremendous difference in the size of the gonads between the breeding and the nonbreeding seasons. These seasonal changes are initiated by secretions of the hypothalamus (Fig. 1). In many birds it is the changing photoperiod that stimulates the hypothalamus, although, according to Farner and Lewis (1971), "it must be emphasized that photoperiodic mechanisms are probably never solely responsible for setting the precise time of reproduction since final adjustments can be made by supplementary and modifying mechanisms."

Photoperiodic control has been demonstrated experimentally in some 50 species in 15 families of birds. For some species (e.g., *Junco hyemalis, Zonotrichia leu-*

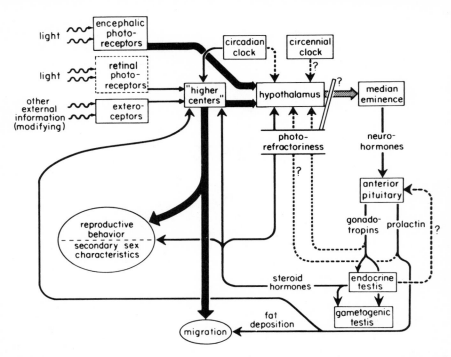

Figure 1. A schematic representation, based primarily on *Zonotrichia leucophrys gambelii*, of neuroendocrine functions and internal information in the control of male reproductive cycles in a photoperiodic species. (By permission of Donald S. Farner and Academic Press.)

cophrys, Fringilla coelebs), photoperiod appears to be the primary environmental factor involved in the control of annual cycles. For others (e.g., *Ploceus philippinus*), day length seems to be less important than other environmental factors. Still other species demonstrate a permissive photoperiodic response. In regard to these species, Farner and Lewis wrote: "Thus far there has been no experimental demonstration that the difference of approximately one hour between the longest and shortest days at a latitude of 10° can serve as the basis for control of an annual reproductive cycle. The changes in gonadal function induced by experimentally altered day lengths, in such species and populations, can only be interpreted as a permissive effect of long days since sufficient changes in day length are not available among the sources of environmental information used in the control of the natural reproductive cycle." Examples of these permissive response species include *Numida meleagris, Quelea quelea, Munia m. malacca,* and *Zonotrichia capensis* (Miller, 1965).

Hamner (1966) demonstrated a circadian rhythm of photic sensitivity in the House Finch (*Carpodacus mexicanus*). He showed that the concept of a "critical day length" is misleading, and he remarked that "a bird which responds to 12 hours of light but not to 11 is apparently stimulated only by the hour of light from 11 to 12, not the total amount of light received. . . . The author believes that most of the interpretive differences which have arisen in the avian photoperiodic literature could be resolved if the concept of critical day length were discarded."

Although there is a reproductive pattern for tropical and subtropical species, it cannot be explained by reference to a single stimulus. Thus Skutch (1950) reported that some species of Central American birds breed when temperatures are rising, others when they are falling; in higher latitudes breeding occurs when the days are becoming longer, but hummingbirds in the western highlands of Guatemala begin to nest when the days are growing shorter; some species begin breeding at the onset of the rainy season, but in other areas many species nest during the height of the dry season. Nearly 80 percent of the nests were found during the 5 months from March to July, the period when food appeared to be most abundant for the majority of species.

At least two species (*Loxops v. virens, Hemignathus wilsoni*) of Hawaiian honeycreepers on the island of Hawaii (20° N.) begin to nest when the days are growing shorter (October-December), although *Loxops v. stejnegeri* of Kauai (22°N) apparently does not initiate nest-building until mid-February (Berger, 1969, 1972; Fig. 2). The Hawaiian Goose (*Branta sandvicensis*) also nests during the fall (Fig. 3).

The Emperor Penguin lays its egg in July in total darkness, and four species of albatrosses and six species of smaller petrels lay when the days are shortening. The Kea (*Nestor notabilis*), the emus (*Dromiceius*), and the Lyrebirds (*Menura*) all begin spermatogenesis while the day length shortens, and breed in winter rather than in summer.

Rain is critical in extreme desert regions where droughts last for months or even for several years. Birds inhabiting Australian deserts may nest during any month of the year; they may breed twice within 6 months after the rains finally come. Keast (1959) estimated that 26 percent of Australian birds are nomadic (Fig. 4); these species move about in search of food, water, and favorable nesting conditions.

Thus, because of the unpredictability of rain in arid environments, natural selection has favored the evolution of a reproductive system in some species that can maintain itself without the usual refractory period common to Temperate Zone species. The Common Weaverbird or Baya (*Ploceus philippinus;* Fig. 5) of Asia has been maintained in an experimental state of sexual readiness with continuous spermatogenesis for a period of 15 months without any indication of regression in the testes.

Figure 2. Nest and eggs of the Hawaii Amakihi (*Loxops v. virens*); Mauna Kea, Hawaii; December 8, 1966.

Rain itself functions as the timer for the breeding cycle in some Australian birds, because they begin to breed before the rain has caused any perceptible change in the food supply, the nesting material, or the general appearance of the environment. In this instance, it is presumed that the sight of rain provides the stimulus, and that the retinal response, in turn, stimulates the hypothalamus. Examples of birds thus affected by rain are the Zebra Finch (*Poephila guttata*), the Black-faced Wood-swallow (*Artamus melanops*), the Budgerigar (*Melopsittacus undulatus*), and the Australian Tree Swallow (*Hylochelidon nigricans*).

Grass-dwelling Old World warblers in Rhodesia and Nyasaland depend on rain for the growth of nesting material and protective cover. The presence of green grass is said to be the stimulus for reproduction in the Red-billed Quelea (*Quelea quelea*) and the Red-bellied Weaver (*Malimbus erythrogaster*) of Africa. "Neither rain itself nor the presence of insect food (another consequence of rain) induces gonadal growth in the absence of green grass; green grass, introduced into the birds' cages, induces gonad growth in the absence of either rain or insect food. Females of this species [*Quelea quelea*] undergo changes of bill color, indicating pituitary stimulation, simply as a result of seeing the male build the nest" (Lehrman, 1964:153).

A population of Abert's Towhees (*Pipilo aberti*) in Arizona may begin to nest 10-14 days after a heavy rain in March or April; sometimes a second nesting peak follows the July rains. The Rufous-winged Sparrow (*Aimophila car-*

Figure 3. Nest and eggs of the Hawaiian Goose; Mauna Loa, Hawaii; January 10, 1973.

Figure 4. Breeding seasons (generalized) of birds in different parts of the Australian continent. Centralward from the coast are extensive intermediate zones where the birds give every evidence of "wanting" to breed in the spring and do so when the spring is fertile. The north of the continent has only one wet season, in summer. (Courtesy of Allen Keast and the editor of *Monographiae Biologicae*.)

Figure 5. Baya Weaverbirds at nests under construction. The male builds the nests; females are perched on the crossbar between the future egg chamber and the entrance tube. (Courtesy of Salim Ali.)

palis) "delays its actual nesting until the summer rains" (Phillips, 1951). Ovulation in the Yellow-headed Blackbird (*Xanthocephalus xanthocephalus*) is said to be delayed if the water over which the birds are building their nests dries up before the nests are completed.

The Red Crossbill depends on a diet of conifer seeds, particularly pine seeds. Its nomadic tendencies, as well as the discovery of nests of this species in every month of the year, strongly suggest that food, rather than light, acts as both the proximate and the ultimate factor in timing the breeding cycle. Experiments have shown that light has some effect in initiating gonadal development, but that under the influence of light the gonads fail to develop fully. Red Crossbills nest only when and where conifers produce a good seed crop.

Although most birds have one breeding season each year, there are several curious exceptions. Sooty Terns on Ascension Island in the Atlantic Ocean breed on a cycle that corresponds to a period of 10 lunar months (295.3 days), and the birds nest twice in some years (Ashmole, 1963a). On Christmas Island (about 2° N. 157° W.) in the Pacific Ocean, however, there are two breeding seasons, and Sooty Terns that fail to raise a chick during one breeding season often return and nest again 6 months later; successful adults, however, do not return to nest until 12 months later (Ashmole, 1963b, 1965) (Fig. 6). Au-

Figure 6. A small portion of a Sooty Tern colony at Cape Manning, Christmas Island (Pacific Ocean), November 14, 1971. It has been estimated that Sooty Tern colonies on this island sometimes contain as many as 800,000 eggs. Ralph W. Schreiber estimated that 250,000 eggs were collected for food from one colony during May and June 1967.

dubon's Shearwater (*Puffinus lherminieri*) on the Galapagos Islands also breeds every 9 months (see Snow and Snow, 1967).

TERMINATION OF THE BREEDING SEASON

The duration of the breeding season depends on the physiological condition of the endocrine organs. Little is known about the possible influence (if any) of external factors on these organs near the end of a species's breeding cycle. Evidence suggests that a bird will not renest, and may even desert a nest containing eggs or young, after a certain period because of a decrease in the internal stimuli responsible for incubation and feeding behavior. September and October nests of American Goldfinches sometimes are deserted for no apparent reason even when they contain young (Berger, 1968).

Farner and Lewis (1971) discussed four possible mechanisms for terminating the reproductive cycle: decreasing day length, negative feedback, photorefractoriness, and photorefractoriness of the "transequatorial type." After the breeding season the gonads pass through a photorefractory stage during which they are insensitive to changes in photoperiod. Farner and Lewis suggested that "the best hypothesis concerning the site of photorefractoriness is that it is at a hypothalamic or higher level," and discussed six possible hypotheses dealing with the cause of photorefractoriness in birds. Among these is the circennial periodicity hypothesis, which has been shown for two species of warblers (*Phylloscopus trochilus, P. sibilatrix*). Birds "maintained on LD 12:12 at constant temperature from September until June underwent prenuptial molt at the same time as controls retained under natural photoperiods either on their breeding or wintering grounds. . . . Furthermore, birds that hatched late in the year terminated molt and developed both autumnal and vernal *Zugunruhe* later than young hatched earlier in the season."

Several phenomena are related to a partial recrudescence of the gonads after the breeding season and the annual molt. Among these is postbreeding song. Sometimes nesting is attempted (Selander and Nicholson, 1962; Berger, 1966).

FUNCTIONS OF COURTSHIP

Courtship may be defined as any behavior pattern that brings the sexes together and leads to copulation. Tinbergen (1954) outlined four main attributes of courtship display.

1. Song, song flights, and other special displays serve an orientation function: they attract a female to the male's territory or to a nest site.

2. Courtship displays serve to suppress nonsexual responses in the mate.

3. Displays serve to synchronize the sexes (Fig. 7). For birds of the Temperate Zone, the effects of photoperiodism lead to a gross synchronization of the sexes. A finer synchronization, leading to copulation, is effected by the signal function of display.

4. Because courtship displays and songs are species specific, they serve as *biological isolating mechanisms;* they tend to ensure that pair-bonds and copulation will occur only between individuals of the same species (Fig. 8).

PAIR FORMATION

The majority of birds are monogamous. A pair remains together during one nesting cycle (or longer), and we say that a pair-bond is formed between the male and the female.

Duration of the Pair-Bond

The length of the pair-bond varies considerably among birds.

1. The sexes meet only at the time of copulation.

a. The males meet at a communal display ground (a "lek"). Examples: Prairie Chicken (*Tympanuchus*), Sharp-tailed Grouse (*Pedioecetes*), Sage Grouse (*Centrocercus*), Ruff (*Philomachus pugnax*), Hermit Hummingbirds (*Phaethornis*), Gould's Manakin (*Manacus vitellinus*), some birds-of-paradise (*Paradisaea*).

b. The males remain isolated and display alone. Examples: Ruffed Grouse (*Bonasa*), Spruce Grouse (*Canachites*), Blue Grouse (*Dendragapus*), bowerbirds (Ptilonorhynchidae).

2. The sexes remain together for only a few days or until the start of incubation. Examples: Ruby-throated Hummingbird (*Archilochus colubris*), Penduline Tit (*Remiz pendulinus*). In cases where the male incubates, the female leaves him at the start of incubation: e.g., hemipode-quails (*Turnix*), Northern Phalarope (*Lobipes lobatus*).

3. The sexes remain together for weeks or months but separate when incubation begins. Examples: most ducks.

4. The sexes remain together throughout the breeding season. Such multiple-brood species as Barn Swallows, House Wrens, Catbirds, and Eastern Bluebirds often change mates for a second brood, but not all individuals do so. Among migratory passerine species, the males often return earlier than the females, and both males and females tend to return to the same territory (or to one nearby) in succeeding years. A pair may remate for several years. We do not know whether there is any carryover in the pair-bond between such a pair.

Figure 7. Postures of Anhingas (*Anhinga a. leucogaster*): 1, adult male in usual perching posture; 2–4, wing-waving of male; 5, bowing display of male; 6 and 7, pointing by male; 8, male in twig-shaking display, female watching; 9, stiff-necked posture of pair formation; 10, bill-rubbing; 11, mock-feeding. (Courtesy of Ted T. Allen and the editor of *Wilson Bulletin*.)

Figure 8. A Glossy Ibis (*Plegadis falcinellus*) on its nest. Ibises, egrets, and herons have many displays during the breeding season, which often involve specialized feather plumes. (Courtesy of Samuel A. Grimes.)

The remating in successive years may be due solely to the tendency to return to the same general area (Fig. 9; Berger, 1951).

5. The sexes pair for life and remain together throughout the year. Available evidence suggests that albatrosses, swans, geese, Common Terns, Ravens, some species of crows, and the Thick-billed Nutcracker, Carolina Chickadee (*Parus carolinensis*), Brown Creeper (*Certhia familiaris*), and Wrentit (*Chamaea fasciata*) form permanent pair-bonds.

Number of Mates

Lack (1968:161) wrote that "well over nine-tenths of nidicolous birds and four-fifths of nidifugous birds are monogamous, primarily because, with a nearly equal sex ratio, each male and each female will, on average, leave most descendents if they share in raising a brood, particularly when they collect the food for the young." He added (p. 150) that "about 2% of the species and 4% of the subfamilies of birds are polygynous, 0.4% of the species and 1% of the

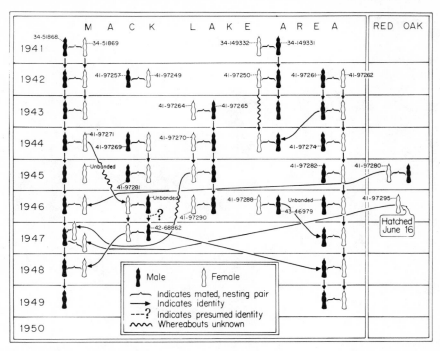

Figure 9. The interrelationships of pairs of Kirtland's Warbler in the Old Mack Lake colony (Sec. 19, T24N, R3E, Oscoda County, Michigan) during the period 1941–1949, inclusive. The chart includes data on all pairs known to have mated in more than one season before 1965. Modified by adding band numbers to Josselyn Van Tyne's original chart. (Courtesy of Andrew J. Berger, Bruce E. Radabaugh, and the editor of *Bird-Banding*.)

subfamilies are polyandrous, and 6% of the species and 3% of the subfamilies are promiscuous," assuming that almost all hummingbirds are promiscuous.

There are two types of polygamy.

1. **Polygyny** (one male fertilizes the eggs of several females) has been reported in most grouse, the Ring-necked Pheasant, the bittern *Botaurus stellaris,* hummingbirds, several icterids (e.g., *Agelaius phoeniceus, A. tricolor, Xanthocephalus xanthocephalus, Icterus wagleri*), weaverbirds (e.g., *Euplectes hordeacea, E. nigroventris, Ploceus philippinus*), and the Corn Bunting (*Emberiza calandra*).

2. **Polyandry** (females copulate with two or more males) occurs in some tinamous (*Crypturus*), hemipode-quails (*Turnix*), painted-snipe (*Rostratula*), and some phalaropes.

Species and Sex Recognition

That birds recognize their own kind is evident from the rarity of interspecific (and intergeneric) matings in the wild. Experiments using recorded songs have shown that males distinguish between the songs of males that occupy adjacent territories (Goldman, 1973). A male Red-headed Woodpecker will attack stuffed males of its own or other species (Fig. 10).

Johnston (1960) wrote that mounting is important in sex recognition for the Inca Dove (*Scardafella inca*); a mounted male actively rids himself of the bird

Figure 10. A Red-headed Woodpecker (*Melanerpes erythrocephalus*) attacking a dummy of its own species (above) and of a dummy Red-bellied Woodpecker (*Centurus carolinus*). (Courtesy of Robert K. Selander and the editor of *Wilson Bulletin*.)

on his back, whereas a female is passively tolerant. A Black Grouse (*Lyrurus tetrix*) will copulate repeatedly for hours with a stuffed hen that is pulled through the male's territory.

One of the implications of imprinting is that the young bird becomes imprinted on its own species, presumably during the nestling and fledgling stages. A puzzling problem, however, is presented by parasitic species. A young Brown-headed Cowbird, for example, may be hatched in the nest of any one of about 200 species of birds, and, if the bird hatches after the adult cowbirds have migrated, it may not see an adult until it reaches the wintering grounds. Even though a young cowbird may never have seen another cowbird, the immature birds gather in premigratory flocks, and they mate and copulate with their own kind the following spring.

Copulation

Copulation is the act of transferring sperm from the male to the cloaca of the female. The transfer is accomplished by cloacal contact as the male balances on the female's back; a cloacal penis assists the process in ratites, ducks, gallinaceous birds, and a few others. Copulation may take place on the ground, in water, in bushes and trees, on telephone wires or fences, or in the air (some swifts). Great Blue Herons usually copulate in or near the nest (Fig. 11). Some birds ignore each other after copulation, but others have a postcopulatory display (Fig. 12). The females of some species (e.g., Red-bellied Woodpecker, Starling, House Finch) may mount the male (*reverse mounting*) as a preliminary step to copulation.

In some species copulation also must serve to maintain the pair-bond. House Sparrows and Red-bellied Woodpeckers, for example, may copulate as early as 2 months before the eggs are laid; in the sparrow copulation may continue throughout the incubation period (Berger, 1957). Black-headed Gulls and Common Terns are said to copulate even after the eggs hatch, and copulation of Starlings has been observed in nearly every month of the year.

Ficken and Dilger (1960) discussed copulation with substitute objects by males of three species (Fisher's Lovebird, *Agapornis fischeri*; Wood Thrush, *Hylocichla mustelina*; and American Redstart, *Setophaga ruticilla*) that were stimulated by females in adjacent cages.

During *courtship feeding* the male offers food to the female; she usually utters distinctive call notes, and her posture often simulates that of a young bird begging for food. Courtship feeding has been observed just before, during, or after copulation in some bitterns, the Laughing Gull, Herring Gull, Rock Dove, Yellow-billed Cuckoo, Roadrunner, and some species of nuthatches, shrikes, and Galapagos finches. It may occur as a part of the courtship period, during incubation, or even when there are young in the nest (Figs. 13, 14).

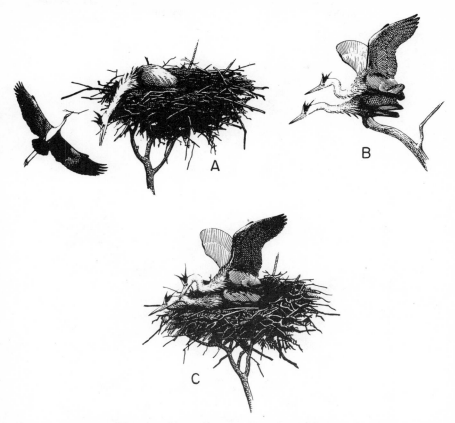

Figure 11. Copulatory behavior of the Great Blue Heron. (*A*) Bird in nest, drooping head and neck over side and shaking bill slowly from side to side; mate arriving with stick. (*B, C*) Copulation on branch and in nest. (Courtesy of W. Powell and Betty D. Cottrille.)

AGE AT SEXUAL MATURITY

Ducks, many gallinaceous birds, pigeons, some owls, and most passerine birds breed the year after they hatch; in most instances they are less than 12 months old. Hens of captive Japanese Quail (*Coturnix*) lay eggs when only 5 or 6 weeks old. The Australian estrildine finch *Poephila guttata* may nest when 3 months old, and a captive bird laid eggs when only 6 weeks of age; the African Fire-finch (*Lagonosticta senegala*) also may breed when 3 months old; some doves "and at least one species of parrot, quail, and button-quail" first breed when less than 6 months old (Lack, 1968:295). Geese, many hawks and

Figure 12. Postcopulatory display of the Fulvous Tree Duck. (Courtesy of Brooke Meanley.)

Figure 13. Courtship feeding in *Agapornis roseicollis*. (Courtesy of William C. Dilger.)

owls, most gulls, and a few passerine species do not breed until 2 years of age; individuals of some species of swifts breed when 1 year old, others not until their second year, and a few not until their third year. Among procellariiform birds, different species do not breed until they are from 3 to 8 or 9 years old, and individual variation typically occurs within a species (Fisher and Fisher, 1969). Fry (1972) found that about two-thirds of Red-throated Bee-eaters (*Merops bulocki*) breed when 1 year old and about a third when 2 years old.

TERRITORIALITY

In 1920 Eliot Howard published a small book, *Territory in Bird Life,* which caught the attention of ornithologists everywhere, and the territory concept became a major guiding factor in bird study.

Figure 14. A male Swainson's Warbler (*Limnothlypis swainsonii*) feeding a female on the nest. (Courtesy of Samuel A. Grimes.)

Territory Defined

By "territory" we mean a limited area defended by a bird, especially against members of its own species and sex during at least part of the breeding cycle. However, so many exceptions and special cases are known that "territory" generally is defined as "any defended area."

A Classification of Territories

Hinde (1956) suggested the following classification of territories.

Type A. Large breeding area within which courtship and copulation, nesting, and food-seeking usually occur. Examples: many species, including American wood-warblers, Old-world warblers, Mockingbirds, Cardinals, sparrows (Figs. 15, 16).

Type B. Large breeding area which, however, does not furnish most of the food. Examples: Willet, Red-winged Blackbird, Yellow-headed Blackbird.

Type C. Small nesting territories of colonial and some noncolonial birds—a small area around the actual nest. Examples: many seabirds, gulls, herons, some doves, some weaverfinches, many species of Hawaiian honey-creepers.

Type D. Pairing territories: small areas (leks, dancing grounds, display arenas) used for copulation but not for nesting. Examples: some grouse, some birds-of-paradise, cotingas, manakins, and hummingbirds.

A bird may defend its mate, its young, a covey, a song or lookout post, or a food supply (Wolf, 1969; Macroberts, 1970). French (1959) concluded that the male Black Rosy Finch (*Leucosticte atrata*) defends the female rather than a territory. The feeding area and the nest may be widely separated in this species; when the female leaves the nest, the male follows her and drives off other males that approach too closely. Family groups of Superb Blue Wrens (*Malurus cyaneus*) in Australia may contain more than one male—a dominant male and one or more supernumerary males; only the dominant male demonstrates aggressive behavior during the winter, but all adults in the family group defend the territory during the breeding season (Rowley, 1963).

A few species apparently exhibit no territorial behavior at all: e.g.,

Figure 15. A Rufous-fronted Longtail Warbler (*Prinia buchanani*; Sylviidae) near its nest; type A territory. (Courtesy of Shivrajkumar of Jasdan, India.)

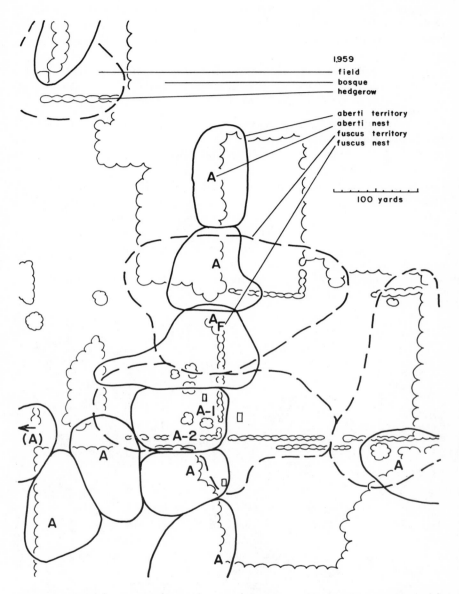

Figure 16. Territories and nests of the Brown Towhee (*Pipilo fuscus*) and Abert's Towhee (*P. aberti*) on the San Xavier Reservation near Tucson, Arizona, in 1959; numerals indicate successive nests of the same pair. (Courtesy of Joe T. Marshall, Jr., and the editor of *Condor*.)

Redshank, *Totanus totanus* (Hale, 1956); Scarlet-Rumped Tanager, *Ramphocelus passerinii* (Skutch, 1954:124); Sharp-tailed Sparrow, *Ammospiza caudacuta* (Woolfenden, 1956; Murray, 1969).

Functions of Territorial Behavior

Several functions are served by territorialism, and the degree of importance of each varies greatly among different species (Hinde, 1956). Furthermore, other factors may exist and may be more important than some of those listed below (see Brown, 1964).

1.   **Food.** Early studies were largely concerned with species of type A above, and Howard and his early disciples emphasized the importance of ensuring an adequate food supply for the young. Modern studies, however, have greatly reduced our belief in the general importance of this interpretation.

2.   **Pair-bond.** Territorial behavior is surely important in many species in the formation and maintenance of the bond between a pair of birds.

3.   **The reduction of interference** by other individuals of the species in reproductive activities is certainly important in most territorial species.

4.   **Regulation of the density** of a species in a favorable habitat is one result of territorial behavior.

5.   **Reduction in loss to predators** doubtless results both from dispersion and from the territorial birds' thorough familiarity with every detail of the terrain.

6.   **Disease prevention** by dispersion has been urged as another function of territorialism, but there is little evidence for this view.

7.   **Social stimulation.** Darling (1952) proposed that "one of the important functions of territory in breeding birds is the provision of *periphery*—periphery being defined as that kind of edge where there is another bird of the same species occupying a territory." He added that, if the hypothesis is correct, "*territorial behavior as a whole is a social phenomenon, and it has survival value.*"

Lewis and Orcutt (1971) redefined the "Fraser Darling effect" as follows: "Birds are, themselves, exteroceptive factors to which other birds respond. One expression of the total response pattern is the facilitation of events of the reproductive cycle." They added: "Social facilitation can operate over a much broader spectrum, temporally and mechanistically, than traditionally thought." (See also Crook, 1964.) In writing about parrots of the genus *Agapornis,* Dilger (1960) said that "if an individual of one of the more social species is deprived of normal social interactions (reciprocal preening, courtship feeding, etc.) in the presence of suitable companions, it will gradually become ill and eventually die (in about six months)." Dilger suggested that some of the problems experienced in such cases might "be partly caused by thyroid difficulties associated with stress."

Methods of Defense

Territories are defended against intruders not only by physical attack or the threat of it, but often merely by song or even by the conspicuous presence of the defending bird. One of the frequently observed phenomena of territorial behavior is that the bird defending a territory is more aggressive than the intruder and usually is successful in driving it away. If, in the course of chasing, two fighting birds should fly into the territory of the intruder, the roles of defender and intruder may become reversed at once.

Smith (1972) found that, if he blackened the brightly colored epaulets of male Red-winged Blackbirds, 64 percent of the birds lost their territories; these males, however, were still able to attract females and mate successfully. In her discussion of the Calfbird (*Perissocephalus tricolor*), a cotingid, Snow (1972) said that the "motionless posture" (Fig. 17) "is the most important display performed by males in their claims to a lek perch." Although this posture is maintained for shorter periods, males sometimes hold it for as long as 50 minutes.

Size of Territory

Many colonial nesting seabirds nest just far enough apart so that the incubating birds cannot reach each other with their bills. Large hawks, eagles, and owls, on the other hand, may have territories occupying several square miles. Among passerine birds, the size of the area defended by type A species varies with many factors. Song Sparrow territories studied by Nice ranged from 0.5 to 1.5 acres, "depending partly on the pugnacity of the owner and partly on the amount of space available." By studying populations of Red-eyed Vireos, Yellow Warblers, and Song Sparrows on a series of islands, Beer *et al.* (1956) found that the "minimum amount of space used by a pair to raise their young successfully may be much smaller when the boundaries are strictly physical barriers rather than invisible lines determined by intraspecific conflict." For example, they found Song Sparrows utilizing areas "as small as one-tenth the minimum size defended by birds on the mainland and in contact with others of their own species." Other examples of territory size are as follows: Robin, 0.11 to 0.6 acre; Red-eyed Vireo, 1.4 to 2.1 acres; Black-capped Vireo (*Vireo atricapilla*), 2.5 to 4.6; Prothonotary Warbler, 1.9 to 6.4; Carolina Chickadee, 10 to 15; Black-capped Chickadee, 8.4 to 17.1.

Figure 17. Display postures of male Calfbirds: (*a*, *b*) motionless postures; (*c*) pouter-pigeon posture; (*d*) fluffed-up posture; (*e*) posture of bird half-way through the "moo" call. (Courtesy of Barbara K. Snow and the editor of *Ibis*.)

NESTS

Some of the larger birds of prey use the same nest for years, adding new material annually (Fig. 18); a Bald Eagle nest in Ohio that had been used for 36 years was about 9 feet wide and 20 feet deep and was estimated to weigh about 2 tons. A few species use the same nest for successive broods in a single breeding season, but the majority of multiple-brooded species build a new nest for each clutch of eggs. The location for the nest may be selected by the male, by the female, or by both together.

Figure 18. A Great-horned Owl nest with two young in cactus; Saguaro National Monument, Arizona; the Santa Catalina Mountains are in the background. (Courtesy of George Olin.)

Nest Locations

It is well known that some birds always nest on the ground and that other species never do so (Figs. 19, 20). Still others (e.g., Rufous-sided Towhee, Field Sparrow, Song Sparrow) typically build early nests on the ground and later ones in low bushes. Many species exhibit geographical differences in their choice of a nest site. The Brown Thrasher (*Toxostoma rufum*), for example, usually builds in low bushes in the eastern part of the United States, but often on the ground in the western part of its range; it nests in bushes in southern Michigan, on the ground in the jack-pine plains of northern Michigan; there is one record of nesting in a tree cavity. Some species that usually nest in trees (sometimes at great heights) may nest on the ground in remote areas or on islands: e.g., Golden Eagle, Osprey, Barred Owl, Great Blue Heron, Common Crow.

Nesting sites may vary considerably within a family, as illustrated by North American species of wood-warblers. Prothonotary and Lucy's warblers nest in tree cavities, and Lucy's Warblers sometimes adopt an old nest of the Verdin.

Figure 19. An Indian Stonecurlew (*Burhinus oedicnemus*) on its nest. (Courtesy of Shivrajkumar of Jasdan, India.)

Figure 20. A Small Indian Pratincole (*Glareola lactea*) on its eggs. (Courtesy of Shivrajkumar of Jasdan, India.)

Other species nest on the ground, among the upturned roots of a fallen tree, in shrubs, and in trees up to heights of 100 feet (Figs. 21, 22).

Role of Sexes in Nest-Building

There are at least seven categories to which species may be assigned according to the roles of the sexes in nest-building.

    1.   Both sexes build the nest.

    a.   Male and female share more or less equally in nest construction or excavation: many seabirds, bee-eaters, kingfishers, woodpeckers, swallows, waxwings, some wrens.

Figure 21. Nest and eggs of the Tennessee Warbler; Schoolcraft County, Michigan, June 26, 1956. (Courtesy of Lawrence H. Walkinshaw.)

Figure 22. Male Black-throated Blue Warbler at its nest; Schoolcraft County, Michigan, June 25, 1952. (Courtesy of Lawrence H. Walkinshaw.)

b. The male builds "dummy" or "cock" nests: many wrens.

2. The female builds, but the male provides the material: African Spoonbill (*Platalea alba*), Wood Pigeon (*Columba palumbus*), Mourning Dove, Ground Dove.

3. The female builds without help from the male: some parrots, some muscicapids, Wagler's Oropendola, the Red-eyed Vireo, Ovenbird, sunbirds, hummingbirds, manakins, and other lek species.

4. The female builds, but both sexes gather the material: Raven, Rook.

5. The male builds, but the female provides the material: frigatebirds.

6. The male alone builds the nest: some shrikes, several weaverbirds (e.g., Baya; Crinson-crowned Bishopbird, *Euplectes hordeacea*).

7. No nest is built: many seabirds, auks, pratincoles, many shorebirds, skimmers, nightjars (Fig. 23).

Time Required to Build Nests

A species that lays its eggs on the ground or on bare rock may spend no time in preparation of the site, whereas some other species work on a nest for weeks. The pace of nest-building tends to be more rapid in the North—especially in the Arctic—where the nesting season is very short. Tropical species may spend weeks constructing a nest, working only a few hours each day. Most of the

Figure 23. A Yellow-wattled Lapwing (*Vanellus malabaricus*) standing over its eggs. (Courtesy of Shivrajkumar of Jasdan, India.)

small passerines, however, build their nests in a few days: e.g., Prothonotary Warbler, 3.3 days; Cedar Waxwing, 5.6 days; American Goldfinch, 9 days. The male and female Elepaio (*Chasiempis sandwichensis*), an endemic Hawaiian genus of the Muscicapidae, build their cup-shaped nest over a period of about 2 weeks. The female Royal Flycatcher (*Onychorhynchus mexicanus*) builds a very long (up to 5 or 6 feet), pensile nest in about 12 days (Skutch, 1960). By contrast, the Red-billed Quelea (*Quelea quelea*) is said to complete a nest in 2 days.

Representatives of a number of bird families excavate nesting chambers in earth banks: e.g., petrels, shearwaters, jacamars, bee-eaters, motmots, some woodpeckers, some furnariids, some swallows. The Turquoise-browed Motmot (*Eumomota superciliosa*) may dig nesting burrows over 5 feet long in less than 5 days. Although the Bank Swallow (*Riparia riparia*) is poorly adapted for digging, Gaunt (1965) found that burrows averaging 27.3 inches in depth were excavated at a rate of about 5 inches per day. Fry (1972) reported that Red-throated Bee-eaters of Africa excavate tunnels about 3 feet deep during November "at the end of the rains, when the ground is still soft," but that the birds do not breed until January, when the earth has become hard.

Comparative Nidification

Among the orders of birds are species that lay their eggs or build their nests in nearly every conceivable location. A multitude of different materials are used to construct nests that vary in diameter from less than 1 inch to 9 feet or more (Fig. 24).

A King Penguin or Emperor Penguin holds a single egg on top of its feet and incubates it in this position; other penguins lay their eggs in burrows or caves. A potoo lays a single egg on the top of a broken tree stub and incubates in an erect position. The conelike nest of the Oilbird (*Steatornis*) is built of seeds and droppings; some species of the swiftlet genus *Collocalia* build their nests entirely of saliva. Tailorbirds (*Orthotomus*) and some wren-warblers (*Prinia*) of the family Sylviidae sew together the edges of one or more leaves and then build their nests in the resulting folds (Fig. 25); the nests of some sunbirds and hummingbirds are sewn to the under side of a large leaf. Although most pigeons and doves build a flimsy platform of twigs, a few nest in tree cavities and burrows in the ground. Skutch (1945) reported that Blue-throated Green Motmots (*Aspatha gularis*) laid their eggs "in burrows which had already been in use for months as dormitories." After the motmots dig new burrows, the old ones often are used for nesting by Cóban Swallows (*Notiochelidon pileatus*).

Most woodpeckers excavate a nest cavity in dead trees, but some use living trees; the Gilded Flicker and the Gila Woodpecker often nest in giant saguaro cacti; the Red-shafted Flicker has been known to excavate into the side of a

Figure 24. Nest and eggs of the White-spotted Fantail Flycatcher (*Rhipidura albogularis*); Baroda, India, August 8, 1964.

haystack, a building, and a dirt bank. Abandoned nesting cavities of wood-peckers are adopted by many other hole-nesting species that do not excavate their own cavities: e.g., Elf, Flammulated, Pygmy, Saw-whet, Screech, and Whiskered owls; Great Crested, Wied's Crested, and Ash-throated flycatchers; Violet-green and Tree swallows; Prothonotary and Lucy's warblers; titmice, starlings, and bluebirds. Great Crested Flycatchers, House Wrens, House Sparrows, and Eastern Bluebirds nest in metal newspaper tubes and mailboxes in rural areas. The Great Crested Flycatcher is noted for using castoff snake skins in its nest, and a similar tendency has been reported for several species of the woodcreeper genus *Synallaxis* (Williams, 1922).

The White Tern (*Gygis alba*) does not build a nest but lays its egg on a rock, on a horizontal branch of a tree or shrub (Fig. 26), or, rarely, on the roof of a building. The Common Nighthawk lays its egg on the ground in wooded areas but on the flat roofs of buildings in towns and cities. The Killdeer, a ground-nesting species, sometimes lays its eggs on the roof of a building.

Figure 25. Nest and eggs of Franklin's Longtail Warbler (*Prinia hodgsonii*); Baroda, India, July 30, 1964.

Several species of weaverbirds (e.g., Chestnut Sparrow, *Sorella eminibey;* Cutthroat Finch, *Amadina fasciata*; Bronze Mannikin, *Lonchura cucullata*; White-throated Munia, *L. malabarica*) frequently use the abandoned nests of another species, lining them with grasses and feathers. The Green Sandpiper (*Tringa ochropus*) and the Solitary Sandpiper (*T. solitaria*) do not build nests but lay their eggs in the abandoned nests of other tree-nesting birds (e.g., Gray Jay, Cedar Waxwing, Robin, Eastern Kingbird). The Bay-winged Cowbird (*Molothrus badius*), some tyrant-flycatchers (notably *Legatus albicollis* and *L. leucophaius*), and some weaverbirds (e.g., Zebra Finch, *Poephila guttata*) sometimes take the nests of other species by force. Roberts (1955) compiled a list of 80 species that habitually or occasionally use the nests of other species.

Little is known about the role of learning in nest construction in any bird family, but Collias and Collias (1964a, 1964b) published photographs of nests built by inexperienced and by experienced male Village Weavers, *Ploceus* (*Textor*) *cucullatus.* They concluded that "practice, channelized by specific response tendencies, but not necessarily tuition by example, is needed for development of the ability to build a normal nest by the male of this species." In several species the males alone build the nests, and each species follows a specific sequence in nest construction (see Fig. 5). Although abnormal nests are common, little is known about the causes (Ali and Ambedkar, 1956; Davis, 1971; Fig. 27).

Figure 26. Above: A White Tern egg on a branch; Keith Island, Eniwetok Atoll, Marshall Islands, July 27, 1971. Below: A White Tern egg on a coral rock; Cook Islet, Christmas Island, Pacific Ocean, November 15, 1971.

Figure 27. Two abnormal nests of the Baya; the nest on the right is 5 feet 6 inches long; Baroda, India, March 6, 1965.

NESTING ASSOCIATES

There are so many different kinds of relationships among nesting birds and other animals that several writers have classified them. The following is a composite classification.

1. Social Nesting among Birds

A. A Single Species. In addition to seabirds and herons, other species defend a type C territory: e.g., bee-eaters, some swallows, many weaverbirds (Fig. 28). Balda and Bateman (1971) described the social interactions of a flock of some 250 Piñon Jays (*Gymnorhinus cyanocephalus*) that nested in an area of about 120 acres (see also Rothstein, 1971).

Two highly specialized forms of social nesting also occur.

(1) *Communal nesting.* From one to four pairs of Groove-billed Anis (*Crotophaga sulcirostris*) build a communal nest, the females lay eggs in it, and all adults take part in incubation and feeding of the young. This unusual form of nesting is found also in the Swamp Hen (*Porphyrio p. melanotus*), the Smooth-billed Ani (*Crotophaga ani*), the Guira Cuckoo (*Guira guira*), the Acorn Woodpecker (*Melanerpes formicivorus*), and at least one timaliid (*Yuhina brunneiceps*).

(2) *Cooperative nesting.* Friedmann (1950) described cooperative nests of the Sociable Weaver (*Philetairus socius*) that may be "as much as 25 feet long and 15 feet wide at the base and from 5 to 10 feet in height" (Fig. 29). Such nests may be built by flocks of 100 or more pairs of birds, each pair having its own nesting chamber. A similar nesting pattern is found in the Red-billed Buffalo-Weaver, *Bubalornis albirostris* (Collias and Collias, 1964b), and in the South American Gray-breasted or Monk Parakeet, *Myiopsitta monachus* (Friedmann, 1927).

"Helpers" at the nest are one or more birds that assist a pair, usually in feeding the young. Helpers may be juvenile birds or adults. The number of helpers varies from 1–3 in the Red-throated Bee-eater to as many as 23 in the Long-tailed Shrike (*Corvinella corvina*). This type of cooperative nesting has been recorded in about 25 percent of African bird families. In Asia the Long-tailed Tit (*Aegithalos caudatus*), many babblers, and possibly Hypocolius (*Hypocolius ampelinus*) nest cooperatively. Cooperative nesting occurs in Australia in such birds as ducks, rails, parrots, swallows, cuckoo-shrikes, nuthatches, honeyeaters, wood-swallows, and babblers.

B. Mixed Colonies. These are of two general types. In one, large numbers of two or more species nest together: e.g., shearwaters, petrels, boobies, frigatebirds, terns, herons, ibises, egrets, and anhingas. In the second type, one

Figure 28. Lower: Portion of a nesting colony of Black Swans (*Cygnus atratus*) on an island in Lake George, N.S.W., Australia. Upper: Nest relief ceremony; the incoming bird is greeted and returns the greeting. (Courtesy of H. J. Frith; photos by Ederic Slater.)

Figure 29.   Cooperative nest of the Sociable Weaver (*Philetairus socius*); South Africa. (Courtesy of Herbert Friedmann and the Smithsonian Institution.)

or more pairs of one species build their nests in the midst of a large colony of other species: jaegers or skuas with penguins or gulls; Tufted Ducks and Turnstones in colonies of terns and Black-headed Gulls.

C.   "Protective" Nesting.   A number of small birds have been found nesting in close association with larger birds, often birds of prey. Grackles, Starlings, and House Sparrows sometimes nest in the side of Osprey nests; Cliff Swallows, near Prairie Falcon nests. Certain African weaverbirds build near nests of the Jackal Buzzard (*Buteo rufofuscus*), and the Pygmy Falcon (*Polihierax semitorquatus*) has been reported to nest in the large cooperative nests of the Sociable Weaver.

D.   Proximity Nesting.   There are many reports of different genera and species building their nests unusually close together, i.e., in the same tree, bush, or clump of vegetation. The distance between nests depends largely on the tolerance of the more strongly territorial species.

2. Social Nesting between Birds and Invertebrate Animals

A.   Birds Nesting in Termitaria.   Chisholm reported that three species of Australian parrots and five species of kingfishers regularly excavate nesting burrows in termite nests (Figs. 30, 31), and the late Loke Wan Tho found similar examples in Asia. Haverschmidt (1960) wrote of the nesting of Brown-throated Parakeets (*Aratinga pertinax surinama*), three species of jacamars (family Galbulidae), the Picine Woodhewer (*Xiphorhynchus picus*), the Gray-breasted Martin (*Progne chalybea*), and the Southern House Wren (*Troglodytes musculus*) in termite nests in South America.

B.   Birds Building Their Nests near the Nests of Hymenopterous Insects. Hindwood (1955) cited these Australian birds as often nesting near wasp or ant nests: six species of bush warblers of the genus *Gerygone* and the Brown-throated Sunbird (*Anthreptes malaccensis*), Banded Finch (*Poephila biche-novii*), Red-browed Finch (*Aegintha temporalis*), Ricebird (*Lonchura punctu-lata*), and Baya.

Janzen (1969) discussed relationships between obligate acacia-ants and six species of birds that nest in swollen-thorn acacias in Central America. Oniki (1970) reported that wasps (*Pison* sp.) build mud cells beneath the lining of nests of the Reddish Hermit (*Phaethornis ruber*) in Brazil. In Hawaii the Ricebird sometimes builds its globular nest in mesquite trees containing nests of the wasp *Polistes exclamans;* both the bird and the wasp are introduced species. In the Old World, most bird species that nest in association with insects build domed nests, but in Central and South America species building open, cup-shaped nests also do so.

C.   Insects Inhabiting Birds' Nests

(1) *Parasites on nestlings or adults.*   The obligatory blood-sucking larvae of several genera of insects have been found on the young of many species of birds; 54 species of birds are known to be parasitized by the blowfly *Protocalliphora metallica* (Nolan, 1955). Biting lice (Mallophaga), fleas (Siphonaptera), and hippoboscid flies (Hippoboscidae), as well as various species of ticks and mites, also are parasitic on birds (Fig. 32).

(2) *Nest-cleaning insects.*   The Golden-shouldered Parrot (*Psephotus chrysopterygius*) digs a nesting burrow in a termitarium and lays its eggs on the bare ant bed. The larvae of a moth (*Neossiosynoeca scatophaga*) feed on the feces

Figure 30. Nesting site of the Blue-winged Kookaburra (*Dacelo leachii*) in a terrestrial termite mound, Northern Territory, Australia, showing old nesting hollows and an occupied nest site (indicated by arrow) drilled through a flange of the mound. (Courtesy of the late K. A. Hindwood and the editor of *Emu*.)

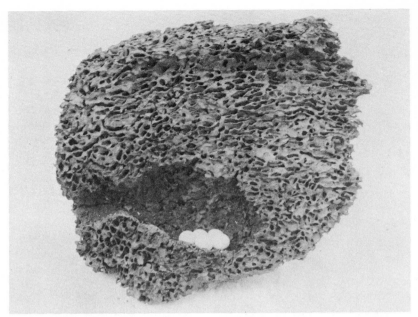

Figure 31. Vertical section of nest site of a Sacred Kingfisher (*Halcyon sancta*) in the nest of the termite *Nasutitermes walkeri,* near Sydney, Australia. (Courtesy of the late K. A. Hindwood and the editor of *Emu.*)

Figure 32. A hippoboscid fly on an incubating Red-tailed Tropicbird (*Phaethon rubricauda*); Christmas Island (Pacific Ocean), November 17, 1971.

of the nestling parrots. Hindwood (1951) reported other genera and species of moths associated with the nests of several Australian birds: Eastern Rosella (*Platycercus eximius*), Red-browed Finch, Zebra Finch, Ricebird, House Sparrow, Spinebill Honeyeater (*Acanthorhynchus brevirostris*), Brown Weebill (*Smicrornis brevirostris*), and Yellow-tailed Thornbill (*Acanthiza chrysorrhoa*). The larvae of the moths eat either the feces or feathers in the nest.

(3) *Miscellaneous insect inhabitants.* Many other insects have been found in birds' nests (over 400 insects representing eight species from a single nest), including insects parasitic on other insects.

D. *Invertebrates Other than Insects.* Nolan found one snail and seven genera of mites (two of which are parasitic on birds) in nine nests of the Prairie Warbler (*Dendroica discolor*).

3. Secondary Uses of Birds' Nests

Nests frequently are appropriated by mice (e.g., *Peromyscus*) and rats (e.g., *Rattus rattus, R. exulans*) and remodeled for their use. Nests also are used as storage sites and feeding platforms by small mammals and as roosting shelters by other birds (Nickell, 1951).

REFERENCES

Ali, S. and V. C. Ambedkar. 1956. Notes on the Baya Weaver Bird, *Ploceus philippinus* Linn. *J. Bombay Nat. Hist. Soc.,* **53**:381–389.

Allen, T. T. 1961. Notes on the breeding behavior of the Anhinga. *Wilson Bull.,* **73**:114–125.

Ambedkar, V. C. 1964. *Some Indian Weaver Birds.* University of Bombay, Bombay, India.

Andrew, R. J. 1961. The displays given by passerines in courtship and reproductive fighting: a review. *Ibis,* **103a**:315–348, 549–579.

Ashmole, N. 1963a. The biology of the Wideawake or Sooty Tern *Sterna fuscata* on Ascension Island. *Ibis,* **103b**:297–364.

——— 1963b. Molt and breeding in populations of the Sooty Tern *Sterna fuscata. Postilla,* **76**:1–18.

——— 1965. Adaptive variation in the breeding regime of a tropical sea bird. *Proc. Natl. Acad. Sci.,* **53**:311–318.

Baerends, G. P., and N. A. Van Der Cingel. 1962. On the phylogenetic origin of the snap display in the Common Heron (*Ardea cinerea* L.). *Symp. Zool. Soc. London,* **8**:7–24.

Balda, R. P., and G. C. Bateman. 1971. Flocking and annual cycle of the Piñon Jay, *Gymnorhinus cyanocephalus. Condor,* **73**:287–302.

Beer, J. R., L. D. Frenzel, and N. Hansen. 1956. Minimum space requirements of some nesting passerine birds. *Wilson Bull.,* **68**:200–209.

Berger, A. J. 1951. Ten consecutive nests of a Song Sparrow. *Wilson Bull.,* **63**:186–188.

——— 1957. Nesting behavior of the House Sparrow. *Jack-Pine Warbler,* **35**:86–92.

——— 1966. Mid-winter nesting of the American Robin in western Pennsylvania. *Auk,* **83**:668.

———— 1968. Clutch size, incubation period, and nestling period of the American Goldfinch. *Ibid.,* **85**:494–498.

———— 1969. The breeding season of the Hawaii 'Amakihi. *Occ. Pap. Bernice P. Bishop Mus.,* **24**:1–8.

———— 1972. *Hawaiian Birdlife.* University Press of Hawaii, Honolulu.

————, and B. E. Radabaugh. 1968. Returns of Kirtland's Warblers to the breeding grounds. *Bird-Banding,* **39**:161–186.

Brown, J. L. 1964. The evolution of diversity in avian territorial systems. *Wilson Bull.,* **76**:160–169.

Chisholm, A. H. 1952. Bird-insect nesting associations in Australia. *Ibis,* **94**:395–405.

Collias, N. E., and E. C. Collias. 1964a. The development of nest-building behavior in a weaverbird. *Auk,* **81**:42–52.

———— 1964b. Evolution of nest-building in the weaverbirds (Ploceidae). *Univ. Calif. Publ. Zool.,* **73**.

Cottrille, W. P., and B. D. Cottrille. 1958. Great Blue Heron: behavior at the nest. *Misc. Publ. Mus. Zool., Univ. Mich.,* No. 102.

Crook, J. H. 1964. The evolution of social organisation and visual communication in the weaverbirds (Ploceinae). *Behaviour,* Suppl. X.

Darling, F. F. 1952. Social behavior and survival. *Auk,* **69**:183–191.

Davis, T. A. 1971. Variation in nest-structure of the common weaverbird *Ploceus philippinus* (L.) of India. *Forma Functio,* **4**:225–239.

Dilger, W. C. 1956. Hostile behavior and reproductive isolating mechanisms in the avian genera *Catharus* and *Hylocichla. Auk,* **73**:313–353.

———— 1960. The comparative ethology of the African parrot genus *Agapornis. Z. Tierpsychol.,* **17**:649–685.

Dixon, K. L. 1963. Some aspects of social organization in the Carolina Chickadee. *Proc. XIIIth Int. Ornithol. Congr.,* **1963**:240–258.

Farner, D. S. 1973. *Breeding Biology of Birds.* Natl. Acad. Sciences, Washington, D.C.

————, and R. A. Lewis. 1971. Photoperiodism and reproductive cycles in birds. In *Photophysiology,* Vol. 6, Academic Press, New York, pp. 325–370.

Ficken, M. S., and W. C. Dilger. 1960. Comments on redirection with examples of avian copulations with substitute objects. *Anim. Behav.,* **8**:219–222.

Fisher, H. I., and M. L. Fisher. 1969. The visits of Laysan Albatrosses to the breeding colony. *Micronesica,* **5**:173–221.

Fisher, J. 1954. Evolution and bird sociality. In *Evolution as a Process,* George Allen & Unwin, London.

Foster, M. S. 1969. Synchronized life cycles in the Orange-crowned Warbler and its mallophagan parasites. *Ecology,* **50**:315–323.

French, N. R. 1959. Life history of the Black Rosy Finch. *Auk,* **76**:159–180.

Friedmann, H. 1927. Notes on some Argentine birds. *Bull. Mus. Comp. Zoöl.,* **68**, No. 4.

———— 1950. The breeding habits of the weaverbirds: a study in the biology of behavior patterns. *Smithson. Rept. for 1949,* **1950**:293–316.

Fry, C. H. 1972. The social organisation of bee-eaters (Meropidae) and co-operative breeding in hot-climate birds. *Ibis,* **114**:1–14.

Gaunt, A. S. 1965. Fossorial adaptations in the Bank Swallow, *Riparia riparia* (Linnaeus). *Univ. Kans. Sci. Bull.,* **46**:99–146.

Gill, F. B., and W. E. Lanyon. 1964. Experiments on species discrimination in Blue-winged Warblers. *Auk,* **81:**53–64.

Goldman, P. 1973. Song recognition by Field Sparrows. *Auk,* **90:**106–113.

Graber, J. W. 1961. Distribution, habitat requirements, and life history of the Black-capped Vireo (*Vireo atricapilla*). *Ecol. Monogr.,* **31:**313–336.

Grubb, T. C., Jr. 1970. Burrow digging techniques of Leach's Petrel. *Auk,* **87:**587–588.

Hale, W. G. 1956. The lack of territory in the Redshank, *Tringa totanus. Ibis,* **98:**398–400.

Hamner, W. M. 1966. Photoperiodic control of the annual testicular cycle in the House Finch, *Carpodacus mexicanus. Gen. Comp. Endocrinol.,* **7:**224–233.

Hann, H. W. 1937. Life History of the Oven-bird in southern Michigan. *Wilson Bull.,* **49:**145–237.

——— 1940. Polyandry in the Oven-bird. *Ibid.,* **53:**69–72.

Haverschmidt, F. 1958. Nesting of a jacamar in a termite nest. *Condor,* **60:**71.

——— 1960. Some further notes on the nesting of birds in termites' nests. *Emu,* **60:**53–54.

Hinde, R. A. 1956. The biological significance of the territories of birds. *Ibis,* **98:**340–369.

Hindwood, K. A. 1951. Moth larvae in birds' nests. *Emu,* **51:**121–133.

——— 1955. Bird/wasp nesting associations. *Ibid.,* **55:**263–274.

Janzen, D. H. 1969. Birds and the ant × acacia interaction in Central America, with notes on birds and other Myrmecophytes. *Condor,* **71:**240–256.

Johnston, R. F. 1960. Behavior of the Inca Dove. *Condor,* **62:**7–24.

Keast, A. 1959. Australian birds: their zoogeography and adaptations to an arid continent. In *Biogeography and Ecology in Australia, Monogr. Biol.* **8:**89–114.

Kilham, L. 1958a. Pair formation, mutual tapping and nest hole selection of Red-bellied Woodpeckers. *Auk,* **75:**318–329.

——— 1958b. Territorial behavior of wintering Red-headed Woodpeckers. *Wilson Bull.,* **70:**347–358.

——— 1960. Courtship and territorial behavior of Hairy Woodpeckers. *Auk,* **77:**259–270.

Kruijt, J. P. 1966. The development of ritualized displays in Junglefowl. *Phil Trans. Roy. Soc. London, B,* **251:**479–484.

———, and J. A. Hogan. 1967. Social behavior on the lek in Black Grouse, *Lyrurus tetrix tetrix* (L.). *Ardea,* **55:**203–240.

Lack, D. 1954. *The Natural Regulation of Animal Numbers.* Oxford University Press, London.

——— 1968. *Ecological Adaptations for Breeding in Birds.* Methuen & Co., London.

Lehrman, D. S. 1964. Control of behavior cycles in reproduction. In *Social Behavior and Organization among Vertebrates,* University of Chicago Press, Chicago, pp. 143–166.

Lewis, R. A., and F. S. Orcutt, Jr. 1971. Social behavior and avian sexual cycles. *Scientia,* **106:**447–472.

McKinney, F. 1961. An analysis of the displays of the European Eider *Somateria mollissima mollissima* (Linnaeus) and the Pacific Eider *Somateria mollissima* v. *nigra* Bonaparte. *Behaviour,* Suppl. **7.**

——— 1965. The spring behavior of wild Steller Eiders. *Condor,* **67:**273–290.

Macroberts, M. H. 1970. Notes on the food habits and food defense of the Acorn Woodpecker. *Condor,* **72:**196–204.

Marshall, J. T., Jr. 1960. Interrelations of Abert and Brown towhees. *Condor,* **62:**49–64.

Masatomi, H. 1959. Attacking behaviour in homosexual groups of the Bengalee, *Uroloncha striata* var. *domestica* Flower. *J. Fac. Sci., Hokkaido Univ.,* **14:**234–251.

Miller, A. H. 1965. Capacity for photoperiodic response and endogenous factors in the reproductive cycles of an equatorial sparrow. *Proc. Natl. Acad. Sci.,* **54:**97–101.

———, and V. D. Miller. 1968. The behavioral ecology and breeding biology of the Andean Sparrow, *Zonotrichia capensis. Caldasia,* **10:**83–154.

Murray, B. G., Jr. 1969. A comparative study of the Le Conte's and Sharp-tailed sparrows. *Auk,* **86:**199–231.

Nice, M. M. 1937–1943. Studies in the life history of the Song Sparrow. I, II. *Trans. Linn. Soc. N. Y.,* **4,** 1937; **6,** 1943.

——— 1941. The role of territory in bird life. *Am. Midl. Nat.,* **26:**441–487.

Nickell, W. P. 1951. Studies of habitats, territory, and nests of the Eastern Goldfinch. *Auk,* **68:**447–470.

——— 1958. Variations in engineering features of the nests of several species of birds in relation to nest sites and nesting materials. *Butler Univ. Bot. Stud.,* **13:**121–140.

——— 1965. Habitats, territory, and nesting of the Catbird. *Am. Midl. Nat.,* **73:**433–478.

——— 1969. Unusual nesting habitats of three bird species in Rondeau Provincial Park, Ontario. *Wilson Bull.,* **81:**454–459.

Nolan, V., Jr. 1955. Invertebrate nest associates of the Prairie Warbler. *Auk,* **72:**55–61.

Oniki, Y. 1970. Nesting behavior of Reddish Hermits (*Phaethornis ruber*) and occurrence of wasp cells in nests. *Auk,* **87:**720–728.

Phillips, A. R. 1951. The molts of the Rufous-winged Sparrow. *Wilson Bull.,* **63:**323–326.

Phillips, R. E., and F. McKinney. 1962. The role of testosterone in the displays of some ducks. *Anim. Behav.,* **10:**244–246.

Pikula, J., and Č. Folk. 1970. Differential breeding in *Corvus monedula, Sturnus vulgaris, Parus major* and *Fringilla coelebs* in woodland and non-woodland habitats. *Zool. Listy,* **19:**261–273.

Radabaugh, B. E. 1972. Polygamy in the Kirtland's Warbler. *Jack-Pine Warbler,* **50:**48–52.

Richardson, F. 1965. Breeding and feeding habits of the Black Wheatear *Oenanthe leucura* in southern Spain. *Ibis,* **107:**1–16.

Roberts, N. L. 1955. A survey of the habit of nest-appropriation. *Emu,* **55:**110–126, 173–184.

Rothstein, S. I. 1971. High nest density and non-random nest placement in the Cedar Waxwing. *Condor,* **73:**483–485.

Rowley, I. 1963. The reaction of the Superb Blue Wren, *Malurus cyaneus,* to models of the same and closely related species. *Emu,* **63:**207–214.

Selander, R. K., and D. R. Giller. 1959. Interspecific relations of woodpeckers in Texas. *Wilson Bull.,* **71:**107–124.

———, and D. J. Nicholson. 1962. Autumnal breeding of Boat-tailed Grackles in Florida. *Condor,* **64:**81–91.

Skutch, A. F. 1945. Life history of the Blue-throated Green Motmot. *Auk,* **62:**489–517.

——— 1950. The nesting seasons of Central American birds in relation to climate and food supply. *Ibis,* **92:**185–222.

——— 1954. Life Histories of Central American birds. *Pacific Coast Avifauna,* No. 31.

——— 1960. Life Histories of Central American birds. II. *Pacific Coast Avifauna,* No. 34.

Smith, D. G. 1972. The role of the epaulets in the Red-winged Blackbird (*Agelaius phoeniceus*) social system. *Behavior*, **41**:251–268.

Snow, B. K. 1970. A field study of the Bearded Bellbird in Trinidad. *Ibis*, **112**:299–329.

——— 1972. A field study of the Calfbird *Perissocephalus tricolor*. *Ibid.*, **114**:139–162.

Snow, D. W., and B. K. Snow. 1967. The breeding cycle of the Swallow-tailed Gull *Creagrus furcatus*. *Ibis*, **109**:14–24.

Stokes, A. W. 1961. Voice and social behavior of the Chukar Partridge. *Condor*, **63**:111–127.

Tinbergen, N. 1954. The origin and evolution of courtship and threat display. In *Evolution as a Process*, J. Huxley *et al.*, eds., George Allen & Unwin, London.

Walkinshaw, L. H. 1953. Life-history of the Prothonotary Warbler. *Wilson Bull.*, **65**:152–168.

Weller, M. W. 1967. Notes on some marsh birds of Cape San Antonio, Argentina. *Ibis*, **109**:391–411.

Williams, C. B. 1922. Notes on the food and habits of some Trinidad birds. *Bull. Dept. Agric. Trinidad Tobago,* **20**:123–185.

Wolf, L. L. 1969. Female territoriality in a tropical hummingbird. *Auk*, **86**:490–504.

Woolfenden, G. E. 1956. Comparative breeding behavior of *Ammospiza caudacuta* and *A. maritima*. *Univ. Kans. Publ., Mus. Nat. Hist.*, **10**:45–75.

CHAPTER ELEVEN

Eggs and Young

The yolk in a chicken egg is composed of alternating bands of white yolk and yellow yolk (Fig. 1). Although the yolks of many bird eggs are yellow, Cott (1954) described other colors: e.g., salmon red in the Gentoo Penguin; deep red to scarlet in the Common Tern, Arctic Tern, and Black-headed Gull; Mandarin red in the White Pelican; fire red in the Spoonbill and Crested Barbet; orange in the Common Loon, Brown Pelican, and Great Black-backed Gull; tangerine orange in many cormorants and in the Cirl Bunting (*Emberiza cirlus*), British Tree Creeper (*Certhia familiaris britannica*), and Puff-backed Shrike (*Dryoscopus cubla*); and orpiment orange in the Yellow Bunting (*Emberiza citrinella*), Wood-lark (*Lullula arborea*), and Spotted Flycatcher (*Muscicapa striata*).

The yolk color of chicken eggs is influenced by diet, and olive-colored and orange-red yolks can be produced by including certain plant materials in the food. It seems doubtful, however, that yolk color is determined solely by diet in all species. The Sooty Tern (*Sterna fuscata oahuensis*) and Brown Noddy (*Anous stolidus pileatus*) in Hawaii have virtually the same diet (fish and squid), but the tern egg has a red yolk and the noddy egg a yellow yolk. Similarly, Ashmole and Ashmole (1967) reported that fish and squid formed the bulk of the diets of shearwaters, terns, and noddies at Christmas Island (Pacific Ocean). Of the species studied by the Ashmoles, Berger found that the yolk is red in the Gray-backed

473

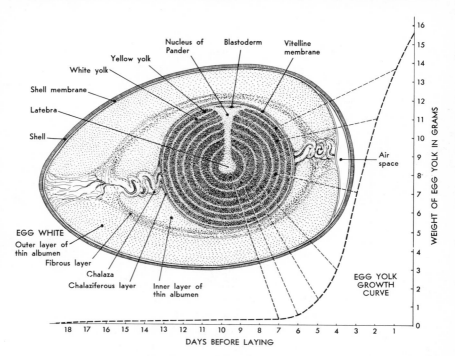

Figure 1. Diagram showing the structure of the hen's egg at the time of laying. The graph indicates the rate of growth of the egg during the 18 days preceding laying. The lines leading from the various layers of the yolk to the growth curve emphasize the times at which these layers were formed. (By permission of the late Bradley M. Patten, from *Foundations of Embryology*, 2nd ed., McGraw-Hill Book Co., copyright 1964; and by permission of the late Emil Witschi, from *Development of Vertebrates*, W. B. Saunders Co., copyright 1956.)

Tern (*Sterna lunata*), orange in the Crested Tern (*Thalasseus bergii*), and yellow in the Blue-gray Noddy (*Procelsterna cerulea*).

The ovary of a 13-day-old Red-winged Blackbird contains about 100,000 ovocytes, about 50 of which will be released during the bird's life (Witschi, 1956). Ovulation consists of the rupture of the mature follicle and the release of the ovum into the body cavity (Fig. 2). Although fertilization is followed at once by cell division, development stops after the egg is laid unless it is incubated. An *air chamber* forms between the inner and outer shell membranes within an hour after the egg is laid.

COLOR AND PATTERN

Many kinds of birds lay white or near-white eggs: grebes, shearwaters, diving-petrels, pelicans, frigatebirds, storks, flamingos, most pigeons and doves, parrots, owls, swifts, hummingbirds, the Coraciiformes except the Upupidae and Phoeniculidae, and all members of the Piciformes.

The eggs of most birds are smooth, with a dull surface. However, the surface is glossy in woodpeckers and most tinamous; pitted in the Ostrich, storks, and

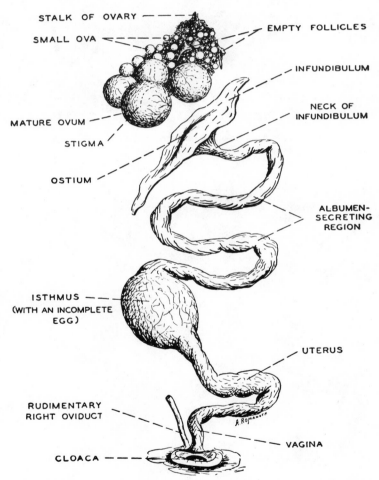

Figure 2. Reproductive organs of the hen. (By permission of Alexis L. Romanoff, from *The Avian Egg*, John Wiley & Sons, copyright 1949.)

Figure 3. The immaculate blue eggs of the Jungle Babbler (*Turdoides striatus*); Baroda, India, February 25, 1965.

Figure 4. Variation in markings and shape of Crested Tern eggs; Cook Islet, Christmas Island, Pacific Ocean, November 15, 1971.

toucans; rough or corrugated in emus and some chachalacas; greasy in ducks; and chalky in grebes, boobies, and anis.

An egg is *immaculate* if the shell is a solid color: white, pale blue (e.g., eggs of the Dickcissel, Indigo Bunting, bluebirds), bluish green (eggs of the Catbird, several American thrushes, some babblers) (Fig. 3).

The majority of birds lay colored eggs, often with elaborate patterns. Variation in amount and pattern of pigment occurs among the members of a species and in the clutch of an individual (Fig. 4). Excellent colored plates of the eggs of world birds are included in the *Handbuch der Oologie* (Schönwetter and Meise).

SIZE AND WEIGHT

Birds' eggs range in length from about 0.25 inch (some hummingbirds) to about 13 inches (extinct *Aepyornis*). Examples of birds that lay large eggs are

Figure 5. A Kiwi and its egg.

the Ostrich, 6.8 by 5.4 inches; Mute Swan, 4.5 by 2.9; California Condor, 4.3 by 2.6; Trumpeter Swan, 4.3 by 2.8; Common Loon and White Pelican, 3.5 by 2.2.

Precocial species lay much larger eggs than altricial species of similar weight. Although the Spotted Sandpiper (*Actitis macularia*) and the Catbird (*Dumetella carolinensis*) are similar in size, the sandpiper's eggs are about twice the size of those of the Catbird. Small birds lay relatively larger eggs than do most large birds. However, the Canvasback (*Aythya valisineria*) and the Ruddy Duck (*Oxyura jamaicensis*) lay eggs of approximately the same size, yet the Ruddy Duck is only about a third as large as the Canvasback. A Kiwi (*Apteryx*) weighing about 4 pounds may lay an egg 5 inches long and weighing 1 pound (Fig. 5).

Pikula (1971) found that the first eggs in a clutch of Song Thrush (*Turdus philomelos*) eggs are smaller than the later ones, and that eggs laid late in the breeding season are larger than those laid at the beginning of the season. Preston (1969) noted that there is much more variation in the length than in the breadth of eggs (Fig. 6).

Kuroda (1963) presented data on total egg weight and on the proportional weights of the eggshell, yolk, and albumen in a series of birds. The eggshells

Figure 6. Nest and eggs of the Prairie Horned Lark, showing an abnormally long egg; Willow Run, Michigan, April 14, 1947.

varied from 7.7 percent of total egg weight in the Rhinoceros Auklet (*Cerorhinca monocerata*) to 14.4 percent in the Common Murre (*Uria aalge*). The shell of Common Tern eggs forms 10.6 percent of total egg weight; that of Roseate Terns, 10.7 percent (Collins and LeCroy, 1972). Kuroda found the thickness of eggshells to vary from 0.2 mm in eggs of *Nycticorax* to 0.75 mm in *Uria*.

The widespread distribution of pesticide contamination is well known; Sladen *et al.* (1966) found DDT residues in Adelie Penguins from Ross Island, Antartica, during 1964. Increased interest in eggshell thickness developed when it was learned that experimental birds given diets containing DDT or its metabolites laid eggs with thinner shells and less calcium than eggs laid by control birds (Longcore *et al.*, 1971; Anderson and Hickey, 1970). In the United States eggshell thinning has been reported in more than 20 species, primarily raptorial and fish-eating birds. "In seven of eight species where shell thinning exceeded 20 percent, there was an associated population decline" (McLane and Hall, 1972).

SHAPE

The majority of birds' eggs are oval, but at least 11 other terms may be used to describe less frequently encountered shapes (Fig. 7). Owls and toucans lay round (spherical) eggs. Swifts, hummingbirds, swallows, and some other birds lay elliptical or long-elliptical eggs. Murres, auks, and shorebirds lay pear-shaped (pyriform) eggs. The eggs of a clutch vary both in size and in shape. Preston (1968, 1969) presented equations for analyzing this variation, and Stonehouse (1966) proposed a formula for determining egg volume from linear dimensions.

CLUTCH SIZE

By "clutch" we mean all the eggs laid by one female for a single nesting. Among the many species of birds, the clutch size varies from 1 to about 20. A single egg is laid by most penguins, albatrosses, shearwaters, petrels, tropic-birds, condors and large eagles, the Crabplover, some auks, potoos, some night-jars, crested-swifts, the Lyrebird, some sunbirds. Two eggs constitute the clutch of some penguins, loons, boobies, gannets, many pigeons, hummingbirds, and many tropical passerine birds; some of these species (e.g., Adelie Penguin, loons, boobies) may typically raise only one young from the 2 eggs laid. Many plovers, sandpipers, avocets, and phalaropes lay a clutch of 4 eggs (Fig. 8). Many small birds in the Temperate Zone lay 4 or 5 eggs, but others (especially

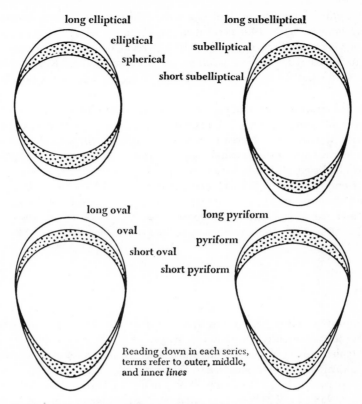

Figure 7. Profiles of 12 egg shapes. (Courtesy of Ralph S. Palmer and Yale University Press.)

wrynecks, wrens, chickadees, titmice, nuthatches) lay 6 to 13. Most ducks and gallinaceous birds lay from 6 to 15 or more eggs, but with the larger clutches it is not always certain that all are laid by one female.

Geographical variation in average clutch size occurs in both nonpasserine and passerine species. Many tropical species lay fewer eggs per clutch than do Temperate Zone species. Two eggs form the clutch for many Central American tyrant-flycatchers, troupials, tanagers, and fringillids, whereas members of these families breeding in the United States usually lay 4 to 6 eggs (Fig. 9).

Seasonal variation in clutch size is demonstrated by the Great Tit (*Parus major*), European Blackbird, European Robin, and Yellowhammer (*Emberiza citrinella*): early and late clutches are slightly smaller than those laid during the middle of the nesting season (Lack, 1954:34).

Coulson *et al.* (1969) showed that the eggs of the Kittiwake (*Rissa tridactyla*) and of the Shag (*Phalacrocorax aristotelis*) decrease in width, shape index, and volume as the breeding season advances but increase in these features as the females become older. Moreover, older birds breed earlier than young ones, so that the volume (rather than the number) of eggs laid throughout the breeding season decreases progressively. These authors proposed, therefore, that "the Shag, Kittiwake, and the Great Skua (*Catharacta skua*) all show a marked decrease in the amount of material females use in egg production as the breeding season advances, and this trend is probably common to many species of seabirds. That this trend results from food shortage is unlikely, and it is suggested that it results from the different degrees of reproductive development reached in seabirds, earlier breeding birds showing a higher reproductive drive."

Figure 8. A Black-winged Stilt (*Himantopus h. himantopus*) and its eggs; Saurashtra, India. (Courtesy of Dharmakumarsinhji of Bhavnagar, India.)

Figure 9.   Eastern Wood Pewee nest and eggs; northern Florida. (Courtesy of Samuel A. Grimes.)

Annual differences in average clutch size have been demonstrated for some birds of prey, in which clutch size seems to vary with the relative abundance of the rodents that serve as prey (Pitelka *et al.,* 1955). Certain birds (e.g., bustards, some starlings and weaverbirds, the Banded Landrail) in Africa and Australia lay smaller clutches in very dry years than in years of ample rainfall. Berger (1968) reported annual differences (from 5.0 to 16.5%) in the number of six-egg clutches of the American Goldfinch.

Significance of Clutch Size

Several theories have been proposed to explain the differences in clutch size among birds.

1.   The number of eggs laid is limited by the physiological capacity of the bird.

2.   A bird lays as many eggs as it can cover with its incubation patch.

3.   Clutch size is correlated with the mortality rate of the species; if the mortality is low, the clutch is small.

4.   For most birds the clutch size represents the largest number of young that the parents can feed (the sturdy-young theory; Skutch, 1967b).

Lack (1954) argued strongly for the fourth theory. Later (1968), however, he revised his ideas because of problems of interpretation raised by certain nidifugous birds in which the young begin to pick up their own food shortly after hatching, but in which there may be a considerable difference in clutch size between tropical and Temperate Zone species.

Ryder (1970) suggested that clutch size in Ross's Goose, which nests in the Canadian arctic, has evolved in relation to the food reserves that the female accumulates before she arrives on the breeding grounds. These food reserves "are allotted to ova and non-ovarian tissues. The number and size of eggs is limited to provide enough reserve food material to the young until they are able to feed themselves, and also to provide the breeding female with food so that she can give maximum protection to the clutch." (See also Cody, 1966.)

Indeterminate and Determinate Laying

Phillips (1887) discovered by removing an egg each day from the nest of a Yellow-shafted Flicker (*Colaptes auratus*) that the bird did not stop when the typical clutch had been laid but laid a total of 71 eggs in a 73-day period. Such species are called indeterminate layers. Other examples of this type are ducks, gallinaceous birds, Budgerigars, wrynecks, the House Wren, the House Sparrow. The ability of domesticated birds to continue to lay has, of course, been exploited to man's advantage, and individual chickens have laid more than 360 eggs per year; ducks, 309; turkeys, more than 200; and Ostriches, 100 (Romanoff and Romanoff, 1949).

Determinate layers, on the other hand, cannot be induced to lay fewer or more eggs than a normal clutch. Pigeons, doves, and the Herring Gull, Ringed Plover, Black-billed Magpie, Barn Swallow, Eastern Bluebird, and Tricolored Blackbird appear to be determinate layers. The many ramifications of determinate and indeterminate laying and of hormonal control of ovulation have been discussed by Kendeigh *et al.* (1956), Fraps (1970), Lehrman (1961), van Tienhoven (1961), Brockway (1968), and Meier and MacGregor (1972).

Time of Laying

The interval between completion of the nest and laying of the egg may be short or long (Figs. 10, 11, 12). Many birds lay the first egg the day after completion of the nest; others wait 2 or 3 days; and for some there is an interval of 1 to 2 weeks (e.g., some antshrikes, some redstarts of the genus *Myioborus,* American Goldfinch). The Wrenthrush (*Zeledonia coronata*) waits 1 week; the Superb Blue Wren (*Malurus cyaneus*), from 5 days to a month after the nest is completed.

Many birds lay one egg each day until the clutch is complete, but the Wrenthrush and the Mountain Elaenia (*Elaenia frantzii*) lay eggs at 48-hour inter-

Figure 10. A Long-tailed Tailorbird (*Orthotomus sutorius maculicollis*) at its nest; Singapore. (Courtesy of the late Loke Wan Tho.)

vals (Hunt, 1971; Skutch, 1967a). The Thrush-like, Blue-crowned, and Orange-collared manakins lay their eggs in the middle of the day and at 48-hour intervals (Skutch, 1969b:162). Other examples of birds that lay at longer intervals are the following: many hawks and owls, 2 or 3 days; European Cuckoo (*Cuculus canorus*) and Smooth-billed Ani, 2 days; Groove-billed Ani, 2 to 4 days; Oilbird, 4 days; megapodes, 2 to 5 days; some hornbills, 5 to 7 days (Lack, 1968).

Egg Dumping

Ducks and Ring-necked Pheasants are noted for laying eggs on the ground, in another hen's nest, or in "dump nests," in which the eggs are not incubated. Weller (1959) reported dump nests containing from 30 to 87 Redhead eggs (Fig. 13).

EGG RECOGNITION BY ADULTS

In view of the wide range in color and markings of birds' eggs, one might assume that birds have a well-developed discriminatory ability for their own eggs. Experiments show, however, that this is rarely the case.

Tinbergen (1953) found that most Herring Gulls were more attracted to the nest site than to their eggs and would incubate on an empty nest while their eggs were in sight a short distance away. Not only would a gull accept the eggs of some other gull, but also it would incubate wooden "eggs" painted blue or

Figure 11. A Black-naped Oriole (*Oriolus chinensis*) at its nest; Singapore. (Courtesy of Salim Ali; photograph by the late Loke Wan Tho.)

Figure 12. A Red-wattled Lapwing (*Hoplopterus indicus*) at its nest. (Courtesy of the late Loke Wan Tho.)

yellow. The importance of tactile stimulation in egg "recognition" is shown by Tinbergen's findings that Herring Gulls would return to nests containing wooden "eggs" of various shapes (rectangular, cylinder, prism) but would not incubate them if they had sharp edges. Tinbergen remarked: "For when the birds made their choice and went to the rectangular blocks, they could truly be said to 'recognize' them as eggs. But when, after touching them, they left them, one could with as much justification say that now they did not 'recognize' them as eggs." (See also Baerends and Drent, 1970; Fredrickson and Weller, 1972; Walker, 1951.)

Peek *et al.* (1972) found that female Red-winged Blackbirds "preferred to remain at the nest site even though the nest, eggs, and young (under 7 days old) were replaced with counterparts from other Redwing nest situations." If the nestlings were at least 10 days old, however, the female went to them regardless of the nest they were in.

The survival of parasitic species depends on the acceptance of their eggs by the host species. Among the few species that typically eject Brown-headed Cowbird eggs from the nest are the Catbird and Robin, both of which lay bluish eggs wholly unlike those of the cowbird; species of thrushes that lay blue eggs, however, apparently do not remove cowbird eggs from their nest. A few species (e.g., Cardinal), whose eggs are similar in size and color pattern to those of the cowbird, are prone to desert parasitized nests, but most host species accept the alien eggs (Fig. 14).

Supernormal Stimuli

Tinbergen used the term "supernormal stimuli" in describing certain behavior reactions. The Ringed Plover, for example, "is more strongly stimulated by a white egg with large black dots than by its own eggs, which are buff with small brownish dots." When given a choice between a real egg and a giant artificial

Figure 13. A dump nest of the Fulvous Tree Duck (*Dendrocygna bicolor*). (Courtesy of Brooke Meanley.)

Figure 14. Lower: Nest of a Cardinal with three host eggs and one egg of the Brown-headed Cowbird; Ann Arbor, Michigan, June 16, 1963. Upper: Indigo Bunting nest with two host eggs and one Cowbird egg; Ann Arbor, Michigan, June 18, 1963.

Figure 15.   Herring Gull rolling a giant wooden egg into its nest. (Courtesy of G. P. Baerends.)

egg, Herring Gulls try to incubate the artificial egg even when its volume is 8 to 20 times that of the normal egg (Fig. 15). Elegant Terns studied by Walker incubated the larger egg of the Royal Tern, and Black Terns accepted chicken eggs "over 35 times the volume of the tern's egg" (Richardson, 1967).

Egg-retrieving

This practice has been found among a number of ground-nesting species: e.g., Red-tailed Tropicbird, Gray Lag-Goose, Ringed Plover, Oystercatcher, Herring, Laughing, and Black-headed gulls, Sooty Tern, Black Tern, European Nightjar, and Common Nighthawk (Fig. 16).

Beer (1962) placed wood models of Black-headed Gull eggs 23 cm from the centers of the nests. His results, he wrote, "indicate that the readiness to perform egg-rolling increases steadily as the date of laying draws closer, remains maximally high during the laying and incubation periods, and declines steadily after hatching."

Richardson (1967) noted that Black Tern eggs placed just outside the nest usually were retrieved but that some terns incubated the displaced eggs. However, "moving a whole nest island up to 23 feet . . . did not cause desertion of the eggs even when the island was moved 15 feet in two steps in two hours."

Howell and Bartholomew (1969) found that Red-tailed Tropicbirds rolled

Figure 16. Female Common Nighthawk rolling a displaced egg under her. (Courtesy of Milton W. Weller and the editor of *Auk*.)

their own eggs back to the nest from a distance of 6 inches. Half of the birds tested rolled an albatross egg (much larger than a tropicbird egg) into the nest, whereas only one tropicbird retrieved a much smaller tern egg. Only one of the tropicbirds retrieved a red plastic egg, and within a few seconds this bird pushed the egg out of the nest. If a red plastic egg was substituted for the bird's own egg in the nest, however, the bird usually incubated it.

Egg Transport

Audubon wrote in 1831 that, if a person touches the eggs, a Chuck-will's-widow will move them to another location, carrying the eggs in the mouth. This statement was long dismissed as a myth, but Rysgaard (1944) told of a Chuck-will's-widow carrying an egg in its left foot, and Kilham (1957) reported similar behavior by a Whip-poor-will. Egg-moving and/or egg-carrying behavior also has been reported for several other species (e.g., Mallard and Pintail ducks, Red-bellied Woodpecker, Yellow-shafted Flicker), and Truslow (1967) presented photographs of Pileated Woodpeckers carrying eggs in their bills from the nest hole after the tree had broken through the upper part of the nesting cavity. Goertz and Rutherford (1972) watched a Carolina Chickadee remove four small young from a nesting cavity.

INCUBATION

Incubation Patches

To incubate is to apply heat to eggs (Fig. 17). Most birds (exceptions: ratites, pelecaniform birds, megapodes, Cassin Auklets) develop incubation (or brood' patches in the skin of the abdominal wall a short time before incubation begins. The patches form almost exclusively in the apteria, but they differ in size, location, and number throughout the orders of birds (Hanson, 1959; Selander and Yang, 1966b). They are formed by the loss of some abdominal feathers and/or simply by an increase in the vascularization of the skin. Down feathers usually are lost; but, in birds lacking down, contour feathers may be shed or plucked. Female ducks and geese pluck breast feathers to line the nest.

In general, only the sex that incubates develops an incubation patch; there are exceptions, however, and the presence or absence of a brood patch for deciding which sex incubates is questionable. Skutch (1957) commented that "among passerines, there appears to be little correlation between incubation by the male and his possession of a brood-patch. Some males which regularly incubate lack it, whereas others which fail to incubate develop it."

Figure 17. An incubating Piping Plover (*Charadrius melodus*); Muskegon County, Michigan. (Courtesy of Lawrence H. Walkinshaw.)

Table 1.   Synopsis of Incubation Patterns.

Parental Incubation[a]

I. Incubation by two parents.
   A. By both sexes simultaneously.
      Red-legged Partridge (*Alectoris rufa*).
   B. By both sexes alternately.
      1. Either sex may cover eggs by night.
         a. Change-overs occur at intervals of about 24 hours, so that the same parent does not often sit for whole successive nights.
            Diving-petrel (*Pelecanoides urinatrix*), Sooty Tern (*Sterna fuscata*), Bridled Tern (*S. anaethetus*), Ringed Kingfisher (*Ceryle torquata*).
         b. Change-overs occur at intervals of much more than 24 hours, so that the same parent may sit for several nights in succession.
            Adelie Penguin (*Pygoscelis adeliae*), Manx Shearwater (*Puffinus puffinus*), Royal Albatross (*Diomedea epomophora*), and many other Procellariiformes.
         c. Change-overs occur at intervals of less than 1 day (sometimes by both daylight and dark).
            Semipalmated Plover (*Charadrius hiaticula*), Herring Gull (*Larus argentatus*), Cape Wagtail (*Motacilla capensis*).
      2. Female covers eggs by night.
         a. Male takes one long session each day.
            Pigeons and doves, Citreoline Trogon (*Trogon citreolus*), Black-throated Trogon (*T. rufus*).
         b. Sexes alternate on nest several times a day, with the diurnal sessions therefore usually shorter than in 2*a*.
            Amazon Kingfisher (*Chloroceryle amazona*), Rufous-tailed Jacamar (*Galbula ruficauda*), antbirds (Formicariidae), many warblers (Sylviidae).
      3. Male covers eggs by night.
         a. Female takes one long session each day.
            Ostrich (*Struthio camelus*), Pale-billed Woodpecker (*Phloeoceastes guatemalensis*).
         b. Sexes alternate on nest several times a day, with the diurnal sessions therefore usually shorter than in 3*a*.
            Many woodpeckers, anis (*Crotophaga*), American Coot (*Fulica americana*).

II. Incubation by one parent only.
   A. By female.
      1. One recess, usually long, is taken each day.
         Bobwhite Quail (*Colinus virginianus*), Marbled Wood Quail (*Odontophorus gujanensis*), Guan (*Pauxi pauxi*).

Table 1 (con't)

2. Several or many recesses are taken each day, sometimes also by night. Redhead Duck (*Nyroca americana*), most hummingbirds, manakins (Pipridae), American flycatchers (Tyrannidae), most songbirds.

3. Female sits continuously for many days.
   a. She fasts while incubating.
      Golden Pheasant (*Chrysolophus pictus*), Great Argus Pheasant (*Argusianus argus*), Eider Duck (*Somateria mollissima*), Blue Goose (*Chen caerulescens*).
   b. She is fed by male.
      Hornbills.

B. By male.

1. One long recess is taken each day.
   Little Tinamou (*Crypturellus soui*), Bonaparte's Tinamou (*Nothocercus bonapartei*).

2. Several or many recesses are taken each day.
   Pheasant-tailed Jacana (*Hydrophasianus chirurgus*), Ornate Tinamou (*Nothoprocta ornata*).

3. Male incubates continuously, fasting, for a number of days.
   Emperor Penguin (*Aptenodytes forsteri*), Kiwi (*Apteryx*), Ému (*Dromiceius n. hollandiae*).

III. Incubation by more than two birds at a single nest.

A. Eggs laid by one female.

1. Several adults assist the female.
   Bush-tit (*Psaltriparus minimus*).

B. Eggs laid by two or more females.

1. Several birds of both sexes participate in incubation.
   Anis (*Crotophaga*), ?Acorn Woodpecker (*Melanerpes formicivorus*).

Substitute Incubation

IV. Incubation by birds of other species.
Many cuckoos, cowbirds (Icteridae), honeyguides (Indicatoridae), Black-headed Duck (*Heteronetta atricapilla*).

V. Incubation without animal heat.
Megapodes, Egyptian Plover (*Pluvianus aegyptius*).

[a] From Skutch (1957).

Incubation Period

This is the elapsed time between the laying of the last egg in the clutch and the hatching of that egg when all the eggs hatch (Nice, 1954a).[1] The most precise method for determining the onset of incubation is to place thermistors between the eggs in the nest. However, even when all external conditions are apparently the same, there is a normal range of variation in the incubation time for a species's eggs. Although the eggs usually hatch in the order laid, there are exceptions.

Role of the Sexes

The eggs may be incubated by the female only, by the male only, by both, or by neither (e.g., megapodes, Egyptian Plover, parasitic species) (Fig. 18). For a number of families there is no reliable information on which sex incubates; indeed, many species of birds exist whose eggs have never been described. For the families for which information is available, we find that incubation is performed by both sexes in about 54 percent of the cases; by the female alone in 25 percent; by the male alone in 6 percent; and by the male, the female, or both in 15 percent. Skutch (1957) suggested that incubation by both sexes was the primitive method among birds, and he presented a synopsis of the incubation pattern (Table 1).

Length of Incubation Period

Among different birds the incubation period varies from a minimum of about 10.5 days to almost 12 weeks. The presumed 10-day incubation period for the Brown-headed Cowbird and the Cape White-eye has been shown to be incorrect (Nice, 1953; Winterbottom, 1955). At the other extreme, the eggs of the Royal Albatross require an incubation period of 77.5 to 80.25 days; those of the Black-footed Albatross, 63 to 68 days.

Parmelee (1970) reported incubation times for the Sanderling (Fig. 19) varying from 24 days, 6 hours, 15 minutes ($\pm$30 minutes) to 31 days, 16 hours, 46

---

[1] Kendeigh (1963b) proposed that the incubation period be measured on the basis of energy transformation because "the energy for embryonic development does not come from the imput of heat from the outside but from oxidation of the fats, carbohydrates, and, perhaps, proteins in the yolk and albumen of the egg itself. The rate and amount of energy that is released for maintenance and growth depends on the amount of oxygen absorbed and this is readily measured." The rate of oxygen absorption is "directly proportional to the temperature and the length of time that the egg is maintained at that temperature. The degree-hour, therefore, indicates the mobilization of a definite amount of energy and a definite amount of embryonic growth." Kendeigh found that the incubation period of the House Wren is 14 days or 366 $\pm$ 3.5 hours, as measured in time units. The sum total of heat above a temperature threshold he estimated to be 5867 $\pm$ 144 degree-hours or 32.9 kcal.

Figure 18. A Phoenix Petrel (*Pterodroma alba*) and its egg; Motu Tabu Islet, Christmas Island, Pacific Ocean, November 13, 1971.

minutes (±34 minutes). The nest in which the maximum incubation period was determined was unusual in that an interval of 5 days and 16 hours elapsed between completion of the clutch and the onset of incubation. Parmelee also determined that there is an interval of 26 to 29 hours between the laying of each egg in a clutch, and that, although only one adult incubates the eggs and attends the young, this may be either the male or the female!

Incubation Temperatures

The optimal incubation temperature is said to be from 2 to 4°C lower than the body temperature of the incubating bird (Witschi, 1956:224). Irving and Krog (1956) reported that 74 percent of nest temperatures recorded during incubation were between 33° and 37°C. Prolonged exposure to high temperatures kills the embryos. Embryos are much better adapted to survive lowered temperatures, and this is especially true before incubation has begun (Parmelee, 1970; Norton, 1972). Greenwood (1969) wrote of a Mallard egg hatching successfully even though the shell had been cracked by freezing.

The megapodes exhibit interesting differences from other birds. The hens lay the eggs in mounds of vegetation, in holes dug in sand, or in volcanic material. Volcanic heat provides the necessary temperature for incubation in the last of

Figure 19. A Sanderling incubating its clutch of eggs; Bathurst Island, Canadian Arctic Archipelago, June 28, 1969. (Courtesy of David F. Parmelee.)

these cases; the sun and the heat of decaying organic materials supply the heat required in the other nesting sites. Natural incubation periods vary from 50 to 96 days, although two chicks were hatched in 44 days at an artificial temperature of 37.7°C. The adults tend the mounds, and the males, especially, periodically check the temperature (apparently by temperature receptors in the tongue) in the mound. By adding or removing materials, the birds are said to maintain a fairly constant temperature (from about 32° to 35.5°C) in the mounds (Frith, 1962; Clark, 1964a, 1964b).

Egg-turning

During the incubation period egg-turning is practiced by most birds. The yolk rotates each time that the egg is turned, so that the developing embryo lies uppermost in the shell. Baerends and Drent (1970:55) wrote that an early adherence of the embryo to the shell causes "disruption in the development of the extra-embryonic membranes and subsequently death, whereas in the later stages immobility of the egg leads to adhesion of allantois and yolk, preventing the retraction of the yolk into the abdomen prior to hatching." Rolle (1963) said that the Red-legged Thrush may turn the eggs as often as 11 times per hour. Hartshorne (1962) used the term "tremble-thrusts" to describe the egg-turning behavior of Eastern Bluebirds.

Figure 20. A female Southern Small Minivet (*Pericrocotus c. cinnamomeus*: Campephagidae) incubating. (Courtesy of Shivrajkumar of Jasden, India.)

Constancy of Incubation

The time that an incubating bird spends on the eggs is called a *session* or the *attentive period*; the time off the eggs is the *recess* or *inattentive period* (Skutch, 1962). The rhythm of sessions and recesses varies widely among different species of birds (Fig. 20). Some female hornbills remain in their nesting cavity throughout the incubation period, so that the eggs are covered nearly 100 percent of the time. This is true also of some species in which both sexes incubate, and of species that nest in February, March, or April in northern parts of the world: e.g., some hawks, owls, the Killdeer and other shorebirds, and the Mourning Dove, Horned Lark, Clark's Nutcracker, Vesper Sparrow, and Song Sparrow.

Norton (1972) studied the incubation rhythms at nests of the Dunlin, Baird's Sandpiper, and Pectoral Sandpiper (*Calidris melanotos*) at Barrow, Alaska, by making continuous telemetric temperature recordings from egg-laying to hatch-

Figure 21. A Clapper Rail (*Rallus longirostris*) repairing nest while incubating. (Courtesy of Samuel A. Grimes.)

Table 2. Incubation of Infertile or Spoiled Eggs[a]

Species	Length of Attendance (days)	Incubation Period (days)	Authority
Black-crowned Night Heron[b] (*Nycticorax nycticorax*)	40, 49, 51	22–24	Noble and Wurm (1942)
Wood Duck (*Aix sponsa*)	62	About 30	Leopold (1951)
Bobwhite (*Colinus virginianus*)	56	23	Stoddard (1946)
Sarus Crane[b] (*Grus antigone*)	70–72	About 32	Walkinshaw (1947)
Smooth-billed Ani (*Crotophaga ani*)	24	13–15	Davis (1940)
Black-chinned Hummingbird (*Archilochus alexandri*)	24	14	Bené (1946)
White-tailed Trogon (*Trogon viridis*)	51	About 17	Original
Yellow-shafted Flicker (*Colaptes auratus*)	30	11–12	Sherman (1952)
Common Crow (*Corvus brachyrhynchos*)	21, 22, 24 26, 28, 32	16–18	Emlen (1942)
Carolina Chickadee (*Parus carolinensis*)	24	12–13	Odum (1942)
Blue Tit (*Parus caeruleus*)	25	13–15	Gibb (1950)
European Wren (*Troglodytes troglodytes*)	25, 26, 51	15–17	Armstrong (1955)
Gray's Thrush (*Turdus grayi*)	17–18, 19	12	Original
European Robin (*Erithacus rubecula*)	35, 48	13–15	Lack (1953)
Eastern Bluebird (*Sialia sialis*)	21, 21, 21	13–14	Laskey (1940)
	33		Thomas (1946)
Chestnut-capped Brush-finch (*Atlapetes brunnei-nucha*)	19	About 15	Original
American Goldfinch (*Spinus tristis*)	23	13	Berger (1953)

[a] From Skutch (1962).
[b] In captivity.

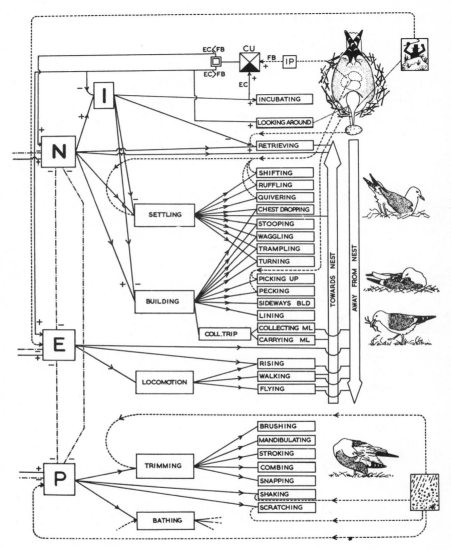

Figure 22. A model of the functional organization of incubation behavior. I, The tendency to incubate; N, the tendency to settle on nest, to incubate, to retrieve, and to build in response to eggs in the nest; E, the tendency to escape; P, a processing unit for evaluating the different aspects of feedback information from the nest (analagous to releasing mechanisms); CU, a comparison unit for feedback information from incubating; FB > EC, a positive output of CU and transferral to different channels; FB < EC, a negative output of CU. (Courtesy of G. P. Baerends and the editor of *Behaviour.*)

ing. He found that the eggs were covered 97.55 percent of the time in Dunlin nests, 96.50 percent in Baird's Sandpiper nests, and 84.99 percent in Pectoral Sandpiper nests. Both sexes of the Dunlin and Baird's Sandpiper incubate, whereas only the female Pectoral Sandpiper performs this function.

The incubation period for Adelie Penguin eggs varies from about 34 to 38 days; the male assumes incubation duties first while the female returns to sea for food. The male incubates for 2 to 2.5 weeks; then the female returns and takes over the incubation. Because the birds arrive at their breeding grounds as much as 3.5 weeks before egg-laying, males may fast as long as 40 days (Sladen, 1958). The Common Eider incubates for four weeks, during which period the female leaves the nest for periods of only 10 or 15 minutes every second or third day, and then merely for water; she fasts for the entire period.

Skutch (1962) reviewed the subject of incubation patterns. He noted that "in some species both parents together devote less time to the nest than do other birds incubating alone," and he added that "incubation or non-incubation by the male appears to be largely indifferent to the welfare of many species, so that we may regard this as a non-adaptive character, resulting from chance mutations, and not closely controlled by natural selection."

Some birds add to their nests or repair them during the incubation period: e.g., boobies, hummingbirds, rails (Fig. 21), the House Sparrow.

Prolonged Incubation Behavior

When the eggs fail to hatch, prolonged incubation is often observed. Skutch wrote that most birds tend to incubate their eggs for an interval at least 50 percent longer than is normally required to hatch them (Table 2).

Functional Organization of Incubation Behavior

In their study of Herring and Black-headed gulls, Baerends and his colleagues (1970) determined egg-laying behavior, clutch size, incubation rhythm, and hatching, as well as heat production by the embryo, heat loss from eggs, energy requirements for incubation, and behavior of the incubating birds. Baerends then proposed an ethological model (Fig. 22) that included all behavior during incubation in terms of the bird's orientation toward the nest and away from it.

THE YOUNG

Egg-Tooth

The fully developed embryo cuts the eggshell with an *egg-tooth*; during this activity, the egg is said to be "pipped." In most birds the egg-tooth is located

Figure 23. A Common Tern (*Sterna hirundo*) hatching;Detroit, Michigan, June 30, 1962.

near the tip of the upper mandible; in several families (e.g., Haematopodidae, Charadriidae, Recurvirostridae, Burhinidae, Bucerotidae), it is said to occur only on the lower mandible; and in some birds (e.g., some alcids, Red-headed and Red-bellied woodpeckers), egg-teeth are found on both mandibles (Parkes and Clark, 1964; Sealy, 1970). The egg-tooth falls off, or is resorbed, during the first week after hatching in most birds, but later than this in some penguins, alcids, hawks, owls, and musophagids. Two young Pigeon Guillemots observed by Sealy lost the egg-tooth between the 27th and 30th days after hatching, whereas the egg-tooth disappears in about 24 hours in the downy Wilson's Phalarope (Höhn, 1967).

The time required for the young bird to hatch varies considerably among species (Fig. 23). From 10 to 20 hours is required for the young of many small species to break open the eggshell. The egg of the Royal Albatross may be pipped for more than 4 days before the chick emerges, and the Laysan Albatross chick may require as long as 6 days to break out of the shell (Fisher, 1971).

Hatching Muscle

The *hatching muscle* (M. complexus) aids the chick in breaking out of the shell. The muscle "is apparent at 7 days of incubation [in the chicken], reaches its maximum size at 20 or 21 days of incubation, and then gradually disappears" (Fisher, 1958). The increase in size of the muscle is related largely to the accumulation of lymph (Bock and Hikida, 1969). Brooks and Garrett (1970) observed the actual hatching of several chickens through sealed "windows" cut in the eggs. They concluded that "the pip is apparently produced by one of a series of strong whole-body convulsions that presses the beak tip through the shell." Fisher (1966) reported that there is "a gross correlation between size of hatching muscle and thickness of egg shell" in grebes, but no correlation in ducks.

Altricial and Precocial Young

Altricial young are helpless at hatching; they require complete parental care for periods ranging from about 1 month to nearly 8 months (Fig. 24). At the time of hatching, altricial young may be *psilopaedic* (either naked or having a scant covering of hairlike down on the crown and back), or *ptilopaedic* (clothed with down; e.g., herons, bitterns, ibises, vultures, hawks, owls). Long, sharp claws on the toes and a strong grasping reflex enable young White Terns to hold onto swaying branches where the egg is laid (Fig. 25).

Precocial young are ptilopaedic at hatching and can run (or swim) as soon as their down dries (e.g., megapodes, ducks, geese, quail, pheasants, grouse, rails, shorebirds). Precocial young usually leave the nest within 48 hours after hatching; the young of many species are never fed by the adults, although the latter may lead the young to food. Megapode chicks dig out of their nests and are independent of parental care.

Alvarez del Toro (1971) reported that the male American Finfoot (*Heliornis fulica*) has a shallow brood pouch on each side of the body below the wings. The newly hatched chick is blind, nearly naked, and helpless, and the male presumably puts a chick into each brood pouch and carries it therein.

There are intermediate conditions of the young, however, and Nice (1962) outlined these in tabular form "according to manner of getting food, activity, amount of down, and development of sight" (Table 3). The eyes of altricial and semialtricial species are closed at hatching. The Song Sparrow opens its eyes on the third or fourth day after hatching; Painted Bunting, on the third day; American Goldfinch, at about 3.5 days; Red-legged Thrush, "for short periods of time" on the fifth day; Black Woodpecker (*Dryocopus martius*), at 12 days; Red-headed Woodpecker, at 13 or 14 days.

Directive Marks or Targets

These special markings in the oral cavity are presumed to aid in coordinating the gaping of the young with the feeding response of the adults. The simplest type of directive mark is the light-colored (often yellow), swollen corners of the mouth that Clark (1969) referred to as transitory *oral flanges*.

Figure 24. Above: Three nestling Japanese White-eyes (*Zosterops japonica*) less than 12 hours old; Honolulu, Hawaii, April 19, 1973. Below: Three nestling Red-crested Cardinals (*Paroaria coronata*); one bird has just hatched and its down is still wet; the largest bird is 18 hours old; Honolulu, Hawaii, April 28, 1973.

Figure 25. A White Tern (*Gygis alba*) chick on the branch where it hatched; Cook Islet, Christmas Island, Pacific Ocean, November 15, 1971.

The color of the roof of the mouth varies among nestling birds: rose-pink to crimson in some Hawaiian honeycreepers; orange-yellow in the Red-faced Warbler and Painted Redstart; red and blue in the Hawfinch (*Coccothraustes*); pink and yellow in crossbills (*Loxia*); mauve or reddish purple in some corvids. However, information is available for only a small percentage of birds, and Ficken (1965) pointed out that the color may change within a few days after hatching; e.g., the Tree Pipit (*Anthus trivialis*) has a deep orange palate at hatching, but this changes to crimson at 4 days of age.

More striking are directive marks on the palate that contrast sharply with the background color, and are characteristic of many ploceid finches and cuckoos (Fig. 26). In color these frequently are black, blue, or white; some are semiluminous.

Table 3. Classification of Maturity at Hatching in Birds[a]

Feed Selves

Precocials—eyes open, down covered, leave nest first day or two.

    Precocials 1. Are independent of parents—e.g., megapodes.
    Precocials 2. Follow parents but find own food—e.g., ducks, shorebirds.
    Precocials 3. Follow parents and are shown food—e.g., quail, chickens.

Fed by Parents

Precocials 4. Follow parents and are fed by them—e.g., grebes, rails.

Semiprecocials—eyes open, down covered, stay at nest though able to walk—e.g.,
        gulls, terns.
Semialtricials—down covered, unable to leave nest.

    Semialtricials 1. Eyes open—e.g., herons, hawks.
    Semialtricials 2. Eyes closed—e.g., owls.

Altricials—eyes closed, little or no down, unable to leave nest—e.g., passerines.

[a] From Nice (1962).

Heel-pads

These horny pads, studded with strong tubercles, cover the joint between the tibiotarsus and the tarsometatarsus. They are found in the Quetzal (Trogonidae), some coraciiform birds (kingfishers, todies, bee-eaters, hornbills), and piciform birds (jacamars, puffbirds, barbets, honeyguides, toucans, woodpeckers, and wrynecks) (Figs. 27, 28).

Length of Nestling Period

A young bird while still in the nest is a *nestling* (Fig. 29). A *fledgling* is a bird that has left the nest (Fig. 30). In general, "the nestling periods of small altricial birds—except hole-nesters and long-winged species like swallows and swifts—are of about the same length as their incubation periods" (Skutch, 1945). Nestling periods shorter than the incubation periods characterize some genera of passerine families: e.g., antbirds, larks, some wood-warblers, and fringillids. Among passerine species, however, the nestling period varies considerably: about 8 days for Ovenbirds; 21 to 22 days for the Iiwi (*Vestiaria*); 28 to 35 days for the Purple Martin; 29 to 35 days for the Common Mynah (*Acridotheres*). Some barbets, toucans, and woodpeckers have nestling periods two or three times longer than the incubation periods. The flightless period for the Laysan Albatross chick averages 165 days; that of the Royal Albatross may be as long as 243 days.

PARENTAL CARE

Tinbergen concluded that egg-turning behavior stops after the eggs are pipped, presumably as a response to movements or call notes of the hatching young, which cause the parent "to treat the pipped egg as a chick."

The Egg-Young Stage

At least two different kinds of sounds (clicking and tapping) are produced by hatching birds: the egg-young stage. After watching hatching young of the

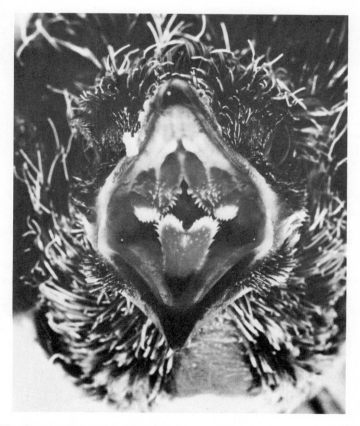

Figure 26. Gape markings of an 11-day-old Roadrunner (*Geococcyx californianus*). (Courtesy of Constance P. Warner, Kerry A. Muller, and the editor of *Wilson Bulletin*.)

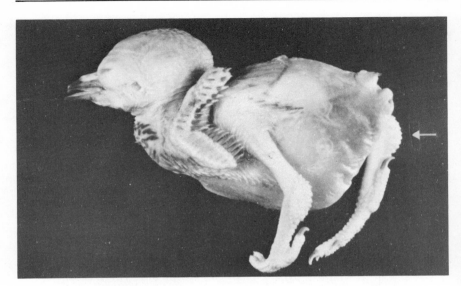

Figure 27. A nestling Puerto Rican Tody (*Todus mexicanus*), estimated to be 3 to 5 days old; the arrow points to the heel-pad. (Courtesy of Francis J. Rolle and the editor of *Auk*.)

Common Eider through openings cut in the shell, Driver (1965) concluded that "the evidence suggests that 'clicking' is a side-effect of the newly functioning respiratory system. It is distinct from the sounds of the shell rupture and vocalization that are also produced by egg-young at this stage of development." Hatching birds also utter soft *peep* call notes, and the adult may "talk" to its egg-young (Fig. 31).

Egg-shell Disposal

Many precocial species leave the egg shells in the nest and lead their young away. Others do not (Fig. 32); shorebirds, for example, remove the shells, even though the young stay at the nest for only a day or two, and Green and Solitary sandpipers (which lay their eggs in the tree nests of other birds) also carry the shells away. A few altricial species (e.g., some cuckoos) leave the shells in the nest, but most altricial species either eat the shells or carry them away.

Brooding Behavior

Eggs are incubated; nestlings are brooded. Brooding behavior is a continuation of incubation behavior. The speed with which adults can change their behavior during the incubation period is demonstrated by some experiments conducted by Graber (1955), who used young from incubator-hatched eggs. He removed the eggs from a Bell's Vireo nest shortly after the last egg of the clutch was laid and then put a young Cardinal in the nest. He reported that the vireos brought larvae to the nest less than 1 minute later. In another instance, Graber put two hatchling Painted Buntings and a Cardinal in a Painted Bunting nest, from which he then removed the eggs. The three young birds were fed by the foster parents and were fledged from the nest.

Many species continue to brood the young at night after daytime brooding has ceased. A female Painted Bunting "brooded three young constantly for eight nights and then abandoned them on the ninth—on the eve of fledging" (Parmelee, 1959). "At lower altitudes, young Scaly-breasted Hummingbirds

Figure 28. A 6-hour-old Red-headed Woodpecker (*Melanerpes erythrocephalus*), showing the use of the well-developed heel-pads. (Courtesy of Jerome A. Jackson.)

Figure 29. Two nestling Secretarybirds, 10 days old. (Courtesy of W. T. Miller.)

may sleep without the maternal coverlet when only 8 to 11 days old and still almost naked; but at greater heights where nights are frosty, White-eared Hummingbirds are brooded until they are 17 or 18 days old and clothed with feathers" (Skutch, 1967a:50). Both the male and the female Rufous-fronted Thornbird spend the night in their domed nest during the incubation and nestling periods (Skutch, 1969a).

Nest Sanitation

Although a few birds (e.g., motmots, the Hoopoe, many woodpeckers) apparently do not remove fecal sacs, the majority of passerine species dispose of them. The sacs often are eaten during the first days of the nestling's life, but later they are carried away; they may be dropped in flight or deposited on a branch. The Green Violet-ear (*Colibri thalassinus*) and the White-eared Hummingbird (*Hylocharis leucotis*) stand on the nest rim, pick up the droppings (not enclosed in a gelatinous sac), and toss them out by sideward jerks of the

Figure 30. Above: A Blue Jay (*Cyanocitta cristata*) on the day it left the nest. Below: A fledgling Brown-headed Cowbird (left) and a fledgling Rose-breasted Grosbeak (*Pheucticus ludovicianus*).

Figure 31.   A Laysan Albatross "talking" to its egg. (Courtesy of Harvey I. Fisher and the editor of *Living Bird*.)

head (Skutch, 1967a). Adult Green Woodpeckers eat the feces inside the nesting cavity during the first 12 to 14 days of the nestling period; for the next 4 or 5 days, they are either eaten or carried away; and during the last 3 or 4 days there is no nest sanitation by the female, although the male removes sacs for an additional day (Tuft, 1956).

Many species of weaverbirds (especially carduelines and estrildines) remove a decreasing number of fecal sacs as the young grow older, so that the nest rim becomes covered with them before nest-leaving time (Fig. 33). Some African and Australian ploceids are said not to remove the feces at all. Skutch found that some female hummingbirds allow the feces to accumulate on the nest rim as the young grow older.

Both aluminum and colored bands on the legs of nestlings are treated as foreign materials in the nest. Berger (1953a) described the behavior of three different female Horned Larks at nests in which the young had been banded. At one nest the female picked up a fully feathered nestling by the leg and flew some 8 yards with the bird before dropping it; she retrieved it and then flew an additional 5 yards before dropping the young bird again. Other species (e.g., Veery, Red-eyed Vireo, Purple Sunbird) have been observed to remove banded young from the nest. Rolle (1963) wrote that after a nestling Red-legged Thrush was banded, "it would receive harsh physical treatment if either parent detected the presence of the band. The parent bird would then pick up the nestling by its banded tarsus and vigorously shake the young bird in an effort to dislodge the band."

Care of Fledglings

In species in which both parents feed the nestlings, each adult may take charge of part of the brood after the young leave the nest, or the brood may be kept together and both adults feed all of the young (Fig. 34). When the interval

Figure 32. A female Redhead (*Aythya americana*) eating the eggshell of a newly hatched duckling. (Courtesy of Milton W. Weller.)

Figure 33. Nest of the House Finch (*Carpodacus mexicanus frontalis*), showing the accumulation of fecal sacs on the rim of the nest; Honolulu, Hawaii, March 25, 1972.

Figure 34. A Palm Warbler (*Dendroica palmarum*) feeding a fledgling; Schoolcraft County, Michigan, June 25, 1956. (Courtesy of Lawrence H. Walkinshaw.)

Figure 35. Recently hatched Stanley Cranes (*Anthropoides paradisea*); Natal, South Africa, December 16, 1961. (Courtesy of Lawrence H. Walkinshaw.)

between nestings is short, the male may take charge of the young while the female is incubating the second set of eggs. Thus Parmelee (1959) wrote that a male Painted Bunting "took over the brood just before egg-laying, and thereafter the female had nothing to do with the brood, so far as known." Among double-brooded American Goldfinches, the male usually takes care of the fledglings of the first brood.

Precocial species keep their downy young (Fig. 35) together and brood them; Wilcox (1959) said that Piping Plover chicks sometimes are brooded until they are 20 days old. Aquatic precocial species (swans, sheldgeese, some ducks, coots, loons, grebes) carry small young on their backs while swimming (Johnsgard and Kear, 1968). Some rails (e.g., Clapper and Virginia rails) have been observed to move their young with their bills, and the African Jacana (*Actophilornis africanus*) and the Australian Lotusbird (*Irediparra gallinacea*) carry chicks by holding them under their wings.

A different type of parent-young relationship is found among certain colonial nesting birds (e.g., some penguins, Shelduck, eider ducks, Sandwich Tern, flamingos) in which the young leave the nests after a few days and herd together to form a *crèche*. Young flamingo chicks leave their nests when 3 or 4 days old and join a crèche in which all of the young are approximately the same age.

Helpers at the nest have been described for a wide variety of world birds (Skutch, 1961; Short, 1964; Harrison 1969), and there is at least one record of a male Cardinal feeding goldfish.

Parent-young Recognition

The evidence suggests that adult passerine and other altricial species do not recognize their own young. This has been demonstrated by transferring nestlings from one nest to another. The new nestlings, of either the same or a different species, are generally accepted and fed, whereas the adults may remove their own banded young from the nest. Most New World passerines accept and rear young cowbirds, and Old World species accept young cuckoos.

Davies and Carrick (1962) found that Crested Terns do not recognize their own eggs or newly hatched chicks but do recognize their own chicks by the time they are 2 days old. Black-footed Albatrosses do not recognize their own chicks until they are about 10 days old; after that, the adults will feed only their own young. Red-tailed Tropicbirds studied by Howell and Bartholomew (1969), however, accepted substitute chicks even though they differed considerably in size and color (Fig. 36).

In his work with ducks, Gottlieb (1971) demonstrated that "the maternal call is the primary basis on which the young birds are able to discriminate a parent of their own species from a parent of another species." Peking Ducks (Fig. 37) usually hatch on the 27th day of incubation, and Gottlieb states: "On Day 26–27, the bill of all normally positioned embryos is in the air space at the large end of the egg, and the action of the bill has pipped the shell over the air space. On Day 25 the bills of most embryos are well into the air space and some embryos may have pipped the shell. On Day 24 about one-half of the embryos have begun to penetrate into the air space, and none have pipped the shell. In the usual course of events, it is at this time (Day 24) that pulmonary respiration is fairly well established and the embryos can vocalize." Gottlieb inserted electrodes into the lower mandible and into the skin subdermally near the heart in order to record bill-clapping and heartbeat; he also recorded vocalizations via a microphone; some of the embryos were devocalized. Gottlieb showed that the embryos "were overtly unresponsive" to the Mallard call on Days 20 and 21; "on Days 22 and 23 they showed an inhibition (decrease) in their bill-clapping rate; and on Days 24 to 27 they showed an increase in their rate of bill-clapping when exposed to the mallard call. . . . At all ages tested the embryos were behaviorally unresponsive to the maternal calls of other species; thus, the embryos' behavioral response to the mallard call is a discriminative reaction, and it occurs as early as Day 22 (five days before hatching). . . . Embryos which have been devocalized and kept in auditory isolation show a delay in the perfection of their postnatal discriminative response to the

Figure 36. A Red-tailed Tropicbird brooding a chick under its wing, a behavior pattern presumably found only in tropicbirds and cracids. Christmas Island, Pacific Ocean, November 6, 1971.

species maternal call. Whereas sham-operated, vocal isolated embryos show a perfect discriminative response to the mallard call in the mallard vs. wood duck call test at 24 hours after hatching, mute-isolated embryos do not show a perfect discriminative response to the mallard call in this test until 48 hours after hatching."

Mortality

The eggs and the young are subject to so many hazards that a book could be written on this topic. Nice (1957) wrote that the success rate of open nests of altricial birds in the North Temperate Zone ranged from "38 to 77 percent, averaging 49," and that "in 29 studies involving 21,951 eggs, fledgling success ranged from 22 to 70 percent, averaging 46." The survival rate for hole-nesting species averages 66 percent, but a wide variation from this average is seen for individual species: 25.7 percent for Prothonotary Warblers and 93.7 percent for Tree Swallows. Moreover, Walkinshaw (1941) reported the survival of

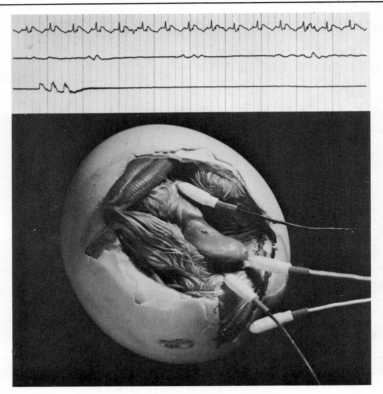

Figure 37. A Peking Duck embryo (day 25–27) with recording electrodes in place. Top line shows heartbeat, middle line depicts bill-clapping, and bottom line shows occurrence of three vocalizations (recorded via microphone above the embryo). (Courtesy of Gilbert Gottlieb and the editor of *The Quarterly Review of Biology.*)

young to be 25.7 percent for 121 Prothonotary Warbler nests over an 11-year period in Michigan, but 61.3 percent for 36 nests over a 2-year period in Tennessee. The low success for the Michigan nests was attributed largely to competition with the House Wren for nest sites. (See also Ricklefs, 1968, 1969.)

BREEDING PARASITISM

Some species have become so specialized in their reproductive behavior that they no longer build nests, incubate eggs, or care for their young but instead lay their eggs in the nests of other species, called the "hosts" or "fosterers." These parasitic species are termed *obligate parasites*. Still other species (*nonobligate parasites*) typically incubate their eggs and take care of their young but either rarely or frequently lay one or more eggs in the nest of some other bird.

Some authors use the term *nest parasitism* as a synonym for breeding parasitism, thus including both obligate and nonobligate parasites.

Nonobligate Parasites

Nonobligate parasites may be divided into two groups: *egg parasites* and *nest parasites*. Nest parasites appropriate the nests of other species in which to lay and incubate their eggs. Nest parasitism has been reported in a wide variety of species: e.g., Piratic Flycatcher (*Legatus leucophaius*), Jinete Flycatcher (*Machetornis rixosa*), Melodious Troupial (*Icterus icterus*), Banded Finch (*Poephila bichenovii*), Yellow-tailed Thornbill (*Acanthiza chrysorrhoa*), Black-headed Pardalote (*Pardalotus melanocephala*), Black-backed Magpie (*Gymnorhina tibicen*). Payne (1969) reported that "Chestnut Sparrows *Passer eminiby* near Magadi, Kenya, built no nests of their own but usurped nests newly built by Grey-capped Social Weavers *Pseudonigrita arnaudi*."

Eggs of certain species of ducks have been found so often in the nests of other species that the former have been referred to as "semiparasitic" species. Egg parasitism reaches its highest development in the Redhead (Fig. 38) and the Ruddy Duck. Weller (1959) found that as many as 13 different Redheads laid

Figure 38. A Redhead egg in the nest of an American Bittern; Delta, Manitoba. (Courtesy of Milton W. Weller.)

eggs in the same nest. Moreover, he concluded that "three distinct types of nest behavior are found in different individual females: normal nesting, semiparasitism, and complete or obligate parasitism." He estimated that from 5 to 10 percent of female Redheads are nonparasitic and nest early in the season. "All other Redhead females apparently lay eggs parasitically at some time. More than half of these hens are semiparasitic and nest after parasitizing while the remainder are probably completely parasitic."

Obligate Parasites

Obligate parasites are found in five bird families: Anatidae, Cuculidae, Indicatoridae, Icteridae, Ploceidae. Although most of the parasitic species inhabit the Old World, a few are found in Central and South America (a duck and several species of cuckoos and cowbirds). The Brown-headed and Bronzed cowbirds are the obligate parasites of North America.

A. Anatidae. The Black-headed Duck (*Heteronetta atricapilla*) of South America is entirely parasitic, laying its eggs in the nests of the Red-fronted Coot (*Fulica rufifrons*), Rosybill (*Netta peposaca*), Red-gartered Coot (*Fulica armillata*), White-faced Ibis (*Plegadis falcinellus*), and other inhabitants of marshes (Figs. 39, 40). The ducklings are remarkably independent and leave the parent host at less than 2 days of age. The Black-headed Duck "is distinctive among the five families with parasitic members because it is the only precocial species and appears to be the least damaging to its host. In this sense it is the most perfected of the brood parasites. The parasitic female does no damage to the eggs or nest of the host and, based on limited data, barely influences clutch size or nest success. Its young do not take food intended for the hosts' young. In fact, its behavior borders on commensalism rather than parasitism. . . . The Black-headed Duck has achieved success not by specializations in laying behavior or egg color but by the random placement of eggs in nests containing eggs of any color. Survival of the young in the nests of these divers is possible because the young rear themselves after only a brief period of parental care. Because it is the least damaging to the host, it may be considered the most perfect of avian parasites" (Weller, 1968).

B. Cuculidae. Although the Yellow-billed and Black-billed cuckoos occasionally lay an egg in the nest of another species, they and several other cuckoos are nonparasitic (Fig. 41). Several genera of cuckoos, however, are obligatory parasites: e.g., *Cacomantis, Cercococcyx, Chrysococcyx, Clamator, Cuculus, Eudynamis, Dromococcyx, Pachycoccyx, Surniculus,* and *Tapera.*

The eggs of many parasitic cuckoos are noted for their extreme variation in color and pattern, and in many cases the birds lay in the nests of hosts whose

Figure 39. A Black-headed Duck egg in the nest of the Red-fronted Coot. (Courtesy of Milton W. Weller and the editor of *Living Bird*.)

eggs are similar. It was suggested, therefore, that the European Cuckoo consists of "races or gentes, inherently parasitical upon particular species. Thus we have meadow-pipit cuckoos, reed-warbler cuckoos, wagtail cuckoos, etc., etc." (Chance, 1940:108; Fig. 42). Many parasitic cuckoos also lay smaller eggs, relative to body size, than do nonparasitic species.

The young of most parasitic species (but not *Clamator*) forcibly eject from the nest the eggs or other nestlings (including other cuckoos usually) so that only one cuckoo is raised in a nest (Fig. 43). Friedmann (1967) proposed the term *alloxenia* "to describe the situation wherein two or more related species of parasites tend to use different species of hosts," and *homoxenia* for the use "of the same hosts by different species of parasites." For discussions of differences in behavior among parasitic cuckoos see Friedmann (1948, 1958, 1964b, 1968b, 1969).

C. Indicatoridae. Varying amounts of information are available on the parasitic habits of 6 of the 11 species of honeyguides (Friedmann, 1955,

Figure 40. A Black-headed Duckling (lower left) in a nest with a downy White-faced Ibis. (Courtesy of Milton W. Weller and the editor of *Living Bird*.)

1968a). Some of the species differ from most parasitic birds in that they lay in the nests of hole-nesting species (bee-eaters, the Hoopoe, kingfishers). Friedmann summarized by saying that the typical honeyguides (*Indicator* and probably *Melichneutes*) "victimize picarian and coraciiform birds primarily, and only such passerine forms as utilize old nesting sites of these birds or other tunnels fairly similar to theirs." *Prodotiscus* is known to parasitize only small passerine birds. The nestlings of *Indicator* have well-developed "mandibular hooks," with which the Lesser Honeyguide, at least, may kill the host young; the hooks and the egg-tooth drop off before the nestlings are 2 weeks old. The Greater Honeyguide nestling has been observed to eject its nest mates. These species have very long nestling periods (37 to 40 days).

D. Icteridae. Some cowbirds are parasitic; others are not. The cowbirds are of special interest because the habits of the several species suggest a way by which parasitism may have evolved. The nonparasitic Bay-winged Cowbird (*Molothrus badius*) sometimes builds its own nest but usually appropriates the nest of some other species in which to lay and incubate its eggs. The parasitic Shiny Cowbird (*M. bonariensis*) exhibits what appears to be nest-building behavior at times (but rarely attempts to build), and it has "the parasitic habit very poorly developed, wasting large numbers of its eggs"; many eggs are found on the ground, and a large number of nests contain more than one cowbird egg—from 15 to 37 cowbird eggs have been found in a single nest of certain ovenbirds (Friedmann, 1929:351); one of the most common hosts is the Rufous-collared Sparrow. The Screaming Cowbird (*M. rufo-axillaris*), so far as is known, is parasitic almost exclusively on the Bay-winged Cowbird. The Giant Cowbird (*Scaphidura oryzivora;* formerly *Psomocolax*) is parasitic on other icterids (*Gymnostinops, Zarhynchus, Psarocolius, Cacicus;* Skutch, 1954:316; Smith, 1968).

Parasitic cowbirds do not develop incubation patches or broodiness behavior. In nonparasitic species, these events are stimulated by prolactin secreted by the

Figure 41. A Sirkeer Cuckoo (*Taccoucua leschenaultii*) at its nest. (Courtesy of Shivrajkumar of Jasdan, India.)

Figure 42. A Meadow Pipit (*Anthus pratensis*) feeding a fledgling European Cuckoo (*Cuculus canorus*). (Courtesy of Eric Hosking.)

pituitary gland. Höhn (1959) demonstrated that "the prolactin content of cowbirds' and red-winged blackbirds' pituitaries is fairly similar," and he concluded that "failure to respond to prolactin rather than lack of prolactin production is involved in the failure of female cowbirds to form brood patches." Selander and Kuich (1963) extended this research and confirmed the conclusion that "the integument of the cowbird is refractory to hormonal stimulation." They also were unsuccessful in stimulating nest-building and incubation behavior by hormonal injection. They concluded that "refractoriness of the integument and of presumed neural mechanisms mediating nesting and incubation behavior to hormones is regarded as a consequence of natural selection once birds are in an environmental situation in which failure to incubate is adaptive. The hypothesis that 'endocrine imbalance' is a causal factor in the evolution of brood parasitism is rejected." Hamilton and Orians (1965) presented a stimulating paper on the evolution of parasitism and suggested the types of research that are needed for a better understanding of brood parasitism.

Eggs of the Bronzed Cowbird (*Molothrus aeneus;* formerly *Tangavius*) have been found in the nests of some 62 species of other birds; little is known of its habits. The Brown-headed Cowbird (*Molothrus ater*) is known to have parasitized more than 200 species (350 species and subspecies) of birds (Friedmann, 1963, 1966, 1971). Some are abnormal or accidental hosts (e.g.,

Figure 43. A nestling European Cuckoo ejecting eggs from the nest of a Tree Pipit (*Anthus trivialis*). (Courtesy of Eric Hosking.)

Blue-winged Teal, Spotted Sandpiper, Killdeer, Wilson's Phalarope, Mourning Dove). For the majority of the remaining species, little information is available regarding the incidence of parasitism or the effect on the host species (Fig. 44); not all parasitized species raise the cowbirds to the fledging stage, but here again the data are scanty (Friedmann, 1963:39).

Female Brown-headed Cowbirds apparently lay 1 egg each day and two or more clutches during a breeding season. Walkinshaw (1949) believed that individual cowbirds laid 25 eggs in one season. Payne (1965) studied sections of a large number of ovaries and concluded that "cowbirds lay in clutches of one to six eggs," and that "most cowbirds lay several clutches in a season. Times between clutches range from a few days to a few weeks"; "they lay an average

Figure 44. Three eggs of the Slaty Vireo (*Neochloe brevipennis*) and an egg of the Brown-headed Cowbird, collected near Cuernavaca, Mexico, on June 13, 1958, by J. Stuart Rowley; these were the first eggs of this vireo to be described. (Courtesy of J. Stuart Rowley and the editor of *Condor.*)

Figure 45. A female Brown-headed Cowbird at the nest of an Ovenbird (*Seiurus aurocapillus*), photographed at 4:57 A.M., May 25, 1939, by the late Harry W. Hann; this was the first photograph taken of a cowbird in the process of egg-laying. (Courtesy of the late Harry W. Hann.)

of 10 to 12 eggs during the breeding season. Variation between birds is great, as some lay no eggs while others lay at least 15." How accurately such microscopical studies reveal how many eggs have actually been laid, however, remains to be determined. In his work with sectioned ovaries of the Starling, Davis (1958) found that "for 32 birds the number of ovulations was the same as the clutch-size in only 7 cases. The ovulations were greater in 16 cases and smaller in 9."

Female cowbirds find nests by watching other birds and by actively searching for nests. They usually parasitize open nests but sometimes lay in the nests of cavity-nesting species. The cowbird usually lays her eggs before sunrise; she often removes an egg from the nest on the same day or on the next day, usually during midmorning but occasionally in the afternoon (Hann, 1941; Fig. 45). Cowbirds often lay asynchronously, i.e., before or after the egg-laying period of the host species; they may lay during the nest-building stage or after the young have hatched. Two or more cowbirds frequently lay in the same nest, and as many as eight cowbird eggs have been found in a single nest (that of a Black-and-White Warbler).

Five passerine families provide most of the cowbird's hosts: tyrant-flycatchers, thrushes, vireos, wood-warblers, finches. Most of the species have not evolved a "defense" against cowbird parasitism. Three reactions exhibited by some hosts, however, are detrimental to the cowbird: ejecting cowbird eggs from the nest, deserting a parasitized nest, and building a new floor over cowbird eggs. The Yellow Warbler commonly covers cowbird eggs (as well as its own eggs) by building a new lining over them; the Willow Flycatcher, American Goldfinch, and a few other species rarely cover cowbird eggs. Double- and triple-storied Yellow Warbler nests are not unusual; Berger found a six-storied nest that held 11 cowbird eggs. In his intensive study of the Yellow Warbler, Daniel S. McGeen found that 86.5 percent of nonsynchronized cowbird eggs were deserted or covered but that only 22.5 percent of synchronized eggs were so treated. Despite the loss of cowbird eggs because of poor synchronization with the host's laying period and the defensive behavior of some hosts, the cowbird is a very successful species. Destruction of host young results primarily from the rapid growth of the cowbird. Hann reported that the weight of an Ovenbird is 81 percent of the cowbird's weight at hatching, but only 53 percent at nest-leaving time. The average weight of the newly hatched cowbird is about 3.25 grams, and a newly hatched Kirtland's Warbler weighs approximately 1.5 grams; when 3 days old, however, the cowbird weighs about 12.7 grams, about the same as an adult warbler.

The incubation period of many hosts is longer than that of the cowbird (average about 11.6 days). The young of most hosts have little chance of fledging unless they hatch before the cowbird or on the same day, as happens when the cowbird lays after the host has completed her clutch; host nestlings sometimes survive when a single cowbird hatches first. Cowbirds, on the other hand, often survive even though they hatch several days after the host young. As a rule, only one or two cowbirds are fledged from a single nest, but there is a record of four cowbirds fledging from a Song Sparrow nest. Several observers have suggested that, in successful parasitized nests, each cowbird is raised at the expense of one host young, but this must be accepted as the broadest of generalizations. In an evolutionary sense, the best hosts for the cowbird should be those that are able to raise both cowbirds and host young.

The amount of cowbird pressure on a species depends on the size of the cowbird population and on the total breeding population of all potential hosts in a given area (McGeen, 1972). It depends also on the relationship of the host's breeding season to that of the cowbird; early and late nests of multibrooded species may escape parasitism entirely. Single-brooded species, as well, may succeed in raising a brood of their own young by renesting after the destruction of a parasitized nest. Hence a simple statement of the percentage of parasitized nests of a species tells little about the effect of parasitism on the production of host young by that species, which is the critical factor. Moreover, one must consider the effects of parasitism over the entire breeding range of a species as well as on local populations.

E. Ploceidae. The weaverbirds include several parasitic genera, about which relatively little is known (Friedmann, 1962; Payne, 1973). Most taxonomists place three of the genera (*Hypochera, Vidua, Steganura*) in the subfamily Viduinae (some recognize only the genus *Vidua*) and the other genus (*Anomalospiza*) in the subfamily Ploceinae. "Whereas the parasitic viduine weavers confine their attention largely to related estrildine species, the cuckoo finch [*Anomalospiza*], insofar as is known, victimizes grass warblers exclusively. Undoubtedly additional host species will be added with further field studies, but the records available to date clearly point to the sylviine genera *Cisticola* and *Prinia* as the mainstays of the cuckoo finch" (Friedmann, 1960:45).

REFERENCES

Alvarez del Toro, M. 1971. On the biology of the American Finfoot in southern Mexico. *Living Bird,* **1971**:79–88.

Amadon, D. 1964. The evolution of low reproductive rates in birds. *Evolution,* **18**:105–110.

Ames, P. L. 1966. DDT residues in the eggs of the Osprey in the north-eastern United States and their relation to nesting success. *J. Appl. Ecol.,* Suppl., **3**:87–97.

Anderson, D. W., and J. J. Hickey. 1970. Oological data on egg and breeding characteristics of Brown Pelicans. *Wilson Bull.,* **82**:14–28.

Ashmole, N. P., and M. J. Ashmole. 1967. Comparative feeding ecology of sea birds of a tropical oceanic island. *Peabody Mus. Nat. Hist. Bull.* No. 24.

———, and Humberto Tovar S. 1968. Prolonged parental care in Royal Terns and other birds. *Auk,* **85**:90–100.

Baerends, G. P., and R. H. Drent, eds. 1970. The Herring Gull and its egg. *Behaviour,* Suppl. 17.

Bartholomew, G. A., and T. R. Howell. 1964. Experiments on nesting behavior of Laysan and Black-footed albatrosses. *Anim. Behav.,* **12**:549–559.

Barton, A. J. 1958. A releaser mechanism in the feeding of nestling Chimney Swifts. *Auk,* **75**:216–217.

Beer, C. G. 1962. The egg-rolling of Black-headed Gulls *Larus ridibundus. Ibis,* **104**:388–398.

Bellrose, F. C., T. G. Scott, A. S. Hawkins, and J. B. Low. 1961. Sex ratios and age ratios in North American ducks. *Ill. Nat. Hist. Surv. Bull.,* **27**:391–474.

Berger, A. J. 1953a. Reaction of Horned Larks to banded young. *Bird-Banding,* **24**:19–20.

——— 1953b. Green frog catches young Phoebe. *Ibid.,* **24**:67–68.

——— 1953c. Three cases of twin embryos in passerine birds. *Condor,* **55**:157–158.

——— 1967. Traill's Flycatcher in Washtenaw County, Michigan. *Jack-Pine Warbler,* **45**:117–123.

——— 1968. Clutch size, incubation period, and nestling period of the American Goldfinch. *Auk,* **85**:494–498.

—— 1972. *Hawaiian Birdlife*. University Press of Hawaii, Honolulu.

Bock, W. J., and R. S. Hikida. 1969. Turgidity and function of the hatching muscle. *Am. Midl. Nat.*, **81**:99–106.

Brockway, B. F. 1968. Budgerigars are not determinate egg-layers. *Wilson Bull.*, **80**:106–107.

Broekhuysen, G. J. 1963. The breeding biology of the Orange-breasted Sunbird *Anthobaphes violacea* (Linnaeus). *Ostrich*, **34**:187–234.

Brooks, W. S., and S. E. Garrett. 1970. The mechanism of pipping in birds. *Auk*, **87**:458–466.

Chance, E. P. 1940. *The Truth about the Cuckoo*. Country Life, London.

Clark, G. A., Jr. 1961. Occurrence and timing of egg teeth in birds. *Wilson Bull.*, **73**:268–278.

—— 1964a. Ontogeny and evolution in the megapodes (Aves: Galliformes). *Postilla*, **78**:1–37.

—— 1964b. Life histories and the evolution of megapodes. *Living Bird*, **1964**:149–167.

—— 1969. Oral flanges of juvenile birds. *Wilson Bull.*, **81**:270–279.

Cody, M. L. 1966. A general theory of clutch size. *Evolution*, **20**: 174–184.

Collins, C. T., and M. LeCroy. 1972. Analysis of measurements, weights, and composition of Common and Roseate tern eggs. *Wilson Bull.*, **84**:187–192.

Coppinger, R. P. 1970. The effect of experience and novelty on avian feeding behavior with reference to the evolution of warning coloration in butterfies. II. Reactions of naïve birds to novel insects. *Am. Nat.*, **104**:323–335.

Cott, H. B. 1951, 1954. The palatability of the eggs of birds. . . . *Proc. Zool. Soc. London*, **121**:1–41; **124**:335–463.

—— 1953, 1954. The exploitation of wild birds for their eggs. *Ibis*, **95**:409–449, 643–675; **96**:129–149.

Coulson, J. C., G. R. Potts, and J. Horobin. 1969. Variation in the eggs of the Shag (*Phalacrocorax aristotelis*). *Auk*, **86**:232–245.

Coulter, M. W. 1957. Predation by snapping turtles upon aquatic birds. *J. Wildl. Manage.*, **21**:17–21.

Davies, S. J. J. F., and R. Carrick. 1962. On the ability of Crested Terns, *Sterna bergii*, to recognize their own chicks. *Aust. J. Zool.*, **10**:171–177.

Davis, D. E. 1958. Relation of "clutch-size" to number of ova ovulated by Starlings. *Auk*, **75**:60–66.

Driver, P. M. 1965. "Clicking" in the egg-young of nidifugous birds. *Nature*, **4981**:315.

Eddinger, C. R. 1970. The White-eye as an interspecific feeding helper. *Condor*, **72**:240.

Ellis, J. A., and R. F. Labisky. 1966. Soft-shelled eggs in a Bobwhite nest. *Wilson Bull.*, **78**:229.

Ficken, M. S. 1965. Mouth color of nestling passerines and its use in taxonomy. *Wilson Bull.*, **77**:71–75.

Fisher, H. I. 1958. The "hatching muscle" in the chick. *Auk*, **75**:391–399.

—— 1966. Hatching and the hatching muscle in some North American ducks. *Trans. Ill. State Acad. Sci.*, **59**:305–325.

—— 1969. Eggs and egg-laying in the Laysan Albatross, *Diomedea immutabilis*. *Condor*, **71**:102–112.

—— 1971. The Laysan Albatross: its incubation, hatching, and associated behaviors. *Living Bird*, **1971**:19–78.

Fraps, R. M. 1970. Photoregulation in the ovulation of the domestic fowl. In *La Photorégulation de la Reproduction chez les Oiseaux et les Mammiferes*, Editions du Centre National de la Recherche Scientifique, Paris.

Fredrickson, L., and M. W. Weller. 1972. Responses of Adelie Penguins to colored eggs. *Wilson Bull.,* **84:**309–314.

Friedmann, H. 1929. *The Cowbirds.* Charles C Thomas, Springfield, Ills.

—— 1948. The parasitic cuckoos of Africa. *Wash. Acad. Sci. Monogr.* No. 1, Washington, D.C.

—— 1955. The honey-guides. *U.S. Natl. Mus. Bull.,* **208.**

—— 1958. Further data on African parasitic cuckoos. *Proc. U.S. Natl. Mus.,* **106:**377–408.

—— 1960. The parasitic weaverbirds. *U.S. Natl. Mus. Bull.,* **223.**

—— 1962. The problem of the Viduinae in the light of recent publications. *Smithson. Misc. Collect.,* **145,** No. 3.

—— 1963. Host relations of the parasitic cowbirds. *U.S. Natl. Mus. Bull.,* **233.**

—— 1964a. The history of our knowledge of avian brood parasitism. *Centaurus,* **10:**282–304.

—— 1964b. Evolutionary trends in the avian genus *Clamator. Smithson. Misc. Collect.,* **146,** No. 4.

—— 1966. Additional data on the host relations of the parasitic cowbirds. *Ibid.,* **149,** No. 11.

—— 1967. Allonexia in three sympatric African species of *Cuculus. Proc. U.S. Natl. Mus.,* **124,** No. 3633.

—— 1968a. Additional data on brood parasitism in the Honey-guides. *Ibid.,* **124,** No. 3648.

—— 1968b. The evolutionary history of the avian genus *Chrysococcyx. U.S. Natl. Mus. Bull.,* **265.**

—— 1969. Additions to knowledge of the Yellow-throated Glossy Cuckoo, *Chrysococcyx flavigularis. J. Ornithol.,* **110:**176–180.

—— 1971. Further information on the host relations of the parasitic cowbirds. *Auk,* **88:**239–255.

Frith, H. J. 1962. *The Mallee-fowl.* Angus and Robertson, Sydney.

Gavrilov, E. L. 1972. On the sex ratio in the Spanish Sparrow, *Passer hispaniolensis* Temm. *Int. Stud. Sparrows,* Warsaw, **6:**11–23.

George, J. C., and P. T. Iype. 1963. The mechanism of hatching in the chick. *Pavo,* **1:**52–56.

Gibb, J. A. 1968. The evolution of reproductive rates: are there no rules? *Proc. N. Z. Ecol. Soc.,* **15:**1–6.

—— 1970. The turning down of marked eggs by Great Tits. *Bird-Banding,* **41:**40–41.

Goertz, J. W., and K. Rutherford. 1972. Adult Carolina Chickadee carries young. *Wilson Bull.,* **84:**205–206.

Gottlieb, G. 1971. *Development of Species Identification in Birds.* University of Chicago Press, Illinois.

Graber, R. R. 1955. Artificial incubation of some non-galliform eggs. *Wilson Bull.,* **67:**100–109.

Greenwood, R. J. 1969. Mallard hatching from an egg cracked by freezing. *Auk,* **86:**752–754.

Hamilton, W. J., III, and G. H. Orians. 1965. Evolution of brood parasitism in altricial birds. *Condor,* **67:**361–382.

Hammar, B. 1970. The karyotypes of thirty-one birds. *Hereditas,* **65:**29–58.

Hann, H. W. 1941. The Cowbird at the nest. *Wilson Bull.,* **53:**209–221.

Hanson, H. C. 1959. The incubation patch of wild geese: its recognition and significance. *Arctic,* **12:**139–150.

Harrison, C. J. O. 1969. Helpers at the nest in Australian birds. *Emu,* **69:**30–40.

Hartshorne, J. M. 1962. Behavior of the Eastern Bluebird at the nest. *Living Bird,* **1962:**131–149.

Heath, R. G., J. W. Spann, and J. F. Kreitzer. 1969. Marked DDE impairment of Mallard reproduction in controlled studies. *Nature,* **224:**47–48.

Hickey, J. J. 1955. Some American population research on gallinaceous birds. In *Recent Studies in Avian Biology,* University of Illinois Press, Urbana, pp. 326–396.

———— 1969. *Peregrine Falcon Populations, Their Biology and Decline.* University of Wisconsin Press, Madison.

Höhn, E. O. 1959. Prolactin in the cowbird's pituitary in relation to avian brood parasitism. *Nature,* **184:**2030.

———— 1967. Observations on the breeding biology of Wilson's Phalarope (*Steganopus tricolor*) in central Alberta. *Auk,* **84:**220–244.

Howell, T. R., and G. A. Bartholomew. 1969. Experiments on nesting behavior of the Red-tailed Tropicbird, *Phaethon rubricauda. Condor,* **71:**113–119.

Hunt, J. H. 1971. A field study of the Wrenthrush, *Zeledonia coronata. Auk,* **88:**1–20.

Irving, L., and J. Krog. 1956. Temperature during the development of birds in Arctic nests. *Physiol. Zool.,* **29:**195–205.

Jackson, J. A. 1970. Observations at a nest of the Red-headed Woodpecker. *Niobrara 1968–1969,* University of Kansas Museum of Natural History, pp. 3–10.

Johnsgard, P. A., and J. Kear. 1968. A review of parental carrying of young by waterfowl. *Living Bird,* **1968:**89–102.

Kendeigh, S. C. 1952. Parental care and its evolution in birds. *Ill. Biol. Monogr.* No. 22, University of Illinois Press, Urbana.

———— 1963a. Thermodynamics of incubation in the House Wren, *Troglodytes aedon. Proc. XIIIth Int. Ornithol. Congr.,* pp. 884–904.

———— 1963b. New ways of measuring the incubation period of birds. *Auk,* **80:**453–461.

————, T. C. Kramer, and F. Hamerstrom. 1956. Variations in egg characteristics of the House Wren. *Ibid.,* **73:**42–65.

Kilham, L. 1957. Egg-carrying by the Whip-poor-will. *Wilson Bull.,* **69:**113.

Kuroda, N. 1963. A comparative study on the chemical constituents of some bird eggs and the adaptive significance. *Misc. Rept. Yamashina Inst. Ornithol. Zool.,* **3:**311–333. (Japanese; English summary.)

Lack, D. 1954. *The Natural Regulation of Animal Numbers.* Oxford University Press, London.

———— 1968. *Ecological Adaptations for Breeding in Birds.* Methuen & Co., London.

Landauer, W. 1961. The hatchability of chicken eggs as influenced by environment and heredity. *Univ. Conn. Agric. Exp. Sta. Monogr.* No. 1.

Lehrman, D. S. 1961. Hormonal regulation of parental behavior in birds and infrahuman mammals. In *Sex and Internal Secretions,* 3rd ed., Williams & Wilkins Co., Baltimore.

Long, C. A. 1963. Production of sterile eggs in the Dickcissel. *Wilson Bull.,* **75:**456.

Longcore, J. R., F. B. Samson, J. F. Kreitzer, and J. W. Spann. 1971. Changes in mineral composition of eggshells from Black Ducks and Mallards fed DDE in the diet. *Bull. Environ. Contam. Toxicol.,* **6:**345–350.

McGeen, D. S. 1972. Cowbird-host relationships. *Auk,* **89:**360–380.

————, and J. J. McGeen. 1968. The Cowbirds of Otter Lake. *Wilson Bull.,* **80:**84–93.

McKinney, D. F. 1954. An observation on Redhead parasitism. *Wilson Bull.,* **66:**146–148.

McLane, M. A. Ross, and L. C. Hall. 1972. DDE thins Screech Owl eggshells. *Bull. Environ. Contam. Toxicol.*, **8**:65–68.

Masui, K. 1967. *Sex Determination and Sexual Differentiation in the Fowl.* University of Tokyo Press, Tokyo.

Meier, A. H., and R. MacGregor III. 1972. Temporal organization in avian reproduction. *Am. Zool.*, **12**:257–271.

Mewaldt, L. R. 1956. Nesting behavior of the Clark Nutcracker. *Condor*, **58**:3–23.

Nice, M. M. 1949. The laying rhythm of Cowbirds. *Wilson Bull.*, **61**:231–234.

———— 1953. The question of ten-day incubation periods. *Ibid.*, **65**:81–93.

———— 1954a. Problems of incubation periods in North American birds. *Condor*, **56**:173–197.

———— 1954b. Incubation periods throughout the ages. *Centaurus*, **3**:311–359.

———— 1957. Nesting success in altricial birds. *Auk*, **74**:305–321.

———— 1962. Development of behavior in precocial birds. *Trans. Linn. Soc. N.Y.*, **8**.

Nickell, W. P. 1966. Ring-necked Pheasants hatch in nest of Blue-winged Teal. *Wilson Bull.*, **78**:472–474.

Norton, D. W. 1972. Incubation schedules of four species of calidridine sandpipers at Barrow, Alaska. *Condor*, **74**:164–176.

Oring, L. W. 1964. Egg moving by incubating ducks. *Auk*, **81**:88–89.

Parkes, K. C., and G. A. Clarke, Jr. 1964. Additional records of avian egg teeth. *Wilson Bull.*, **76**:147–154.

Parmelee, D. F. 1959. The breeding behavior of the Painted Bunting in southern Oklahoma. *Bird-Banding*, **30**:1–18.

———— 1970. Breeding behavior of the Sanderling in the Canadian High Arctic. *Living Bird*, **1970**:97–146.

Parsons, J. 1972. Egg size, laying date and incubation period in the Herring Gull. *Ibis*, **114**:536–541.

Payne, R. B. 1965. Clutch size and numbers of eggs laid by Brown-headed Cowbirds. *Condor*, **67**:44–60.

———— 1969. Nest parasitism and display of Chestnut Sparrows in a colony of Grey-capped Social Weavers. *Ibis*, **111**:300–307.

———— 1973. Behavior, mimetic songs and song dialects, and relationships of the parasitic indigo-birds (*Vidua*) of Africa. *Ornithol. Monogr.* No. 11.

Peek, F. W., E. Franks, and D. Case. 1972. Recognition of nest, eggs, nest site, and young in female Red-winged Blackbirds. *Wilson Bull.*, **84**:243–249.

Petersen, A. J. 1955. The breeding cycle in the Bank Swallow. *Wilson Bull.*, **67**:235–286.

Phillips, C. L. 1887. Egg-laying extraordinary in *Colaptes auratus*. *Auk*, **4**:346.

Pikula, J. 1971. Die Variabilität der Eier der Population Turdus philomelos, Brehm 1931 in der CSSR. *Zool. Listy*, **20**:69–83. (English summary.)

Pitelka, F. A., P. Q. Tomich, and G. W. Treichel. 1955. Ecological relations of jaegers and owls as lemming predators near Barrow, Alaska. *Ecol. Monogr.*, **25**:85–117.

Pitman, C. R. S. 1957. Further note on aquatic predators of birds. *Bull. Brit. Ornithol. Club.*, **77**:105–110.

Preston, F. W. 1968. The shapes of birds' eggs: mathematical aspects. *Auk*, **85**:454–463.

———— 1969. Shapes of birds' eggs: extant North American families. *Ibid.*, **86**:246–264.

Rahn, H., and A. Ar. 1974. The avian egg: incubation time and water loss. *Condor.* **76**:147–152.

Ratcliffe, F. N. 1958. Factors involved in the regulation of mammal and bird populations. *Aust. J. Sci.,* **21**:79–87.

Rice, D. W., and K. W. Kenyon. 1962. Breeding cycles and behavior of Laysan and Black-footed albatrosses. *Auk,* **79**:517–567.

Richardson, F. 1967. Black Tern nest and egg moving experiments. *Murrelet,* **48**:52–56.

Ricklefs, R. E. 1968. Patterns of growth in birds. *Ibis,* **110**:419–451.

———— 1969. An analysis of nesting mortality in birds. *Smithson. Contrib. Zool.,* No. 9. (See also *Auk,* **87**, 1970:826–828.)

Rolle, F. J. 1963. Life history of the Red-legged Thrush (*Mimocichla plumbea ardosiacea*) in Puerto Rico. *Stud. Fauna Curaçao Other Carribbean Ids.,* **14**:1–40.

Romanoff, A. L., and A. J. Romanoff. 1949. *The Avian Egg.* John Wiley & Sons, New York.

Rowley, I. 1965. The life history of the Superb Blue Wren *Malurus cyaneus. Emu,* **64**:251–297.

Ryder, J. P. 1970. A possible factor in the evolution of clutch size in Ross' Goose. *Wilson Bull.,* **82**:5–13.

Rysgaard, G. N. 1944. A Chuck-wills-widow carrying an egg. *Auk,* **61**:138.

Schönwetter, M., and W. Meise. 1960–1975. *Handbuch der Oologie.* Akademie-Verlag, Berlin. (See *Auk,* **80**, 1963:390–391.)

Sealy, S. G. 1970. Egg teeth and hatching methods in some alcids. *Wilson Bull.,* **82**:289–293.

Selander, R. K. 1960. Sex ratio of nestlings and clutch size in the Boat-tailed Grackle. *Condor,* **62**:34–44.

————, and L. L. Kuich. 1963. Hormonal control and development of the incubation patch in icterids, with notes on behavior of cowbirds. *Ibid.,* **65**:73–90.

————, and S. Y. Yang. 1966a. Behavioral responses of Brown-headed Cowbirds to nests and eggs. *Auk,* **83**:207–232.

————, and ———— 1966b. The incubation patch of the House Sparrow, *Passer domesticus* Linnaeus. *Gen. Comp. Endocrinol.,* **6**:325–333.

Short, L. L., Jr. 1964. Extra helpers feeding young of Blue-winged and Golden-winged warblers. *Auk,* **81**:428–430.

Skutch, A. F. 1945. Incubation and nestling periods of Central American birds. *Auk,* **62**:8–37.

———— 1952. On the hour of laying and hatching of birds' eggs. *Ibis,* **94**:49–61.

———— 1954. Life histories of Central American birds. *Pacific Coast Avifauna,* No. 31.

———— 1957. The incubation patterns of birds. *Ibis,* **99**:69–93.

———— 1961. Helpers among birds. *Condor,* **63**:198–226.

———— 1962. The constancy of incubation. *Wilson Bull.,* **74**:115–152.

———— 1965. Life history of the Long-tailed Silky-Flycatcher, with notes on related species. *Auk,* **82**:375–426.

———— 1967a. Life histories of Central American highland birds. *Publ. Nuttall Ornithol. Club,* No. 7.

———— 1967b. Adaptive limitation of the reproductive rate of birds. *Ibis,* **109**:579–599.

———— 1969a. A study of the Rufous-fronted Thornbird and associated birds. *Wilson Bull.,* **81**:5–43, 123–139.

———— 1969b. Life histories of Central American birds. III. *Pacific Coast Avifauna,* No. 35.

Sladen, W. J. L. 1958. The pygoscelid penguins. *Falkland Isl. Depend. Surv. Sci. Rept.*, No. 17, 97 pp.

——, C. M. Menzie, and W. L. Reichel. 1966. DDT residues in Adelie Penguins and a crabeater seal from Antarctica: ecological implications. *Nature,* **210:**670–673.

Smith, N. G. 1968. The advantage of being parasitized. *Nature,* **219:**690–694.

Stonehouse, B. 1966. Egg volumes from linear dimensions. *Emu,* **65:**227–228.

Switzer, B. and V. Lewin. 1971. Shell thickness, DDE levels in eggs, and reproductive success in common terns (*Sterna hirundo*) in Alberta. *Can. J. Zool.,* **49:**69–73.

Tinbergen, N. 1953. *The Herring Gull's World.* Collins, London.

Truslow, F. K. 1967. Egg-carrying by the Pileated Woodpecker. *Living Bird,* **1967:**227–236.

Tuft, H. R. 1956. Nest-sanitation and fledging of the Green Woodpecker. *Brit. Birds.,* **49:**32–36.

Udakawa, T. 1956. Karyogram studies in birds. VIII. The chromosomes of some species of Turdidae and Troglodytiidae (*sic*) *Jap. J. Zool.,* **12:**105–111.

Van Tienhoven, A. 1961. Endocrinology of reproduction in birds. In *Sex and Internal Secretions,* 3rd ed., Williams & Wilkins Co., Baltimore.

Victoria, J. K. 1972. Clutch characteristics and egg discriminative ability of the African Village Weaverbird *Ploceus cucullatus. Ibis,* **114:**367–376.

Vosburgh, F. G. 1948. Easter egg chickens. *Natl. Geogr. Mag.,* **94:**377–387.

Walker, L. W. 1951. Sea birds of Isla Raza. *Natl. Geogr. Mag.,* **49:**239–248.

Walkinshaw, L. H. 1941. The Prothonotary Warbler, a comparison of nesting conditions in Tennessee and Michigan.*Wilson Bull.,* **53:**3–21.

—— 1949. Twenty-five eggs apparently laid by a Cowbird. *Wilson Bull.,* **61:**82–85.

—— 1952. Chipping Sparrow notes. *Bird-Banding,* **23:**101–108.

Weller, M. W. 1959. Parasitic egg laying in the Redhead (*Aythya americana*) and other North American Anatidae. *Ecol. Monogr.,* **29:**333–365.

—— 1967. Distribution and habitat selection of the Black-headed Duck (*Heteronetta atricapilla*). *El Hornero,* **10:**299–306.

—— 1968. The breeding biology of the parasitic Black-headed Duck. *Living Bird,* **1968:**169–207.

—— 1971. Experimental parasitism of American Coot nests. *Auk,* **88:**108–115.

White, F. N., and J. L. Kinney. 1974. Avian incubation. *Science,* **186:**107–115.

Wiemeyer, S. N., *et al.* 1972. Residues of organochlorine pesticides, polychlorinated biphenyls, and mercury in Bald Eagle eggs and changes in shell thickness,—1969 and 1970. *Pestici. Monit. J.,* **6:**50–55.

Wiens, J. A. 1971. "Egg-dumping" by the Grasshopper Sparrow in the Savannah Sparrow nest. *Auk,* **88:**185–186.

Wilcox, Le Roy. 1959. A twenty year banding study of the Piping Plover. *Auk,* **76:**129–152.

Winterbottom, J. M. 1955. The incubation period of the Cape White-eye. *Wilson Bull.,* **67:**135–136.

Witschi, E. 1956. *Development of Vertebrates.* W. B. Saunders Co., Philadelphia.

CHAPTER TWELVE

Taxonomy and Nomenclature

Taxonomy or systematics is the science of classifying the forms of life. Nomenclature is a system of names; it permits the interchange of information among biologists. Modern systems of classification result from man's attempt to interpret information on extinct and living animals. Insofar as the attempt is successful, the classification will be a natural one, i.e., will express the phylogenetic relationships of animals.

Linnaeus (1707–1778) founded the binominal system of Latin names, and the 10th edition (1758) of his *Systema Naturae Regnum Animale* is the starting point for zoological nomenclature. In this work, Linnaeus adopted two Latin words (in place of many) for naming animals: one name to indicate the general kind (the *genus*) of animal; the other, the particular kind (the *species*).

An International Commission on Zoological Nomenclature was established in 1901. The Commission publishes the *Bulletin of Zoological Nomenclature,* in which the latest decisions are announced. A new International Code of Zoological Nomenclature was published in 1964.

The principal provisions of the International Code may be summarized as follows (for details see Blackwelder, 1967; Mayr, 1969).

1.   Zoological nomenclature is independent of botanical nomenclature.

2.   Family and subfamily names are formed by adding -idae and -inae, respectively, to the stem of the name of the type genus. The 1961 Code recommended that superfamily and tribal names be formed by adding -oidea and -ini, respectively. Before 1960, family and subfamily names had to be changed if the name of the type genus was changed or if, on the basis of priority, a new type genus was designated (e.g., the family name of the American wood-warblers has been, in turn, Mniotiltidae, Compsothlypidae, and Parulidae). This is no longer the rule: the older synonym is now called a *senior synonym*; a younger synonym, the *junior synonym* (see Blackwelder, 1967:197).

3.   Names of subgenera and all higher categories are uninominal; names of species are binominal; of subspecies, trinominal. All names should be Latin or Greek words, but names from other languages or even arbitrary combinations of letters are not invalid. The generic name must be unique in zoology. A specific or subspecific name must be unique in the genus; it should be a Latin or Latinized adjective, noun, or patronym (name of a person). Generic and subgeneric names are capitalized; specific and subspecific names are lower case (formerly, patronyms often were capitalized). The generic, specific, and subspecific names are underlined in manuscripts, and usually italicized in published works or at least presented in a different type from the text.

4.   The *law of priority* states that the generic and specific names of a particular animal shall be the names proposed by Linnaeus in 1758, or the names first proposed after that date. However, a recognizable description or diagnosis of the animal must be published together with a binominal or trinominal name (preferably but not necessarily appropriate).

5.   The author of a name is the one who first publishes it with a suitable description of the animal. The author's name follows the species name without punctuation (and without being italicized) unless the species has been transferred to a different genus, in which event the author's name is enclosed in parentheses.

THE TAXONOMIC HIERARCHY

The framework of classification is a hierarchy consisting of certain obligatory groups: kingdom, phylum, class, order, family, genus, and species. In order to express natural relationships clearly, however, other categories often are used. These are formed by adding the prefixes super-, sub-, and infra-, to certain terms of the basic hierarchy. The classification of the Stanley Crane (Fig. 1) in

Figure 1. A Stanley Crane (*Balearica p. pavonina*) near its nest; Nigeria, August 15, 1965. (Courtesy of Lawrence H. Walkinshaw.)

hierarchic order is:

Kingdom	Animalia
Phylum	Chordata
Class	Aves
Subclass	Neornithes
Superorder	Neognathae
Order	Gruiformes
Family	Gruidae
Genus	*Balearica*
Species	*pavonina*

A systematist "creates" a genus in order to express the similarity among species of birds—a similarity that he believes is due to a common ancestry. Thus several kinds of nuthatches are placed in the genus *Sitta* to suggest that they evolved from a common ancestor and therefore are more closely related to each other than to any other birds. Hence the genus is a concept. The species are the actual populations of animals; species exhibit diversity within a genus.

A species is a population or group of populations of similar individuals occupying a definite (and usually continuous) geographical range and breeding among themselves (or believed able to do so) but normally not breeding with individuals of other species. Within a species, especially one that occupies a large range, differences between the various populations sometimes are so distinct that isolated populations are recognized as *geographic races* or *subspecies* (Mayr, 1963:348; Fig. 2).

CLIMATIC RESPONSES OF POPULATIONS

In a species with an extensive geographical range, morphological characters often exhibit a gradual change from one extremity of the range to the other. Huxley (1942:206) coined the term *cline* to refer to such character gradients.

For expressing other correlations between environment and structure, several "rules" have been proposed. All have exceptions, but support for each rule is found among various subspecies (and, rarely, some species) of birds.

Bergmann's Rule

"Within a polytypic warm-blooded species, the body-size of a subspecies usually increases with decreasing mean temperature of its habitat" (Huxley, 1942:211). In practice, this "decreasing mean temperature" means increasing latitude or altitude (see Kendeigh, 1969; Lasiewski *et al.*, 1964). Examples are found among the subspecies of many well-known birds: Bald Eagle, Red-tailed

Figure 2. Drawings of the heads of North American Peregrine Falcons. (*A*) Arctic Peregrine (*Falco peregrinus anatum*) from Banks Island, Canadian Arctic. (*B*) *F. p. anatum* from Alaska. (*C*) Peale's Falcon (*F. p. pealei*) from Langara Island, British Columbia. (Courtesy of Frank L. Beebe and the editor of *Condor*.)

Hawk, American Kestrel, Bobwhite, Great-horned Owl, Screech Owl, Hairy Woodpecker, Horned Lark, Red-winged Blackbird, Song Sparrow. Exceptions to the rule include the Sandhill Crane and the Phainopepla.

Allen's Rule

"In warm-blooded species, the relative size of exposed portions of the body (limbs, tail, and ears) decreases with decrease of mean temperature" (Huxley, 1942). The bills of birds seem to follow this rule, and it has been reported to apply also to nonmigratory birds' wings. Actually, however, the opposite is true of migratory birds and even some nonmigratory birds (Mayr, 1956).

Kelso's Rule

"The ears [i.e., ear-opening and dermal flap] of northern owls are relatively larger than in southern owls" (Kelso, 1940).

Gloger's Rule

"Intensity of melanin pigmentation tends to decrease with mean temperature; the amount of black pigment tends to increase with increase of humidity; yellowish or reddish brown pigmentation is characteristic of regions of high temperature and aridity." These correlations are very general among birds. Examples are the Bobwhite, Red-tailed Hawk, Screech Owl, Hairy Woodpecker, Horned Lark, Fox Sparrow, Song Sparrow.

Feathering of Legs and Feet

"The amount of feathering on the legs and feet of certain nonmigratory birds, such as owls and grouse, decreases with increasing mean temperature" (Kelso and Kelso, 1936). In populations of the Ruffed Grouse, for example, the tarsus is feathered in North Dakota birds but bare in Virginia birds.

CONFUSING FACTORS

The following problems are faced by the avian systematist.

Plasticity

Why do some (stable) species exhibit so little geographical variation, whereas closely related (plastic) species vary so much? The Swamp Sparrow, *Zonotrichia (Melospiza) georgiana,* has an extensive breeding range extending from northern Manitoba and Newfoundland to Missouri and Delaware, and yet ornithologists recognize but two weakly differentiated races. By contrast, so much differentiation in external characters has occurred in the Song Sparrow that 39 races are recognized (Fig. 3). Striking geographical variation is demonstrated by subspecies of the Golden-fronted Woodpecker (*Centurus a. aurifrons, C. a. dubius*), by the Hawaiian honeycreeper genus *Hemignathus*, and by the drongo *Dicrurus hottentottus* (Mayr, 1963:591; Monroe and Howell, 1966; Snow, 1954).

Individual, seasonal, and age variations in plumage patterns also may cause problems. Mayr *et al.* (1953), for example, commented that several hundred bird synonyms are based on juvenal plumages.

Sexual Dimorphism

Sexual dimorphism is very widespread among birds; in some species it varies geographically. Among the phalaropes and a few other birds the female is more brightly colored than the male. Sexual dimorphism may be expressed in overall

Figure 3. Some geographical races of the Song Sparrow. From top to bottom: *Zonotrichia melodia sanaka,* Aleutian Islands; *Z. m. caurina,* southeastern Alaska; *Z. m. samuelis,* Vallejo, California; *Z. m. montana,* Chiricahua Mountains, Arizona; *Z. m. saltonis,* Oak Creek, Arizona; *Z. m. mexicana,* Lerma, Mexico.

size, bill size, or tail length (Amadon, 1959; Earhart and Johnson, 1970; Jackson, 1971; Johnson, 1966a; Short, 1970).

Sibling Species

Sibling species are at least partly sympatric forms that are morphologically similar and reproductively isolated (Mayr, 1963:34). Thus not all "good" species exhibit well-marked morphological differences. The three small *Empidonax* flycatchers, *virescens*, *traillii*, and *minimus*, are virtually indistinguishable in the field during migration, and yet each species can be identified instantly on the basis of song or nest.

Polymorphism

Polymorphism "is the occurrence in the same population of two or more distinct forms of a species in such proportions that the rarest of them cannot be maintained by recurrent mutation" (Mayr, 1951:111). A good example is the two color phases (red and gray) of the Screech Owl (*Otus asio*) in eastern North America (Owen, 1963). Polymorphism is common in Wedge-tailed Shearwaters, herons, egrets, the Hawaiian Hawk (*Buteo solitarius*), the Parasitic Jaeger (Arctic Skua), the Black Bulbul (*Hypsipetes madagascariensis*), some paradise flycatchers (*Terpsiphone*), the Bananaquit (*Coereba flaveola*), and Brown Jays (O'Donald and Davis, 1959).

Convergence

Convergence in form and color of unrelated species makes it difficult to determine the true relationship of the convergent species. The Great Kiskadee Flycatcher (*Pitangus sulphuratus*), the Boat-billed Flycatcher (*Megarhynchus pitangua*), and the Rusty-margined Flycatcher (*Myiozetes cayanensis*) are probably but distantly related tyrant-flycatchers that have secondarily become very similar in plumage. In this example, the lack of close relationship seems generally accepted, but there are large groups of passerine birds (e.g., Sylviidae, Muscicapidae, Timaliidae; Fig. 4) in which the interrelationships of the species and genera are poorly understood; how much of the difficulty is due to convergence is unknown. Convergence in structure and habit also may occur among species of unrelated families, as illustrated by the American meadowlarks (*Sturnella*) and species of African pipits of the genus *Macronyx*. Both genera have streaked upper parts and yellowish under parts with a black pectoral band; moreover, they occupy the same type of habitat and build semidomed nests on the ground (Friedmann, 1946).

Figure 4. Nestling Rufous-bellied Babblers (*Dumetia h. hyperythra*); Baroda, India, August 5, 1964. The babblers (Timaliidae) are a large and diverse group whose relationships are poorly understood.

Divergence

Divergence or adaptive radiation is most pronounced on isolated islands, and, in general, the degree of divergence is related to the distance of the islands from a continental land mass or intervening islands. Since islands are colonized by long-distance dispersal, the numbers of kinds of plants and animals are typically inversely related to the isolation of the islands. Thus there is a higher degree of endemism and divergence in the Hawaiian Islands (e.g., the honeycreeper family Drepanididae) than in the Galapagos Islands (e.g., the Galapagos finches Geospizinae). So striking are the differences in bill development among the honeycreepers that at one time the several species were assigned to different families: Meliphagidae, Dicaeidae, Fringillidae (Fig. 5).

Figure 5. Divergence in bill structure in Hawaiian honeycreepers (family Drepanididae, subfamily Psittirostrinae). Left to right: *Loxops v. virens, Hemignathus o. obscurus, H. lucidus affinis, H. wilsoni, Pseudonestor xanthophrys, Psittirostra psittacea, P. cantans, P. kona.* (Courtesy of Dean Amadon and the American Museum of Natural History.)

Hybridization

Hybridization occurs most often among species that do not form a lengthy pair-bond: e.g., grouse, hummingbirds, manakins, birds-of-paradise. Interfamily hybrids are exceptional, but a few have been reported among gallinaceous birds.

The successful mating of individuals of different species, genera, or even families indicates that isolating mechanisms break down occasionally. This may involve elimination of ecological isolation. Mayr (1951) reported that hybridization between *Passer domesticus* and *P. hispaniolensis* "occurs where the Willow Sparrow has shifted from river bottom into human settlements.'' Interspecific hybridization occurs commonly between Blue-winged and Golden-winged warblers (Fig. 6). In this instance, the zone of overlap of the two species has increased with the northward extension of the breeding range of the Blue-winged Warbler during the present century, and the frequency of hybridization is very high (Short, 1963; Gill and Lanyon, 1964). Studies of hybridization in flickers, orioles, meadowlarks, tanagers, buntings, and towhees have been published by Lanyon (1966), Rising (1970), Short (1965, 1969), and Sibley and Sibley (1964).

HIGHER TAXONOMIC CATEGORIES

It would be convenient to have a list of taxonomic characters for orders, another list for families, a third list for subfamilies, etc. Unfortunately, it is not possible to prepare such lists. The features that characterize the nuthatches are not the same as those that characterize the crowlike birds: the characteristics of the genera differ from each other. Inasmuch as subfamilies are composed of genera and families are composed of subfamilies, the diagnostic features will

Figure 6. Male and female Blue-winged Warblers at the nest; Calhoun County, Michigan, June 11, 1955. (Courtesy of Lawrence H. Walkinshaw and William A. Dyer.)

vary among the subfamilies, the families, and the orders. The taxonomist can find the diagnostic characters only by studying the birds. Wetmore (1960) recognized 27 orders of recent birds; Stresemann (1959) recognized 48 orders in 1934 and 51 orders in 1959. There is less diversity of opinion on most families of nonpasserine birds; considerable disagreement exists in regard to both the number and the arrangement of passerine families. A few passerine families are admittedly "taxonomic wastebaskets"; others represent convenience and tradition. The crux of the problem is deciding which characters are plastic and which are relatively stable in an evolutionary sense. The subjective character of the taxonomic categories has led some zoologists to suggest numerical bases for determining relationships (Selander, 1971).

TAXONOMIC CHARACTERS

Taxonomic characters fall into five main groups: morphological, physiological, ethological, ecological, and geographical.

Morphological Characters

Nostrils or *Nares*.   The condition of the external openings of the nasal cavities is often cited by taxonomists. Usually the nostrils are open (*pervious*); in most birds they are exposed (*gymnorhinal*), but they are concealed by feathers in grouse and crows. In many penguins, frigatebirds, cormorants, anhingas, and boobies or gannets, the nostrils close during ontogeny and are absent (*impervious* or *obsolete*) in adults; this seems to be a variable feature in the Red-footed Booby (*Sula sula*), with some adults having small, slitlike openings.

Commonly the nostrils are placed laterally on the bill and are posterior to the midpoint, but the position ranges from very near the bill tip (uniquely in *Apteryx*) to the posterior rim of the bill (*Ramphastos* toucans, in which the nostrils open to the rear). In outline, nostrils range from round through various oval shapes to linear slits. In albatrosses and other "tubinares" (tube-nosed swimmers) the nostrils are at the ends of horny tubes; in many nightjars (Caprimulgidae) they are at the ends of rather long, soft, flexible tubes. A *nasal operculum* is a flaplike structure along the upper rim of the nostril in Galliformes, Thinocoridae, Columbidae, some hummingbirds (e.g., *Eugenes fulgens*), Rhinocryptidae, Neosittidae, Zosteropidae, and Drepanididae; the operculum is movable in the Rhinocryptidae. In wrynecks (Jyngidae) the flap is attached along the lower rim of the nostril. There are other special conditions; e.g., in falcons (*Falco, Polihierax*) and some cuckoos (*Chrysococcyx*) the nostril is circular but has a conspicuous central tubercle.

In many birds the two nasal cavities are separated by a complete internasal septum (a condition called *imperforate*), but in others (cranes, American vultures, rails) there is a central opening in the partition (*perforate*).

Nasal Bones. Garrod (1873a) described the two principal arrangements of the bony structure forming the nasal openings in the skulls of birds (Fig. 7) and proposed the following terms.

Holorhinal: the posterior outline of the opening is fairly rounded. A line drawn over the culmen connecting the posterior ends of the two nasal openings will pass across the nasal processes of the premaxilla. Examples: rails, grouse, pheasants.

Schizorhinal: the posterior outline of the nasal opening forms a deep slit. In most cases a line drawn between the openings as noted above will pass behind the nasal processes of the premaxilla. Examples: gulls, cranes.

Two additional categories were proposed later.

Pseudoschizorhinal: a modification of the holorhinal type, found in many ovenbirds (Furnariidae). The posterior outline of the nasal opening is rounded, but the openings extend far back—even posterior to the nasal processes of the premaxilla.

Amphirhinal: having two bony nostril openings (one anterior to the other) on each side. Feduccia (1967) found amphirhinal nostrils among representatives of more than 15 passerine familes, but for some genera the condition was present in some species but not in others. Consequently, he commented that "the apparent ease with which the amphirhinal condition has arisen in so many passerine families, plus the fact of its occurrence in some species but not in others of reasonably well-defined genera is sufficient recommendation for extreme caution with its use, if any, in passerine taxonomy."

Palate. Huxley (1867) used the structure of the bony palate to characterize four major groups of "carinate" birds. He placed "Dromaeus" (the emu) and the other "struthious" birds in a separate division, the Ratitae.

1. **Dromaeognathous** (Greek *dromaios*, a runner); Fig. 8. The prevomers or vomers[1] extend far back and articulate with the posterior ends of the palatines and the anterior ends of the pterygoids, separating both from the parasphenoid ("basisphenoid" of some authors). Huxley placed only the tinamous (Tinamidae) in his suborder "Dromaeognathae." Later Pycraft (1901) showed that the palate of the rhea fits the same fundamental description

[1] De Beer (1937:434) wrote that "the prevomers of non-mammalian vertebrates are totally different structures from the vomer in mammals." Jollie (1957:423), however, considers the prevomers of reptiles and birds to be homologous with the vomer of mammals.

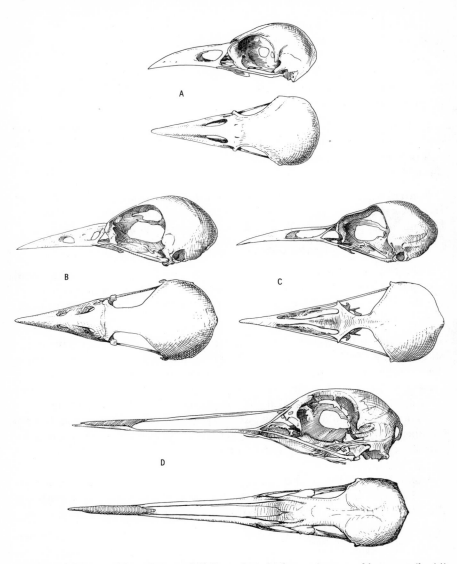

Figure 7. Dorsal and lateral views of skulls to show the four main types of bony nostrils. (*A*) Holorhinal—*Corvus brachyrhynchos* (Corvidae). (*B*) Amphirhinal—*Gymnopithys leucaspis* (Formicariidae). (*C*) Pseudoschizorhinal—*Cinclodes antarcticus* (Furnariidae). (*D*) Schizorhinal—*Grus canadensis* (Gruidae).

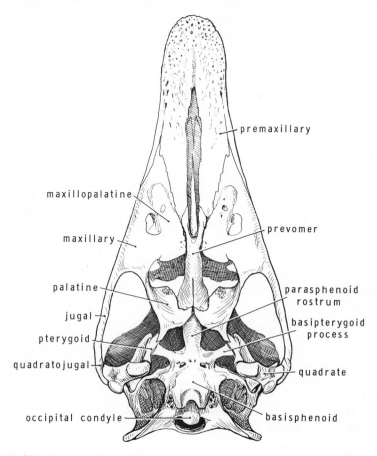

Figure 8. Ventral view of skull of *Rhea americana* to illustrate the dromaeognathous palate.

and must likewise be called dromaeognathous. He went further and brought all other struthious birds (emu and cassowaries, Ostrich, apteryx, the moas, and aepyornis) into one group and called them "paleognathous" birds. Bock (1963b) also considered that all ratites have an homologous palate type, and that "the main mass of available evidence suggests that the palaeognathous palate evolved from the neognathous palate as an adaptation for some new, but still unknown, method of feeding." He suggested, therefore, that "the ratites be reunited as a single taxonomic unit, but the rank to be assigned left as an open question for further study."

2. **Schizognathous** (Greek *schizo*, I cleave). Prevomers completely fused and fairly large or small. Maxillopalatines do not meet along the central line.

Palatines and pterygoids articulate with the parasphenoid rostrum. Characteristic of the Galliformes, Gruiformes, Charadriiformes, Piciformes, etc.

3. **Desmognathous** (Greek *desmos*, a bond). Prevomers fused or absent. Maxillopalatines meet in the midline (and, in many cases, fuse). Palatines and pterygoids articulate with the parasphenoid rostrum. Characteristic of Anseriformes, Ciconiiformes, Pelecaniformes, Falconiformes, and many others.

4. **Aegithognathous** Greek *aigeithalos*, some small bird). Prevomers large, completely fused and truncated in front, separating the maxillopalatines. Characteristic of passerine birds and swifts.

The last three palate types are shown in Fig. 9.

Further study has shown that the second, third, and fourth palate types merge into each other, and that the dromaeognathous type is not always sharply demarcated. (The young of many birds actually develop a dromaeognathous palate first and later acquire a palate of type 2, 3, or 4.)

Columella (Fig. 10).   The columella is the single ear bone in the middle ear cavity of birds. Feduccia (1974) described the structure of this bone in passerine and nonpasserine species and found that its morphology "in the major groups of both Old and New World suboscines represents a unique, derived character state, and argues strongly for monophyly."

Cervical Vertebrae.   Unlike mammals, which regularly have 7 cervical vertebrae, birds of different groups have numbers that range from 13 (e.g., certain passerines, some cuckoos) to 25 (some swans). The most frequent number is 14 or 15. Despite the fact that F. A. Lucas gave 13 as the minimum number of cervical vertebrae as long ago as 1902, many modern works state (erroneously) that some birds have fewer.

The cervical vertebrae are defined as those between the skull and the trunk. One or more of the posteriormost cervical vertebrae commonly bear short, incomplete, cervical or cervicodorsal ribs; they should always be included in the counts of cervical vertebrae for taxonomic diagnoses. A peculiar feature of the atlas in birds is that its body contains either a foramen or a notch in which the *odontoid* (toothlike) *process* of the *axis* articulates. Hence one speaks of either a *perforated* or a *notched atlas.*

Sternum (Fig. 11).   The condition of the sternum—unkeeled or keeled—has long been used to define two major groups, the "ratite" and "carinate" birds.[2]

In **carinates**, the body of the sternum bears a downward-projecting *carina*

---

[2] After it was realized that the tinamous (Tinamidae), a strongly keeled group, were more closely related to the "ratite" (flat-sternumed) birds than to other "carinate" (keel-sternumed) birds, these terms were no longer used for taxonomic units, but they remain as important ornithological adjectives.

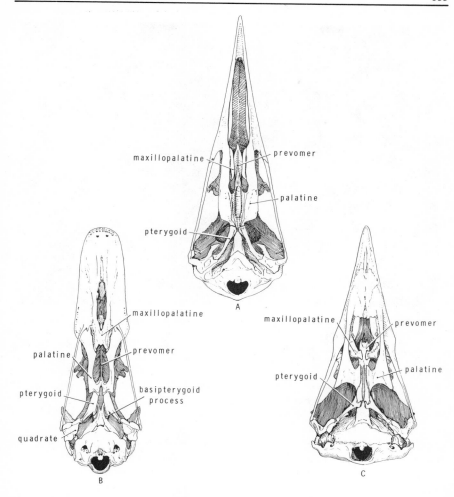

Figure 9. The three other "classic" types of palate. (*A*) Schizognathous—*Larus atricilla* (Laridae). (*B*) Desmognathous—*Branta canadensis* (Anatidae). (*C*) Aegithognathous—*Corvus corax* (Corvidae).

or keel which serves as an area of origin for two of the flight muscles (supra-coracoideus and pectoralis); the relative development of the carina is directly related to the development of these muscles. At its anterior end, the sternum has two midline projections, the *spina interna* (toward the inside of the bird) and the *spina externa* (ventral to spina interna). In some birds, such as pheasants, these two fuse to form a *spina communis* (also *rostrum, spina*

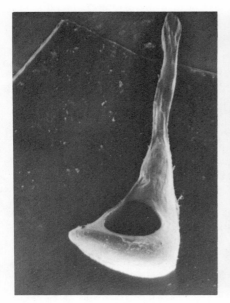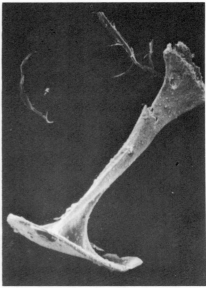

Figure 10. Scanning electron micrographs of the columella of (left) *Laniocera rufescens* (Cotingidae) and (right) *Cincinnurus regina* (Paradisaeidae); actual length approximately 2 mm. The vast majority of birds, including the oscines, possess a columella of the reptilian type, with a flat footplate and straight bony shaft (right). The major groups of suboscines (Eurylaimidae, Pittidae, Furnariidae, Formicariidae, Rhinocryptidae, Cotingidae, Pipridae, Tyrannidae, and Phytotomidae) have a columella characterized by a large bulbous footplate perforated by one large fenestra. (Courtesy of J. Alan Feduccia.)

sternalis; but not "manubrium," because no part of the bird sternum is homologous to the mammalian manubrium; Gadow, 1896:910). On each side of the sternum, caudal to the rib facets, there is usually a *posterolateral process*; in many galliform birds an *oblique process* is also present. In some groups, there is a *lateral metasternal process* between the posterolateral process and the median metasternum, thus producing a double-notched posterior border of the sternum (see Heimerdinger and Ames, 1967; no part of the bird sternum is homologous with the mammalian xiphoid process). In other birds, especially those of strong flight (e.g., hawks), the spaces between the processes are largely filled in, leaving only small openings (fenestra, windows); in some the spaces are obliterated so that there is a continuous bony plate. These different types of sterna are called *notched* (either single or double-notched), *fenestrate*, and *entire*. Heimerdinger and Ames examined the sterna of nearly 1000 suboscine passerines from 11 families and found six different patterns that varied from entire to double-notched. They recommended caution, therefore,

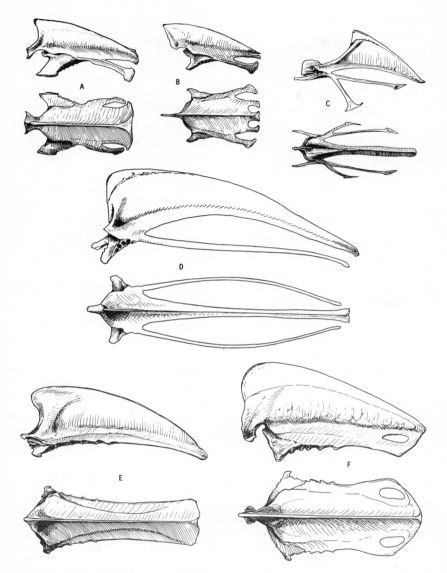

Figure 11. Selected examples of the avian sternum. (*A*) *Psarocolius angustifrons* (Icteridae). (*B*) *Megaceryle alcyon* (Alcedinidae). (*C*) *Callipepla squamata* (Phasianidae). (*D*) *Tinamus major* (Tinamidae). (*E*) *Aramus guarauna* (Aramidae). (*F*) *Ara macao* (Psittacidae).

555

"in the taxonomic use of sternal configurations until their range of variation has been determined, and their functional and adaptive significance can be understood." (See also Feduccia, 1972.) In **ratite** birds and in some flightless carinate birds, such as the New Zealand parrot *Strigops* and the rail *Notornis*, there is virtually no keel on the sternum. The flightless parrot and rail are obviously degenerate members of well-known families of carinate, flying birds, and it now appears that the ratite birds also are descended from flying (and therefore keeled) birds.

Tarsometatarsus ("Tarsus"). The tarsometatarsus shows a variety of characters that have been used in taxonomy. In some species, such as grouse, sandgrouse, and many owls, it is partly or wholly feathered, but in most other birds it is unfeathered. The unfeathered sheath (*podotheca*) of the tarsometatarsus may be any of the following types.

Scutellate (with large, scalelike segments).
Reticulate (with a network of small, scalelike segments).
Scutellate-reticulate (scaled in front and reticulate behind).
Scaleless (the sheath soft, undivided, bare skin): e.g., the American green kingfishers, *Chloroceryle*.

In passerines the horny sheath of the tarsometatarsus is obviously divisible into an anterior segment called the *acrotarsium* (Greek *akron*, top) and a posterior segment or *planta* ("the sole").

The shape of the cross section of the tarsus is of two sorts in the higher songbirds (Oscines). The larks, family Alaudidae, have the posterior part of the tarsus rounded, and they are called "latiplantar oscines"; Rand (1959) reported this condition in one species (but not in others) of the vireo genus *Hylophilus* and in the Australian Treecreeper *Climacterus*. All other oscines thus far examined have the posterior part of the tarsus sharp-angled and are called "acutiplantar oscines."

Several types of passerine tarsal scutellation have been described (Ridgway, 1901–1911; Fig. 12).

Pycnaspidean (*pycnos*, dense): with rear (plantar) surface of the tarsus densely covered with small scales or granules.

Exaspidean: with anterior, scutellated segment of tarsal sheath extending across the lateral side of the tarsus.

Endaspidean (the reverse of the preceding): with anterior, scutellated segment of tarsal sheath extending across the medial surface of the tarsus.

Holaspidean: with rear surface of the tarsus covered by a single series of broad, rectangular scales.

Taxaspidean: with rear surface of the tarsus covered by two (or sometimes three) series of small, rectangular (or hexagonal) scales.

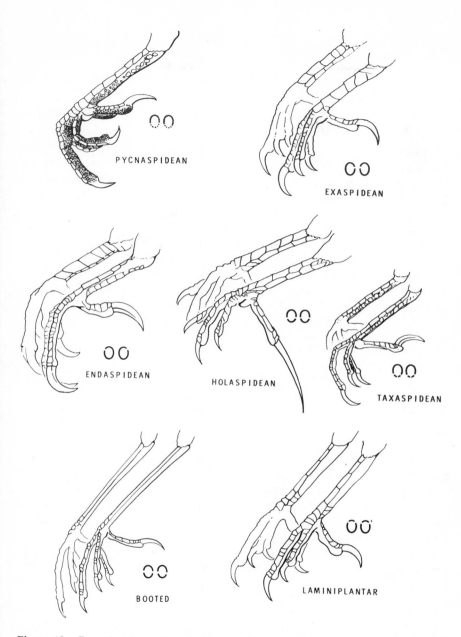

Figure 12. Examples of tarsal scutellation in passerine birds. Pycnaspidean—*Phytotoma rara* (Phytotomidae; after Küchler, 1936). Exaspidean—*Muscivora forficata* (Tyrannidae). Endaspidean—*Dendrocolaptes platyrostris* (Dendrocolaptidae). Holaspidean—*Alauda arvensis* (Alaudidae). Taxaspidean—*Acropternis orthonyx* (Rhinocryptidae). Booted—*Turdus migratorius* (Turdidae). Laminiplantar—*Cardinalis cardinalis* (Fringillidae). The oval figures show schematically the pattern of scales and scutes but not the cross-sectional shapes of the tarsi.

Booted ("ocreate"): with scutella fused into a single smooth sheath or "boot," except, in some cases, at the very lower end.[3]

Laminiplantar: with a smooth, undivided plantar surface, but a scutellate acrotarsal (anterior) surface.

Leonhard Stejneger wrote in the *Riverside Natural History* in 1888 that "the horny covering of the tarsus still plays a great role in the classification of the Passeres." Ridgway described the different patterns and used them as taxonomic characters. He wrote (1907:336), however, that "they have disappointed me . . . they seem of little value beyond the definition of genera (even sometimes failing here!) or minor supergeneric groups; indeed, it has been found that each of them is more or less variable within what appears to be proper generic limits." Nevertheless, most taxonomists have continued to use these terms. In his survey of tarsal scutellation in a large number of passerines, however, Rand (1959) showed that its taxonomic value is doubtful in most groups. He concluded that "the pattern of scutellation of the tarsal envelope is of limited importance as a character indicating relationships within the oscines, once the larks, Alaudidae, are excepted, nor will it separate oscines from nonoscines in all cases. The three main types of scutellation occur in one family (Corvidae)."

Toes. The foot of the bird has, typically, four toes: a hind toe (hallux) and three front toes. The toes are numbered as follows: I, hallux; II, inner front toe; III, central front toe; IV, outer front toe.

The number of bones (phalanges) in the toes of birds is extraordinarily constant: toe I typically has two bones, II has three, III has four, and IV has five. (Note that each toe has one more phalanx than its ordinal number.) This invariance helps us to ascertain the homology of bird toes even when in the course of evolution toes have been lost or turned to different positions. A few swifts, nightjars, and some other birds have a reduced number of phalanges in some toes (Forbes, 1882).

In most birds the hallux is functional and is at the same level as the other toes (a condition called *incumbent*), but in some species the hallux is *elevated* and may not touch the ground at all (e.g., in cranes and many rails). The hallux in larks is notable because it typically has a long, straight, and sharp claw; *Lessonia rufa* (Tyrannidae), a terrestrial species, also has a larklike hind toe and claw. Some species have a pectinate (comblike) margin on the claw of

---

[3] Unfortunately, German and French authors have not only used other names (i.e., *knemidophorer* or *cnémidophore*) for this type of tarsal sheath, but also have transferred Sundevall's term "ocreate" to the type of tarsus (laminiplantar) characteristic of most higher passerine birds. Moreover, the descriptions and/or illustrations in Reichenow, Stresemann, and Boetticher are incorrect. The term "booted" is used in poultry literature to designate feathered shanks; the preferred term is *ptilopody* (feathered foot).

the third toe: examples are the Red-footed Booby, frigatebirds, herons (including the Boat-billed Heron, *Cochlearius cochlearius*), the Black-tailed Godwit (*Limosa limosa*), the Crabplover (*Dromas ardeola*), pratincoles (*Glareola* spp.), the Inca Tern (*Larosterna inca*), barn owls, and the Caprimulgidae (Fig. 13).

Some birds have only three toes: examples are Rheidae, Casuariidae, Dromiceiidae, some Tinamidae, Diomedeidae, Pelecanoididae, Turnicidae, Otididae, Haematopodidae, many Charadriidae, one member of the Scolopacidae (*Calidris alba*), Recurvirostridae (except *Recurvirostra*, hallux vestigial), Burhinidae, Glareolidae (except *Glareola*), most Alcidae, the Kittiwake (*Rissa tridactyla*), the Andean Flamingo (*Phoenicoparrus andinus*), James's Flamingo (*P. jamesi*), the Three-toed Jacamar (*Jacamaralcyon tridactyla*), the Three-toed Swiftlet (*Collocalia papuensis*), and some woodpeckers (*Picoides*). In a single passerine species, *Paradoxornis paradoxa* of western China, there are only three functional toes, IV having been reduced to

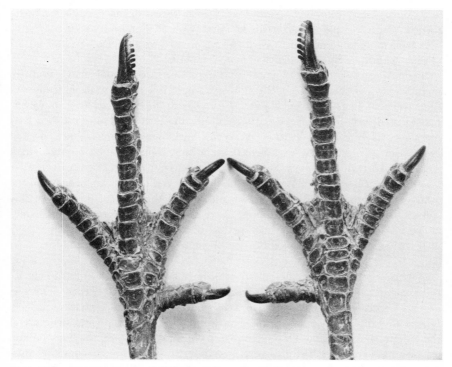

Figure 13. Feet of a Poor-will (*Phalaenoptilus nuttallii*), showing the pectinate margin on the medial border of the middle toe. (Courtesy of Joseph Brauner and the editor of *Condor*.)

a vestigial stub; the possible significance of the reduction is unknown. In most three-toed birds it is the hallux that has been lost, but in certain kingfishers (*Alcyone, Ceyx*), toe II is lacking. Tridactylism, therefore, is found in some running birds, in tree-climbing birds, and in certain birds that swim under water primarily with their wings (diving-petrels and the auk family). Only the Ostrich is didactyl; toes I and II have been lost.

Most birds have *anisodactyl* feet: three toes turned forward and one turned backward. There are several variations, however, from the usual toe arrangement, the most frequent being the *zygodactyl* or yoke-toed foot characteristic of woodpeckers, toucans, cuckoos, and parrots, in which toes I and IV are turned backward. Owls, touracos, and *Pandion* are zygodactyl but can move the outer toe freely from back to front. Although the zygodactyl foot, adapted to climbing and grasping, is found mainly in arboreal groups, several genera of cuckoos (e.g., *Geococcyx, Morococcyx, Carpococcyx, Centropus*) have become cursorial birds.

Heterodactyl feet—in which toes I and II are turned backward—are peculiar to trogons.

Pamprodactyl feet—having all toes turned forward (or capable of being turned forward)—are characteristic of colies (Coliidae) and most swifts.

A less common arrangement is called *syndactyl*: two toes are fused for a part of their length, as e.g., in kingfishers and hornbills.

Swimming birds (or birds whose ancestors swam) commonly have adaptive modifications of the feet. A well-known example is a duck, with its front toes fully webbed; the terrestrial Hawaiian Goose (*Branta sandvicensis*), however, has reduced webbing between the toes. Gulls and terns also have webbed feet, but some, like the White Tern (*Gygis alba*), have the webs much "incised" and reduced in area. Many wading birds (e.g., herons, shorebirds) have only a small web between two of the toes. Other birds, such as cormorants and pelicans, are *totipalmate*, the web including even the hallux. Grebes and phalaropes, on the other hand, swim by means of broad lobes along the sides of the toes. A few species, like the gallinule (*Gallinula*), have only a narrow lobe along the sides of the toes, and other species, such as the dippers (*Cinclus*), swim remarkably well without any apparent modification of the simple perching type of foot.

Jaw Muscles (Fig. 14).   The jaw muscles have been used in attempts to clarify relationships among passerine birds, and further studies are needed. There is general agreement that the relative development of the jaw muscles is largely related to feeding habits, as are the shape of the bill, the shape and structure of the tongue, and presumably, the pattern of both the horny and the bony palate; conversely, some skull features (e.g., tubercles, fossae) are reflections of muscular development. Consequently, the investigator is faced by many difficulties in his attempt to separate adaptive from phylogenetic features in the jaw

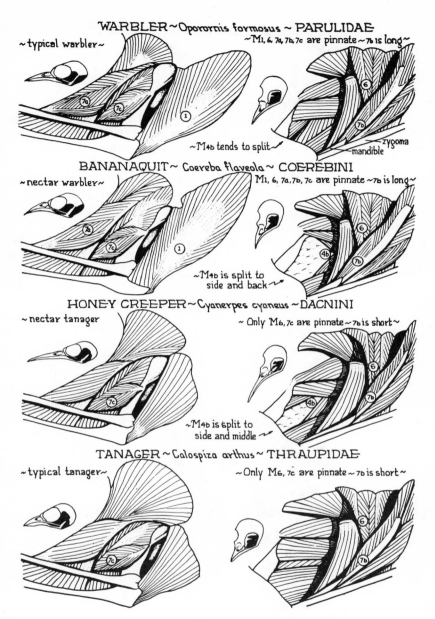

Figure 14. Comparison of the jaw muscle patterns in typical warblers (Parulidae) and tanagers (Thraupidae) with those of their nectar-adapted forms. Key to the numbers: 1, M. depressor mandibulae; 2, protractor quadrati; 3, pterygoideus dorsalis, (a) anterior, (b) posterior; 4, pterygoideus ventralis, (a) anterior, (b) posterior; 5, pseudotemporalis profundus; 6, pseudotemporalis superficialis; 7, adductor mandibulae: (*a*) externus superficialis, (*b*) externus medialis, (*c*) externus profundus, (*d*) posterior. (Courtesy of W. J. Beecher and the editor of *Wilson Bulletin*.)

muscles and related structures (see Beecher, 1953; Tordoff, 1954; Bock, 1963a; Zusi, 1967).

The Tensor Patagii Muscles (Fig. 15).   There are two muscular slips derived phylogenetically from M. deltoideus major: Mm. tensor patagii longus and tensor patagii brevis. In some birds (e.g., passerines) there are two separate muscles, but in others a single belly arises from the medial surface of the apex of the furcula (occasionally also from the acromion); in either case, two tendons are formed. The tendon of the "long" tensor runs outward in the leading edge of the anterior skin fold (propatagium) of the wing, to insert primarily on the extensor process of the carpometacarpus (accessory insertions on other parts of the manus or wrist are common). Shortly after its formation, the tendon may be reinforced by the biceps slip, pars propatagialis of M. pectoralis, or by a tendon attached to the deltoid crest of the humerus. The tendon of the "short" tensor inserts by one or more branches on some part of the muscles (usually M. extensor metacarpi radialis) or bones of the forearm. There is a single tendon of insertion in the Ramphastidae and in some passerine birds. The most complicated pattern of insertion is seen in the Laridae and Alcidae, where multiple tendons of insertion are found (Beddard, 1898:figs. 171–175). Although Garrod, Forbes, and Beddard placed considerable emphasis on the taxonomic value of the pattern (simple or complex) formed by the tendon of insertion of M.

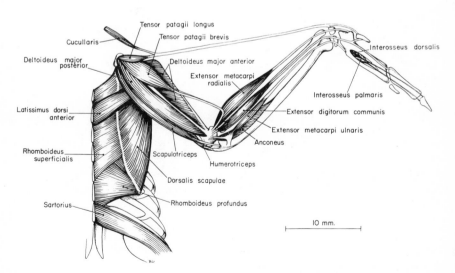

Figure 15.   Dorsal view of the superficial wing muscles of the Kirtland's Warbler (*Dendroica kirtlandii*). In this species the tensor patagii brevis muscle has a simple insertion by a single tendon. (Courtesy of Andrew J. Berger and the editor of *Auk*.)

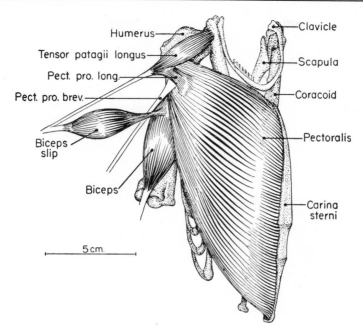

Clavicle

Humerus

Tensor patagii longus

Pect. pro. long.

Pect. pro. brev.

Scapula

Coracoid

Biceps
slip

Biceps

Pectoralis

Carina
sterni

5 cm.

Figure 16. Ventral view of certain wing muscles of *Goura victoria* to show relationships of the biceps slip. (By permission of Andrew J. Berger, from *Biology and Comparative Physiology of Birds,* Vol. 1, Academic Press, copyright 1960.)

tensor patagii brevis, little attention has been given to this anatomical feature during this century.

The Biceps Slip (Fig. 16).   This is a fleshy fasciculus (also called the "biceps propatagialis" and the "tensor accessorius") given off from M. biceps brachii. It inserts (either by fleshy or tendinous fibers) on the tendon of M. tensor patagii longus (rarely also on forearm muscles). The biceps slip, widely used in taxonomic diagnoses, has been reported in the Gaviidae, Podicipedidae, Procellariiformes, Anatidae, many Galliformes, Gruidae, Rallidae, Charadriidae, Laridae, Alcidae, Columbidae (in which the biceps slip is very large), Caprimulgidae, and some other nonpasserine genera (e.g., *Phaethon, Anhinga, Phoenicopterus*; see George and Berger, 1966:330–331).

The Expansor Secundariorum Muscle (Figs. 17, 18).   Garrod (1876a:193) first described this peculiar wing muscle, and Fürbringer (1886:124) later stated that it was composed of smooth muscle. The muscle is present in both passerine and nonpasserine birds (Berger, 1956; George and Berger,

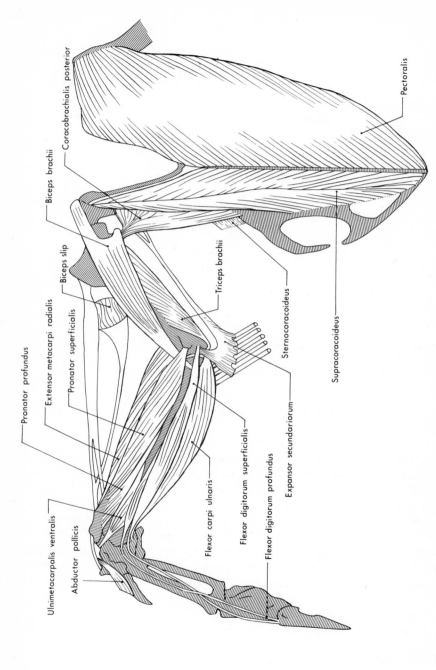

Figure 17. Ventral view of the superficial wing muscles of the Rock Dove (*Columba livia*), showing the two parts of the expansor secundariorum muscle in this species. (By permission of Olin Sewall Pettingill, Jr., from *Ornithology in Laboratory and Field*, 4th ed., Burgess Publishing Co., copyright 1970.)

Coracobrachialis posterior

Biceps brachii

Biceps slip

Pronator profundus

Extensor metacarpi radialis

Pronator superficialis

Triceps brachii

Sternocoracoideus

Supracoracoideus

Pectoralis

Ulnimetacarpalis ventralis

Abductor pollicis

Flexor carpi ulnaris

Flexor digitorum superficialis

Flexor digitorum profundus

Expansor secundariorum

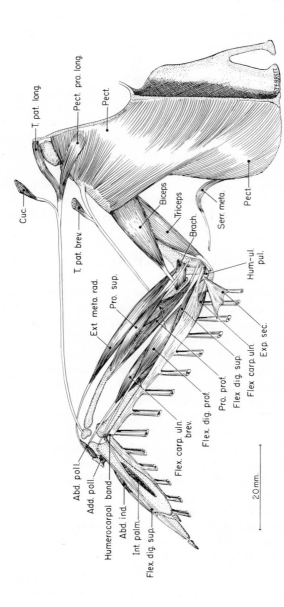

Figure 18. Ventral view of the wing muscles of the extinct Bourbon-crested Starling (*Fregilupus varius*) to show the form of M. expansor secundariorum (Exp. sec.) in passerine birds. (Courtesy of Andrew J. Berger and the American Museum of Natural History.)

1966:338–344). In passerine birds and in certain unrelated nonpasserine genera, the muscle arises by a tendon from the distal end of the humerus and/or the humeroulnar pulley. In other nonpasserines there are two tendons: one is attached to the distal end of the humerus, the other to the pectoral girdle (scapula, coracoid, or sternum) and/or to one or more of the following muscles: dorsalis scapulae, subcoracoideus, coracobrachialis posterior, sternocoracoideus, pectoralis. The fleshy belly, located at the elbow, inserts on the calami of two (rarely) or more (six in *Grus canadensis* and *Aceros undulatus*) of the proximal secondaries (occasionally on some of the distal tertials). Although this muscle is widely used in technical diagnoses, its value for such purposes cannot be known until further information is available regarding its presence or absence and its configuration throughout the orders of birds. Buckley and Wheater (1968) studied the pharmacology of the expansor secundariorum muscle and discovered that the isolated muscle of the fowl exhibited spontaneous activity, which they concluded was not mediated by nerve impulses; tentative conclusions were that the muscle contains adrenergic nerve endings but not cholinergic endings.

Thigh Muscles (Fig. 19). Garrod (1873–1874) studied the muscles of the thigh in birds and decided that five were so variable in their occurrence among different groups of birds that a concise statement regarding them might be useful in taxonomy. He arbitrarily assigned the letters A, B, X, and Y to four of these variable muscles and indicated the presence or absence of the ambiens muscle[4] by placing a plus or minus sign at the end of the series of letters. Hudson (1937:59) proposed that two other muscles, designated as C and D, should be added to the muscle formula, that the ambiens muscle should be represented by the symbol Am, and that the vinculum (a tendinous band) often connecting the tendons of Mm. flexor perforatus digiti III and flexor perforans et perforatus digiti III should be represented by the letter V. Later, Berger (1959) proposed three additions (E, F, G) based on study of both Old World and New World genera. Both Hudson and Berger considered, but rejected, the idea of adding Mm. peroneus longus and peroneus brevis to the formula (see George and Berger, 1966:429); Kurochkin (1968), however, proposed that they be added as M and N.

The expanded list of formula muscles is given in Table 1. Obviously, a great deal of work will be required to obtain data on the expanded formula for all orders and families of birds. George and Berger (1966) presented in tabular form the muscle formulae (excluding M and N) for representatives of 40 bird

---

[4] Garrod considered the ambiens to be of such critical importance that he proposed a division of birds into two subclasses, the Homalogonatae ("with typical knees") and the Anomalogonatae ("with aberrant knees"), according to the presence or absence of this muscle. It soon became apparent, however, that these categories were unnatural ones, and the terms are now obsolete.

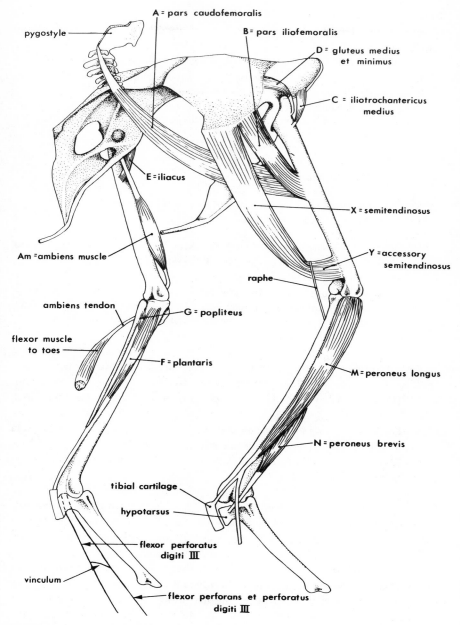

Figure 19. A generalized illustration of the formulae muscles.

Table 1. Symbols for Formula Muscles

Code Letter	Name of Muscle
A	Piriformis pars caudofemoralis (= femorocaudal)
B	Piriformis pars iliofemoralis (= accessory femorocaudal)
C	Iliotrochantericus medius
D	Gluteus medius et minimus (= "piriformis" of Fisher)
E	Iliacus (= "psoas" of Fisher)[a]
F	Plantaris
G	Popliteus
M	Peroneus longus
N	Peroneus brevis
X	Semitendinosus (= flexor cruris lateralis of Fisher)
Y	Accessory semitendinosus
Am	Ambiens
V	Vinculum (between the tendons of Mm. flexor perforatus digiti III and flexor perforans et perforatus digiti III)[b]

[a] M. iliacus of Fisher equals M. iliotrochantericus anterior of Hudson and Berger.
[b] This vinculum (a tendinous connection) is not to be confused with the vinculum between the tendons of Mm. flexor digitorum longus and flexor hallucis longus.

families. Examples of the expanded formulae are as follows: some grebes, BCEFGMX; most Galliformes (not *Opisthocomus* or turkeys), ABCDEFGMNXYAmV; the Osprey (*Pandion*), ADEGNAm; the Nighthawk (*Chordeiles minor*), *AEFGMXY;* the Chimney Swift (*Chaetura pelagica*), AEN; the Eastern Kingbird (*Tyrannus tyrannus*), ACEFXYMN (this formula has been found in most passerine birds studied, but the formula is ACEFXMN in the African River Martin, *Pseudochelidon eurystomina,* and ACEXMN in typical swallows; see Gaunt, 1969).

The Deep Plantar Tendons (Fig. 20).   The value to the bird taxonomist of the variations in these tendons was first shown by the Swedish ornithologist Sundevall, and the idea was then further developed by the English anatomist Garrod (1875).

All birds have two deep flexor muscles of the toes (i.e., muscles which close the digits of the foot): *flexor hallucis longus* and *flexor digitorum longus*. The fleshy bellies of these muscles lie in the leg (above the tarsometatarsus); the long tendons (*plantar tendons*) pass through the hypotarsus (when present) and down the posterior surface of the tarsometatarsus, where the two tendons can be distinguished because the flexor hallucis longus is superficial to the flexor digitorum longus.

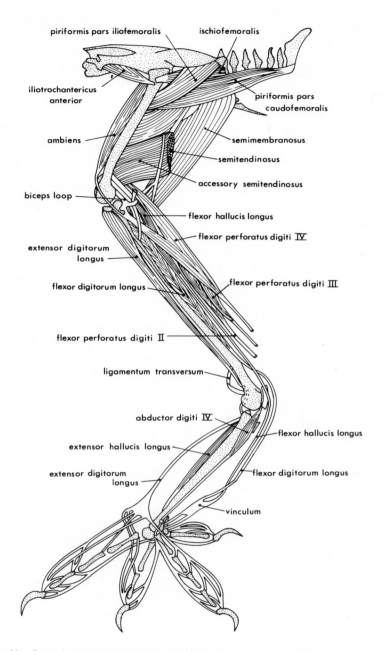

Figure 20. Lateral view of the left leg of the Blue Coua (*Coua caerulea*) to show the relationships of some of the deeper muscles and especially those of the flexor hallucis longus and flexor digitorum longus, whose "deep plantar" tendons are interconnected by a large vinculum.

In all passerines studied except the broadbills (Eurylaimidae), these two ten-dons are entirely separate;[5] in most other birds, however, the two are connected by a fibrous band (a vinculum, but not to be confused with the vinculum in muscle formulae). There are also differences in the insertion of these tendons. In one common type, the flexor digitorum inserts on digits II, III, and IV, whereas the flexor hallucis inserts only on digit I. In another type, the flexor digitorum inserts only on digits III and IV, and the flexor hallucis on digits I and II. In a few birds (e.g., *Gavia, Podiceps, Hydrophasianus, Larus*), the tendon of the flexor hallucis does not insert on the hallux at all, but fuses with the tendon of the flexor digitorum. In tridactyl birds (hallux absent), also, the hallucis tendon fuses with the tendon of M. flexor digitorum longus. Twelve major types of arrangement of the plantar tendons have been described, and a good many intermediate conditions are recorded. These arrangements are sometimes referred to as "type III," "type VIII," etc., following the designa-tions of Gadow (1896:615–618). The terms "antiopelmous," "desmopelmous," "schizopelmous," and "synpelmous" are also sometimes used to designate types of arrangement.

The relationships of the deep plantar (and other) tendons to the hypotarsus (see Fig. 19) have not been examined extensively; but, because of the many pat-terns found, a careful study might reveal points of taxonomic interest. The hypotarsus itself exhibits considerable variation among the orders of birds. It may be a well-developed bony process containing from one to five bony canals, or it may take the form of one or more vertical ridges and grooves and a single (or no), bony canal. All seven of the tendons of flexor muscles to the toes pass through bony canals in *Gavia*, the Picidae, and Passeriformes. All of the ten-dons except that of M. flexor hallucis longus pass through bony canals in *Indi-cator variegatus*. In some genera only the deep plantar tendons pass through bony canals. (See George and Berger, 1966:433.)

Syrinx (Fig. 21).   Three types occur.

1. **Bronchial**—a modification of several rings in each bronchus (such birds actually have two syringes). This type is found in most cuckoos, nightjars and allied families, and some owls.

2. **Tracheal**—a modification of the lower end of the trachea; found in certain New World passerines (ovenbirds, woodcreepers, antbirds, antpipits, tapaculos), which are sometimes called "tracheophones."

3. **Tracheobronchial**—involving both the trachea and the bronchi; found in most birds.

---

[5] Forbes (1880:390–391) proposed that passerine birds be separated into two major divisions: the Desmodactyli, those in which the plantar vinculum is present (the Eurylaimidae), and the Eleu-therodactyli, those in which this vinculum is absent.

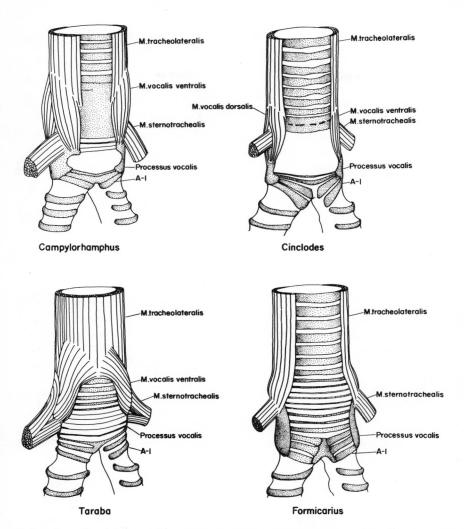

Figure 21. The syringes of four passerine species. Red-billed Scythebill (*Campylorhamphus trochilirostris*; Dendrocolaptidae); Bar-winged Cinclodes (*Cinclodes fuscus*; Furnariidae); Great Antshrike (*Taraba major*; Formicariidae); Black-faced Antthrush (*Formicarius analis*; Formicariidae). (Courtesy of Peter L. Ames and the Peabody Museum of Natural History.)

Authors have described as many as eight pairs of intrinsic syringeal muscles, but there are myological "splitters" and "lumpers"; some authors presumably count muscle slips as separate muscles. Functional syringeal muscles are absent in some ratite birds, in storks, and in American vultures. Such birds "grunt," "boom," and make mechanical noises but in general are called "voiceless."

The different arrangements of the intrinsic muscles were first used in the classification of passerine birds by Müller (1878; see Gadow, 1896:937–942). For many years, the passerines, exclusive of the broadbills (Eurylaimidae), were divided into two groups.

Mesomyodian passerines ("Clamatores"), which have the syringeal muscles attached to the middle of the bronchial semirings.

Acromyodian passerines (Passeres or Oscines; Lyrebirds, scrubbirds), which have the syringeal muscles attached to the extremities of the bronchial semirings. Coues (1903:246) pointed out that, with few exceptions (notably the Alaudidae), a bird possessing an acromyodian type of syrinx also has a laminiplantar tarsus (bilaminate, trilaminate, or booted). A scutellate tarsus is associated with the mesomyodian type of syrinx.

Ames (1971) examined the syringes of nearly 1000 specimens, representing 65 passerine families, and said that "it is probable that the present syringeal structure was fully evolved prior to the radiation of the oscines into their present diversity." He concluded that "taken in conjunction with other anatomical characters, syringeal morphology suggests that the Passeriformes be divided into five suborders: Eurylaimi, Furnarii, Tyranni, Menurae, and Passeres (Oscines)."

Carotid Arteries (Fig. 22).   Rather than having two equal carotid arteries (one on each side of the neck), some birds have a large artery on one side and a small one on the opposite side; in still other birds, there is a single carotid artery; and, in some, both of the deep carotid arteries are absent in the adult and blood is carried to the head by more superficial *cervical arteries*. Some of these variations were used in taxonomy by Garrod (1873b). Beddard (1898:52–54) recorded eight major patterns of the carotid arteries (see also Gadow, 1896:76–77). Passerine birds were said to be "aves laevocarotidinae" because only a left dorsal carotid artery persists in the adult. Glenny (1955:544–548) described 20 different patterns of the dorsal carotid arteries based on the presence of one (*unicarotid,* Class B) or two (*bicarotid,* Class A) dorsal carotid arteries and on the persistence of various embryological remnants. These ligamentous vestiges, however, may represent individual variation and, if so, would be of doubtful value in classification (see Fisher, 1955; Baumel, 1964).

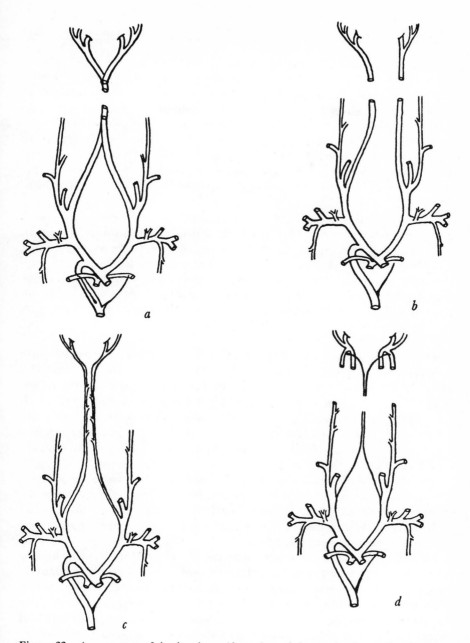

Figure 22. Arrangements of the dorsal carotid arteries and the associated cervical and thoracic arteries in Aves bicarotidinae (ventral views): (*a*) bicarotidinae normales; (*b*) bicarotidinae abnormales, left vessel superficial; (*c*) bicarotidinae infranormales; (*d*) ligamenti carotidinae normales. (Courtesy of Fred H. Glenny and the United States National Museum.)

Arteries of the Thigh.   Branches of two main arteries carry blood to each pelvic appendage in birds: *external iliac artery* and *external ischiatic artery.* The *femoral artery,* a terminal branch of the external iliac artery, is large and carries most of the blood to the thigh in some birds. The external ischiatic (also ischiadic of authors) artery leaves the pelvis by passing through the ilioischiatic (ischiadic) foramen in company with the sciatic (ischiadic) nerve; the artery and nerve leave the foramen dorsal to M. obturator internus. The external ischiatic artery gives off five named branches plus a series of muscular branches in the thigh and then becomes the popliteal artery near the knee joint (Nishida, 1963).

Garrod (1876b) reported that the main artery of the thigh is the femoral artery in certain passerine birds (Pipridae and Cotingidae, except *Rupicola*). He proposed, therefore, that the Mesomyodian passerines be subdivided into the *Heteromeri* (those with the femoral artery predominant) and the *Homoeomeri* (those with the ischiatic artery predominant). Among nonpasserine genera, *Dacelo,* some genera of touracos, and *Centropus phasianinus* (but not other species of that genus) are said to have the enlarged femoral artery. A modern study of these arteries is needed.

Intestinal Convolutions (Fig. 23).   The pattern of arrangement of the much coiled and looped intestine in the bird has been studied, and the data used with some success as one more criterion in the diagnosis of genera and families. Gadow (1889) gave the classic account, describing seven general types, which he called *isocoelous, anticoelous,* etc., but which in practice are commonly designated as type I, type II, etc. Gadow also provided two general descriptive terms, *orthocoelous* (straight-gutted, i.e., with long loops parallel to the long axis of the body) and *cyclocoelous* (with coils in some of the loops).

The use of the pattern of the intestinal convolutions for taxonomic purposes has been much criticized. The fact remains, however, that, if we arrange the groups of birds according to their intestinal patterns, the resulting series parallels arrangements based on other anatomical criteria, and in many groups the type is quite consistent throughout.

Although the correlation between food habits and length of intestine has not been adequately investigated, it is well established that the intestine is relatively longer in plant- and grain-eating birds than in fruit- and meat-eating ones; it tends to be long in fish-eating birds, however. Among herbivorous birds there are some interesting differences. Blue Grouse (*Dendragapus obscurus*) and Spruce Grouse (*Canachites canadensis*) have a diet composed largely of the low-grade food of conifer needles, and their small intestines are about 28 percent longer, relatively, than those of quail. Even among populations of the same species, a significant variation in intestinal length exists when food habits differ. Hence, in the California Quail (*Lophortyx californicus*), the race that

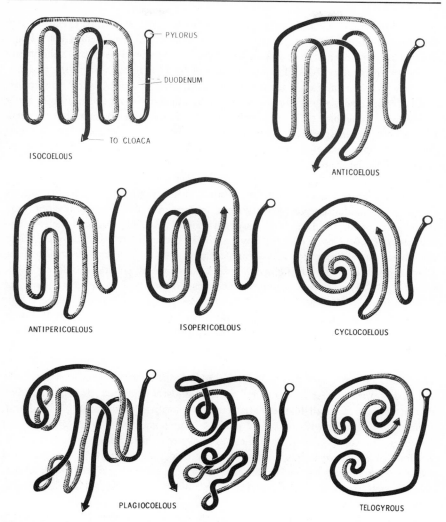

PYLORUS

DUODENUM

TO CLOACA

ISOCOELOUS

ANTICOELOUS

ANTIPERICOELOUS

ISOPERICOELOUS

CYCLOCOELOUS

PLAGIOCOELOUS

TELOGYROUS

Figure 23. Patterns of the intestinal loops (after Gadow). The ascending limb of each loop is shaded; the descending limb, black.

lives along the humid coastal regions of northern California (*L. c. brunnescens*) eats considerably more green food and has an 11 percent relatively longer small intestine than another race (*L. c. californicus*) that inhabits arid, interior valleys. Hence phylogenetic relations may be obscured by extreme diet specialization.

Ceca. Intestinal ceca (Latin *caecum,* blind gut; singular: cecum) are diverticulae found at the junction of the small and large intestine. Their absence, or presence and degree of development, has been used by taxonomists in the technical diagnoses of major groups of birds. The ceca are large and apparently have a major digestive function in the Ostrich, rhea, tinamous, gallinaceous birds, many ducks, shorebirds, etc. Browsing gallinaceous birds have longer ceca than their seed-eating relatives, and it is possible that bacteria in the ceca are responsible for the chemical breakdown of cellulose. Grouse (browsers) have ceca that are as much as 136 percent longer than those of quail (seed-eaters). Moreover, in the two races of the California Quail already mentioned, the ceca are 19 percent longer in the race that eats more green food.

In many birds the ceca are small but seemingly still functional. In others (e.g., herons, gulls, passerines), the ceca are vestigial and presumably nonfunctional (except for the lymphoid tissue in their walls). Herons and at least one African cuckoo (*Chrysococcyx*) have a single cecum rather than a pair. Ceca are absent or rudimentary in anhingas, parrots, some pigeons, plantain-eaters, kingfishers, some hummingbirds, swifts, and woodpeckers (Gadow, 1896; Markus, 1964). Markus reported the presence of ceca in several species of Columbidae: the Rock Pigeon (*Columba guinea*), Cape Turtle Dove (*Streptopelia capicola*), Laughing Dove (*S. senegalensis*), Red-eyed Turtle Dove (*S. semitorquata*); ceca are absent, however, in many species of the family.

Oil Gland (Uropygial Gland). The oil gland shows some variations that have led taxonomists to use its condition as a diagnostic character. In many birds a distinct circlet of small feathers surrounds the orifice (or orifices) of the gland, forming a brushlike tuft (*tufted* oil gland); in a considerable number of species, however, the tuft is absent, and these species are said to have a *nude* oil gland. Miller (1924) showed that, although many bird families may be characterized by the invariable presence or absence of a tuft of feathers on the oil gland, others, such as the Rallidae, the Ardeidae, and the Strigidae, show much variation among the species.

The oil gland is absent in the Ostrich, emu, cassowaries, bustards, frogmouths (Podargidae), the mesite (*Mesoenas*), some parrots (e.g., *Amazona, Pionus, Brotogeris*), some pigeons, and some woodpeckers (e.g., *Campethera*). In some breeds of the domestic pigeon, glandlessness is caused by a recessive gene.

Elder (1954) discussed the chemistry and function of the oil gland: "In waterfowl the secretion is essential for maintenance of feather structure from one molt to the next. . . . It seems unlikely that a bird rendered glandless could survive in the wild."

Wing Formula (Fig. 24). Regardless of how total wing length may vary among individuals of a species, the lengths of the remiges (especially the primaries) relative to each other are usually constant. Description of the relative lengths is called the "wing formula" of a particular species. In certain groups, such as harriers (*Circus*), Old-world warblers (Sylviidae), tyrant-flycatchers (Tyrannidae), and vireos (Vireonidae), the comparison of wing formulae is often one of the best ways of identifying specimens of closely similar species.

Feather Tracts (Figs. 25, 26). Differences among feather tracts were first noted in 1840, but only recently have intensive studies been conducted to determine the value of such data for indicating phylogenetic relationships. Clench (1970) examined 190 specimens of the genus *Passer* and found "no marked individual variation" in *Passer domesticus*. She added that "patterns may be studied confidently from a limited number of specimens per species. The overall patterns also seem to be remarkably consistent and evolutionarily conservative within the class Aves as a whole; the pterylographic differences between taxa are thus believed to be highly significant indications of relationships, especially at higher taxonomic levels." She found two types of variation that appear to be important for taxonomic use: patterns and numbers of feathers.

Morlion (1971) examined 179 specimens representing 19 genera and 40 species of three subfamilies of the Ploceidae: Ploceinae, Estrildinae, Viduinae. She reported that "on comparing different species of the same genus, a nearly identical spinal tract is found. Among the various genera, however, this presents a distinctive pattern. The cervical and the interscapular regions are

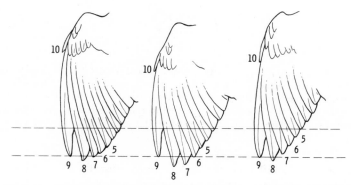

Figure 24. The wing formulae of three species of Old-world warblers (Sylviidae). Left: Reed Warbler (*Acrocephalus scirpaceus*). Center: Blyth's Reed Warbler (*A. dumetorum*). Right: Marsh Warbler (*A. palustris*). (By permission of Roger Tory Peterson, from *A Field Guide to the Birds of Britain and Europe,* Houghton Mifflin Co., copyright 1954.)

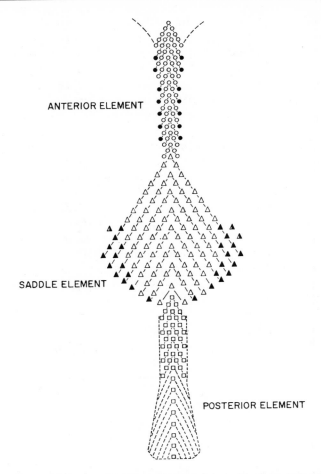

Figure 25. Dorsal feather tract of the House Sparrow (schematic). Open circles: feathers of the anterior element. Open triangles: saddle element. Open squares: posterior element. Solid circles and triangles: basic plumage feathers not present in a 7-day-old nestling. Partly solid triangles: examples of the extra lateral row of feathers present in some specimens. Dashed lines in posterior element: irregular rows of feathers. (Courtesy of Mary H. Clench and the editor of *Auk*.)

very similar, even in different subfamilies. In contrast, the dorsal and the pelvic regions show different patterns in the genera examined."

Pattern of Downy Young. Delacour and Mayr (1945) first used the color of downy young to determine relationships among the Anatidae. In his study of downy grebes, Storer (1967) noted that "the pattern of the young of *occipitalis* is the simplest form known within the genus [*Podiceps*]. What pattern there is resembles that of young *nigricollis*. This similarity supports the conclusions based on behavioral, ecological, and morphological evidence that *occipitalis* and its very recent offshoot, *taczanowskii,* are geographical representatives of *nigricollis.*" Similarly, Jehl (1968) used the color of downy charadriiform birds in delineating relationships in that group.

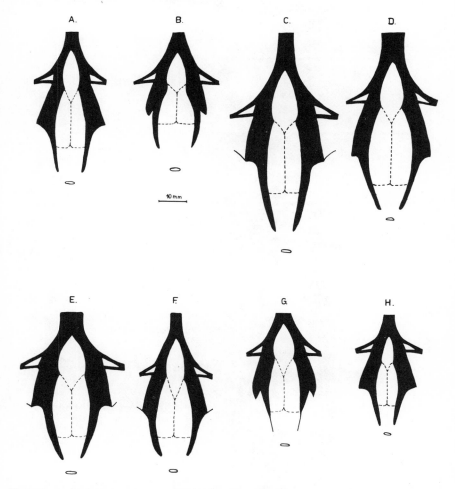

Figure 26. Ventral feather tracts in estrildine finches. (*A*) *Lonchura c. cucullata.* (*B*) *Nigrita f. fusconota.* (*C*) *Spermophaga poliogenys.* (*D*) *Pytilia p. phoenicoptera.* (*E*) *Clytospiza dybowskii.* (*F*) *Nesocharis capistrata.* (*G*) *Lagonosticta larvata nigricollis.* (*H*) *Estrilda atricapilla graueri.* (Courtesy of Maria Morlion and the editor of *Verhandeligen van de Koninklijke Vlaasmse Academie voor Wetenschappen*, Letteren en Schone Kunsten van Belgie.)

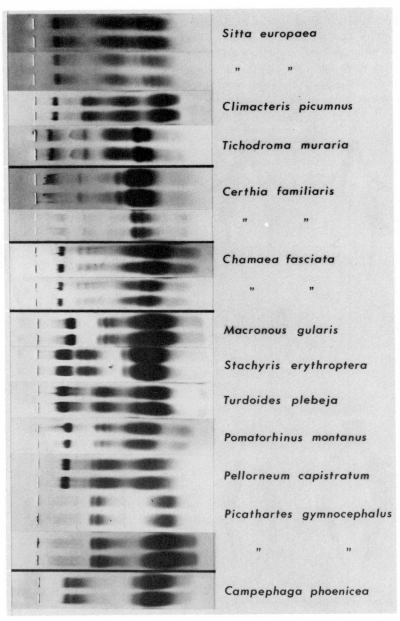

Sitta europaea

"   "

Climacteris picumnus

Tichodroma muraria

Certhia familiaris

"   "

Chamaea fasciata

"   "

Macronous gularis

Stachyris erythroptera

Turdoides plebeja

Pomatorhinus montanus

Pellorneum capistratum

Picathartes gymnocephalus

"   "

Campephaga phoenicea

Figure 27. Starch gel electrophoretic patterns of representatives of the following families: Sittidae, Climacteridae, Certhiidae, Sylviidae (*Chamaea*), Timaliidae, and Campephagidae. (Courtesy of Charles G. Sibley and the Peabody Museum of Natural History.)

Physiological Characters

Selander (1971) discussed the physiological and biochemical techniques that have been used in the search for taxonomic clues. Sibley investigated the egg-white proteins (Fig. 27) of a wide variety of birds, and proposed changes in classification among both passerine and nonpasserine groups (Sibley, 1968, 1970, 1973; Sibley and Ahlquist, 1972). Other biochemical approaches have been investigated by Anderson and Haslewood (1957; bile salts), Arnheim and Wilson (1967; lysozymes), Corbin (1967; ovalbumin peptides), Ferrell (1966; blood groups), Kitto and Wilson (1966; malate dehydrogenase), Mainardi (1962; immunology), Sibley and Brush (1967; eye-lens proteins), Sibley and Hendrickson (1970; plasma proteins), and Wilson *et al.* (1964; lactic dehydrogenase).

Ethological Characters

Ficken and Ficken (1966) discussed the use of behavior in taxonomic studies (Table 2). Other examples in which behavior and voice were used in an attempt to determine closeness of relationship can be found in the papers by Crook (1963), Davis (1965), Dilger (1960), Hardy (1964), Johnsgard (1963,

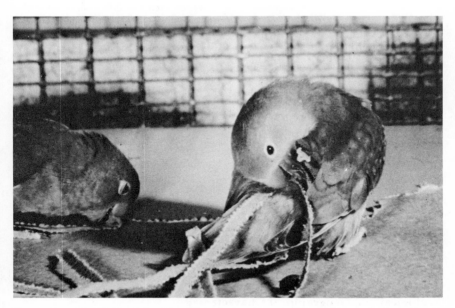

Figure 28. A female Peach-faced Lovebird (*Agapornis roseicollis*) tucking a strip of nesting material into her rump feathers. She will then carry the material to her nest. Some other species of the genus carry nesting material in the bill. (Courtesy of William C. Dilger.)

Table 2. Examples of the Use of Behavioral Characters in Taxonomy[a]

Activity	Level Used	Taxonomic Discrimination
Head-scratching	Familial	Shows that *Recurvirostra, Himantopus,* and *Haematopus* are related to charadriids rather than scolopacids (Simmons, 1957b)
	Subfamilial	Suggest Psittacinae may be polyphyletic (Brereton and Immelmann, 1962)
Holding food with the foot	Familial	Separates *Icteria virens* from Parulidae (Ficken and Ficken, 1962b)
Dust bathing	Subfamilial	Supports relationship of *Passer* to ploceids (Mayr *et al.,* 1953:120)
Water bathing	Familial	Separates timaliids from most other passerines (Simmons, 1963)
Oiling feathers	Familial	Separates timaliids from most other passerines (Simmons, 1963)
Visual agonistic displays	Generic	Splits *Hylocichla mustelina* from *Catharus* spp. (Dilger, 1956b)
	Generic	Often useful at this level in Anatidae (Johnsgard, 1961)
Song	Tribal	Useful at this level in estrildine finches (Delacour, 1943)
	Intergeneric	Shows relationship of *Setophaga ruticilla* and *Dendroica* (Ficken and Ficken, 1962a)
	Intrageneric	Shows close affinities of *Vermivora pinus* and *V. chrysoptera* and of *V. peregrina* and *V. ruficapilla* (Saunders, 1951)
Flight call notes	Subfamilial	Suggest relationship of *Fringilla* and carduelines (Marler, 1957)
Length of pair-bond	Subfamilial	Separates estrildines from ploceids (Steiner, 1955, *in* Mayr, 1958)

1964), Lanyon (1963), Smith (1966), Stein (1963), and Vaurie and Schwartz (1972) (Fig. 28).

Ecological and Geographical Characters

Although Mayr wrote about the "ecological factors in speciation" more than 25 years ago and later (1963) devoted a chapter to "the ecology of speciation,"

Table 2 (con't)

Activity	Level Used	Taxonomic Discrimination
	Subfamilial	Separates Anatinae from some other subfamilies (Johnsgard, 1961)
Bower construction	Subfamilial and familial	Subdivides Ptilonorhynchidae into two subfamilies and places the catbirds in a separate family (Ailuroedidae) (Marshall, 1954:183)
Form of courtship display	Subfamilial	Shows affinity between *Fringilla* and carduelines (Andrew and Hinde, 1956)
	Interfamilial	Presence of tail quivering suggests possible relationship among corvids, estrildines, and ploceids (Andrew, 1961a:561)
Nest construction	Ordinal	Evidence for relationship between hummingbirds and swifts (Pearson, 1953; Amadon, 1959)
	Intergeneric	Indicates close relationship between *Hirundo* and *Ptyonoprogne* (Mayr and Bond, 1943)
	Familial	Separates *Peucedramus* from Parulidae (George, 1962)
Time of initiation of incubation	Subfamilial	Separates estrildines from ploceids (Steiner, 1955, *in* Mayr, 1958)
Participation of sexes in parental care	Subfamilial	Separates *Anseranas semipalmata* from other Anatidae (Johnsgard, 1961)
Feeding of nestlings	Subfamilial	Separates estrildines from ploceids (Steiner, 1955, *in* Mayr, 1958)
Nest sanitation	Familial	Separates *Peucedramus* from Parulidae (George, 1962)

[a] From Ficken and Ficken (1966), by permission of Robert W. and Millicent S. Ficken and the editor of *Auk*.

Selander (1971) stated that "avian taxonomy is still largely based on the morphological features of museum skins, although increasing attention is being given to behavioral, ecological, physiological, and biochemical characters." Orians (1971) reviewed the status of research in behavioral ecology (ethoecology).

Lack (1971) discussed three main ecological isolating mechanisms in birds:

separation by range, by habitat, and by feeding habits. Stallcup (1968) showed that certain nuthatches and woodpeckers living in the same forest were ecologically isolated because of differences in foraging behavior. The sexes of a species may be isolated in their foraging behavior, and the isolation may be related to morphological differences, as in Hammond's Flycatcher (Johnson, 1966b) and in the Huia of New Zealand.

Figure 29.   Displays of *Cassidix*. (*A*) Ruff-out of male *C. mexicanus,* directed to another male at close quarters. (*B*) A male *C. major* in ruff-out display begins a wing-flip as another male returns head-up display from cock posture. (*C*) Solicitation display of male *C. mexicanus,* directed to a female dummy. (*D*) Male *C. major* soliciting a female of the same species. (*E*) Head-up display of *C. mexicanus.* (*F*) Two males of *C. major* exchanging head-up displays. (Courtesy of Robert K. Selander and the editor of *Condor.*)

Figure 30. Swallow-tailed Gulls. (Courtesy of Margaret H. Hundley and John Henry Dick.)

Species-specific behavioral characteristics typically are pronounced in sympatric species. Selander and Giller (1961) studied comparative ecology and ethology in their analysis of Great-tailed and Boat-tailed grackles in a 100-mile zone of sympatry. They concluded that failure of the two species to interbreed "indicates that they are good species, notwithstanding the fact that they have long been considered conspecific by morphological taxonomists," and they suggested that "hybridization is prevented by an ethological isolating mechanism involving the females' selection of homospecific mates on the basis of major differences in the displays and vocalizations of the males" (Fig. 29). Similarly, Gill (1971) studied two species of white-eye (*Zosterops*) that are sympatric on the Mascarene Islands of Mauritius and Réunion, and found differences involving structure, behavior, feeding habits, and vocalizations.

Ecological and geographical adaptations may involve other morphological or behavioral features. The Nihoa race of the Millerbird (*Acrocephalus familiaris kingi*), which inhabits a volcanic island, is darker than the Laysan race (*A. f. familiaris*), which inhabits a sand and coral island. Similarly, the Lava Gull (*Larus fuliginosus*) of the Galapagos Islands has a dark plumage that is cryptic on old lava, on which the birds spend much of their time; it has been suggested that this adaptation reduces competition with other sea scavengers (e.g., frigatebirds and the White-vented Storm-Petrel, *Oceanites gracilis*). By contrast, the Swallow-tailed Gull (*Creagrus furcatus*) of the Galapagos Islands has developed nocturnal habits, presumably in response to daytime predation by frigatebirds (Hailman, 1964; Nelson, 1970; Harris, 1970; Fig. 30).

REFERENCES

Amadon, D. 1959. The significance of sexual differences in size among birds. *Proc. Am. Phil. Soc.,* **103**:531–536.

———1968. Further remarks on the superspecies concept. *Sys. Zool.,* **17**:345–346.

Ames, P. L. 1971. The morphology of the syrinx in passerine birds. *Peabody Mus. Nat. Hist. Bull.* No. 37.

———, M. A. Heimerdinger, and S. L. Warter. 1968. The anatomy and systematic position of the antpipits *Conopophaga* and *Corythopis. Postilla,* **114.**

Anderson, I. G., and G. A. D. Haslewood. 1957. Comparative studies of "bile salts." 10. *Biochem. J.,* **67**:323–328.

Arnheim, N. Jr., and A. C. Wilson. 1967. Quantitative immunological comparison of bird lysozymes. *J. Biol. Chem.,* **242**:3951–3956.

Baumel, J. J. 1964. Vertebral-dorsal carotid artery interrelationships in the pigeon and other birds. *Anat. Anz.,* **114**:113–130.

Beddard, F. E. 1898. *The Structure and Classification of Birds.* Longmans, Green, and Co., London.

Beecher, W. J. 1953. A phylogeny of the oscines. *Auk,* **70**:270–333.

Berger, A. J. 1956. The expansor secundariorum muscle, with special reference to passerine birds. *J. Morphol.,* **99**:137–168.

——— 1959. Leg-muscle formulae and systematics. *Wilson Bull.,* **71**:93–94.

——— 1969. Appendicular myology of passerine birds. *Ibid.,* **81**:220–223.

Blackwelder, R. E. 1967. *Taxonomy: A Text and Reference Book.* John Wiley & Sons, New York.

Bock, W. J. 1963a. Relationships between the birds of paradise and the bower birds. *Condor,* **65**:91–125.

———— 1963b. The cranial evidence for ratite affinities. *Proc. XIII Int. Ornithol. Congr.,* **1962**:39–54.

———— 1969. Comparative morphology in systematics. *Systematic Biology: Proceedings of the International Conference,* National Academy of Sciences, Washington, D.C., pp. 411–448.

Buckley, G. A., and L. E. Wheater. 1968. The isolated expansor secundariorum—a smooth muscle preparation from the wing of the domestic fowl. *J. Pharm. Pharmacol.,* Suppl., **20**:114S–121S.

Clench, M. H. 1970. Variability in body pterylosis, with special reference to the genus *Passer. Auk,* **87**:650–691.

Cooch, G. 1961. Ecological aspects of the Blue-Snow Goose complex. *Auk,* **78**:72–89.

Corbin, K. W. 1967. Evolutionary relationships in the avian genus *Columba* as indicated by ovalbumin tryptic peptides. *Evolution,* **21**:355–368.

Coues, E. 1903. *Key to North American Birds,* 5th ed., Dana Estes & Co., Boston.

Crook, J. H. 1963. The Asian weaver birds: problems of co-existence and evolution with particular reference to behaviour. *J. Bombay Nat. Hist. Soc.,* **60**:1–48.

Davis, L. I. 1965. Acoustic evidence of relationship in *Ortalis* (Cracidae). *Southwest. Nat.,* **10**:288–301.

De Beer, G. R. 1937. *The Development of the Vertebrate Skull.* Oxford University Press, Oxford.

Delacour, J., and E. Mayr. 1945–1946. The family Anatidae. *Wilson Bull.,* **57**:3–55; **58**:104–110.

Dilger, W. C. 1960. The comparative ethology of the African parrot genus *Agapornis. Z. Tierpsychol.,* **17**:649–685.

Dixon, K. L. 1961. Habitat distribution and niche relationship in North American species of *Parus.* In *Vertebrate Speciation,* University of Texas, Austin.

Earhart, C. M., and N. K. Johnson. 1970. Size dimorphism and food habits of North American owls. *Condor,* **72**:251–264.

Elder, W. H. 1954. The oil gland of birds. *Wilson Bull.,* **66**:6–31.

Feduccia, J. A. 1967. The amphirhinal condition in the Passeriformes. *Wilson Bull.,* **79**:453–455.

———— 1972. Variation in the posterior border of the sternum in tree-trunk foraging birds. *Ibid.,* **84**:315–328.

———— 1974. Morphology of the bony stapes in New and Old World suboscines: new evidence for common ancestry. *Auk,* **91**:427–429.

Ferrell, G. T. 1966. Variation in blood group frequencies in populations of Song Sparrows of the San Francisco Bay region. *Evolution,* **20**:369–382.

Ficken, R. W., and M. S. Ficken. 1966. A review of some aspects of avian field ethology. *Auk,* **83**:637–661.

Fisher, H. I. 1955. Major arteries near the heart in the Whooping Crane.*Condor,* **57**:286–289.

Forbes, W. A. 1880. Contributions to the anatomy of passerine birds. III. *Proc. Zool. Soc. London,* **1880**:387–391.

———— 1882. On the variation from the normal structure of the foot in birds. *Ibis,* **1882**:386–390.

Friedmann, H. 1946. Ecological counterparts in birds. *Sci. Mon,* **63**:395–398.

Fürbringer, M. 1886. Uber Deutung und Nomenklatur der Musculatur des Vogelflugels. *Morphol. Jahrb.,* **11**:121–125.

Gadow, H. 1889. On the taxonomic value of the intestinal convolutions in birds. *Proc. Zool. Soc. London*, **1889**:303–316.

—— 1896. In *A Dictionary of Birds*, A. Newton and H. Gadow, eds. Adam and Charles Black, London.

Garrod, A. H. 1873a. On the value in classification of a peculiarity in the anterior margin of the nasal bones in certain birds. *Proc. Zool. Soc. London*, **1873**:33–38.

—— 1873b. On the carotid arteries of birds. *Ibid.*, **1873**:457–472.

—— 1873–1874. On certain muscles of the thigh of birds and on their value in classification. *Ibid.*, **1873**:626–644; **1874**:111–123.

—— 1875. On the disposition of the deep plantar tendons in different birds. *Ibid.* **1875**:39–348.

—— 1876a. On the anatomy of *Chauna derbiana*, and on the systematic position of the screamers (Palamedeidae). *Ibid.*, **1876**:189–200.

—— 1876b. On some anatomical characters which bear upon the major divisions of the passerine birds. I. *Ibid.*, **1876**:506–519.

Gaunt, A. S. 1969. Myology of the leg in swallows. *Auk*, **86**:41–53.

George J. C., and A. J. Berger. 1966. *Avian Myology*. Academic Press, New York.

Gill, F. B. 1971. Ecology and evolution of the sympatric Mascarene white-eyes, *Zosterops borbonica* and *Zosterops olivacea*. *Auk*, **88**:35–60.

——, and W. E. Lanyon. 1964. Experiments on species discrimination in Blue-winged Warblers. *Ibid.*, **81**:53–64.

Glenny, F. H. 1955. Modifications of pattern in the aortic arch system of birds and their phylogenetic significance. *Proc. U. S. Natl. Mus.*, **104**:525–621.

Hailman, J. P. 1964. The Galápagos Swallow-tailed Gull is nocturnal. *Wilson Bull.*, **76**:347–354.

Hall, B. P., and R. E. Moreau. 1969. *An Atlas of Speciation in African Passerine Birds*. British Museum (Natural History), London.

Hardy, J. W. 1964. Behavior, habitat, and relationships of the jays of the genus *Cyanolyca*. *Occas. Pap. C. C. Adams Center Ecol. Stud.*, no. 11.

Harris, M. P. 1970. Breeding ecology of the Swallow-tailed Gull, *Creagrus furcatus*. *Auk*, **87**:215–243.

Heimerdinger, M. A., and P. L. Ames. 1967. Variation in the sternal notches of suboscine passeriform birds. *Postilla*, **105**:1–44.

Hudson, G. E. 1937. Studies on the muscles of the pelvic appendage in birds. *Am. Midl. Nat.*, **18**:1–108.

Huxley, J. 1942. *Evolution: The Modern Synthesis*. Harper & Brothers, New York.

Huxley, T. H. 1867. On the classification of birds; and on the taxonomic value of the modifications of certain of the cranial bones observable in that class. *Proc. Zool. Soc. London*, **1867**:415–472.

Jackson, J. A. 1971. The adaptive significance of reversed sexual dimorphism in tail length of woodpeckers: an alternate hypothesis. *Bird-Banding*, **42**:18–20.

Jehl, J. R., Jr. 1968. Relationships in the Charadrii (shorebirds): a taxonomic study based on color patterns of the downy young. *Soc. Nat. Hist., San Diego, Mem.* No. 3.

Johnsgard, P. A. 1963. Behavioral isolating mechanisms in the family Anatidae. *Proc. XIII Int. Ornithol. Congr.*, **1962**:531–543.

—— 1964. Comparative behavior and relationships of the eiders. *Condor*, **66**:113–129.

Johnson, N. K. 1966a. Bill size and the question of competition in allopatric and sympatric populations of Dusky and Gray flycatchers. *Sys. Zool.,* **15**:70–87.

—— 1966b. Morphologic stability versus adaptive variation in the Hammond's Flycatcher. *Auk,* **83**:179–200.

Jollie, M. T. 1957. The head skeleton of the chicken and remarks on the anatomy of this region in other birds. *J. Morphol.,* **100**:389–436.

Kelso, L. 1940. Variation of the external ear-opening in the Strigidae. *Wilson Bull.,* **52**:24–29.

——, and E. H. Kelso. 1936. The relation of feathering of feet in American owls to humidity of environment and to life zone. *Auk,* **53**:51–56.

Kendeigh, S. C. 1969. Tolerance of cold and Bergmann's rule. *Auk,* **86**:13–25.

Kitto, G. B., and A. C. Wilson. Evolution of malate dehydrogenase in birds. *Science,* **153**:1408–1410.

Kurochkin, E. N. 1968. Locomotion and morphology of the pelvic extremities in swimming and diving birds (in Russian). *Acad. Sci. Ukr. SSR Inst. Zool.,* Moscow, 18 pp.

Lack, D. 1970. Island birds. *Biotropica,* **2**:29–31.

—— 1971. *Ecological Isolation in Birds.* Harvard University Press, Cambridge, Mass.

Lanyon, W. E. 1963. Experiments on species discrimination in *Myiarchus* flycatchers. *Am. Mus. Novit.,* No. 2126.

—— 1966. Hybridization in meadowlarks. *Bull. Am. Mus. Nat. Hist.,* **134**:1–26.

Lasiewski, R. C., S. H. Hubbard, and W. R. Moberly. 1964. Energetic relationships of a very small passerine bird. *Condor,* **66**:212–220.

Mainardi, D. 1962. Studio immunogenetico sulle parentele filogenetiche nell' ordine dei Galliformi. *Ist. Lombardo Accad. Sci. Lett.,* **96**:131–140.

Markus, M. B. 1964. Intestinal caeca in the South African Columbidae. *Bull. Brit. Ornithol. Club,* **84**:137–138.

Mayr, E. 1951. Speciation in birds: Progress report on the years 1938–1950. *Proc. 10th Int. Ornithol. Congr.,* **1950**:91–131.

—— 1956. Geographical character gradients and climatic adaptations. *Evolution,* **10**:105–108.

—— 1963. *Animal Species and Evolution.* Harvard University Press, Cambridge, Mass.

—— 1969. *Principles of Systematic Zoology.* McGraw-Hill Book Co., New York.

—— 1970. *Populations, Species and Evolution.* Harvard University Press, Cambridge, Mass.

——, E. G. Linsley, and R. L. Usinger. 1953. *Methods and Principles of Systematic Zoology.* McGraw-Hill Book Co., New York.

——, and L. L. Short. 1970. Species taxa of North American birds. *Publ. Nuttall Ornithol. Club,* No. 9.

Mengel, R. M. 1970. The North American central plains as an isolating agent in bird speciation. In *Pleistocene and Recent Environments of the Central Great Plains, Dept. Geol. Univ. Kans. Spec. Publ. No. 3.*

Miller, W. DeW. 1924. Further notes on ptilosis. *Bull. Am. Mus. Nat. Hist.,* **50**:305–331.

Monroe, B. L., Jr., and T. R. Howell. 1966. Geographic variation in Middle American parrots of the *Amazona ochrocephala* complex. *Occas. Pap. Mus. Zool. La. State Univ.,* No. 34.

Morlion, M. 1971. *Vergelijkende Studie van de Pterylosis in Enkele Afrikaanse Genera van de Ploceidae.* Verhand. Koninkl. Acad. Vet., Let. en Schonee Kunster van Belgie, 119, 2 vols.

Müller, J. 1878. *On Certain Variations in the Vocal Organs of the Passeres That Have Hitherto Escaped Notice.* Clarendon Press, Oxford, 74 pp.

Nelson, J. B. 1970. The relationships between behaviour and ecology in the Sulidae with reference to other sea birds. *Oceanogr. Marine Biol.,* **8:**501–574.

Nishida, T. 1963. Comparative and topographical anatomy of the fowl. X. The blood vascular system of the hind-limb in the fowl (in Japanese). *Jap. J. Vet. Sci.,* **25:**93–106.

O'Donald, P., and P. E. Davis. 1959. The genetics of the colour phases of the Arctic Skua. *Heredity,* **13:**481–486.

Orians, G. 1971. Ecological aspects of behavior. In *Avian Biology,* Vol. 1, Academic Press, New York, pp. 513–546.

Owen, D. F. 1963. Polymorphism in the Screech Owl in eastern North America. *Wilson Bull.,* 75:183–190.

Payne, R. B. 1971. Paradise Whydahs *Vidua paradisaea* and *V. obtusa* of southern and eastern Africa, with notes on differentiation of the females. *Bull. Brit. Ornithol. Club,* **91:**66–76.

Peters, J. A. 1971. A new approach in the analysis of biogeographic data. *Smithson. Contrib. Zool.,* No. 107.

Pycraft, W. P. 1901. Some points in the morphology of the palate of the Neognathae. *J. Linn. Soc. London, Zool.,* **28:**342–357.

Rand, A. L. 1959. Tarsal scutellation of song birds as a taxonomic character. *Wilson Bull.,* **71:**274–277.

Ridgway, R. 1901–1911. The birds of North and Middle America. *U.S. Natl. Mus. Bull.,* **50,** Pts. 1–5.

Rising, J. D. 1970. Morphological variation and evolution in some North American orioles. *Sys. Zool.,* **19:**315–351.

Selander, R. K. 1971. Systematics and speciation in birds. In *Avian Biology,* Vol. 1, Academic Press, New York, pp. 57–147.

————, and D. R. Giller. 1961. Analysis of sympatry of Great-tailed and Boat-tailed grackles. *Condor,* **63:**29–86.

Short, L. L. 1963. Hybridization in the wood warblers *Vermivora pinus* and *V. chrysoptera. Proc. XIIIth Int. Ornithol. Congr.,* **1962:**147–160.

———— 1965. Hybridization in the flickers (*Colaptes*) of North America. *Bull. Am. Mus. Nat. Hist.,* **129:**307–428.

———— 1969. Taxonomic aspects of avian hybridization. *Auk,* **86:**84–105.

———— 1970. Reversed sexual dimorphism in tail length and foraging differences in woodpeckers. *Bird-Banding,* **41:**85–92.

Sibley, C. G. 1968. The relationships of the "Wren-thrush," *Zeledonia coronata* Ridgway. *Postilla,* **125.**

————1970. A comparative study of the egg-white proteins of passerine birds. *Peabody Mus. Nat. Hist. Bull.* No. 32.

———— 1973. The relationships of the silky flycatchers. *Auk,* **90:**394–410.

————, and J. E. Ahlquist. 1972. A comparative study of egg white proteins of non-passerine birds. *Peabody Mus. Nat. Hist. Bull.* No. 39.

————, and A. H. Brush. 1967. An electrophoretic study of avian eye-lens proteins. *Auk,* **84:**203–219.

————, and H. T. Hendrickson. 1970. A comparative electrophoretic study of avian plasma proteins. *Condor,* **72:**43–49.

————, and L. L. Short, Jr. 1959. Hybridization in the buntings (*Passerina*) of the Great Plains. *Auk,* **76:**443–463.

————, and F. C. Sibley. 1964. Hybridization in the Red-eyed Towhees of Mexico. . . . *Ibid.*, **81**:479–504.

Smith, W. J. 1966. Communication and relationships in the genus *Tyrannus*. *Publ. Nuttall Ornithol. Club*, No. 6. (See also *Auk*, **84**, 1967:606–609.)

Snow, D. W. 1954. Trends in geographical variation in Palaearctic members of the genus Parus. *Evolution*, **8**:19–28.

————, and B. K. Snow. 1967. The breeding cycle of the Swallow-tailed Gull *Creagrus furcatus*. *Ibis*, **109**:14–24.

Stallcup, P. L. 1968. Spatio-temporal relationships of nuthatches and woodpeckers in ponderosa pine forests of Colorado. *Ecology*, **49**:831–843.

Stein, R. C. 1963. Isolating mechanisms between populations of Traill's Flycatcher. *Proc. Amer. Phil. Soc.*, **107**:21–50.

Storer, R. W. 1967. The patterns of downy grebes. *Condor*, **69**:469–478.

———— 1969. What is a tanager? *Living Bird*, **1969**:127–136.

———— 1971. Classification of birds. In *Avian Biology*, Vol. 1, Academic Press, New York, pp. 1–18.

Stresemann, E. 1959. The status of avian systematics and its unsolved problems. *Auk*, **76**:269–280.

Tordoff, H. B. 1954. Relationships in the New World nine-primaried oscines. *Auk*, **71**:273–284.

Vaurie, C., and P. Schwartz. 1972. Morphology and vocalizations of *Synallaxis unirufa* and *Synallaxis castanea* (Furnariidae, Aves), with comments on other *Synallaxis*. *Am. Mus. Novit.*, No. 2483.

Wetmore, A. 1960. A classification for the birds of the world. *Smithson. Misc. Collect.*, **139**.

Wilson, A. C., *et al.* 1964. Evolution of lactic dehydrogenase. *Fed. Proc.*, **23**:1258–1266.

Zusi, R. L. 1967. The role of the depressor mandibulae muscle in kinesis of the avian skull. *Proc. U.S. Nat. Mus.*, **123**, No. 3607: 1–28.

———— 1971. Functional anatomy in systematics. *Taxon*, **20**:75–84.

————, and J. T. Marshall. 1970. A comparison of Asiatic and North American sapsuckers. *Nat. Hist. Bull. Siam Soc.*, **23**:393–407.

CHAPTER THIRTEEN

The Classification of World Birds by Families

We now recognize about 8600 species of birds in the world. No museum contains specimens of all of them, and no ornithologist can immediately recognize every one, even as a specimen in the hand. Only by grouping like species into families can we begin to deal with them with any success.

Most ornithologists divide world birds into about 170 families, each of which contains from 1 to more than 300 species. By definition, a family is a monophyletic group (i.e., it stems from some single ancestral species), and many families are so well marked that there is not the slightest disagreement among ornithologists about which species should be included in these families. Among such families are the parrots, the hummingbirds, and the toucans. Other families, however, are not sharply demarcated, and there is considerable disagreement among ornithologists about where the boundary lines should be drawn.

As we pass from the more ancient types of birds (such as the tinamous and struthious birds) to modern birds (i.e., the "songbirds"), the differences between the family groups become less and less. A subdivision is forced upon us as a matter of convenience by the great evolutionary proliferation of passerine birds, which constitute more than half of all the species now known. Yet if we rigidly maintained the same standards for the songbirds as for the nonpasserine families, we would probably place all passerine birds (or at least the songbirds—the suborder Passeres) in one family. Since this is obviously inexpedient, taxonomists have divided these songbirds into 44 to 53 families. Insofar as this has been done correctly, the birds in each family are related more closely to each other than to the species in any other family.

There are two well-considered schools of thought on which groups of passerines (fringillids or corvids) should be placed last in the classification, i.e., are the highest evolved birds. Not until far more is known about the structure, behavior, and ecology of these groups will we be certain which group should occupy that position.

Some taxonomists reduce the status of certain apparently related families to that of subfamilies of a single, all-inclusive family. It has been known for a long time that there are degrees of relationship between our family categories—so many that we cannot hope to express them all in our linear arrangement of families. Since this is the case, we prefer to maintain many traditional families, for we agree with the philosophy expressed in the *Handbook of British Birds* (Vol. 2, 1938, p. 1): ". . . we consider it preferable to recognize the *Muscicapidae, Sylviidae,* and *Turdidae* as distinct families rather than to merge them into one huge and unwieldy family [= 1109 spp.], as was done by Hartert. It may be repeated that the above three groups evidently do represent the main lines of divergence within this admittedly allied assemblage, and the several stocks are not less worthy of recognition because a small number of more generalized types tend to link them up and do not fit readily into simple definitions. In any case definition is not rendered any easier by lumping them together."

Another example is that of the vireos (Vireonidae), pepper-shrikes (Cyclarhidae), and shrike-vireos (Vireolaniidae), which some taxonomists lump into a single family. It seems not at all certain, however, that they belong together. In 1907 Pycraft pointed out anatomical characters for separating these groups, and we are unaware of any later anatomical study of these birds. Very little is known of the life histories of pepper-shrikes and shrike-vireos, or of many of the vireos. When more complete information is available on all three groups, we may find that they should be combined. At present, however, there appears to be no good basis for doing so. Similarly, the Ibisbill (*Ibidorhynca*) has been placed in the avocet family, in the sandpiper family, and in a separate group. Lowe apparently is the only man to have had the chance to study the anatomy of *Ibidorhynca,* and his paper contains very few details. The point is that, if a taxonomist places a puzzling genus into the nearest likely-looking family (where it may finally prove *not* to belong), he or she may set up a whole series of misleading or false impressions. Other workers will commonly assume that there was a good reason for the action and therefore adopt the placement. Consequently, in time, there develops an imposing series of "authorities" who have placed genus X in a certain family. We believe, therefore, that a very good argument can be made for leaving with separate family status all really doubtful cases, particularly as an aid to students concerned with learning about world birds. This chapter is not a taxonomic treatise.

FAMILY DESCRIPTIONS

These statements of family attributes have been greatly condensed, and a few explanatory remarks are probably needed.

Physical characteristics. *Length* is, of course, the distance from the tip of the bill to the tip of the tail. This is at best an inexact measurement because it depends greatly on how much the curves in the bird's spinal column are straightened out when the measurement is made. For this reason collectors and taxonomists rarely even record it, and as a result we often have had to use the even less accurate measurement taken from a museum specimen. Nevertheless, there seems to be no other measurement that will serve as well to tell the student the general size of a bird, and therfore we have used "length" figures to describe the size range within each bird family and to give the scale for birds illustrated.

Range. The breeding range only is given. Also, man-made changes are ignored.

Habits. In many cases the data on habits, especially breeding habits, are necessarily based on very scanty information. As our knowledge increases, some of these statements will surely be greatly modified. However, for the sake of brevity we have not modified all of these statements with phrases like "as far as is now known," although in most families this qualification is implied.

Breeding. Egg number is the usual number (as far as is known), ignoring rare individual extremes.

Technical diagnosis. Space limitations prohibit a recital of the technical characters of each family (facts that cannot possibly be remembered to any important extent). Therefore we have given a reference to a good published statement.

References. For each family we provide references to several published studies on the habits, especially breeding habits. References to works commonly cited in the Technical diagnosis, Classification, and Reference items are given in brief to save space. Thus Murphy, **1**:329–471, means Vol. 1 of R. C. Murphy's *Oceanic Birds of South America,* published in 1936, pp. 329–471.

Alexander, W. B., *Birds of the Ocean,* 2nd ed. G. P. Putnam, New York, 1954.

Baker, E. C. Stuart, *Fauna of British India: Birds.* Taylor & Francis, London, 8 vols., 1922–1930.

Baker, E. C. Stuart, *The Nidification of Birds of the Indian Empire.* Taylor & Francis, London, 4 vols., 1932–1935.

Bannerman, D. A., *The Birds of Tropical West Africa.* Crown Agents for the Colonies, London, 7 vols., 1930–1949.

Beddard, F. E., *The Structure and Classification of Birds.* Longmans, Green, New York, 1898.

Bent, A. C., Life histories of North American birds. *U.S. Natl. Mus. Bull.,* 1919–1968.

Cory, C. B., B. Conover, and C. E. Hellmayr, Catalogue of birds of the Americas. *Field Mus. Nat. Hist., Zool. Ser.,* Vol. 13. Part 1, 1942–1949, is by C. E. Hellmayr and B. Conover. Part 2, 1918, is by C. B. Cory. Parts 3–5, 1924–1927, are by C. B. Cory and C. E. Hellmayr. Parts 6–11, 1929–1938, are by C. E. Hellmayr.

Mathews, G. M., *The Birds of Australia.* H. F. & G. Witherby, London, 12 vols., 1910–1927.

Peters, J. L., *et al. Check-list of Birds of the World.* Harvard University Press, Cambridge, Mass., 13 vols., 1931–1970.

Ridgway, R., and H. Friedmann, Birds of North and Middle America. *U.S. Natl. Mus. Bull.,* Vol. 50. Parts 1–8, 1901–1919, are by R. Ridgway. Parts 9 and 10, 1941 and 1946, are by R. Ridgway and H. Friedmann. Part 11, 1950, is by H. Friedmann.

Witherby, H. F., *et al., Handbook of British Birds.* H. F. & G. Witherby, London, 5 vols., 1938–1941.

Illustrations

The drawings for this chapter are by George M. Sutton.

ORDERS AND FAMILIES OF LIVING BIRDS OF THE WORLD

TINAMIFORMES

Tinamidae—Tinamou family

STRUTHIONIFORMES

Struthionidae—Ostrich family

RHEIFORMES

Rheidae—Rhea family

CASUARIIFORMES

Casuariidae—Cassowary family

Dromiceiidae—Emu family

APTERYGIFORMES

Apterygidae—Kiwi family

PODICIPEDIFORMES

Podicipedidae—Grebe family

GAVIIFORMES

Gaviidae—Loon family

SPHENISCIFORMES

    Spheniscidae—Penguin family

PROCELLARIIFORMES

    Diomedeidae—Albatross family

    Procellariidae—Shearwater family

    Hydrobatidae—Storm-petrel family

    Pelecanoididae—Diving-petrel family

PELECANIFORMES

    Phaethontidae—Tropicbird family

    Sulidae—Booby family

    Phalacrocoracidae—Cormorant family

    Anhingidae—Anhinga family

    Pelecanidae—Pelican family

    Fregatidae—Frigatebird family

CICONIIFORMES

    Ardeidae—Heron family

    Cochleariidae—Boat-billed Heron family

    Balaenicipitidae—Whale-billed Stork family

    Scopidae—Hammerhead family

    Ciconiidae—Stork family

    Threskiornithidae—Ibis family

    Phoenicopteridae—Flamingo family

ANSERIFORMES

    Anhimidae—Screamer family

    Anatidae—Duck family

FALCONIFORMES

    Cathartidae—American Vulture family

    Sagittariidae—Secretarybird family

    Accipitridae—Hawk family

    Pandionidae—Osprey family

    Falconidae—Falcon family

GALLIFORMES

    Megapodiidae—Megapode family

    Cracidae—Curassow family

    Phasianidae—Pheasant family

    Tetraonidae—Grouse family

    Numididae—Guineafowl family

Meleagrididae—Turkey family

Opisthocomidae—Hoatzin family

GRUIFORMES

Turnicidae—Hemipode-quail family

Pedionomidae—Collared-hemipode family

Gruidae—Crane family

Aramidae—Limpkin family

Rallidae—Rail family

Psophiidae—Trumpeter family

Heliornithidae—Finfoot family

Eurypygidae—Sunbittern family

Rhynochetidae—Kagu family

Cariamidae—Cariama family

Otididae—Bustard family

Mesoenatidae—Mesites family

CHARADRIIFORMES

Jacanidae—Jaççana family

Rostratulidae—Painted-snipe family

Haematopodidae—Oystercatcher family

Scolopacidae—Sandpiper family

Phalaropodidae—Phalarope family

Recurvirostridae—Avocet family

Charadriidae—Plover family

Dromadidae—Crabplover family

Burhinidae—Thick-knee family

Glareolidae—Pratincole family

Thinocoridae—Seedsnipe family

Chionididae—Sheathbill family

Stercorariidae—Skua family

Laridae—Gull family

Rynchopidae—Skimmer family

Alcidae—Auk family

COLUMBIFORMES

Pteroclidae—Sandgrouse family

Columbidae—Pigeon family

PSITTACIFORMES

Psittacidae—Parrot family

MUSOPHAGIFORMES

    Musophagidae—Touraco family

CUCULIFORMES

    Cuculidae—Cuckoo family

STRIGIFORMES

    Tytonidae—Barn owl family

    Strigidae—Typical owl family

CAPRIMULGIFORMES

    Steatornithidae—Oilbird family

    Podargidae—Frogmouth family

    Nyctibiidae—Potoo family

    Aegothelidae—Owlet-frogmouth family

    Caprimulgidae—Nightjar family

APODIFORMES

    Apodidae—Swift family

    Hemiprocnidae—Crested-swift family

    Trochilidae—Hummingbird family

COLIIFORMES

    Coliidae—Coly family

TROGONIFORMES

    Trogonidae—Trogon family

CORACIIFORMES

    Alcedinidae—Kingfisher family

    Todidae—Tody family

    Momotidae—Motmot family

    Meropidae—Bee-eater family

    Coraciidae—Roller family

    Leptosomatidae—Cuckoo-roller family

    Upupidae—Hoopoe family

    Phoeniculidae—Woodhoopoe family

    Bucerotidae—Hornbill family

PICIFORMES

    Galbulidae—Jacamar family

    Bucconidae—Puffbird family

    Indicatoridae—Honeyguide family

    Ramphastidae—Toucan family

    Capitonidae—Barbet family

Picidae—Woodpecker family

Jyngidae—Wryneck family

PASSERIFORMES

Suborder Eurylaimi

Eurylaimidae—Broadbill family

Suborder Furnarii

Dendrocolaptidae—Woodcreeper family

Furnariidae—Ovenbird family

Formicariidae—Antbird family

Rhinocryptidae—Tapaculo family

Suborder Tyranni

Cotingidae—Cotinga family

Pipridae—Manakin family

Phytotomidae—Plantcutter family

Tyrannidae—Tyrant-flycatcher family

Oxyruncidae—Sharpbill family

Pittidae—Pitta family

Xenicidae—New Zealand wren family

Philepittidae—Asity family

Suborder Menurae

Menuridae—Lyrebird family

Atrichornithidae—Scrubbird family

Suborder Passeres ("Oscines" of many authors)

Alaudidae—Lark family

Hirundinidae—Swallow family

Campephagidae—Cuckoo-shrike family

Corvidae—Crow family

Cracticidae—Bellmagpie family

Ptilonorhynchidae—Bowerbird family

Paradisaeidae—Bird-of-paradise family

Grallinidae—Mudnest-builder family

Paridae—Titmouse family

Certhiidae—Creeper family

Rhabdornithidae—Philippine creeper family

Climacteridae—Australian treecreeper family

Sittidae—Common nuthatch family

Neosittidae—Australian nuthatch family

Timaliidae—Babbler family

Dicruridae—Drongo family

Oriolidae—Oriole family

Pycnonotidae—Bulbul family

Irenidae—Leafbird family

Troglodytidae—Wren family

Mimidae—Mockingbird family

Cinclidae—Dipper family

Turdidae—Thrush family

Sylviidae—Old-world warbler family

Muscicapidae—Old-world flycatcher family

Prunellidae—Hedge-sparrow family

Zosteropidae—White-eye family

Motacillidae—Pipit family

Bombycillidae—Waxwing family

Ptilogonatidae—Silky-flycatcher family

Dulidae—Palmchat family

Artamidae—Wood-swallow family

Hyposittidae—Coral-billed nuthatch family

Vangidae—Vanga-shrike family

Laniidae—Shrike family

Prionopidae—Wood-shrike family

Callaeidae—Wattlebird family

Sturnidae—Starling family

Meliphagidae—Honeyeater family

Nectariniidae—Sunbird family

Dicaeidae—Flowerpecker family

Cyclarhidae—Pepper-shrike family

Vireolaniidae—Shrike-vireo family

Vireonidae—Vireo family

Ploceidae—Weaverbird family

Zeledoniidae—Wrenthrush family

Parulidae—American wood-warbler family

Icteridae—Troupial family
Tersinidae—Swallow-tanager family
Thraupidae—Tanager family
Catamblyrhynchidae—Plush-capped Finch family
Drepanididae—Hawaiian honeycreeper family
Fringillidae—Finch family

TINAMIDAE

Tinamou Family (45 species ±)

Figure 1. Rufous-winged Tinamou (*Rhynchotus rufescens*). Total length 17 in. (432 mm).

Physical characteristics. Length—203 to 534 mm (8 to 21 in.). Plumage—tawny, brown, and gray, usually streaked, spotted, or barred in cryptic patterns. Erectile crest in some spp. Bill weak, somewhat elongated and curved, or short, chickenlike. Body compact. Wings short and rounded; tail very short (hidden by contour feathers in some spp.). Legs strong, short to moderately long; hallux elevated (or lacking). Sexes very similar in color; ♀ usually larger.

Range. Neotropical. Mainland from s. Mexico to s. Argentina. Habitat—forest or brushland, and (in S. America) grassland. Nonmigratory.

Habits. Solitary or (some spp.) gregarious. Many spp. crepuscular. Terrestrial; cursorial. Flight strong but not prolonged. Crouch in presence of danger. Voice—whistles and trills, often loud but very mellow in tone—flutelike.

Food. Mainly fruit and seeds; some insects.

Breeding. Polyandrous (at least some spp.). Nest on ground, unlined hollow to substantial structure of sticks.

> *Eggs.* 1 to 10 (or more?); glossy; immaculate; green, blue, yellow, or purplish brown. Incubated by ♂.
>
> *Young.* Nidifugous. Downy. Cared for by ♂.

Technical diagnosis. Beddard, 485–493.

Classification. Hellmayr and Conover, Pt. l, No. 1, 1942:6–114.

References. W. H. Hudson, *Birds of La Plata,* Vol. 2, London, 1920:219–230.

> A. Wetmore, Observations on the birds of Argentina, Paraguay, Uruguay, and Chile, *U. S. Natl. Mus. Bull.,* **133,** 1926:27–42.
>
> A. K. and O. P. Pearson, Natural history and breeding behavior of the Tinamou, *Nothoprocta ornata, Auk,* **72,** 1955:113–127.
>
> A. Wetmore, *The Birds of the Republic of Panamá,* 1, *Smithson. Misc. Collect.,* **150,** 1965:5–24.
>
> F. Haverschmidt, *Birds of Surinam,* Edinburgh, 1968:1–3.
>
> R. Meyer de Schauensee, *A Guide to the Birds of South America,* Wynnewood, Pa., 1970:3–9.

Synonym: Crypturidae.

STRUTHIONIDAE

Ostrich Family (1 species)

Figure 2. Ostrich (*Struthio camelus*).
Total length 72 in. (1829 mm).

Physical characteristics. Length—about 1829 mm (72 in.). The largest extant bird, attaining 300 lb (136 kg) in weight, 8 ft (2.44 m) in height. Plumage—soft and loose-webbed (black in ♂, brownish gray in ♀) on body, sparse on head and most of neck. Thighs bare. Remiges and rectrices (the "plumes" of commerce) very numerous, white in ♂, brownish gray in ♀. Bill short and flat; eyes very large; head small and neck long. Legs long and powerful; 2 toes (3rd and 4th). Sexes unlike.

Range. Africa (exc. c. West Africa), Arabia, and s. Syria. Habitat—open arid country. Nonmigratory.

Habits. Somewhat gregarious. Cursorial; run swiftly. Flightless. Crouch in presence of danger. Voice—a loud hiss; a booming roar (♂ in breeding season).

Food. Succulent plants, berries, seeds; some animal food.

Breeding. Polygamous? Nest a depression scraped in ground (by ♀).

    Eggs. 10 to 20 or 25 (two or more ♀♀ often lay in one nest); with glossy, pitted surface; cream-colored. Incubated by ♂ (at night only) and ♀.

    Young. Nidifugous. Downy. Cared for by ♂ and ♀.

Technical diagnosis. Beddard, 495–496.

Classification. Peters, **1**, 1931:3–4.

References. B. Laufer, Ostrich egg-shell cups of Mesopotamia and the Ostrich in ancient and modern times, *Field Mus. Nat. Hist., Anthropol. Leafl.*, **23**, 1926.

    L. S. Crandall, The struthious birds. II. The Ostriches and Rheas, *Bull. N. Y. Zool. Soc.*, **32**, No. 5, 1929:193–212.

    F. J. Jackson, *The Birds of Kenya Colony and the Uganda Protectorate,* Vol. 1, London, 1938:4–9.

    K. M. Schneider, Vom Brutleben des Strausses (*Struthio*) in Gefangenschaft. In G. Creutz, ed., *Beiträge zur Vogelkunde,* Leipzig, 1949:169–272.

    E. G. F. Sauer and E. M. Sauer, The behavior and ecology of the South African Ostrich, *Living Bird,* **1966**:45–75.

RHEIDAE

Rhea Family (2 species)

Figure 3. Common Rhea (*Rhea americana*). Total length 52 in. (1321 mm).

Physical characteristics. Length—914 to 1321 mm (36 to 52 in.). The heaviest New World bird—weighs about 44 lb (20 kg). Plumage—loose-webbed; brownish gray, with some areas of paler, and some of darker, feathers on head, neck, and under parts; body and wing feathers white-tipped in 1 sp. Wide, flat bill. Neck long. Wings short, with long, soft remiges. No rectrices. Legs long and powerful; 3 toes (2nd, 3rd, and 4th). Sexes very similar—♂ slightly larger and darker.

Range. Eastern Brazil (Rio Grande do Norte), s. Bolivia, and s.e. Peru s. to Straits of Magellan. Habitat—grassland and open-brush country. Nonmigratory.

Habits. Gregarious. Cursorial; run swiftly. Flightless. Swim well. Crouch in presence of danger. Voice—adult, a deep boom; young, a whistle.

Food. Grass, roots, and seeds; also insects, mollusca.

Breeding. Polygamous. Nest a large hollow scraped in ground (by ♂), thinly lined with grass.

    Eggs. 20 to 30 (or even more), laid by several ♀♀ in one nest; golden yellow or deep green (fading rapidly). Incubated by ♂.

    Young. Nidifugous. Downy. Cared for by ♂.

Technical diagnosis. T. Salvadori, Cat. of Birds of Brit. Mus., **27**, 1895:577.

Classification. Hellmayr and Conover, Pt. 1, No. 1, 1942:1–6.

References. W. H. Hudson, *Birds of La Plata*, Vol. 2, London, 1920:230–236.

    A. Wetmore, Observations on the birds of Argentina, Paraguay, Uruguay, and Chile, *U.S. Natl. Mus. Bull.*, **133**, 1926:23–27.

    G. Steinbacher, Zur Brutbiologie des Nandus, *Rhea americana* L., *Zool. Garten*, **18**, 1951:127–137.

    D. F. Bruning, Social structure and reproductive behavior in the Greater Rhea, *Living Bird*, **1974**:251–294.

CASUARIIDAE

Cassowary Family (3 species)

Figure 4. One-wattled Cassowary (*Casuarius unappendiculatus*). Total length 65 in. (1651 mm).

Physical characteristics. Length—1321 to 1651 mm (52 to 65 in.). Plumage—harsh, hairlike, and drooping; black in adults, brown in immatures. Bill short, strong, and laterally compressed. Head and neck featherless, carunculated, particolored in brilliant shades, with heavy casque on forehead and crown, and wattles on lower neck. Wings extremely reduced, remiges represented by only a few bare shafts. No rectrices apparent. Legs long and robust; 3 toes (2nd, 3rd, and 4th), the 2nd with very long, sharp claw. Sexes alike, but ♀ larger.

Range. New Guinea and neighboring islands; Australia (n. Queensland). Habitat—forests. Nonmigratory.

Habits. Somewhat gregarious. Cursorial; run swiftly, even through dense brush. Flightless. Swim well. Pugnacious. Voice—harsh, guttural croaks, or "snorting, grunting, and bellowing."

Food. Chiefly fruits and berries; also insects and spiders.

Breeding. Shallow nest on ground, built of sticks and leaves.

    Eggs. 3 to 8; light green, with coarsely granulated surface. Incubated by ♂.

    Young. Nidifugous. Downy. Cared for by ♂ (and ♀ ?).

Technical diagnosis. T. Salvadori, Cat. of Birds of Brit. Mus., **27,** 1895:585, 590.

Classification. Peters, **1,** 1931:5–9.

References. W. Rothschild, A monograph of the genus *Casuarius, Trans. Zool. Soc. London,* **15,** 1900:109–148, pls. 22–41.

    A. J. Campbell, *Nests and Eggs of Australian Birds,* Vol. 2, Sheffield, England, 2, 1901:1069–1072.

    N. W. Cayley, *What Bird Is That?,* 4th ed., Sydney, 1966:5.

    A. L. Rand and E. T. Gilliard, *Handbook of New Guinea Birds,* New York, 1967:21–25.

DROMICEIIDAE

Emu Family (2 species)

Figure 5. Common Emu (*Dromiceius n. hollandiae*). Total length 78 in. (1981 mm).

Physical characteristics. Length—1397 to 1981 mm (55 to 78 in.). Plumage—rather coarse, hairlike, and drooping; blackish brown above, lighter brown below. Bill short, stout, and somewhat flattened. Sides of head and neck bare and bluish white. Wings much reduced, without real quills. No distinguishable retrices. Legs long and robust; tarsus and tibiotarsal joint bare. Three toes (2nd, 3rd, and 4th); claws short and strong. Sexes alike or nearly so.

Range. Australia and Tasmania. Habitat—open, arid country. Nonmigratory.

Habits. Somewhat gregarious. Cursorial. Flightless. Swim well. Voice—hissing, grunting, or booming sounds.

Food. Vegetable matter, esp. fruits.

Breeding. Nest a hollow in ground, sometimes lined with leaves and twigs.

    Eggs. 7 to 12; dark green. Incubated mainly by ♂.

    Young. Nidifugous. Downy. Cared for by ♂ and ♀.

Technical diagnosis. L. Brasil, Fam. Dromaiidae, Wytsman's *Genera Avium,* Pt. 25, 1914:5 pp., col. pl.

Classification. Peters, **1,** 1931:9–10.

References. A. J. Campbell, *Nests and Eggs of Australian Birds,* Vol. 2, Sheffield, England, 1901:1058–1069.

    D. W. Gaukrodger, The Emu at home, *Emu,* **25,** 1925:53–57, pls. 12–18.

    D. Fleay, Nesting of the Emu, *Emu,* **35,** 1936:202–210, pls. 17–20.

    H. J. Frith, ed., *Birds in the Australian High Country,* Sydney 1969:37–39.

Synonyms: Dromaeidae, Dromaiidae.

APTERYGIDAE

Kiwi Family (3 species)

Figure 6. South. Island Kiwi (*Apteryx australis*). Total length 27 in. (686 mm).

Physical characteristics. Length—482 to 838 mm (19 to 33 in.). Plumage—coarse and furlike; brown or grayish, streaked or barred, esp. above. Bill long, slender, and slightly curved; nostrils near tip (lateral and somewhat ventral). Very long, thin bristles about base of bill. Head small, neck rather long. Wings extremely reduced; no discernible rectrices. Legs rather short and very stout; tarsi bare. Four toes; hallux short, elevated; claws long and sharp. Sexes similar, but ♀ larger.

Range. New Zealand. Habitat—humid forest. Nonmigratory.

Habits. Somewhat gregarious. Pugnacious. Nocturnal or crepuscular. Cursorial. Flightless. Often feed by probing ground with bill. Voice—shrill whistle or scream.

Food. Worms, insects, berries, young shoots of plants.

Breeding. Nest a hole in ground, sometimes lined with leaves and grass.

    Eggs. 1 or 2; white or slightly greenish; extremely large. Incubated by ♂.

    Young. Nidicolous? Downy. Cared for by ♂?

Technical diagnosis. T. Salvadori, Cat. of Birds of Brit. Mus., **27**, 1895:603.

Classification. Peters, **1**, 1931:11–12.

References. W. L. Buller, *A History of the Birds of New Zealand,* Vol. 2, 2nd ed., London, 1888:308–332; Suppl., Vol. 1, 1905:1–30.

    W. R. B. Oliver, *New Zealand Birds,* Wellington, 1930:55–62.

    H. R. Haeusler, Notes on the habits of the North Island Kiwi (*Apteryx mantelli*), *Emu,* **22**, 1923:175–179.

    R. J. Anderson, The Kiwi, New Zealand's wonder bird, *Natl. Geogr. Mag.,* Sept. 1955:394–398.

PODICIPEDIDAE

Grebe Family (18 species)

Figure 7. Pied-billed Grebe (*Podilymbus podiceps*). Total length 13 in. (330 mm).

Physical characteristics. Length—222 to 603 mm (8.75 to 23.75 in.). Plumage—satiny; upper parts dark gray or black; under parts in some solidly reddish brown, but usually glistening white, immaculate or mottled with brown; throat and foreneck rich reddish brown in some; head usually with lateral tufts of feathers. Bill sharply pointed (exc. *Podilymbus*), relatively longer in larger spp. Wings short; tail rudimentary. Tarsi laterally compressed; toes lobed; claws extremely flattened. Sexes alike.

Range. Worldwide (exc. extreme n. and some oceanic islands). Habitat—lakes and ponds with much emergent vegetation (some to seacoast in winter). Northern species migratory.

Habits. Solitary or slightly gregarious. Swim under water (foot-propelled). Completely aquatic. Flight usually weak and infrequent (*Rollandia* flightless). Voice—great variety of clucks, croaks, whistling notes, and tremulous calls.

Food. Fish, crustacea, mollusca, insects, and some vegetable matter. Eat and even feed to small young many of their own body feathers.

Breeding. Nest a heap of wet, decaying vegetable matter in shallow water or floating in deep water.

Eggs. 3 to 9; dull white; unspotted. Incubated by ♂ and ♀.

Young. Nidifugous. Downy—head and neck boldly streaked and spotted (exc. *Aechmophorus occidentalis,* which is gray, unpatterned). Cared for by ♂ and ♀.

Technical diagnosis. Witherby, 4, 1940:84–85.

Classification. Peters, 1, 1931:35–41.

References. Bent, No. 107:1–47; Witherby, 4, 1940:84–111.

J. S. Huxley, The courtship-habits of the Great Crested Grebe . . . , *Proc. Zool. Soc. London,* 1914:491–562.

A. Wetmore, Food and economic relations of North American grebes, *U.S. Dept. Agric. Dept. Bull.* No. 1196, 1924.

J. A. Munro, Studies of waterfowl in British Columbia; The Grebes, *B. C. Prov. Mus. Occas. Pap.* No. 3, 1941.

R. W. Storer, Observations on the Great Grebe, *Condor,* 65, 1963:279–288.

R. W. Storer, The color phases of the Western Grebe, *Living Bird,* 1965:59–63.

R. W. Storer, The patterns of downy grebes, *Condor,* 69, 1967:469–478.

Synonyms: Colymbidae, Podicipitidae, Podicepidae, Podicipidae.

GAVIIDAE

Loon Family (4 species)

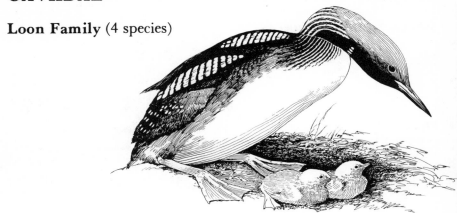

Figure 8. Arctic Loon (*Gavia arctica*).
Total length 27 in. (686 mm).

Physical characteristics. Length—661 to 953 mm (26 to 37.5 in.). Plumage—dense, compact on body; head, neck, and upper parts black or gray, streaked, barred, and spotted with white in bold patterns (in winter gray, unspotted, exc. *Gavia stellata*); under parts white (foreneck rich chestnut in *G. stellata*). Bill strong, tapering, acute. Wings relatively small, pointed; tail well developed but short. Tarsi reticulate, laterally compressed. First 3 toes fully webbed. Sexes alike.

Range. Northern Eurasia and n. N. America. Habitat—lakes and ponds; mainly sea in winter. Migratory.

Habits. Found singly or in pairs, occasionally (in winter) in small, loose flocks. Swim under water (foot-propelled). Go ashore only to nest. Flight strong, direct; large species require a long run along surface of water before rising. Voice—great variety of deep, guttural notes and loud, quavering calls.

Food. Mainly fish; also crustacea, mollusca, insects.

Breeding. Nest a slight depression in ground to large heap of vegetable matter, at edge of water.
  Eggs. 2 (rarely 1 or 3); olive-brown; irregularly spotted with darker brown. Incubated by ♂ and ♀.
  Young. Nidifugous. Downy—uniform gray above, white below. Cared for by ♂ and ♀.

Technical diagnosis. Witherby, **4**, 1940:111.

Classification. Peters, **1**, 1931:34–35.

References. Bent, No. 107:47–82; Witherby, **4**, 1940:111–129.
  R. A. Johnson and H. S. Johnson, A study of the nesting and family life of the Red-throated Loon, *Wilson Bull.,* **47**, 1935:97–103.
  S. T. Olson and W. H. Marshall, The Common Loon in Minnesota, *Minn. Mus. Nat. Hist. Occas. Pap.* No. 5, 1952:vi + 77 pp.
  W. E. Godfry, *The Birds of Canada,* Ottawa, 1966:9–14.
  D. J. Tate and J. Tate, Jr., Mating behavior of the Common Loon, *Auk,* **87**, 1970:125–130.

Synonyms: Colymbidae, of European ornithologists; Urinatoridae.

SPHENISCIDAE

Penguin Family (15 species)

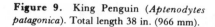

Figure 9. King Penguin (*Aptenodytes patagonica*). Total length 38 in. (966 mm).

Physical characteristics. Length—406 to 1219 mm (16 to 48 in.). Plumage—extremely dense, compact coat of very small glossy feathers; gray, black, or bluish above, white below; specific distinctions in pattern and color confined mainly to head and neck; in some, long superciliary crests, yellow on sides of head or neck, etc. No apteria. Bill short and stout to rather long and curved. Neck short; body stout. Wings paddlelike, with scalelike remiges; do not fold. Tail very short. Legs short and stout, set far back; feet webbed. Sexes alike.

Range. Coasts of Antarctica; subantarctic islands; s. coasts of Australia, Africa, and S. America; n. along w. coast of S. America to the Galapagos (on equator). Some spp. migratory.

Habits. Gregarious. Most completely marine of all birds. Flightless. Swim low in water, using wings alone, frequently leaping and diving porpoise-fashion. On land or ice walk upright, hop, or slide on belly. Voice—loud harsh brays, croaks, barks, or trumpeting calls.

Food. Fish, squid, crustacea.

Breeding. Nest on ground or ice (in some spp. the egg is held, sometimes carried, on top of the feet), or in burrows or caves, sometimes lined with grass or sticks.

Eggs. 1 or 2 (rarely 3); immaculate; white to pale olive-green. Incubated by ♂ and ♀, or ♂ only (Emperor Penguin).

Young. Nidicolous. Downy. Cared for by ♂ and ♀.

Technical diagnosis. Beddard, 396–402 ("Sphenisci").

Classification. Peters, **1,** 1931:29–33.

References. Murphy, **1:**329–471; Alexander, 146–159, pls. 51–57.

L. E. Richdale, Sexual behavior in penguins, Lawrence, Kans., 1951, 316 pp.

W. J. L. Sladen, The pygoscelid penguins, 1, 2, *Falkland Isl. Depend. Surv. Sci. Rept.* No. 17, 1958, London.

O. S. Pettingill, Jr., Penguins ashore at the Falkland Islands, *Living Bird,* **1964:**45–64.

O. L. Austin, Jr., *Antarctic Bird Studies,* Antarctic Res. Ser. 12, American Geophysical Union, 1968.

Synonym: Aptenodytidae.

DIOMEDEIDAE

Albatross Family (13 species)

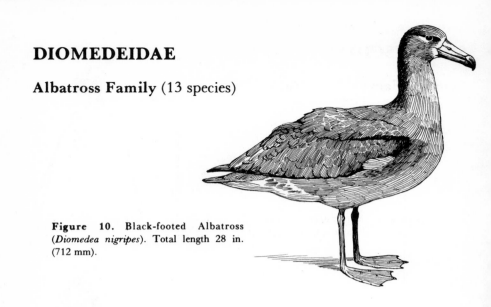

Figure 10. Black-footed Albatross (*Diomedea nigripes*). Total length 28 in. (712 mm).

Physical characteristics. Length—712 to 1346 mm (28 to 53 in.). Plumage—mainly white or sooty brown; tip, edge, or whole of wing black or dark brown; tail and midback black in many spp. Bill very stout, hooked, covered with horny plates; nostrils tubular. Head large, body stout. Wings extremely long and narrow; tail short or moderately long. Legs short; hallux absent or rudimentary; feet webbed. Sexes alike (exc. in *Diomedea exulans*).

Range. Southern oceans, from about 30° S. to Antarctica; n. in the Pacific to the Bering Sea. Most spp. migratory.

Habits. Often gregarious. Pelagic. Exceptional powers of gliding flight. Many spp. follow ships for long periods. Discharge oil from mouth and nostrils when disturbed. Remarkable communal displays or dances. Voice—loud screams, hoarse croaks, loud braying notes.

Food. Fish, squid, and other marine animals.

Breeding. Colonial (exc. *Phoebetria*). Nest a mere scrape in ground or a heap of grass, moss, and earth.

> *Eggs.* 1; white; speckled with brown in some spp. Incubated by ♂ and ♀.
>
> *Young.* Nidicolous. Downy. Fed by ♂ and ♀.

Technical diagnosis. E. Coues, *Key to North American Birds,* Vol. 2, Boston, 1903:1022–1023.

Classification. Peters, **1,** 1931:41–46.

References. Bent, No. 121:1–25; Witherby, **4,** 1940:25, 81–84; Murphy, **1:**471–584; Alexander, 1–15.

> L. E. Richdale, A Royal Albatross nesting on the Otago Peninsula, New Zealand, *Emu,* **38,** 1939:467–488, pls. 60–66 (see also Vol. 41, 1942:169–184, 253–264).
>
> D. W. Rice and K. W. Kenyon, Breeding distribution, history, and populations of North Pacific albatrosses, *Auk,* **79,** 1962:365–386.
>
> Bryan Nelson, *Galapagos Islands of Birds,* New York, 1968.
>
> H. I. Fisher, Eggs and egg-laying in the Laysan Albatross, *Diomedea immutabilis, Condor,* **71,** 1969:102–112
>
> H. I. Fisher, The Laysan Albatross: its incubation, hatching, and associated behaviors, *Living Bird,* **1971:**19–78.

PROCELLARIIDAE

Shearwater Family (53 species)

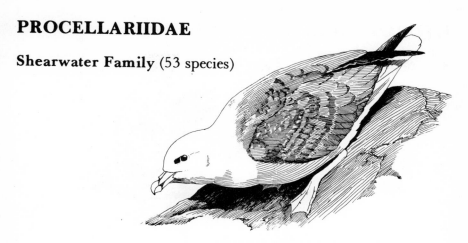

Figure 11. Fulmar (*Fulmarus glacialis*).
Total length 19 in. (482 mm).

Physical characteristics. Length—280 to 914 mm (11 to 36 in.). Plumage—white, gray, brown, or black, or combinations of these. Two color phases in some spp. Bill short and heavy to rather long and slender, hooked, covered with horny plates; nostrils tubular. Wings long and pointed; tail short. Legs short to medium; feet webbed; hallux minute. Sexes alike.

Range. Oceans of the world. Migratory.

Habits. Usually gregarious. Pelagic. Flight rapid and gliding, usually close above surface of water. Pick food from surface in passing flight, or swim about on water to feed. Discharge oil from mouth and nostrils when disturbed. Voice—great variety of shrieks, wails, raucous calls, and cooing sounds.

Food. Fish, squid, crustacea, and other animal matter, including small birds; also carrion.

Breeding. Colonial. Nest in burrows or rock crevices, sometimes lined with vegetable matter. A few large spp. nest on open ground.

    Eggs. 1; white. Incubated by ♂ and ♀.

    Young. Nidicolous. Downy. Cared for by ♂ and ♀.

Technical diagnosis. E. Coues, *Key to North American Birds,* Vol. 2, Boston, 1903:1026–1027, 1031.

Classification. Peters, **1,** 1931:46–68.

References. Bent, No. 121:26–123; Witherby, **4,** 1940:25, 41–80; Murphy, **1:**471–489, and **2:**584–726; Alexander, 16–51.

    L. E. Richdale, The Sooty Shearwater in New Zealand, *Condor,* **46,** 1944:93–107. (See also Vol. 47, 1945:45–62.)

    L. E. Richdale, The Parara or Broad-billed Prion, *Emu,* **43,** 1944:191–217.

    J. Fisher, *The Fulmar,* London, 1952.

    R. S. Palmer, ed., *Handbook of North American Birds,* Vol. 1, New Haven, 1962, pp. 136–217.

    W. B. King and P. J. Gould, The status of Newell's race of the Manx Shearwater, *Living Bird,* 1967:163–186.

    M. P. Harris, The biology of an endangered species, the Dark-rumped Petrel (*Pterodroma phaeopygia*), in the Galapagos Islands, *Condor,* **72,** 1970:76–84.

Synonyms: Fulmaridae, Puffinidae.

HYDROBATIDAE

Storm-petrel Family (22 species)

Figure 12. Wilson's Petrel (*Oceanites oceanicus*). Total length 7 in. (177 mm).

Physical characteristics. Length—139 to 254 mm (5.5 to 10 in.). Plumage—brownish black or gray; with white on rump or under parts of many spp., on face and throat of 3. Bill and legs black (but webs of feet bright yellow in 2 spp.). Bill medium in length, slender, grooved, hooked; nostrils tubular. Neck short, wings long, tail medium to long. Legs slender, medium to long; feet webbed; hallux minute. Sexes alike. Strong musky odor.

Range. Oceans of the world, principally s. of Arctic Circle. Most spp. migratory.

Habits. Usually gregarious. Pelagic. Flight erratic and fluttering. Pick food from surface of sea as they flutter over it, often striking the water with the feet, or (rarely) dive for food. Discharge oil from mouth and nostrils when disturbed. Voice (rarely heard at sea)—great variety of chirping, squealing, and cooing notes used about nesting grounds.

Food. Crustacea, fish, and other small marine animals; also carrion, esp. fatty substances.

Breeding. Colonial. Nest in burrows or rock crevices, sometimes lined.

    Eggs. 1; white; often with red or black spots on the larger end. Incubated by ♂ and ♀.

    Young. Nidicolous. Downy. Cared for by ♂ and ♀.

Technical diagnosis. Witherby, **4**, 1940:25.

Classification. Peters, **1**, 1931:68–75.

References. Bent, No. 121:123–181; Murphy, **1**:471–489, and **2**:726–771; Alexander, 52–64.

    B. Roberts, The life cycle of Wilson's Petrel, *Oceanites oceanicus* (Kuhl), *Brit Graham Land Exped. 1934–37, Sci. Rept.*, **1**, 1940:141–194.

    R. S. Palmer, ed., *Handbook of North American Birds*, Vol. 1, New Haven, 1962:217–254.

    M. P. Harris, The biology of storm petrels in the Galápagos Islands, *Proc. Calif. Acad. Sci.*, **37**, 1969:95–165.

    D. G. Ainley *et al.*, Patterns in the life histories of Storm Petrels on the Farallon Islands, *Living Bird*, **1974**:295–312.

Synonyms: Procellariidae, Thalassidromidae.

PELECANOIDIDAE

Diving-petrel Family (5 species)

Figure 13. Peruvian Diving-petrel (*Pelecanoides garnotii*). Total length 10 in. (254 mm).

Physical characteristics. Length—165 to 254 mm (6.5 to 10 in.). Plumage—black above and white below. Bill short and stout, hooked, covered with horny plates; nostrils tubular (opening upward). Neck, wings, and tail short. Legs short, laterally compressed; feet webbed; hallux absent. Sexes alike.

Range. Southern oceans (between about 35° and 60° S. lat.) and n. along the w. coast of S. America to n. Peru. Largely nonmigratory.

Habits. Somewhat gregarious. Pelagic. Fly with very rapid wing beats. Dive from the wing into the sea to obtain food and to escape enemies. Use wings to swim under water; often emerge flying. Voice—croaking and mewing sounds.

Food. Fish, crustacea, and other animal matter.

Breeding. Somewhat colonial. Nest in burrows or rock crevices with little or no lining.

Eggs. 1; white. Incubated by ♂ and ♀.

Young. Nidicolous. Downy. Cared for by ♂ and ♀.

Technical diagnosis. O. Salvin, Cat. of Birds of Brit. Mus., **25**, 1896:340–342.

Classification. Peters, **1**, 1931:75–77.

References. Murphy, **2**:771–792; Alexander, 65–69, pl. 22.

R. C. Murphy and F. Harper, A review of the Diving Petrels, *Am. Mus. Nat. Hist. Bull.*, **44**, 1921:495–554, pls. 21–24.

L. E. Richdale, The Kuaka or Diving Petrel, *Emu, 43*, 1943:24–48, 97–107.

L. E. Richdale, Supplementary notes on the Diving Petrel, *Trans. Roy. Soc. N. Z.*, **75**, 1945:42–53.

Synonym: Pelecanoidae.

PHAETHONTIDAE

Tropicbird Family

(3 species)

Figure 14. White-tailed Tropicbird (*Phaethon lepturus*). Total length (without central tail feathers) 17 in. (432 mm).

Physical characteristics. Length—406 to 482 mm (16 to 19 in.), excluding the attenuate middle tail feathers, which may add as much as 535 mm (21 in.) to length. Plumage—white, with varied amounts of black on side of head (as orbital line), on wing, and on shafts of tail feathers. Webs of elongated tail feathers red in *Phaethon rubricauda*. Bill yellow or orangered; rather long, stout, slightly decurved, and pointed. Head large, neck short. Wings long and pointed; tail wedge-shaped and of moderate length, except for extremely long and narrow central feathers. Legs extremely short; feet webbed (including all 4 toes). Sexes alike.

Range. Tropical seas; n. in Atlantic to Bermuda and s. in Pacific to s. Australia. Migratory in part.

Habits. Somewhat gregarious. Range daily far out to sea (60 to 80 mi). Fly with quick, steady wing beats high above water. Hover and plunge from height of 50 ft or more for food. On ground can only shuffle along. Voice—shrill whistle, rasping calls.

Food. Fish, squid, crustacea.

Breeding. Somewhat colonial. Lay eggs on bare rock, in shallow caves, or on ground under bushes.

    Eggs. 1; buff; heavily marked with brown and black. Incubated by ♂ and ♀.

    Young. Nidicolous. Downy. Cared for by ♂ and ♀.

Technical diagnosis. E. Coues, *Key to North American Birds,* Vol. 2, Boston, 1903:971.

Classification. Peters, **1,** 1931:77–79.

References. Bent, No. 121:181–193; Murphy, **2**:796–807; Alexander, 222–225.

    B. Stonehouse, The tropic birds (genus *Phaethon*) of Ascension Island, *Ibis,* **103b,** 1962:124–161.

    R. H. Palmer, ed., *Handbook of North American Birds,* Vol. 1, New Haven, 1962:255–264.

    T. R. Howell and G. A. Bartholomew, Experiments on nesting behavior of the Red-tailed Tropicbird, *Phaethon rubricauda, Condor,* **71,** 1969:113–119.

Synonym: Phaëtontidae.

SULIDAE

Booby Family (7 species)

Figure 15. Blue-faced Booby (*Sula dactylatra*). Total length 30 in. (762 mm).

Physical characteristics. Length—661 to 1029 mm (26 to 40.5 in.). Plumage—typically white in adult, with primaries or whole wing brownish black. Tail of some spp. partly or wholly dark in color. Head and neck brown, streaked with white in *Sula nebouxii,* brownish black in *S. leucogaster.* Bill, bare skin about face and throat, and feet usually brightly colored. Bill stout, conical, pointed, and slightly curved toward the tip; nostrils obsolete. Neck moderate in length; body stout. Wings long and pointed; tail rather long and wedge-shaped. Legs short; feet large and fully webbed. Sexes alike or nearly so.

Range. Tropical and temperate seas, exc. n. Pacific. Spp. of high latitudes are migratory.

Habits. Gregarious. Flight strong, with rapid, regular wing beats, interrupted by occasional glides. Feed by diving from the wing (from 30 to 100 ft in air); pursue fish under water, even to considerable depths. Voice—silent at sea; soft guttural notes, loud grunts, croaks, and whistles used about nesting grounds.

Food. Fish, squid.

Breeding. Colonial. Nest variable, from slight hollow in ground to structure of sticks in a tree.
    Eggs. 1 to 3; pale blue, with chalky, white surface. Incubated by ♂ and ♀.
    Young. Nidicolous. Naked at hatching, downy later. Cared for by ♂ and ♀.

Technical diagnosis. Witherby, **4,** 1940:14.

Classification. Peters, **1,** 1931:82–85.

References. Bent, No. 121:193–229; Murphy, **2:**827–870; Mathews, **4,** 1914–15:199–235; Alexander, 192–199.
    James Fisher and H. G. Vevers, The breeding distribution, history and population of the North Atlantic Gannet (*Sula bassana*), *J. Anim. Ecol.,* **12,** 1943:173–213; **13,** 1944:49–62.
    K. A. Wodzicki and C. P. McMeekan, The Gannet on Cape Kidnappers, *Trans. Proc. Roy. Soc. N. Z.,* **76,** 1947:429–452.
    C. B. Kepler, Breeding biology of the Blue-faced Booby, *Sula dactylatra personata,* on Green Island, Kure Atoll, *Publ. Nuttall Ornithol. Club,* No. 8, 1969.

PHALACROCORACIDAE

Cormorant Family (30 species)

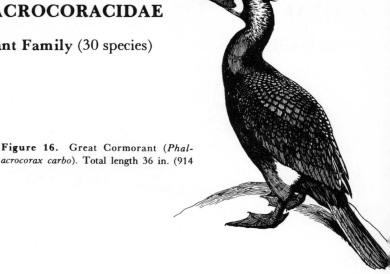

Figure 16. Great Cormorant (*Phalacrocorax carbo*). Total length 36 in. (914

Physical characteristics. Length—482 to 1016 mm (19 to 40 in.). Plumage—largely black in most spp.; in spp. of S. Hemisphere the throat or whole under parts may be white, or the plumage largely gray and feet brightly colored. Eyes, bill, and skin of face usually brightly colored. In breeding plumage of many spp. there are light-colored plumes on head, neck, and flanks. Bill cylindrical, rather slender, hooked; nostrils obsolete. Neck and body long. Wings short or of moderate length; tail long and stiff. Legs short; feet large and fully webbed. Sexes very similar.

Range. Seacoasts and large lakes and rivers of the world, exc. n.c. Canada and n. Asia. Absent from islands of central Pacific. Spp. of higher latitudes migratory.

Habits. Gregarious. Littoral. Fly with steady wing beats, usually close to surface of the water. One species (of the Galapagos Islands) flightless. Feed largely under water, diving (from surface) to considerable depths. Voice—rarely heard exc. at nest site, where a variety of grunts and guttural notes is used.

Food. Fish, crustacea, amphibia.

Breeding. Colonial (most spp.). Nest of seaweed, sticks, etc., on low rocks, on cliffs, or in trees.

> *Eggs.* 2 to 4 (or 6); pale blue or pale green, with chalky surface (blotched with brown in *P. nigrogularis*). Incubated by ♂ and ♀.
>
> *Young.* Nidicolous. Naked at hatching, downy later. Cared for by ♂ and ♀.

Technical diagnosis. Witherby, **4**, 1940:1–2.

Classification. Peters, **1**, 1931:85–94.

References. Bent, No. 121:236–282; Murphy, **2**:870–919; Alexander, 200–221.

> L. Williams, Display and sexual behavior of the Brandt Cormorant, *Condor,* **44,** 1942:85–104.
>
> R. H. Palmer, ed., *Handbook of North American Birds,* Vol. 1, New Haven, 1962:315–357.
>
> A. M. Bailey and J. H. Sorensen, Subantarctic Campbell Island, *Proc. Denver Mus. Nat. Hist.,* No. 10, 1962:235–241.
>
> H. J. Frith, ed., *Birds in the Australian High Country,* Sydney, 1969:47–68.

Synonym: Graculidae (which has also been used as name for Sturnidae).

ANHINGIDAE

Anhinga Family (4 species)

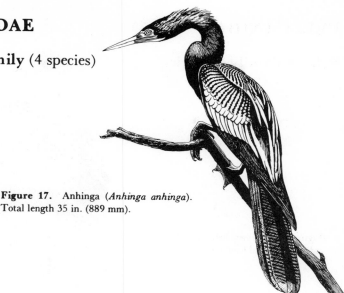

Figure 17. Anhinga (*Anhinga anhinga*).
Total length 35 in. (889 mm).

Physical characteristics. Length—864 to 914 mm (34 to 36 in.). Plumage—black, dotted and streaked with white on upper back, scapulars, and wing coverts; head and neck brown (black in ♂ *A. anhinga*). Heavy transverse fluting on central tail feathers and innermost tertials. Light-colored plumes on sides of head and neck of breeding ♂. Bill long, slender, sharply pointed, finely serrate; nostrils obsolete. Head small; neck very long and slender; body elongate. Wings long and pointed; tail long and stiff. Legs very short; feet large, fully webbed. Sexes unlike.

Range. Southeastern U.S., Middle America, S. America (exc. s. third); Africa s. of Sahara; s. Asia; East Indies, Philippines; Australia and New Zealand. Habitat—lakes and rivers with wooded shores. Migratory in U.S.

Habits. Gregarious during migration and nesting. Fly well, soar with ease, but alight and take off with difficulty. Enter water only to feed; then perch in sun for long periods with wings spread. Feed by swimming under water and spearing fish with beak; at surface often swim with all but head and neck submerged (hence called "snake birds"). Voice (rarely heard)—harsh, grating calls and whistling notes when nesting.

Food. Fish, crustacea, amphibia, insects.

Breeding. Colonial. Bulky nests of sticks in trees or bushes near or over water.

Eggs. 3 to 6; bluish white, with chalky surface. Incubated by ♂ and ♀.

Young. Nidicolous. Naked at hatching, downy (exc. head) later. Cared for by ♂ and ♀.

Technical diagnosis. E. Coues, *Key to North American Birds,* Vol. 2, Boston, 1903:968.

Classification. Peters, **1**, 1931:94–95.

References. Bent, No. 121:229–236.

F. J. Jackson, *The Birds of Kenya Colony and the Uganda Protectorate,* Vol. 1, London, 1938:23–25.

B. Meanley, Nesting of the Water-Turkey in Eastern Arkansas, *Wilson Bull.,* **66,** 1954:81–88.

R. S. Palmer, ed., *Handbook of North American Birds,* Vol. 1, New Haven, 1962:357–365.

H. J. Frith, ed., *Birds in the Australian High Country,* Sydney, 1969:45–47.

Synonym: Plotidae.

PELECANIDAE

Pelican Family (6 species)

Figure 18. Brown Pelican (*Pelecanus occidentalis*). Total length 50 in. (1270 mm).

Physical characteristics. Length—1270 to 1829 mm (50 to 72 in.). Plumage—white (tinged with pink in some spp.), brown, or gray; primaries dark to black. Bill, face, and gular pouch brightly colored in some spp. Some spp. crested. Bill very long, straight, and hooked; nostrils obsolete; gular pouch very large. Neck long; body stout. Wings extremely large; tail short. Legs short and stout; feet large and fully webbed. Sexes alike.

Range. North America (from Great Slave Lake, s. Mackenzie) s. through C. America to the Galapagos; the w. coast (to about 40° S.) and part of the n.e. coast of S. America; Africa (exc. n.w. part), s.e. Europe, s. Asia, the East Indies, and Australia. Habitat—large lakes and the seacoast. Some spp. migratory.

Habits. Gregarious. Can fly strongly and even soar. Feed while swimming on surface, or (*P. occidentalis*) dive for food from the wing. Voice—adults usually silent, rarely uttering grunts and croaks; young grunt and chatter noisily.

Food. Fish; sometimes crustacea.

Breeding. Colonial. Nest of sticks in trees, or of reeds, grass, or mud on the ground.
 Eggs. 1 to 4; white. Incubated by ♂ and ♀.
 Young. Nidicolous. Naked at hatching, downy later. Cared for by ♂ and ♀.

Technical diagnosis. E. Coues, *Key to North American Birds,* Vol. 2, Boston, 1903:956.

Classification. Peters, **1,** 1931:79–81.

References. Bent, No. 121:282–306; Murphy, **2:**807–827; Baker, *Fauna,* **6,** 1929:270–275; Alexander, 184–191.

E. Raymond Hall, Pelicans versus fishes in Pyramid Lake, *Condor,* **27,** 1925:147–160.

R. S. Palmer, ed., *Handbook of North American Birds,* Vol. 1, New Haven, 1962:264–280.

V. E. M. Burke and L. H. Brown, Observations on the breeding of the Pink-backed Pelican *Pelecanus rufescens, Ibis,* **112,** 1970:499–512.

FREGATIDAE

Frigatebird Family (5 species)

Figure 19. Magnificent Frigatebird (*Fregata magnificens*). Total length 38 in. (966 mm).

Physical characteristics. Length—787 to 1041 mm (31 to 41 in.). Plumage—brownish black, with some iridescence. White areas on under parts of ♀ (and of ♂ in 2 spp.). Throat bare—that of breeding males greatly inflated at times, even in flight. Bill long and strongly hooked, rounded in cross section, the culmen convex; nostrils obsolete. Wings very long and pointed, with greatest area, relative to weight, of any bird; tail long and deeply forked. Tarsi extremely short and feathered. Feet small; web includes all 4 toes but only basally. Claws strong; middle claw pectinate. ♀ larger than ♂. Head white in juvenal plumage.

Range. Pan-tropical. Oceanic islands and adjacent seas of all parts of the world. Nonmigratory, but immature birds may wander over ocean.

Habits. Gregarious. Very light and graceful in flight—the most aerial of all water birds (apparently never voluntarily alight on water). Capture food from surface of sea while swooping or hovering; also rob boobies, terns, and other birds of food, especially during stress of feeding young. Voice (rarely heard)—a harsh croak; the young chatter.

Food. Fish, squid, crustacea, jellyfish, turtles, young birds.

Breeding. Nest of sticks in trees, or bushes, or on rocks or ground; built by ♂ of material brought by ♀.

Eggs. 1 (rarely 2); white. Incubated by ♂ and ♀.

Young. Nidicolous. Naked at first, then downy (white). Fed by ♂ and ♀.

Technical diagnosis. Baker, *Fauna*, 6, 1929:295.

Classification. Peters, 1, 1931:95–97.

References. Bent, No. 121:306–315; Murphy, 2:919–940; Alexander, 178–183.

R. S. Palmer, ed., *Handbook of North American Birds,* Vol. 1, New Haven, 1962:365–380.

B. and S. Stonehouse, The Frigate Bird *Fregata aquila* of Ascension Island, *Ibis,* 103b, 1963:409–422.

A. Wetmore, *The Birds of the Republic of Panamá, Smithson. Misc. Collect.,* 150, 1965:72–78.

R. W. Schreiber and N. P. Ashmole, Sea-bird breeding seasons on Christmas Island, Pacific Ocean, *Ibis,* 112, 1970:363–394.

Synonym: Tachypetidae.

ARDEIDAE

Heron Family (63 species)

Figure 20. Great Blue Heron (*Ardea herodias*). Total length 41 in. (1041 mm).

Physical characteristics. Length—280 to 1422 mm (11 to 56 in.). Plumage—loose-textured; white, gray, blue, purple, or brown, usually in simple patterns. Some spp. all white; some speckled or barred above and/or streaked below. Many spp. have long filamentous plumes (esp. in breeding plumage) on back, lower foreneck, or head. Lores bare. Bill long and spearlike. Neck long. Wings large and broad; tail short. Legs and toes long and slender; tibiae partly bare. Claw of middle toe pectinate. Sexes alike or nearly so in most spp.

Range. Worldwide exc. n. N. America and Asia and some oceanic islands. Habitat—shores and marshes. Migratory exc. in tropics.

Habits. Solitary or (usually only while breeding) gregarious. Flight direct; neck carried in S-shape. Typically feed while standing or wading in shallows, darting at prey with the bill. Swallow prey whole, regurgitating undigested portions in pellets. Voice—varied squawks and raucous calls; a few rather musical notes.

Food. Fish; also amphibians, reptiles, crustacea, insects, mollusca, rodents, young birds.

Breeding. Colonial. Nest (in most spp.) shallow, made of sticks; in trees or bushes, or on ground.

> *Eggs.* 3 to 6 or 7; blue, white, buff, or even yellow; typically unspeckled. Incubated by ♂ and ♀.
>
> *Young.* Nidicolous. Downy. Cared for by ♂ and ♀.

Technical diagnosis. Witherby, 3, 1939:125.

Classification. Peters, 1, 1931:97–125.

References. Bent, No. 135:72–219; Witherby, 3, 1939:125–162; Baker, *Fauna,* 6, 1929:335–371.

> W. P. and B. D. Cottrille, Great Blue Heron: behavior at the nest, *Misc. Publ. Mus. Zool. Univ. Michigan,* No. 102, 1958.
>
> A. J. Meyerriecks, Comparative breeding behavior of four species of North American herons, *Publ. Nuttall Ornithol. Club,* No. 2, 1960.
>
> R. W. Dickerman and G. Gavino T., Studies of a nesting colony of Green Herons at San Blas, Nayarit, Mexico, *Living Bird,* **1969:**95–111.
>
> W. McVaugh, Jr., The development of four North American Herons, *Living Bird,* **1972:**155–173.

COCHLEARIIDAE

Boat-billed Heron Family (1 species)

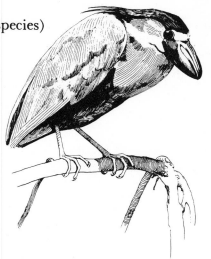

Figure 21. Boat-billed Heron (*Cochlearius cochlearius*). Total length 21 in. (534 mm).

Physical characteristics. Length—508 to 534 mm (20 to 21 in.). Plumage—loose-textured. Crown, nuchal feathers (which are broad and elongated), anterior part of back, and flanks bluish black; rest of head and neck pale buff and gray; wings, tail, and back pale gray; ventral region brown. Lores bare. Eyes very large. Bill extremely broad, flat, and shovel-like; slightly hooked. Gular pouch bare, somewhat distensible. Neck rather long. Wings large and broad; tail short. Legs and toes moderately long and slender; tibiae partly bare; claw of middle toe pectinate. Sexes nearly alike.

Range. Mexico (from Sinaloa and Tamaulipas) to Peru and s. Brazil. Habitat—shores and marshes. Nonmigratory.

Habits. Usually gregarious. Largely nocturnal. Terrestrial, but perch in trees. Flight direct, not strong. Voice—froglike calls, barking or squawking notes; also rattle or clap mandibles.

Food. Fish, crabs, amphibians, mice.

Breeding. Usually colonial. Nest a shallow structure of sticks placed in trees.

    Eggs. 2 to 4; bluish white; slightly speckled with brown. Incubated by ♂ and ♀.

    Young. Not recorded.

Technical diagnosis. R. Ridgway, Studies of the American Herodiones, *Bull. U. S. Geol. Geogr. Surv. Terr.,* **4,** No. 1, 1878:220.

Classification. Peters, 1, 1931:125.

References. D. R. Dickey and A. J. Van Rossem, The birds of El Salvador, *Field Mus. Nat. Hist., Zool. Ser.,* **23,** 1938:84–86.

    A. Wetmore, *The Birds of the Republic of Panamá,* 1, *Smithson. Misc. Collect.,* **150,** 1965:114–119.

    F. Haverschmidt, *Birds of Surinam,* Edinburgh, 1968:25–26.

    C. G. Sibley and J. E. Ahlquist, A comparative study of the egg white proteins of non-passerine birds, *Peabody Mus. Nat. Hist. Bull.* No. 39, 1972.

Synonym: Cancromidae.

BALAENICIPITIDAE

Whale-billed Stork Family (1 species)

Figure 22. Whale-billed Stork (*Balaeniceps rex*). Total length 46 in. (1168 mm).

Physical characteristics. Length—about 1168 mm (46 in.). Plumage—gray, darker, with slight greenish gloss above. Remiges and rectrices grayish black. Short, bushy crest. Bill very large, broad, and flattened, with massive hook at tip of maxilla. Neck moderately long. Wings large and broad; tail short. Legs and toes long; tibiae partly bare. Sexes alike.

Range. Africa: from the valley of the White Nile in s. Sudan to n. Uganda and e. Belgian Congo. Habitat—marshes. Nonmigratory.

Habits. Sometimes gregarious. Largely nocturnal. Move with slow direct flight; sometimes soar. Voice—usually silent, but have a "shrill kitelike cry" and a laughing note; clatter bill.

Food. Fish, frogs, snakes, mollusca, carrion.

Breeding. Nest a large platform of rushes, lined with grass, on ground.

Eggs. 2; white, with slightly bluish tinge.

Young. Nidicolous. Downy.

Technical diagnosis. Beddard, 433–434.

Classification. Peters, 1, 1931:125.

References. Bannerman, 1, 1930:lxviii-lxix, 86–89.

> J. P. Chapin, The Birds of the Belgian Congo, I, *Am. Mus. Nat. Hist. Bull.,* **65,** 1932:447–449.
>
> F. J. Jackson, *The Birds of Kenya Colony and the Uganda Protectorate,* Vol. 1, London, 1938:62–66.
>
> E. T. Gilliard, *Living Birds of the World,* Garden City, 1958:70.
>
> C. W. Mackworth-Praed and C. H. B. Grant, *Birds of the Southern Third of Africa,* Vol. 1, London, 1962:77–78.
>
> C. W. Benson *et al., The Birds of Zambia,* London, 1971:44.

SCOPIDAE

Hammerhead Family (1 species)

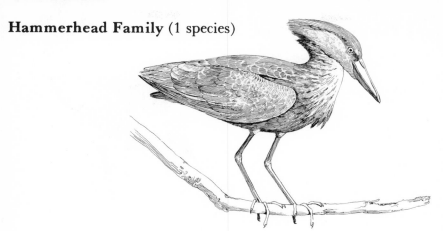

Figure 23. Hammerhead (*Scopus umbretta*). Total length 19.75 in. (502 mm).

Physical characteristics. Length—502 mm (19.75 in.). Plumage—brown, with slight purplish iridescence on upper parts and flight feathers; throat faintly streaked with gray; tail faintly barred with dark brown. Bill long, straight, laterally compressed; maxilla slightly hooked. Head large; long crest carried horizontally. Neck short. Wings and tail rather long. Legs rather long, with tibiae partly bare. Feet slender; toes long. Sexes alike.

Range. Southwestern Arabia, Africa, and Madagascar. Habitat—tree-bordered streams, esp. in grassland. Nonmigratory.

Habits. Usually found in pairs. Slow flight, with neck only slightly curved (not drawn in on shoulders). Voice—harsh, metallic croak; a call, *taket taket . . .,* given "continually in flight" (Rand).

Food. Amphibians, fish, insects, crustacea.

Breeding. Massive roofed nest of sticks and mud, placed in branches of large trees.

> *Eggs.* 3 to 6; white. Incubated apparently by ♂ and ♀.
>
> *Young.* Nidicolous. Downy. Cared for by ♂ and ♀.

Technical diagnosis. Beddard, 420–422; Bannerman, **1**, 1930:89–93.

Classification. A. L. Rand, *Am. Mus. Novit.,* No. 827, 1936.

References. R. B. Cowles, The life history of *Scopus umbretta bannermani* C. Grant in Natal, South Africa, *Auk,* **47**, 1930:159–176.

J. P. Chapin, The Birds of the Belgian Congo, I, *Am. Mus. Nat. Hist. Bull.,* **65**, 1932:449–452,

A. L. Rand, The distribution and habits of Madagascar Birds, *Am. Mus. Nat. Hist. Bull.,* **72**, Art. 5, 1936:339–340.

C. W. Mackworth-Praed and C. H. B. Grant, *Birds of the Southern Third of Africa,* London, 1962:75–77.

P. A. Clancey, *The Birds of Natal and Zululand,* Edinburgh, 1964:42.

CICONIIDAE

Stork Family (17 species)

Figure 24. White Stork (*Ciconia ciconia*). Total length 40 in. (1016 mm).

Physical characteristics. Length—762 to 1524 mm (30 to 60 in.). Plumage—in most spp. boldly black and white, or black (the black often with green, blue, or purple reflections). Remiges black (exc. in 3 spp.). Face or whole head and neck bare in some spp. Bill long, massive, ungrooved; may be recurved or decurved (both in *Anastomus*), or straight. Neck long. Wings long and broad. Tail short; under tail coverts greatly developed in some spp. Legs very long; tibiae partly bare; toes of moderate length, webbed at base. Sexes alike.

Range. Southeastern U. S. and Mexico, C. and S. America, Africa, Eurasia (to about 60° N.), East Indies, Philippines, and Australia (exc. s. third). Northern spp. migratory.

Habits. Gregarious at times. Fly strongly and even soar. Most spp. fly with neck extended. Some spp. have a remarkable dance display. Voice—nearly voiceless (rarely give low grunts or hisses), but rattle or snap mandibles.

Food. Mainly animal matter: fish, amphibia, insects, snails, carrion.

Breeding. Nest a platform of sticks in trees, on cliffs, or on buildings.

    Eggs. 3 to 5, or even 6; white. Incubated by ♂ and ♀.

    Young. Nidicolous. Sparse hairlike prepennae at hatching. Cared for by ♂ and ♀.

Technical diagnosis. Witherby, **3**, 1939:112.

Classification. Peters, **1**, 1931:126–131.

References. Bent, No. 135:57–72; Witherby, **3**, 1939:112–118.

    Fr. Haverschmidt, *The Life of the White Stork,* Leiden, 1949.

    C. W. Mackworth-Praed and C. H. B. Grant, *Birds of the Southern Third of Africa,* Vol. 1, London, 1962:78–86.

    S. Ali and S. D. Ripley, *Handbook of the Birds of India and Pakistan,* Vol. 1, Bombay, 1968:91–109.

    M. P. Kahl, Social behavior and taxonomic relationships of the storks, *Living Bird,* **1971**:151–170.

THRESKIORNITHIDAE

Ibis Family (28 species)

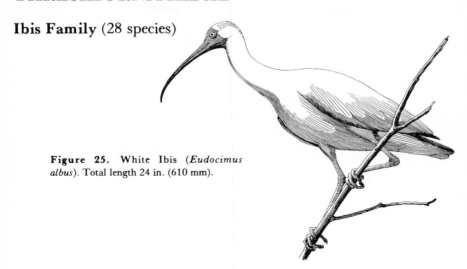

Figure 25. White Ibis (*Eudocimus albus*). Total length 24 in. (610 mm).

Physical characteristics. Length—482 to 1067 mm (19 to 42 in.). Plumage—in most spp. white, grayish brown, or greenish black (1 sp. pink, 1 scarlet); typically little pattern. Some spp. crested; neck feathers or inner secondaries ornamentally modified in some spp. Bill long, slender, grooved, and decurved; or extremely flattened and spatulate at tip. Face and throat, or whole head and neck, bare in some spp. Neck rather long. Wings long; tail short. Legs and toes moderately long; tibiae partly bare; toes webbed at base. Sexes alike or nearly so.

Range. Southern U. S., Middle and S. America, Africa (incl. Madagascar), Eurasia (exc. n. part), and Australia. Habitat—shores and marshes. Northern spp. migratory.

Habits. Usually gregarious. Fly strongly, with neck extended. Voice—harsh croaking or cackling.

Food. Fish, crustacea, reptiles, insects; some grain and other vegetable matter.

Breeding. Nest of sticks, or of rushes and grass; in trees, on cliffs, on rocky islands, or on ground in marshes.

Eggs. 2 to 5; white or blue; immaculate or spotted. Incubated by ♂ and ♀.

Young. Nidicolous. Downy. Cared for by ♂ and ♀.

Technical diagnosis. Witherby, 3, 1939:118.

Classification. Peters, 1, 1931:131–139.

References. Bent, No. 135:23–57; Witherby, 3, 1939:118–125.

Robert P. Allen, The Roseate Spoonbill, *Natl. Audubon Soc. Res. Rept.* No. 2, 1942.

A. Wetmore, *The Birds of the Republic of Panamá*, 1, *Smithson. Misc. Collect.*, **150**, 1965:122–128.

S. Ali and S. D. Ripley, *Handbook of the Birds of India and Pakistan*, Vol. 1, Bombay, 1968:109–118.

H. J. Frith, ed., *Birds in the Australian High Country*, 1969:76–87.

R. P. ffrench and F. Haverschmidt, The Scarlet Ibis in Surinam and Trinidad, *Living Bird*, **1970**:147–165.

Synonyms: Ibidae, Ibididae, Plegadidae, Tantalidae. The Plataleidae are included.

PHOENICOPTERIDAE

Flamingo Family (6 species)

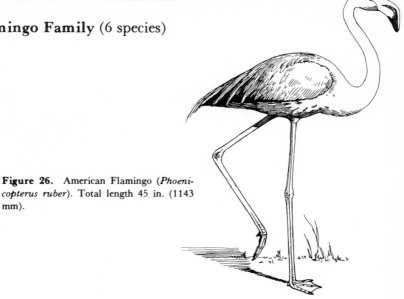

Figure 26. American Flamingo (*Phoenicopterus ruber*). Total length 45 in. (1143 mm).

Physical characteristics. Length—914 to 1219 mm (36 to 48 in.). Plumage—pinkish white or light vermilion, the color strongest on upper wing coverts. Remiges black. Bill thick, lamellate, bent sharply downward at midpoint. Face bare. Neck extremely long. Wings large; tail short. Legs extremely long; tibiae largely bare; toes short and webbed. Sexes alike or nearly so.

Range. West Indies, Yucatán, S. America (n. of 40°), Africa, s. Europe, s.w. Asia to India. Habitat—shallow lagoons and lakes. Some spp. migratory.

Habits. Gregarious. Fly with neck extended. Feed in shallow water with head immersed and bill in inverted position. Voice—a resonant honking and other gooselike notes.

Food. Mollusca, crustacea, insects, fish; also blue-green algae and diatoms.

Breeding. Colonial. Nest a truncated cone of mud built on a mud flat.

Eggs. 1 or 2; white. Incubated by ♂ and ♀.

Young. Nidifugous. Downy. Cared for by ♂ and ♀.

Technical diagnosis. Witherby, **3**, 1939:162–163.

Classification. Peters, **1**, 1931:140–142.

References. Bent, No. 135:1–12; Witherby, **3**, 1939:162–166.

F. M. Chapman, A contribution to the life history of the American Flamingo (*Phoenicopterus ruber*), with remarks upon specimens, *Bull. Am. Mus. Nat. Hist.*, **21**, 1905:53–77.

E. Gallet, *The Flamingos of the Camargue,* Oxford, 1950.

Robert P. Allen, The flamingos: their life history and survival, *Natl. Audubon Soc. Res. Rept.* No. 5, 1956.

C. W. Mackworth-Praed and C. H. B. Grant, *Birds of the Southern Third of Africa,* London, 1962:91–94.

S. Ali and S. D. Ripley, *Handbook of the Birds of India and Pakistan,* Vol. 1, Bombay, 1968:118–122.

ANHIMIDAE

Screamer Family

(3 species)

Figure 27. Horned Screamer (*Anhima cornuta*). Total length 33 in. (838 mm).

Physical characteristics. Length—712 to 914 mm (28 to 36 in.). Plumage—black or gray above, light gray or white below; little pattern; head and neck feathers short and downy or sparse and attenuate; crest (2 spp.) or long, horny frontal spike. Bill short, henlike. Head small and slender; neck rather short. Wings large and broad, with 2 long, sharp spurs on forward edge of manus; tail short and broad. Legs short and heavy; feet very large, slightly palmate; toes and claws long. Sexes alike or nearly so. Subcutaneous air cells well developed.

Range. South America s. to Uruguay and n. Argentina. Habitat—marshes and well-watered grassland. Nonmigratory.

Habits. Gregarious at times. Largely terrestrial and aquatic, but some perch in trees. Able to walk on mats of floating vegetation. Swim occasionally. Fly slowly; rise laboriously from the ground but soar for hours at great heights. Voice—adult, a loud trumpeting; young, a low piping.

Food. Vegetable matter.

Breeding. Shallow nest of rushes, etc., on marshy ground.

> *Eggs.* 3 to 6; white. Incubated by ♂ and ♀.
>
> *Young.* Nidifugous. Downy. Cared for by ♂ and ♀.

Technical diagnosis. Beddard, 451–456 ("Palamedae").

Classification. Peters, **1,** 1931:142.

References. W. H. Hudson, *Birds of La Plata,* Vol. 2, London, 1920:130–133.

> A. Wetmore, Observations on the birds of Argentina, Paraguay, Uruguay, and Chile, *U. S. Natl. Mus. Bull.,* **133,** 1926:67–69.
>
> C. R. Stonor, Notes on the breeding habits of the common Screamer (*Chauna torquata*), *Ibis,* **1939:**45–49.
>
> M. W. Weller, Notes on some marsh birds of Cape San Antonio, Argentina, *Ibis,* **109,** 1967:391–411.
>
> F. Haverschmidt, *Birds of Surinam,* Edinburgh, 1968:32–34.
>
> R. Meyer de Schauensee, *A Guide to the Birds of South America,* Wynnewood, Pa., 1970:30.

Synonym: Palamedeidae.

ANATIDAE

Duck Family (147 species ±)

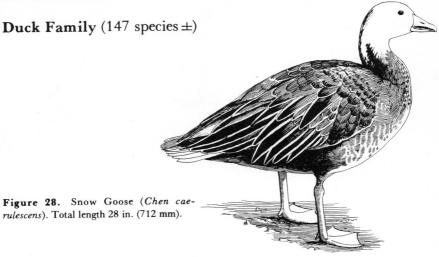

Figure 28. Snow Goose (*Chen cae-rulescens*). Total length 28 in. (712 mm).

Physical characteristics. Length—292 to 1524 mm (11.5 to 60 in.). Plumage—white, black, gray, brown, etc., often in pronounced patterns. Many spp. have colored "speculum" or a white area on the wing. Many spp. crested. Bill typically lamellate, broad, flat, and rounded at the tip. Neck medium to very long. In most spp., wings rather narrow and pointed, tail short. Legs medium to very short, feet webbed. Sexes alike or unlike.

Range. Worldwide. Most spp. migratory

Habits. Many spp. gregarious exc. in nesting season. Typically aquatic. Most spp. fly well; some are flightless. Most spp. dive, regularly or occasionally, for food or to escape enemies. Voice—quack, croak, cackle, whistle, or hiss.

Food. Great variety of aquatic animals and plants; some spp. browse or eat grain on upland.

Breeding. Nest on ground; on rocky ledges, stumps, etc.; in holes in ground; in trees. Nest often lined with female's own down.

> *Eggs.* 2 to 16; white, buff, or greenish; immaculate. Incubated by ♀, by ♂, or by both (*Heteronetta* is parasitic).

> *Young.* Nidifugous. Downy. Cared for by ♀, by ♂, or by both.

Technical diagnosis. Witherby, **3,** 1939:167.

Classification. J. Delacour and E. Mavr. *Wilson Bull., 57,* 1945:3–55.

References. Bent, Nos. 126, 130; Baker, *Fauna,* **6,** 1929:377–475; Murphy, **2:**941–972.

J. Delacour *et al., The Waterfowl of the World,* London, 4 vols., 1954–1964.

P. A. Johnsgard, *Handbook of Waterfowl Behavior,* Ithaca, 1965, 392 pp.

H. J. Frith, *Waterfowl in Australia,* Sydney, 1967, 328 pp.

S. Ali and S. D. Ripley, *Handbook of Birds of India and Pakistan,* Vol. 1, Bombay, 1968:122–210.

P. A. Johnsgard, *Waterfowl: Their Biology and Natural History,* Lincoln, Neb., 1968, 138 pp., 59 col. pls.

CATHARTIDAE

American Vulture Family (6 species)

Figure 29. King Vulture (*Sarcoramphus papa*). Total length 31 in. (787 mm).

Physical characteristics. Length—635 to 1118 mm (25 to 44 in.). Plumage—typically brownish black, with a light area on the under surface of the wing. (Body plumage of *Sarcoramphus* white and cream.) Ruff of lanceolate or downy feathers about base of neck of larger spp. Head bare and carunculated; black, red, or yellow. (*Sarcoramphus* has the head elaborately patterned with black hairlike feathers.) Bill medium to very thick, rounded, hooked; nostrils perforate. Neck short. Wings very long and broad; tail short to long. Legs medium or short; toes rather long; claws weakly hooked (not used for grasping). Sexes alike or nearly so.

Range. South America; N. America to s. Canada. Populations in high latitudes migratory.

Habits. Typically solitary; some spp. use colonial roosts. Have remarkable soaring flight, but larger spp. have difficulty getting off the ground. Disgorge food when disturbed on the ground. Voice (rarely heard)—low croaks or hissing sounds.

Food. Mainly carrion. Some spp. occasionally attack living animals, esp. newborn young.

Breeding. Nest in caves or hollow trees, usually without lining.

>*Eggs.* 1 to 3; white or pale gray-green; immaculate or spotted with brown. Incubated by ♂ and ♀.

>*Young.* Nidicolous. Downy. Cared for by ♂ and ♀.

Technical diagnosis. Friedmann, Pt. 11, 1950:3–6.

Classification. Hellmayr and Conover, Pt. 1, No. 4, 1949:1–14.

References. Bent, No. 167:1–44.

>C. B. Koford, The California Condor, *Natl. Audubon Soc. Res. Rept.* No. 4, 1953.

>K. E. Stager, The role of olfaction in food location by the Turkey Vulture (*Cathartes aura*), *Contrib. Sci.,* **81,** 1964:1–63.

>A. Wetmore, *The Birds of the Republic of Panamá*, 1, *Smithson. Misc. Collect.,* **150,** 1965:153–171.

Synonym: Sarcoramphidae. The Cathartidae were formerly included in the "Vulturidae."

SAGITTARIIDAE

Secretarybird Family

(1 species)

Figure 30. Secretarybird (*Sagittarius serpentarius*). Total length 59 in. (1499 mm).

Physical characteristics. Length—1270 to 1499 mm (50 to 59 in.). Plumage—largely pale gray on body; remiges and thighs black; crest black and gray; rump black, barred with white; upper tail coverts white; tail gray, with white tip and black subterminal band. Conspicuous crest of long, paddle-shaped feathers on nape. Bill short and hooked; face bare, the skin bright orange. Neck rather long. Wings large; tail long, with extremely long pair of central feathers. Legs extremely long and slender. Toes short and partly webbed; claws long. Sexes alike, but ♀ slightly smaller.

Range. Africa: Senegambia; s. Sudan to the Cape of Good Hope (not Madagascar). Habitat—open plains. Some migration reported.

Habits. Solitary. Terrestrial; largely cursorial. Fly well but infrequently. Roost in trees. Voice (rarely heard)—loud, strident *doo-doo-dut*, "a deep raucous groan."

Food. Reptiles, small mammals, insects, young birds.

Breeding. Bulky nest of sticks and sod placed in a bush or tree.

    Eggs. 2 or 3; bluish white. Incubated by ♀.

    Young. Nidicolous. Downy. Cared for by ♂ and ♀.

Technical diagnosis. Friedmann, Pt. 11, 1950:60–61.

Classification. Peters, 1, 1931:192.

References. Bannerman, 1, 1930:165–169.

    C. D. Priest, *The Birds of Southern Rhodesia,* Vol. 1, London, 1933:180–189,

    F. J. Jackson, *The Birds of Kenya Colony and the Uganda Protectorate,* Vol. 1, London, 1938:123–127.

    L. H. Brown, On the biology of the large birds of the Emu district, Kenya Colony, *Ibis,* **94,** 1952:592–595; see also *Ibis,* **97,** 1955:43–48.

    C. W. Mackworth-Praed and C. H. B. Grant, *Birds of the Southern Third of Africa,* London, 1962:123–124.

Synonyms: Gypogeranidae, Serpentariidae.

ACCIPITRIDAE

Hawk Family

(208 species)

Figure 31. Crowned Hawk-eagle (*Stephanoaetus coronatus*). Total length 39 in. (991 mm).

Physical characteristics. Length—200 to 1143 mm (8 to 45 in.). Plumage—typically of blended grays and browns, lighter below. Many spp. have under parts barred or streaked and/or tail barred. Bill short and strongly hooked. Cere and orbital skin bare and usually brightly colored. Neck short. Wings large and, in most spp., rather rounded; tail medium to long. Legs strong; medium to rather long. Claws stout and hooked. Sexes usually alike, but ♀ larger.

Range. Nearly worldwide (absent from Antarctic, n. Arctic, and many oceanic islands). Most spp. migratory.

Habits. Usually solitary. Diurnal. Fly strongly; many soar. Capture prey with their feet. Voice—loud screams, whistles, cackling notes.

Food. Mammals, birds, reptiles, amphibians, fish, mollusca, various invertebrates, carrion.

Breeding. Nest of sticks in trees or on cliffs; or of grass and weed stalks, on ground.

    Eggs. 1 to 6; white; immaculate or marked with brown. Incubated by ♀, or by ♂ and ♀.

    Young. Nidicolous. Downy. Cared for by ♂ and ♀.

Technical diagnosis. Witherby, **3,** 1939:1, 38.

Classification. Peters, **1,** 1931:192–274 (not including the Pandioninae).

References. Bent, No. 167:44–352; Witherby, **3,** 1939:38–106.

    C. W. Mackworth-Praed and C. H. B. Grant, *Birds of the Southern Third of Africa,* Vol. 1, London, 1962:124–131, 147–195.

    A. Wetmore, *The Birds of the Republic of Panamá,* 1, *Smithson. Misc. Collect.,* **150,** 1965:171–256.

    S. Ali and S. D. Ripley, *Handbook of the Birds of India and Pakistan,* Vol. 1, Bombay, 1968:210–335.

    L. Brown and D. Amadon, *Eagles, Hawks and Falcons of the World,* New York, 2 vols., 1969, 945 pp., 165 col. pls.

Synonyms: Aquilidae, Vulturidae (Cathartidae plus "Aegypiidae"). The Pandionidae are sometimes included in the Accipitridae. The Aegypiidae and Buteonidae are included here.

PANDIONIDAE

Osprey Family (1 species)

Figure 32. Osprey (*Pandion haliaetus*).
Total length 24 in. (610 mm).

Physical characteristics. Length—559 to 610 mm (22 to 24 in.). Plumage—brownish black above, with white edgings; head white, with black on crown and black line through eye and auricular region; under parts white, usually spotted across breast; tail barred. Bill short and strongly hooked. Wings long and pointed; tail medium and narrow. Tarsus reticulate; outer toe reversible; claws large; soles of feet studded with spines. Sexes alike.

Range. Europe and Asia s. to Spain, n. Africa, and s. China; East Indies, Australia, and some s.w. Pacific islands; N. America s. to Sinaloa and to Gulf Coast; also Bahamas, Yucatán, and British Honduras. Migrate in winter to S. Africa, India, s. S. America.

Habits. Always near water. Capture fish by hovering and then plunging, feet first, with half-closed wings, often submerging completely. Fish is carried head foremost by both feet (unless very small) to nest or perch. Voice—usually a series of short, shrill whistles, dropping in pitch at end.

Food. Fish. (When pressed by hunger: small mammals, wounded birds, or domestic fowl.)

Breeding. Sometimes loosely colonial. Nest of sticks and finer materials (as available) on rocks, in trees, or on ground. Built by ♂ and ♀.

    Eggs. 3 (rarely 2 or 4); white to fawn color; strongly marked with rich brown. Incubated mainly by ♀.

    Young. Nidicolous. Downy. Food brought by ♂, fed to young by ♀.

Technical diagnosis. Witherby, **3**, 1939:106–107.

Classification. Peters, **1**, 1931:275 ("Pandioninae").

References. Bent, No. 167:352–379; Witherby, **3**, 1939:106–111.

    J. B. May, *The Hawks of North America,* New York, 1935:93–95.

    S. Ali and S. D. Ripley, *Handbook of the Birds of India and Pakistan,* Vol. 1, Bombay, 1968:335–338.

    C. J. Henny and H. M. Wight, An endangered Osprey population: estimates of mortality and production, *Auk,* **86,** 1969:188–198.

    J. G. Reese, Reproduction in a Chesapeake Bay Osprey population, *Auk,* **87,** 1970:747–759.

Synonymy: Sometimes included in Accipitridae.

FALCONIDAE

Falcon Family (58 species)

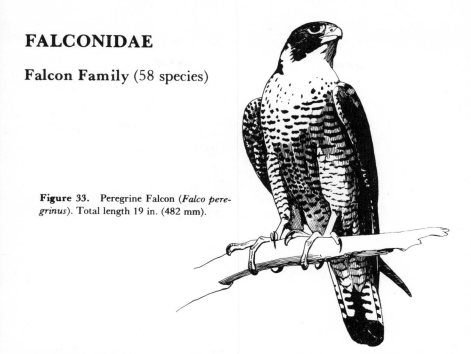

Figure 33. Peregrine Falcon (*Falco pere-grinus*). Total length 19 in. (482 mm).

Physical characteristics. Length—152 to 635 mm (6 to 25 in.). Plumage—gray or brown with white or buff, streaked and/or barred, esp. below; or with solid contrast of black and white. Bill short, strongly hooked, and (in most spp.) toothed. Cere and orbital skin bare and usually brightly colored. Neck short. Wings of most spp. long and pointed; tail medium to long. Legs strong, medium to rather long. Toes long; claws strong and hooked. Sexes usually alike, but ♀ larger.

Range. Nearly worldwide (absent from Antarctic and from most oceanic islands). Many spp. migratory.

Habits. Usually solitary. Diurnal. Fly swiftly, soar infrequently. Capture prey with feet. Voice (little heard exc. at nest)—loud screams, squeals, and chattering notes.

Food. Mammals, birds, reptiles, amphibians, crustacea, insects, mollusca, carrion.

Breeding. Nest on cliffs, on ground, in holes in trees, in old nests of other hawks, etc.; or nest of sticks in a tree.

    Eggs. 2 to 6; white; heavily blotched and spotted with brown. Incubated by ♂ and ♀.

    Young. Nidicolous. Downy. Cared for by ♂ and ♀.

Technical diagnosis. Witherby, **3**, 1939:1–2.

Classification. Peters, **1**, 1931:276–306.

References. Bent, No. 170:1–139; Baker, *Fauna*, **5**, 1928:29–67.

    A. Wetmore, *The Birds of the Republic of Panamá*, 1, *Smithson. Misc. Collect.*, **150**, 1965:259–292.

    S. Ali and S. D. Ripley, *Handbook of the Birds of India and Pakistan,* Vol. 1, Bombay, 1968:338–369.

    L. Brown and D. Amadon, *Eagles, Hawks and Falcons of the World,* New York, 2 vols., 1969, 945 pp., 165 col. pls.

    J. J. Hickey, *Peregrine Falcon Populations, Their Biology and Decline,* Madison, 1969, 596 pp.

Synonymy: The Polyboridae are included.

MEGAPODIIDAE

Megapode Family (10 species)

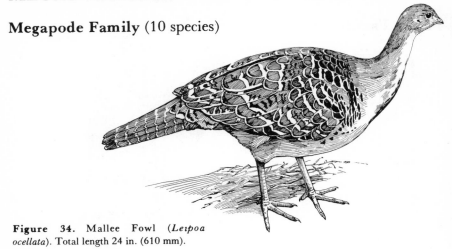

Figure 34. Mallee Fowl (*Leipoa ocellata*). Total length 24 in. (610 mm).

Physical characteristics. Length—254 to 648 mm (10 to 25.5 in.). Plumage—solidly dark brown or slate (with white under parts in some spp.); or gray and brown, streaked and barred with buff or chestnut. One sp. crested, the others with head and neck partly or wholly bare (colored red and yellow or olive-drab). Some spp. have long wattles on nape and foreneck (*Aepypodius*) or a casque (*Macrocephalon*). Bill small and henlike to rather large and massive. Head small; neck medium. Wings large and rounded; tail medium to rather long, vaulted in several spp. Legs, feet, and claws very large and strong, esp. in *Megapodius*. Sexes alike or nearly so.

Range. Australia; Nicobars; islands off n. and e. Borneo, Philippines, Marianas; Celebes and eastward through New Guinea, Solomons, and New Hebrides to c. Polynesia. Habitat—forested areas, esp. near coast, and dry scrub. Nonmigratory.

Habits. Most spp. gregarious. Largely terrestrial. Some fly to offshore islands to roost. Voice—loud, harsh calls and cackling notes.

Food. Insects, worms, snails, seeds, berries.

Breeding. Eggs buried in sand, warm volcanic ash, or mounds of decaying vegetation (usually mixed with soil).

Eggs. 6 to 24; buffy white; immaculate. Not incubated by parents (nests serve as natural incubators).

Young. Nidifugous. Downy. Extremely precocial—not cared for by parents.

Technical diagnosis. Ridgway and Friedmann, Pt. 10, 1946:5.

Classification. Peters, **2**, 1934:3–9.

References. Baker, *Fauna,* **5**, 1928:436–439.

F. Lewis, Notes on the breeding habits of the Mallee-Fowl, *Emu,* **40**, 1940:97–110.

H. J. Frith, Breeding habits in the family Megapodiidae, *Ibis,* **1956**:620–640.

G. A. Clark, Jr., Life histories and the evolution of Megapodes, *Living Bird,* **1964**:149–168.

S. Ali and S. D. Ripley, *Handbook of the Birds of India and Pakistan,* Vol. 2, Bombay, 1969:1–3.

J. E. duPont, *Philippine Birds,* Greenville, Del., 1971.

CRACIDAE

Curassow Family

(44 species)

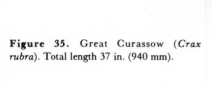

Figure 35. Great Curassow (*Crax rubra*). Total length 37 in. (940 mm).

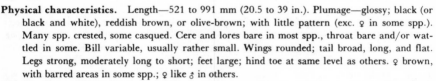

Physical characteristics. Length—521 to 991 mm (20.5 to 39 in.). Plumage—glossy; black (or black and white), reddish brown, or olive-brown; with little pattern (exc. ♀ in some spp.). Many spp. crested, some casqued. Cere and lores bare in most spp., throat bare and/or wattled in some. Bill variable, usually rather small. Wings rounded; tail broad, long, and flat. Legs strong, moderately long to short; feet large; hind toe at same level as others. ♀ brown, with barred areas in some spp.; ♀ like ♂ in others.

Range. Neotropical. Mainland, s. Texas and Sonora to Paraguay. Habitat—forests. Nonmigratory.

Habits. Often gregarious. *Nothocrax urumutum* said to be nocturnal. Arboreal (may feed on ground but when alarmed take refuge in trees). Flight heavy and direct. ♂ in *Penelope* and *Penelopina* has a slow, "drumming" display flight. Voice—loud and often harsh (trachea of ♂ in some spp. elongated and looped between pectoral muscles and skin).

Food. Fruits, leaves.

Breeding. Simple nest of sticks or leaves in branches of trees.

    Eggs. 2 to 3, rarely 4; white. Incubated by ♀ only.

    Young. Nidifugous. Downy (striped on head).

Technical diagnosis. Ridgway and Friedmann, Pt. 10, 1946:5–8.

Classification. C. Vaurie, Taxonomy of the Cracidae (Aves), *Am. Mus. Nat. Hist. Bull.,* **138,** 1968:131–260.

References. Bent, No. 162:345–352.

    G. M. Sutton and O. S. Pettingill, Jr., Birds of the Gomez Farias region, southwestern Tamaulipas, *Auk,* **59,** 1942:10–12.

    E. Schäfer, Estudio bio-ecologico comparativo sobre algunos Cracidae del norte y centro de Venezuela, *Bol. Soc. Venezol. Cienc. nat.,* **15,** No. 80:30–63.

    A. Wetmore, *The Birds of the Republic of Panamá,* 1, *Smithson. Misc. Collect.,* **150,** 1965:293–310.

    H. Sick and D. Amadon, Notes on Brazilian Cracidae, *Condor,* **72,** 1970:106–108.

    J. Delacour and D. Amadon, *Curassows and Related Birds,* American Museum of Natural History, 1973.

PHASIANIDAE

Pheasant Family
(174 species)

Figure 36. Silver Pheasant (*Lophura nycthemera*). Total length 44 in. (1118 mm).

Physical characteristics. Length—127 to 1981 mm (5 to 78 in.). Plumage—ranging from brown and black in cryptic patterns to multicolored in conspicuous patterns (bright yellow, red, green, blue, white). Colored wattles or areas of bare skin on heads of some spp. Some spp. crested. Bill henlike, small or moderate in size. Neck short to fairly long. Wings strong, rounded; tail extremely short to extremely long. Legs strong, short to rather long; spurred in some spp. Sexes unlike in most spp.

Range. Mainland of: extreme s. Canada, s. through n. two-thirds of S. America; Old World, exc. n. Scandinavia and n. Asia. A few spp. migratory.

Habits. Gregarious or, in some spp., solitary. Terrestrial, but some roost in trees. Flight strong but not prolonged. Many spp. scratch or dig in ground with bill for food. Males of many spp. display plumage or whir wings in courtship. Voice—crowing, clucking, and cackling notes.

Food. Seeds, grains, berries, insects, worms, etc.

Breeding. Some spp. polygamous. Nest on ground or (*Tragopan*) in trees.

> *Eggs.* 2 to 22; white to buff or olive-green; speckled or immaculate. Incubated by ♀, or by ♂ and ♀.

> *Young.* Nidifugous. Downy. Cared for by ♀, or by ♂ and ♀.

Technical diagnosis. Ridgway and Friedmann, Pt. 10, 1946:231.

Classification. Peters, **2,** 1934:42–133.

References. Bent, No. 162:1–90; Witherby, **5,** 1941:233–254.

> A. F. Skutch, Life history of the Marbled Wood-Quail, *Condor,* **49,** 1947:217–232.

> J. Delacour, *The Pheasants of the World,* London, 1951.

> A. Wetmore, *The Birds of the Republic of Panamá,* 1, *Smithson. Misc. Collect.,* **150,** 1965:310–333.

> S. Ali and S. D. Ripley, *Handbook of the Birds of India and Pakistan,* Vol. 2, Bombay, 1969:4–127.

> P. A. Johnsgard, Experimental hybridization of the New World quail (Odontophorinae), *Auk,* **88,** 1971:264–275.

> P. A. Johnsgard, *Grouse and Quails of North America,* Lincoln, Neb., 1973, 553 pp., 52 col. pls.

Synonymy: The Odontophoridae and Perdicidae are included.

TETRAONIDAE

Grouse Family (18 species)

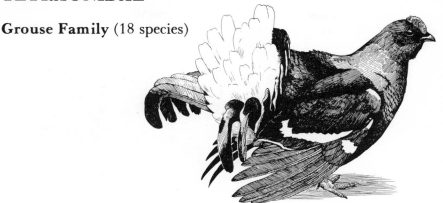

Figure 37. Black Grouse (*Lyrurus te-trix*). Total length 20 in. (508 mm).

Physical characteristics. Length—305 to 889 mm (12 to 35 in.). Plumage—mainly brown, gray, or black, usually paler (or with white) below; most spp. barred, mottled, or scaled in cryptic patterns (3 spp. assume white plumage in winter). Bill short and strong; nostrils feathered. Head entirely feathered (exc. bare orbital line in some spp.). Neck rather short, with (in some spp.) lateral inflatable air sacs and/or tufts of erectile feathers. Wings short and rounded; tail diverse in form, usually large. Legs short to medium (not spurred); tarsi (and, in some spp., toes) feathered; feet rather large. Sexes alike or unlike.

Range. North America, n. Europe and Asia. Most spp. nonmigratory.

Habits. Most spp. solitary. Terrestrial (some spp. somewhat arboreal). Flight strong but not prolonged. Males, alone or in groups, have (mainly in breeding season) various displays involving strutting, whirring the wings, and (sometimes) vocal accompaniment. Voice—whistling, cooing, cackling, and clucking notes.

Food. Buds, leaves, berries, and a variety of other vegetable matter; insects.

Breeding. Some spp. polygamous. Nests on ground (exceptionally in a tree).

> *Eggs.* 6 to 16; white, buffy, brown, cinnamon-red; immaculate, speckled, or heavily blotched with brown, black, or red. Incubated by ♀.

> *Young.* Nidifugous. Downy. Cared by by ♀.

Technical diagnosis. Ridgway and Friedmann, Pt. 10, 1946:63–65.

Classification. Peters, **2,** 1934:24–42.

References. Bent, No. 162:91–345; Witherby, **5,** 1941:209–233.

> W. E. Godfrey, *The Birds of Canada,* Ottawa, 1966:106–115.

> D. F. Parmelee *et al.,* The birds of southeastern Victoria Island and adjacent small islands, *Natl. Mus. Can. Bull.* No. 222, 1967:71–79.

> S. D. MacDonald, The courtship and territorial behavior of Franklin's race of the Spruce Grouse, *Living Bird,* **1968:**5–25.

> S. D. MacDonald, The breeding behavior of the Rock Ptarmigan, *Living Bird,* 1970:195–238.

NUMIDIDAE

Guineafowl Family (7 species)

Figure 38. Vulturine Guineafowl
(*Acryllium vulturinum*). Total length 27 in.
(686 mm).

Physical characteristics. Length—432 to 750 mm (17 to 29.5 in.). Plumage—black, uniformly spotted (exc. in *Phasidus*) with white, the spots becoming lines or bars on the remiges (*Acryllium* has the lower neck and breast bright blue, with long, pointed, black and white hackles; *Agelastes* has the forepart of body white). Head and neck largely bare (colored red, yellow, blue, or slate), marked with casque, a conspicuous vertical crest, or a band of short feathers, and (in most spp.) with wattles or loose folds of skin. Bill short and stout. Wings moderate and very rounded; tail medium (long and pointed in *Acryllium*), typically drooping and heavily overlain with coverts. Legs strong and feet large; spurs lacking exc. in *Phasidus* and *Agelastes*. Sexes alike or nearly so.

Range. Madagascar and Africa s. of the Sahara. Habitat—forest, brush, exceptionally grassland. Nonmigratory.

Habits. Most spp. gregarious. Largely terrestrial, but roost in trees. Flight strong but not prolonged. Notably wary. Voice—loud, harsh, and monotonously repetitive.

Food. Insects, snails, seeds, tubers, leaves.

Breeding. Nest, with little lining, placed on ground.
> *Eggs.* 7 to 20; white to pale reddish brown; immaculate or lightly speckled. Incubated by ♀.
> *Young.* Nidifugous. Downy. Cared for by ♂ and ♀.

Technical diagnosis. Ridgway and Friedmann, Pt. 10, 1946:430–431.

Classification. Peters, **2,** 1934:133–139.

References. J. P. Chapin, The Birds of the Belgian Congo, I, *Am. Mus. Nat. Hist. Bull.,* **65,** 1932:656–683.

> F. J. Jackson, *The Birds of Kenya Colony and the Uganda Protectorate,* Vol. 1, London, 1938:275–285.

> A. Roberts, *The Birds of South Africa,* London, 1940:78–80.

> A. R. Lee, The Guinea Fowl, *U. S. Dept. Agric. Farmers' Bull.* No. 1391, 1940.

> C. W. Mackworth-Praed and C. H. B. Grant, *Birds of the Southern Third of Africa,* Vol. 1, London, 1962:230–235.

MELEAGRIDIDAE

Turkey Family (2 species)

Figure 39. Ocellated Turkey (*Agriocharis ocellata*). Total length 41 in. (1041 mm).

Physical characteristics. Length—838 to 1092 mm (33 to 43 in.). Plumage—dark metallic brown or green, barred with black; wings barred with white; rectrices brown or gray, barred with black, tipped with a two-color band; upper tail coverts in *Agriocharis* broadly banded with copper, black, and metallic blue. Bill short. Skin of head and neck blue and red—bare and carunculated. Neck long. Wings and tail broad and rounded. Legs rather long, spurred (very slightly in ♀); feet large. Sexes similar, but ♀ smaller and duller.

Range. Eastern U. S. and s.w. to Arizona; Mexico, including the Yucatán Peninsula; n. Guatemala; British Honduras. Habitat—wooded country. Nonmigratory.

Habits. Gregarious. Terrestrial, but roost in trees. Flight strong but not prolonged. Voice—many "gobbling" and clucking sounds.

Food. A great variety of vegetable material; some insects and other animal matter.

Breeding. Polygamous. Nest a hollow, with little lining, in the ground.

Eggs. 8 to 18; buff; speckled with brown. Incubated by ♀.

Young. Nidifugous. Downy. Cared for by ♀.

Technical diagnosis. Ridgway and Friedmann, Pt. 10, 1946:436–437.

Classification. Ridgway and Friedmann, Pt. 10, 1946:436–463.

References. Bent, No. 162:323–345.

E. A. McIlhenny, *The Wild Turkey and Its Hunting,* New York, 1914.

H. S. Mosby and C. O. Handley, *The Wild Turkey in Virginia,* Richmond, Va., 1943.

J. S. Ligon, History and management of Merriam's Wild Turkey, *Univ. N. Mex. Publ. Biol.* No. 1, 1946.

A. S. Leopold, The Wild Turkeys of Mexico, *Trans. 13th N. Am. Wildl. Conf.,* 1948:393–400.

A. W. Schorger, *The Wild Turkey—Its History and Domestication,* Norman, Okla., 1966.

OPISTHOCOMIDAE

Hoatzin Family

(1 species)

Figure 40. Hoatzin (*Opisthocomus hoazin*). Total length 24 in. (610 mm).

Physical characteristics. Length—about 610 mm (24 in.). Plumage—loose-webbed; dark brown above, with olive reflections and conspicuous white streaks; top of head reddish brown; tail broadly tipped with pale buff; pale buff on throat and upper breast to rich chestnut on thighs and crissum. Bill short, stout, and laterally compressed. Long crest of stiff, narrow feathers; head otherwise scantily feathered. Neck long and slender. Wings and tail large and rounded. Legs short; feet rather large. Sexes alike.

Range. Northern S. America. (Distributed irregularly in Atlantic drainage from the Amazon system northward.) Habitat—wooded river banks. Nonmigratory.

Habits. Gregarious. Arboreal. Flight very weak. Voice—a harsh hissing screech; a harsh, monotonous *ca cherk, ca cherk.*

Food. Leaves and fruit, esp. of arum.

Breeding. Nest a platform of sticks 4 to 15 (rarely up to 50) ft above water in trees.

Eggs. 2 to 3; white; speckled with brown. Incubated (by ♂ and ♀?).

Young. Seminidicolous. Almost naked at hatching. Cared for (by ♂ and ♀?).

Technical diagnosis. Ridgway and Friedmann, Pt., 10, 1946:2–4.

Classification. Peters, **2,** 1934:141.

References. [C.] W. Beebe, *Tropical Wildlife in British Guiana,* New York, 1917: 155–182.

J. L. Grimmer, Strange little world of the hoatzin, *Natl. Georg. Mag.,* **122,** 1962:391–401.

H. Sick, Hoatzin. In *A New Dictionary of Birds,* London, 1964:369–371.

C. G. Sibley and J. E. Ahlquist, A comparative study of the egg white proteins of nonpasserine birds. *Peabody Mus. Nat. Hist. Bull.* No. 39, 1972.

TURNICIDAE

Hemipode-quail Family (15 species)

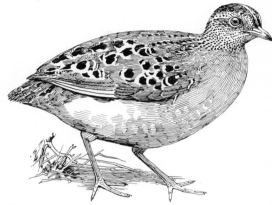

Figure 41. Spotted Hemipode-quail (*Turnix ocellata*). Total length 7.5 in. (190 mm).

Physical characteristics. Length—114 to 190 mm (4.5 to 7.5 in.). Plumage—brown, gray, and black; cryptically patterned above; plain (in most spp.) below; barred or spotted on sides. Bill short, stout to slender and pointed (*Ortyxelos*). Neck short; body compact. Wings short and rounded; tail feathers soft and extremely short (exc. in *Ortyxelos*). Legs strong but short; 3 toes (2nd, 3rd, and 4th). ♂ somewhat smaller and duller than ♀.

Range. Southern Spain and Portugal, Africa, India, Burma, Thailand, Indochina, s. and e. China, Malaysia, Papuan Region, Philippines, Solomons, Australia. Habitat—grassland, scanty brush, or open forest. Nonmigratory (exc. in China).

Habits. Solitary. Terrestrial. Very secretive. Flight brief and infrequent. Pugnacious (esp. the ♀ during courship). Voice—a booming note (♀); cooing or purring notes.

Food. Seeds, insects.

Breeding. Polyandrous. Nest of grass on the ground.

    Eggs. Usually 4 (2 in *Ortyxelos*); white, gray, or olive; spotted with brown. Incubated by ♂.

    Young. Nidifugous. Downy. Cared for by ♂.

Technical diagnosis. Ridgway and Friedmann, Pt. 9, 1941:2–4.

Classification. Peters, **2**, 1934:142–149.

References. Bannerman, **2**, 1931:303–312.

    R. C. Pendleton, Field observations on the Spotted Button-quail on Guadalcanal, *Auk,* **64,** 1947:417–421.

    B. E. Smythies, *The Birds of Burma,* 2nd ed., Edinburgh, 1953:451–452.

    C. W. Mackworth-Praed and C. H. B. Grant, *Birds of the Southern Third of Africa,* London, 1962:367–371.

    S. Ali and S. D. Ripley, *Handbook of the Birds of India and Pakistan,* Vol. 2, Bombay, 1969:127–135.

PEDIONOMIDAE

Collared-hemipode Family (1 species)

Figure 42. Collared-hemipode (*Pedionomus torquatus*). Total length 6.5 in. (165 mm).

Physical characteristics. Length—152 to 171 mm (6 to 6.75 in.). Plumage of ♀—reddish brown, buff, and black in cryptic patterns above. A broad white collar, spotted with black. Upper breast (and nape) chestnut; rest of under parts pale buff, marked along sides with dark brown and black. Bill medium, rather slender. Neck of moderate length; body compact. Wings short and rounded; tail very short. Legs fairly long; tibiae partly bare; 4 toes. ♂ smaller and duller in coloration than ♀.

Range. Australia (New S. Wales, Victoria, s. Australia). Habitat—open plains. Migratory in part.

Habits. Solitary. Very secretive. Terrestrial. Flight weak. Often rise on toes and peer about. Voice—a hollow, "tapping" sound.

Food. Seeds, insects.

Breeding. Nest a hollow, lined with grass, in the ground.

   Eggs. Usually 4; yellowish or greenish white; spotted with olive and gray. Incubated (by ♂?).

   Young. Nidifugous (?). Downy (?). Cared for (by ♂?).

Technical diagnosis. Mathews, **1,** 1911:96.

Classification. Peters, **2,** 1934:150.

References. Mathews, **1,** Pts. 1–2, 1910–11:96–99.

   E. T. Gilliard, *Living Birds of the World,* Garden City, 1958:145.

   N. W. Cayley, *What Bird Is That?*, 4th ed., Sydney, 1966:211–212.

   H. J. Frith, ed., *Birds in the Australian High Country,* Sydney, 1969:149.

   P. Slater *et al., A Field Guide to Australian Birds,* Wynnewood, Pa., 1971:266.

GRUIDAE

Crane Family (14 species)

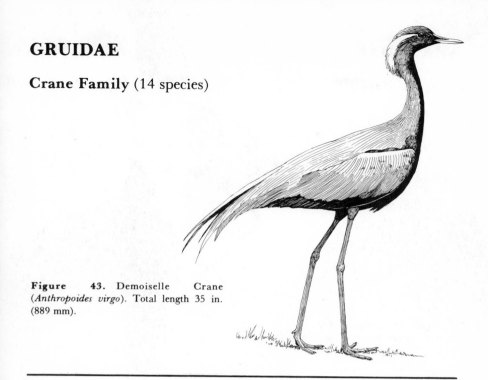

Figure 43. Demoiselle Crane (*Anthropoides virgo*). Total length 35 in. (889 mm).

Physical characteristics. Length—787 to 1524 mm (31 to 60 in.). Plumage—largely brown, gray, or white; bare red areas or ornamental plumes on head. Bill straight and rather long. Wings large; secondaries (used in display) usually modified—curled or elongated. Tail short. Legs long, with tibiae partly bare; hind toe elevated. Sexes alike or nearly so.

Range. North America s. to Cuba (to c. Mexico in winter) and the Old World exc. the Malayan Archipelago, Polynesia, and New Zealand. Most spp. migratory.

Habits. Gregarious (exc. in breeding season). Fly with neck extended; often soar. The elaborate "dance" characteristic of most cranes is apparently performed by some spp. at all seasons of the year. Voice—loud and resonant (trachea is elongated and coiled in sternum in most spp.).

Food. Omnivorous.

Breeding. Nest a sparse to bulky mass of vegetation on ground or in shallow water.

> *Eggs.* 2 to 3; dull white to brown or olive-brown; spotted with darker shades (exc. those of *Balearica,* which are immaculate pale blue). Incubated by ♂ and ♀.
>
> *Young.* Nidifugous. Downy. Cared for by ♂ and ♀.

Technical diagnosis. Ridgway and Friedmann, Pt. 9, 1941:1–6.

Classification. Peters, **2,** 1934:150–154.

References. Bent, No. 135:219–259, pls. 65–71; Witherby, **4,** 1940:449–455.

> L. H. Walkinshaw, The Sandhill Cranes, *Cranbrook Inst. Sci. Bull.* No. 29, 1949.
>
> R. P. Allen, The Whooping Crane, *Natl. Audubon. Soc. Res. Rept.* No. 3, 1952.
>
> L. H. Walkinshaw, The Wattled Crane *Bugeranus carunculatus* (Gmelin), *Ostrich,* **36,** 1965:73–81.
>
> S. Ali and S. D. Ripley, *Handbook of the Birds of India and Pakistan,* Vol. 2, Bombay, 1969:135–148.
>
> L. H. Walkinshaw, *Cranes of the World,* New York, 1973.

Synonyms: Balearicidae, Megalornithidae, "Psophiidae" of Mathews (a name now used for the Trumpeters).

ARAMIDAE

Limpkin Family

Figure 44. Limpkin (*Aramus guarauna*).
Total length 25 in. (635 mm).

Physical characteristics. Length—584 to 712 mm (23 to 28 in.). Plumage—dark olive-brown with greenish iridescence on upper part of body; feathers of head, neck, and under parts (also of back and wing coverts in some subspp.) broadly streaked with white. Bill long and laterally compressed (tip slightly decurved, usually somewhat twisted). Neck long and slender. Wings broad and rounded; tail broad and of moderate length. Legs long, with tibiae partly bare. Toes long; claws long and sharp. Sexes alike.

Range. Southern Georgia, Florida, Greater Antilles, s. Mexico, tropical C. and S. America s. to c. Argentina (Buenos Aires Prov.). Habitat—swamps, wooded or open, or arid brush (West Indies). Nonmigratory.

Habits. Solitary or slightly gregarious. Somewhat crepuscular or nocturnal. Fly slowly and infrequently. Vociferous—loud wails and screams; a variety of clucking notes.

Food. Large snails (esp. *Ampullaria*); a few insects and seeds.

Breeding. Shallow nest of rushes or sticks, in marsh vegetation just above water, or in bushes or trees (as much as 17 ft up).

    Eggs. 4 to 8; pale buffy; blotched and speckled with light brown. Incubated by ♂ and ♀.

    Young. Nidifugous. Downy—dark brown, unmarked. Cared for by ♂ and ♀.

Technical diagnosis. Ridgway and Friedmann, Pt. 9, 1941:28.

Classification. Hellmayr and Conover, Pt. 1, No. 1, 1942:301–308.

References. Bent, No. 135:254–259.

    D. J. Nicholson, Habits of the Limpkin in Florida, *Auk,* **45,** 1928:305–309.

    A. H. Howell, *Florida Bird Life,* New York, 1932:199–202.

    C. Cottam, Food of the Limpkin, *Wilson Bull.,* **48,** 1936:11–13.

    A. Wetmore, *The Birds of the Republic of Panamá,* 1, *Smithson. Misc. Collect.,* **150,** 1965:334–338.

    F. Haverschmidt, *Birds of Surinam,* Edinburgh, 1968:82–83.

RALLIDAE

Rail Family (132 species)

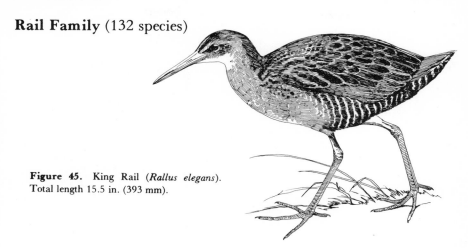

Figure 45. King Rail (*Rallus elegans*).
Total length 15.5 in. (393 mm).

Physical characteristics. Length—139 to 508 mm (5.5 to 20 in.). Plumage—black or soft shades of blue, gray, brown, or green; some spp. plain, others streaked above and/or barred below (esp. on sides and thighs). Bill strong; short and stout to long and curved. Some spp. have a bright-colored horny frontal shield. Neck medium to long. Body laterally compressed. Wings short and rounded; tail short and (usually) of soft feathers. Legs and toes long (toes lobed in *Fulica*); tibiae partly bare. Sexes alike or nearly so in most spp.

Range. Worldwide (exc. high latitudes). Habitat—marshes (typically), woods, or dry plains. Many spp. migratory.

Habits. Most spp. solitary. Secretive. Many spp. crepuscular. Terrestrial, but swim well. Flight appears weak, but some spp. make long migratory flights; many island spp. are flightless. Voice—great variety of screams, cackles, etc.; one note may be rapidly repeated in a long series.

Food. A great variety of animal and vegetable matter.

Breeding. Open nest on ground or slightly above it in a bush; in a burrow; on piles of vegetation in the water; or globular nest of grass with side entrance.

    Eggs. 2 to 16; white to buff or olive; immaculate or speckled. Incubated by ♂ and ♀.

    Young. Nidifugous. Downy. Cared for by ♂ and ♀ (and sometimes by young of a previous brood).

Technical diagnosis. Witherby, **5**, 1941:173–174.

Classification. Peters, **2**, 1934, 157–213.

References. Bent, No. 135:260–371; Witherby, **5**, 1941:173–208.

    A. O. Gross and J. Van Tyne, The Purple Gallinule (*Ionornis martinicus*) of Barro Colorado Island, Canal Zone, *Auk,* **46,**1929:431–446.

    Paul H. Baldwin, The life history of the Laysan Rail, *Condor,* **49,** 1947:14–21.

    A. Wetmore, *The Birds of the Republic of Panamá,* 1, *Smithson. Misc. Collect.,* **150,** 1965:338–365.

    B. Meanley, Natural history of the King Rail, *North Am. Fauna,* No. 67, 1969.

    S. Ali and S. D. Ripley, *Handbook of the Birds of India and Pakistan,* Vol. 2, Bombay, 1969:148–183.

647

PSOPHIIDAE

Trumpeter Family (3 species)

Figure 46. White-winged Trumpeter (*Psophia leucoptera*). Total length 21 in. (534 mm).

Physical characteristics. Length—432 to 534 mm (17 to 21 in.). Plumage—very soft and loose-webbed, velvetlike on head and neck; outer webs of tertials and secondaries modified to long, hairlike strands. Largely black, with purple, green, or bronze reflections, esp. on lower neck and wing coverts. Secondaries and tertials white, gray, or brown. Bill short, stout, and somewhat curved. Neck long. Wings very rounded; tail short; tail coverts long and full. Legs long; tibiae partly bare; feet of moderate size. Sexes alike.

Range. South America: s.e. Venezuela, the Guianas, and the Amazon Valley. Habitat—humid forests. Nonmigratory.

Habits. Gregarious. Largely terrestrial, but roost in trees. Run swiftly; fly only when forced. Carriage "humped" like that of guineahen (*Numida*). Voice—a loud, deep-toned cry; a prolonged cackle.

Food. Vegetable material, insects.

Breeding. Nest in hole in a tree or in crown of a palm.

> *Eggs.* 6 to 10; white or green; immaculate. Incubated by ♀.
> *Young.* Probably nidifugous. Downy. Cared for (by ♀?).

Technical diagnosis. Beddard, 374–377.

Classification. Hellmayr and Conover, Pt. 1, No. 1, 1942:308–314.

References. F. P. Penard and A. P. Penard, *De Vogels van Guyana*, Vol. 1, Paramaribo, [1908]:221–225.

C. Chubb, *The Birds of British Guiana*, Vol. 1, London, 1916:144–146.

[C.] W. Beebe, *Tropical Wild Life in British Guiana*, New York, 1917:247–252.

K. Plath, Trumpeters, *Aviculture*, **2** (Ser. 2), 1930:107–108.

F. Haverschmidt, *Birds of Surinam*, Edinburgh, 1968:83–84.

R. Meyer de Schauensee, *A Guide to the Birds of South America*, Wynnewood, Pa., 1970:64.

Note: The "Psophiidae" of Mathews = Gruidae.

HELIORNITHIDAE

Finfoot Family (3 species)

Figure 47. African Finfoot (*Podica senegalensis*). Total length 24.5 in. (623 mm).

Physical characteristics. Length—305 to 623 mm (12 to 24.5 in.). Plumage—brown or greenish black above (spotted with white in *Podica*), buffy white below. Head and neck partly black, with white stripes (faint in *Podica*). Bill, legs, and feet green, yellow, or red. Bill strong, rather long and tapered. Neck fairly long. Wings short and rounded; tail rather long and broad. Legs very short; feet broadly lobed. Sexes alike or ♀ slightly smaller and duller.

Range. Central America and n. half of S. America; c. and s. Africa (exc. Madagascar); n.e. India, Burma, Thailand, Malay Peninsula, Sumatra. Habitat—streams, lakes, marshes. Nonmigratory.

Habits. Solitary. Very shy. Largely aquatic, but run rapidly on land (*Heliopais, Podica*). Perch on branches over water. Flight fairly strong but not prolonged. Fly along surface of water when alarmed. Swim with head-bobbing motion; dive well. Voice—a bark of one to three notes; a guttural growl or quack; a bubbling sound.

Food. Insects, crustacea, snails, fish, amphibia; seeds.

Breeding. Nest of reeds or sticks in brush or trees, 1 to 12 ft above water.

    Eggs. 2 to 7; cream-colored or greenish white, speckled and blotched. Incubated by ♂ and ♀.

    Young. Nidicolous. Sparse down at hatching. Cared for (by ♂)?

Technical diagnosis. Ridgway and Friedmann, Pt. 9, 1941:224–226.

Classification. Peters, **2**, 1934:213–214.

References. Baker, *Fauna, 6,* 1929:36–38; Bannerman, **2**, 1931:xvi–xvii, 36–40.

    J. P. Chapin, The Birds of the Belgian Congo, II, *Bull. Am. Mus. Nat. Hist.,* **75,** 1939:34–40.

    C. W. Mackworth-Praed and C. H. B. Grant, *Birds of the Southern Third of Africa,* London, 1962:256–257.

    S. Ali and S. D. Ripley, *Handbook of the Birds of India and Pakistan,* Vol. 2, 1969:183–185.

    M. Alvarez del Toro, On the biology of the American Finfoot in southern Mexico, *Living Bird,* **1971:**79–88.

EURYPYGIDAE

Sunbittern Family (1 species)

Figure 48. Sunbittern (*Eurypyga helias*).
Total length 18 in. (457 mm).

Physical characteristics. Length—457 mm (18 in.). Plumage—full and soft (but feathers of neck very short); mottled and barred: black, gray, brown, and white; with spread wing showing bold pattern of black, chestnut, pale olive, yellow, gray, and white; 2 broad bands of black and chestnut on tail. Bill long. Head large; neck long, slender. Wings broad; tail long and broad. Legs long, with tibiae partly bare; toes rather long. Sexes alike.

Range. Neotropical. Tabasco and Chiapas through C. America to e. Peru and Bolivia and to s.e. Brazil. Habitat—about streams and ponds in tropical forest. Nonmigratory.

Habits. Solitary. Wading, terrestrial birds, but take refuge in trees when alarmed. Flight light and graceful. "Dance" with wings and tail outspread. Voice—soft, long-drawn whistle and plaintive piping; rattle bill.

Food. Insects, crustacea.

Breeding. Bulky nest of sticks and mud in tree, 12 to 20 ft above the ground, built by ♂ and ♀.

    Eggs. 2 to 3; nearly oval; buff; spotted and blotched with brown and purplish gray. Incubated by ♂ and ♀.

    Young. Nidicolous. Downy. Fed by ♂ and ♀.

Technical diagnosis. Ridgway and Friedmann, Pt. 9, 1941:232.

Classification. Peters, **2**, 1934:215–216.

References. A. Wetmore, *The Birds of the Republic of Panamá*, 1, *Smithson. Misc. Collect.,* **150**, 1965:369–372.

    F. Haverschmidt, *Birds of Surinam,* Edinburgh, 1968:93.

    R. Meyer de Schauensee, *A Guide to the Birds of South America,* Wynnewood, Pa., 1970:70–71.

RHYNOCHETIDAE

Kagu Family (1 species)

Figure 49. Kagu (*Rhinochetos jubatus*).
Total length 22 in. (559 mm).

Physical characteristics. Length—about 559 mm (22 in.). Plumage—loose-webbed; pale gray, shaded with brown; wings and tail somewhat darker and narrowly barred with dark gray. Spread wing conspicuously barred with black and chestnut; marked with white. Long bushy crest. Bill rather long and sharp; slightly flattened and decurved. Neck medium. Wings broad and rounded; tail fairly long and rounded. Legs long; tibiae partly bare; hind toe elevated. Sexes nearly alike.

Range. New Caledonia (e. of Australia). Habitat—forests. Nonmigratory.

Habits. Solitary. Crepuscular. Terrestrial. Apparently flightless. Has remarkable dancelike displays. Voice—guttural, rattling notes, piercing cries.

Food. Worms, snails, insects, frogs.

Breeding. Nest of sticks and leaves on ground.

Eggs. 1; buff; speckled with brown and gray. Incubated by ♂ and ♀.

Young. Seminidicolous. Downy. Cared for by ♂ and ♀.

Technical diagnosis. Ridgway and Friedmann, Pt. 9, 1941:2–3, 4.

Classification. Peters, **2**, 1934:215.

References. A. J. Campbell, The Kagu of New Caledonia, *Emu,* **4,** 1905:166–168.

A. J. Campbell, A Kagu chick, *Emu,* **5,** 1905:32. (See also footnote, *Emu,* **5:**95.)

H. E. Finckh, Notes on Kagus (*Rhinochetus jubatus*), *Emu,* **14,** 1915:168–170.

D. W. Warner, The present status of the Kagu, *Rhynochetos jubatus,* on New Caledonia, *Auk,* **65,** 1948:287–288.

J. C. Greenway, Jr., *Extinct and Vanishing Birds of the World,* New York, 1958:253–256.

J. Fisher *et al., Wildlife in Danger,* New York, 1969:234–235.

Synonym: Rhinochetidae.

CARIAMIDAE

Cariama Family (2 species)

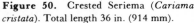
Figure 50. Crested Seriema (*Cariama cristata*). Total length 36 in. (914 mm).

Physical characteristics. Length—762 to 914 mm (30 to 36 in.). Plumage—soft and loose-webbed; rather light grayish brown, finely vermiculated with buff or white; under parts paler, becoming buffy white on belly and under tail coverts. Nuchal, as well as frontal, crest; long feathers on throat and lower neck. Remiges boldly barred with buff; tail tipped with gray or white. Bill short, broad, decurved. Wings short and rounded; tail long. Legs very long; tibiae extensively bare; feet small, semipalmate. Sexes alike or nearly so.

Range. East-central Brazil s. to n. Argentina. Habitat—grassland (*Cariama*) and forests (*Chunga*). Nonmigratory.

Habits. Occur in pairs or small groups. Largely cursorial; run rapidly. Fly weakly. Voice—barking and screaming cries, often given in chorus.

Food. Insects, reptiles, small mammals; also berries, seeds, leaves.

Breeding. Nest of sticks in trees or bushes.

Eggs. 2 to 3; white; sparsely spotted with brown. Incubated by ♂ and ♀.

Young. Nidicolous. Down with some pattern, very long and shaggy, especially on head. Cared for (by ?).

Technical diagnosis. Beddard, 373–374.

Classification. Hellmayr and Conover, Pt. 1, No. 1, 1942:427–430.

References. H. S. Boyle, Field notes on the Seriema (*Chunga burmeisteri*), *Auk,* **34,** 1917:294–296.

O. Heinroth, Die Jugendentwicklung von *Cariama cristata, J. Ornithol.,* **72,** 1924:119–124.

E. T. Gilliard, *Living Birds of the World,* New York, 1958:155–156.

R. Meyer de Schauensee, *A Guide to the Birds of South America,* Wynnewood, Pa., 1970:71.

Synonym: Dicholophidae.

OTIDIDAE

Bustard Family (23 species)

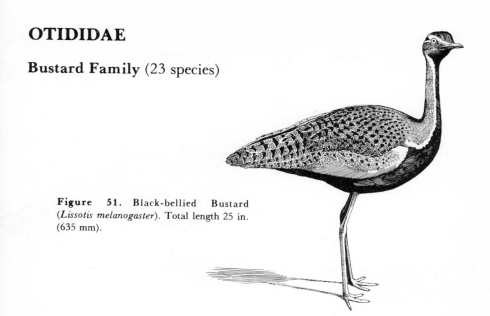

Figure 51. Black-bellied Bustard (*Lissotis melanogaster*). Total length 25 in. (635 mm).

Physical characteristics. Length—368 to 1321 mm (14.5 to 52 in.). Plumage—buff and gray above, barred and vermiculated with black in cryptic patterns; pale buff, white, or solid black below; bold black and white patterns on head and neck in a few spp.; some spp. have concealed white areas on the wing which become very conspicuous in flight. Some spp. crested; some with long, bristly feathers on sides of head or neck. Bill rather short, stout, and flattened. Neck long. Wings broad; tail short to medium. Legs strong and fairly long; tibiae extensively bare; 3 toes (2nd, 3rd, and 4th), short and broad. Sexes unlike.

Range. Southern Europe, Africa, s. Asia, Australia. Habitat—open plains. Some spp. migratory.

Habits. Rather gregarious. Terrestrial; cursorial. Flight strong but infrequent. Crouch in presence of danger. ♂ has remarkable courtship flight and posturing and strutting displays. Voice—a variety of booming, crowing, clucking, and other notes.

Food. Omnivorous.

Breeding. Nest a slight depression, with little or no lining, in ground.

    Eggs. 1 to 5; reddish brown to olive-green; speckled and blotched with brown. Incubated by ♀.

    Young. Nidicolous. Downy. Cared for by ♀.

Technical diagnosis. Witherby, **4**, 1940:436.

Classification. Peters, **2**, 1934:217–225.

References. Bannerman, **2**, 1931:49–67.

    R. S. Dharmakumarsinhji, *Birds of Saurashtra India,* Bombay, 1954:145–153.

    C. W. Mackworth-Praed and C. H. B. Grant, *Birds of the Southern Third of Africa,* London, 1962:261–275.

    S. Ali and S. D. Ripley, *Handbook of the Birds of India and Pakistan,* Bombay, 1969:185–198.

    H. J. Frith, ed., *Birds in the Australian High Country,* Sydney, 1969:150–151, 154.

Synonym: Otidae.

MESOENATIDAE

Mesites Family (3 species)

Figure 52. Brown Mesites (*Mesoenas unicolor*). Total length 10 in. (254 mm).

Physical characteristics. Length—254 to 266 mm (10 to 10.5 in.). Plumage—chestnut-brown, or gray and olive-brown, above. One or more white streaks on side of head or neck. Throat and breast pale buff or white, spotted or barred (exc. in *Mesoenas unicolor*). Bill slender, moderate in length and straight to long (40 mm) and sickle-shaped (*Monias*). Nostrils long, operculate slits. Wings short and rounded; clavicles rudimentary; tail rather long and rounded. Legs slender or stout; moderate in length. ♀ unlike ♂ in *Monias,* like in others. Five pairs of powder down patches.

Range. Madagascar (in forests or brushy areas). Nonmigratory.

Habits. Usually found in pairs (*Monias* sometimes in flocks of 30 or more). Terrestrial. Run well. Fly only when forced to do so (*Monias* apparently flightless). Voice (of *Monias*)—loud, explosive *nak nak nak . . .* ; a low *creeu.*

Food. Fruits, seeds, insects.

Breeding. Some evidence of polyandry. Nest of sticks and leaves, 2 to 6 ft above the ground, in low trees or brush.

> *Eggs.* 1 to 3; white, greenish white, or buffy gray. Incubated by ♂ or by ♀ (or both? —recorded for ♂ in *Monias,* for ♀ in *Mesoenas*).

> *Young.* Nidifugous. Downy. Apparently (in *Monias*) cared for by ♂.

Technical diagnosis. P. R. Lowe, *Proc. Zool. Soc. London,* **1924**:1131–1152.

Classification. Peters, **2**, 1934:141–142.

References. J. Delacour, Les oiseaux de la Mission Zoologique Franco-Anglo-Americaine á Madagascar, *L'Ois. Rev. Franç. Ornithol.,* **2**, 1932:30–31.

> A. L. Rand, The distribution and habits of Madagascar birds, *Bull. Am. Mus. Nat. Hist.,* **72**, Art. 5, 1936:364–368.

> A. L. Rand, The nests and eggs of *Mesoenas unicolor* of Madagascar, *Auk,* **68**, 1951:23–26.

> E. T. Gilliard, *Living Birds of the World,* New York, 1958:127–128.

> O. L. Austin, Jr., *Birds of the World,* New York, 1961:101.

Synonyms: Mesitidae, Mesoenidae, Mesitornithidae.

JACANIDAE

Jaçana Family (7 species)

Figure 53. American Jaçana (*Jacana spinosa*). Total length 10 in. (254 mm).

Physical characteristics. Length—165 to 534 mm (6.5 to 21 in.). Plumage—reddish or greenish brown and black above; greenish black, brown, or white below. One sp. with white crown, 1 with white rump; 5 spp. with yellow area on neck; 3 spp. with spread wing boldly marked with white or yellow. Bare forehead and crown, frontal wattle, or rictal lappets (exc. in *Microparra* and *Hydrophasianus*). Bill of moderate length and nearly straight. Wings broad, with metacarpal spur or knob; tail short and weak (very long and narrow in *Hydrophasianus*). Legs long; tibiae extensively bare; toes and esp. claws extremely long. Sexes alike, but ♀ larger than ♂.

Range. Mexico, C. America, S. America (exc. s.w. quarter), Africa (s. of Sahara), s. Asia, Philippines, Malaysia, Papuan Region, n. Australia. Habitat—marshy shores of lakes and streams. Largely nonmigratory.

Habits. Some spp. gregarious in winter. Largely cursorial; walk habitually across water on floating lily pads. Swim readily and dive to escape danger. Flight slow and labored. Voice—loud, harsh cries, mewing, chattering notes.

Food. Insects, snails, fish, seeds of water plants.

Breeding. Nest of water weeds, usually on floating vegetation.

> *Eggs.* 3 to 6, usually 4; brown; immaculate or heavily marked with dark brown or black. Incubated by ♂ (or by ♂ and ♀?).
>
> *Young.* Nidifugous. Downy. Cared for by ♂ (or by ♂ and ♀?).

Technical diagnosis. Ridgway, Pt. 8, 1919:1–6.

Classification. Peters, **2**, 1934:226–230.

References. C. W. Mackworth-Praed and C. H. B. Grant, *Birds of the Southern Third of Africa,* London, 1962:278–280.

> A. Wetmore, *The Birds of the Republic of Panamá,* 1, *Smithson. Misc. Collect.,* **150,** 1965:372–378.
>
> J. B. D. Hopcraft, Some notes on the chick-carrying behavior in the African jacana, *Living Bird,* **1968:**85–88.
>
> S. Ali and S. D. Ripley, *Handbook of the Birds of India and Pakistan,* Vol. 2, Bombay, 1969:198–202.

Synonym: Parridae.

ROSTRATULIDAE

Painted-snipe Family (2 species)

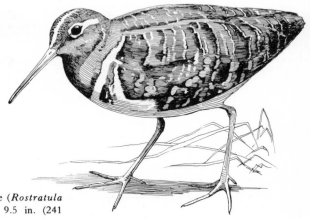

Figure 54. Painted-snipe (*Rostratula benghalensis*). Total length 9.5 in. (241 mm).

Physical characteristics. Length—190 to 241 mm (7.5 to 9.5 in.). Plumage—brown, olive-brown, gray, black, and white; cryptically patterned above; throat and breast dark, belly buffy white. Bill long, somewhat curved, and slightly swollen at tip. Neck short. Wings broad; tail short and weak. Legs moderately long; tibiae partly bare; toes long. ♂ smaller and duller than ♀ (*Rostratula*); sexes alike (*Nycticryphes*).

Range. Southern S. America, Africa (s. of Sahara), Egypt, s. Asia, Japan, Malaysia, Australia, Philippines. Habitat—marshes. Nonmigratory (exc. for local movements).

Habits. Solitary. Crepuscular. Terrestrial. Flight strong but not prolonged. Voice—a booming call; whistling, purring, and hissing notes.

Food. Insects, mollusca, worms, grain.

Breeding. Apparently polyandrous. Nest of grass, placed on ground and usually concealed by surrounding vegetation.

Eggs. 2 to 5; buff; heavily marked with dark brown and black. Incubated by ♂.

Young. Nidifugous. Downy. Cared for by ♂.

Technical diagnosis. P. R. Lowe, *Ibis,* **1931**:507–530, 532; **1932**:390–391.

Classification. Peters, **2**, 1934:230.

References. A. Wetmore, Observations on the birds of Argentina, Paraguay, Uruguay, and Chile, *U.S. Natl. Mus. Bull.,* **133,** 1926:163–164.

P. A. Clancey, *Gamebirds of Southern Africa,* Johannesburg, 1967:213–214.

S. Ali and S. D. Ripley, *Handbook of the Birds of India and Pakistan,* Vol. 2, Bombay, 1969:325–328.

H. J. Frith, ed., *Birds in the Australian High Country,* Sydney, 1969:168–170.

G. M. Henry, *A Guide to the Birds of Ceylon,* New York, 1971:325–326.

HAEMATOPODIDAE

Oystercatcher Family (6 species)

Figure 55. American Oystercatcher (*Haematopus palliatus*). Total length 17.5 in. (445 mm).

Physical characteristics. Length—381 to 508 mm (15 to 20 in.). Plumage—dark brown or blackish brown exc. head and neck, which are black, and (in some spp.) the rest of under parts, rump, lower back, and large area on wing, which are white. Bill bright red, legs pink. Bill long, stout, and much compressed laterally. Neck short. Wings long and pointed; tail short. Legs and feet very stout; feet slightly webbed; 3 toes (2nd, 3rd, and 4th). Sexes alike.

Range. Most temperate and tropical seacoasts; n. to Iceland and the Aleutians; s. to Cape Horn, Argentina, and Tasmania. Also breed on some inland waters of Europe and Asia. Some forms migratory.

Habits. Gregarious. Flight strong and direct. Excitable, vociferous. Voice—a variety of loud, shrill notes; "song" a long series of monotonous piping notes, ending in a trill.

Food. Oysters, mussels, limpets, chitons, and other "shellfish"; sandworms (*Nereis*).

Breeding. Nest a hollow, usually lined (with grass, moss, or rock flakes), in the ground.

   Eggs. 2 or 3, sometimes 4; buff; blotched and speckled with black and brown. Incubated by ♂ and ♀.

   Young. Nidifugous. Downy. Cared for by ♂ and ♀.

Technical diagnosis. Ridgway, Pt. 8, 1919:26–27.

Classification. Peters, **2**, 1934:231–234.

References. Bent, No. 146:305–323; Baker, *Fauna,* **6,** 1929:164–167; Witherby, **4,** 1940:414–421; Murphy, **2:**973–993.

   J. S. Huxley and F. A. Montague, Studies on the courtship and sexual life of birds. V. The Oyster-catcher (*Haematopus ostralegus* L.), *Ibis,* **1925:**868–897.

   I. R. Tompkins, Life history notes on the American Oyster-catcher, *Oriole,* **19,** 1954:37–45.

   G. F. Makkink, Contribution to the knowledge of the behavior of the Oyster-Catcher (*Haematopus ostralegus* L.), *Ardea,* **31,** 1942:23–74.

   A. Wetmore, *The Birds of the Republic of Panama,* 1, *Smithson. Misc. Collect.,* **150,** 1965:378–382.

   N. W. Cayley, *What Bird Is That?*, Sydney, 1966:296–297.

SCOLOPACIDAE

Sandpiper Family (82 species)

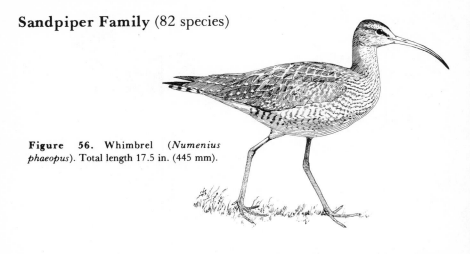

Figure 56. Whimbrel (*Numenius phaeopus*). Total length 17.5 in. (445 mm).

Physical characteristics. Length—127 to 610 mm (5 to 24 in.). Plumage—pale buff or gray to chestnut brown (black in a few spp.), and white; the back cryptically patterned in most spp., the under parts largely white in many, barred or spotted with black or brown in some; the winter plumage in many spp. paler, grayer, and less marked. Bill slender, moderately to extremely long; straight, decurved, or recurved; spatulate in 1 sp. Neck medium to long. Wings long; tail short to medium. Legs short to long; tibiae partly bare in many spp. Toes rather long (hallux lacking in Sanderling). Sexes alike or nearly so in most spp.

Range. Worldwide. Habitat—open, bare areas, usually near water. Most spp. migratory.

Habits. Rather gregarious. Terrestrial. Most spp. are waders. Flight strong. Many have elaborate courtship flights and displays, usually with vocal or instrumental (i.e., sounds made by flight feathers) accompaniment. Voice—a great variety of sharp, harsh callnotes, musical trills, whistling, and other sounds.

Food. Variety of animal matter; some vegetable material.

Breeding. Nest on ground, in vacant tree nests, or (*Coenocorypha*) in ground burrows of other birds.

Eggs. 2 to 4; buff or olive; heavily marked with brown and black. Incubated by ♂ and ♀, by ♀ alone, or largely by ♂.

Young. Nidifugous. Downy. Cared for by ♂ and ♀, by ♀ alone, or by ♂ alone.

Technical diagnosis. Ridgway, Pt. 8, 1919:143–144; P. R. Lowe, *Ibis,* **1931**:747–750 (for subfamily Arenariinae).

Classification. Peters, **2,** 1934:258–288.

References. Bent, No. 142:54–359, and No. 146:1–143, 269–305; Witherby, **4,** 1940:152–346.

W. G. Sheldon, *The Book of the American Woodcock,* Boston, 1967.

H. J. Frith, ed., *Birds in the Australian High Country,* Sydney, 1969:185–192.

S. Ali and S. D. Ripley, *Handbook of the Birds of India and Pakistan,* Vol. 2, Bombay, 1969:241–321.

D. F. Parmelee, Breeding behavior of the Sanderling in the Canadian Arctic, *Living Bird,* **1970**:97–146.

Synonymy: The Aphrizidae are included.

PHALAROPODIDAE

Phalarope Family (3 species)

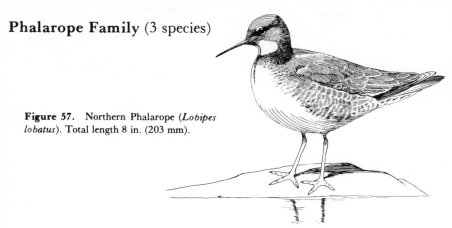

Figure 57. Northern Phalarope (*Lobipes lobatus*). Total length 8 in. (203 mm).

Physical characteristics. Length—190 to 254 mm (7.5 to 10 in.). Plumage—soft and dense; the ♀ with bold patterns of black, white, gray, and rich brown on head and neck; back dark gray, streaked with buff or rich brown; white on rump area; under parts white or reddish brown. Winter plumage—black (or gray) and white. Bill moderate to very long and slender. Wings long and pointed; tail moderate. Tarsi rather long (exc. *Phalaropus*) and laterally compressed; toes lobed, semipalmate. ♂ smaller and much duller in coloration than ♀.

Range. Arctic and subarctic regions; *Steganopus* s. in N. America to about 40° N. lat. Migratory; winter on ocean in restricted areas in tropics (*Steganopus* inland in s. S. America).

Habits. Aquatic. Float buoyantly; swim erratic course, bobbing head back and forth. Notably tame. Flight swift. Often feed by swimming rapidly and continuously in very small circle, snatching from water the small animal life thus stirred up. Voice—low, musical calls and whistles.

Food. Crustacea, insects, some vegetable matter.

Breeding. Nest on open ground near water, often in small colonies. Substantial nest of grass or scantily lined hollow.

> *Eggs.* 4 (rarely 3 or 5); olive-buff; heavily blotched and speckled with brown and black. Incubated typically by ♂ only.

> *Young.* Nidifugous. Downy—light brown, streaked and spotted with light gray and black. Cared for by ♂ (sometimes assisted by ♀).

Technical diagnosis. Ridgway, Pt. 8, 1919:416–417.

Classification. Peters, **2**, 1934:292–293.

References. Bent, No. 142:1–37; Witherby, **4**, 1940:213–222.

W. E. Godfrey, *The Birds of Canada,* Ottawa, 1966:166–170.

E. O. Höhn, Observations on the breeding biology of Wilson's Phalarope (*Steganopus tricolor*) in central Alberta, *Auk,* **84**, 1967:220–244.

D. F. Parmelee *et al.,* The birds of southeastern Victoria Island and adjacent small islands, *Natl. Mus. Can. Bull.* No. 222, 1967.

G. P. Dement'ev *et al., Birds of the Soviet Union,* Vol. 3, Washington, 1969:289–303.

RECURVIROSTRIDAE

Avocet Family (7 species)

Figure 58. Black-necked Stilt (*Himantopus mexicanus*). Total length 14 in. (355 mm).

Physical characteristics. Length—292 to 482 mm (11.5 to 19 in.). Plumage—black (sometimes glossed with green) or brownish gray, and white above (uniform gray in *Ibidorhyncha*); mostly white below; chestnut pectoral shield in 1 sp., reddish brown head and neck in 2 spp. Bill long, very slender, straight or recurved (strongly decurved in *Ibidorhyncha*). Head small; neck moderately long. Wings rather long and pointed; tail rather short and square. Legs moderate to extremely long and slender; feet slightly to strongly webbed (slightly lobed in *Ibidorhyncha*); hallux vestigial or absent. Sexes alike or nearly so.

Range. Africa (incl. Madagascar), c. and s. Europe and Asia, Malaysia, Papuan Region, Australia, New Zealand, Philippines, Hawaiian Ids.; w. and s. U.S., West Indies, C. America, Galapagos Ids., and most of S. America. Habitat—usually near water. Northern forms migratory.

Habits. Gregarious. Terrestrial (waders). Swim readily. Flight strong. Voice—loud, harsh barks and calls; softer, more musical notes.

Food. Aquatic insects, mollusca, crustacea; fish, frogs, lizards; some vegetable matter.

Breeding. Colonial. Nest a scrape in sand or mud, bare or lined with pebbles or grass; sometimes a bulky platform (rarely floating) of sticks, etc.

> *Eggs.* 2 to 5 (usually 4); pale stone color to olive; spotted, scrawled, or blotched with reddish brown to black. Incubated by ♂ and ♀.
>
> *Young.* Nidifugous. Downy. Cared for (by ♂ and ♀)?

Technical diagnosis. Ridgway, pt. 8, 1919:435 (does not include *Ibidorhyncha*).

Classification. Peters, **2**, 1934:288–291.

References. Bent, No. 142:37–54; Witherby, **4**, 1940:403–413.

Jack Jones, The Banded Stilt, *Emu,* **45**, 1945:1–36, 110–118.

C. W. Mackworth-Praed and C. H. B. Grant, *Birds of the Southern Third of Africa,* London, 1962:303–305.

H. J. Frith, ed., *Birds in the Australian High Country,* Sydney, 1969:192–194.

S. Ali and S. D. Ripley, *Handbook of the Birds of India and Pakistan,* Vol. 2, Bombay, 1969:328–336.

CHARADRIIDAE

Plover Family (63 species)

Figure 59. Dotterel (*Eudromias morinellus*). Total length 9 in. (228 mm).

Physical characteristics. Length—152 to 393 mm (6 to 15.5 in.). Plumage—brown, olive, gray, or black, and white. Most spp. with back uniformly colored; many with broad collar and/or contrastingly marked tail or rump. Head and neck boldly marked in many spp. Below: white (or, rarely, solid black), marked in many spp. with one or two broad, contrasting bands. Spread wing with conspicuous pattern. Some (among Vanellinae) have long crest, wattles about face, and/or spurs on wing. Bill straight, of moderate length. Neck rather short. Wings long; tail short to medium. Legs short to long; tibiae partly bare. Toes moderately long; hallux vestigial or lacking. Sexes alike or nearly so.

Range. Worldwide. Habitat—open, bare areas. Most spp. migratory.

Habits. Rather gregarious. Terrestrial. Run swiftly. Flight strong. Voice—shrill cries, melodious whistling notes.

Food. Variety of animal matter; some vegetable material.

Breeding. Nest on ground, with little or no lining.

    Eggs. 2 to 5 (usually 4); buff or gray; heavily marked with black. Incubated by ♂ and ♀.

    Young. Nidifugous. Downy. Cared for by ♂ and ♀.

Technical diagnosis. Ridgway, Pt. 8, 1919:61–62.

Classification. Peters, **2,** 1934:234–258.

References. Bent, No. 146:144–305; Witherby, **4,** 1940:346–403; Baker, *Nidification, 4,* 1935:387–403.

    C. W. Mackworth-Praed and C. H. B. Grant, *Birds of the Southern Third of Africa,* London, 1962:280–301.

    W. G. Conway and J. Bell, Observations on the behavior of Kittlitz's Sandplovers at the New York Zoological Park, *Living Bird,* **1968:**57–70.

    S. Ali and S. D. Ripley, *Handbook of the Birds of India and Pakistan,* Vol. 2, Bombay, 1969:205–241.

    W. D. Graul, Breeding biology of the Mountain Plover, *Wilson Bull.,* **87,** 1975:6–31.

DROMADIDAE

Crabplover Family (1 species)

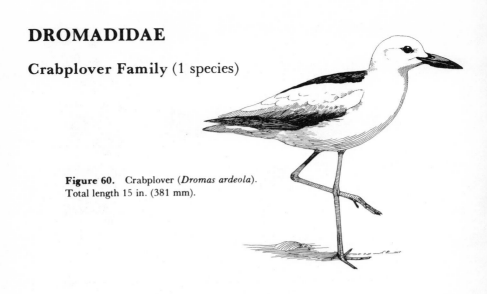

Figure 60. Crabplover (*Dromas ardeola*).
Total length 15 in. (381 mm).

Physical characteristics. Length—381 mm (15 in.). Plumage—white, with a few orbital feathers, central feathers of back, and longer scapulars, black; remiges and greater upper wing coverts brownish black (inner webs of remiges pale brown to white). Bill strong, laterally compressed. Wings long and pointed; tail short. Legs long and bare; middle claw pectinate. ♀ somewhat smaller, with black dorsal feathers shorter, than ♂.

Range. Coasts and island shores of Indian Ocean from Natal and Madagascar to Ceylon; also islands off Burma and the shores of s. portion of Red Sea. Migratory in part.

Habits. Gregarious. Noisy. Crepuscular. Fly well and run swiftly (in short, abrupt dashes). Voice—chatter; raucous cry somewhat like a crow's.

Food. Crabs and other crustacea; mollusca.

Breeding. Nest an unlined depression at the end of a burrow (4 to 8 ft long, near surface of ground).
Eggs. 1; white; extremely large (65 × 46 mm.). Incubation unrecorded.
Young. Nidicolous. Downy. Cared for by ♂ and ♀?

Technical diagnosis. Baker, *Fauna,* **6,** 1929:94.

Classification. Peters, **2,** 1934:293.

References. Baker, *Fauna,* **6,** 1929:94–95; Baker, *Nidification,* **4,** 1935:353–355.

P. R. Lowe, Some notes on the Crab-Plover (*Dromas ardeola* Paykull), *Ibis,* **1916:**317–337.

G. Archer and E. M. Godman, *The Birds of British Somaliland and the Gulf of Aden,* Vol. 2, London, 1937:495–497.

R. Meinertzhagen, *Birds of Arabia,* London, 1954:470–471.

S. Ali and S. D. Ripley, *Handbook of the Birds of India and Pakistan,* Vol. 2, Bombay, 1969:337–338.

G. M. Henry, *A Guide to the Birds of Ceylon,* 2nd ed., New York, 1971:275–277.

BURHINIDAE

Thick-knee Family (9 species)

Figure 61. Stonecurlew (*Burhinus oedicnemus*). Total length 16 in. (406 mm).

Physical characteristics. Length—355 to 521 mm (14 to 20.5 in.). Plumage—brown, gray-brown, or buff, mottled, barred or streaked, with conspicuous wing pattern in some spp.; side of head patterned with broad stripes. Bill stout, short to moderately long. Head large and broad; eyes very large. Wings medium to long and pointed; tail moderately long, graduated. Legs rather long; tarsus bare, tibiotarsal joint thickened. Feet partially webbed, with 3 toes (2nd, 3rd, and 4th). Sexes alike.

Range. Temperate and tropical Europe, Africa, Asia; Australia; tropical America (s. Mexico to n.w. Brazil and s. Peru; Hispaniola). Habitat—stony or sandy ground in semiopen country; some spp. littoral. Northern forms migratory.

Habits. Somewhat gregarious. Crepuscular and nocturnal. Run swiftly. Flight brief and infrequent but strong. Some spp. have remarkable communal display in autumn. Voice—clamorous; loud wailing and croaking cries; variety of shorter notes.

Food. Worms; land and water insects, crustacea, mollusca; mice; frogs; nestlings of other birds.

Breeding. Eggs laid directly on ground or in slight unlined depression—in open or near grass or shrubs.

Eggs. 2 (rarely 3); white to pale buff; speckled and blotched. Incubated by ♀, assisted by ♂ in most spp.

Young. Nidicolous, or may leave nest in 1 day. Downy. Cared for by ♂ and ♀.

Technical diagnosis. Witherby, **4,** 1940:430.

Classification. Peters, **2,** 1934:293–298.

References. Witherby, **4,** 1940:430–436; Baker, *Nidification, 4,* 1935:340–345.

R. S. Dharmakumarsinhji, *Birds of Saurashtra India,* Bombay, 1954:166–169.

C. W. Mackworth-Praed and C. H. B. Grant, *Birds of the Southern Third of Africa,* London, 1962:275–278.

H. J. Frith, ed., *Birds in the Australian High Country,* Sydney, 1969:194–196.

Synonym: Oedicnemidae.

GLAREOLIDAE

Pratincole Family (17 species)

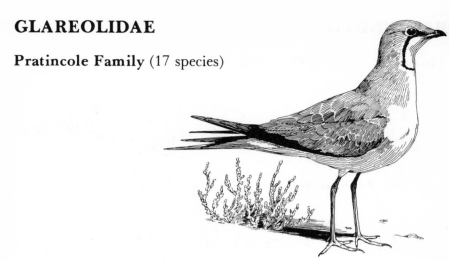

Figure 62. Pratincole (*Glareola pratincola*). Total length 9.5 in. (241 mm).

Physical characteristics. Length—152 to 254 mm (6 to 10 in.). Plumage—olive, brown, gray, and chestnut; plain, with bold markings of black, white, and chestnut (a few spp. with metallic areas or cryptic pattern). Pratincoles have short bills, long narrow pointed wings, forked tails, short legs, 4 toes. Coursers have rather long tapering bills, rather short and broad wings and tails, long legs, 3 toes. Middle toe elongate, with claw usually pectinate in both coursers and pratincoles; basal web between outer and middle toe in most spp. Sexes similar; may differ in size.

Range. Around Mediterranean, n. to British Isles and Germany; Africa, e. through India to Indochina; Australia. Habitat—usually sandy or stony ground, often near water. More northern spp. migratory.

Habits. Usually gregarious. Some spp. crepuscular. Pratincoles aerial, their flight buoyant and erratic (hawk insects). Coursers terrestrial, running swiftly, their flight direct and sometimes swift, but infrequent. Some spp. wade. Voice (most spp. vociferous)—a harsh *hark hark*; *zic zac*; *kik-kikki*; etc.

Food. Land and water insects, lizards, snails, seeds, (?) fish.

Breeding. Usually colonial. Eggs laid directly on sand or rock, sometimes in shallow depression (rarely lined); sometimes buried in sand.

    Eggs. 2 to 4 (rarely 5). Incubated by ♀, assisted by ♂ in some spp.

    Young. Nidicolous (but may leave nest in 1 day?). Downy. Cared for by ♂ and ♀.

Technical diagnosis. Witherby, **4**, 1940:421.

Classification. Peters, **2**, 1934:298–306.

References. Witherby, **4**, 1940:421–429; Bannerman, **2**, 1931:191–220; Baker, *Nidification, **4**,* 1935:345–353.

    C. W. Mackworth-Praed and C. H. B. Grant, *Birds of the Southern Third of Africa,* London, 1962:335–339.

    S. Ali and S. D. Ripley, *Handbook of the Birds of India and Pakistan,* Vol. 3, Bombay, 1969:7–17.

Synonymy: The Cursoriidae are included.

THINOCORIDAE

Seedsnipe Family (4 species)

Figure 63   D'Orbigny Seedsnipe
(*Thinocorus orbignyianus*). Total length
8.25 in. (210 mm).

Physical characteristics.   Length—171 to 280 mm (6.75 to 11 in.). Plumage—cryptic pattern above, brown with pale buff edgings; below: pale buff, marked with darker brown; or white, with broad breast band of gray or mottled brown. Bill strong, short, pointed, rather sparrow-like. Nostrils long slits with protective operculum above. Wings long and pointed; tail short. Legs very short. Sexes unlike.

Range.   Southern half of Argentina and Chile, n. along the higher Andes to n. Ecuador. Habitat—open country. Some forms migratory.

Habits.   Gregarious. Terrestrial. When disturbed, creep away along ground or crouch, and suddenly, with snipelike calls, flush in zigzag flight. Run rapidly. *Thinocorus* has a soaring flight song. Voice—whistling or cooing calls; grating *chairp*.

Food.   Largely vegetable; some insects.

Breeding.   Nest a scantily lined depression or rough aggregate of plant material, on ground.

> *Eggs.*   4; pointed; cream-white, pink, or greenish; finely speckled with darker shades. Incubated by ♀.
>
> *Young.*   Nidifugous. Downy. Cared for by ♀.

Technical diagnosis.   R. B. Sharpe, Cat. of Birds of Brit. Mus., **24,** 1896:714.

Classification.   Peters, **2,** 1934:306–308.

References.   J. D. Goodall *et al., Las Aves de Chile,* Vol. 2, Buenos Aires, 1951:265–273.

> G. L. Maclean, A study of seedsnipe in southern South America, *Living Bird,* **1969:**33–80.
>
> R. Meyer de Schauensee, *A Guide to the Birds of South America,* Wynnewood, Pa., 1970:82–83.
>
> P. S. Humphrey *et al., Birds of Isla Grande (Tierra del Fuego),* Preliminary Smithsonian Manual, Washington, 1970:207–213.

Synonyms: Thinocorythidae, Attagidae.

CHIONIDIDAE

Sheathbill Family (2 species)

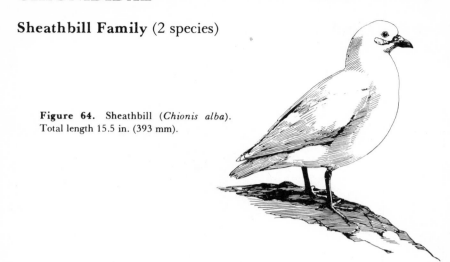

Figure 64. Sheathbill (*Chionis alba*).
Total length 15.5 in. (393 mm).

Physical characteristics. Length—355 to 432 mm (14 to 17 in.). Plumage—white. Bill black or yellow and black, with horny sheath on base of maxilla; legs and feet pale flesh or gray. Face partly bare and flesh-colored, more or less carunculated near base of bill. Eyes small. Bill short and stout. Neck short; body compact. Wings rather long; tail short and somewhat rounded. Legs short and stout; tibiae partly bare. Feet large and powerful, slightly webbed; hallux elevated. Sexes alike or ♀ smaller than ♂.

Range. Islands of the extreme s. Atlantic and w. Indian oceans; tip of Graham Land and southernmost S. America. Habitat—seacoast. Locally migratory.

Habits. Gregarious. Largely terrestrial but make long sea flights. Swim well but infrequently. Flight strong. Pugnacious. Voice—"angry cries"; cooing sound (in courtship).

Food. Omnivorous. Small fish, mollusca, and crustacea; birds' eggs and nestlings; algae and lichens; variety of offal.

Breeding. Nest a bulky collection of feathers, shells, algae, etc., in natural hole, old burrow, or rock crevice.

Eggs. 2 to 3; white; blotched and freckled with brown or black. Incubated by ♂ and ♀.

Young. Nidifugous? Downy. Cared for by ♂ and ♀.

Technical diagnosis. J. H. Kidder and E. Coues, *U.S. Natl. Mus. Bull.*, **3,** 1876:115 ("Chionomorphae").

Classification. Peters, **2,** 1934:308–309.

References. Murphy, **2:**999–1006.

    A. F. Cobb, *Birds of the Falkland Islands,* London, 1933:62–63.

    P. S. Humphrey *et al., Birds of Isla Grande* (*Tierra del Fuego*), Preliminary Smithsonian Manual, Washington, 1970:214–215.

    P. Slater *et al., A Field Guide to Australian Birds,* Wynnewood, Pa., 1971:316.

Synonym: Chionidae.

STERCORARIIDAE

Skua Family (4 species)

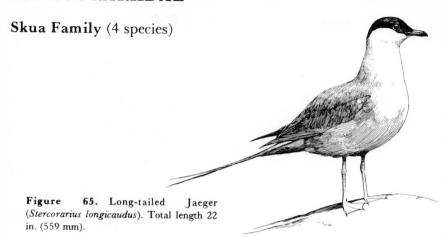

Figure 65. Long-tailed Jaeger (*Stercorarius longicaudus*). Total length 22 in. (559 mm).

Physical characteristics. Length—432 to 610 mm (17 to 24 in.). Plumage (variable, with dark and light phases)—blackish brown above, with some light edgings or streaks; light brown to white below, plain or barred with brown; neck yellow or streaked with buff; white flash on spread wing. Bill strong, medium in length, rounded, and strongly hooked, with horny "cere." Body stout. Wings long, pointed; tail medium to long, wedge-shaped, the central rectrices elongated in *Stercorarius* (twisted in 1 sp.). Legs short; feet stout, webbed; hallux very small; claws strongly hooked. Sexes alike (or ♀ larger).

Range. High latitudes of both S. and N. hemispheres. Habitat—oceans and their coasts; sometimes nest inland. Migratory.

Habits. Somewhat gregarious. Pelagic. Flight swift and powerful (some soar). Aggressive and predatory, robbing other birds of food. Sometimes eat own eggs and young. Some hawk insects. Striking flight display in 1 sp. Voice—quacking and chattering notes; wailing cries; shrieks and screams.

Food. Small mammals; large insects; eggs, young, and even adult birds; fish, crustacea, mollusca; carrion, offal; some vegetable matter.

Breeding. Loosely colonial (generally in or near colonies of other seabirds). Nest a depression, lined or unlined, in vegetation on tundra or pastures.

Eggs. 2 to 4 (usually 2); pale green or blue to dark olive-brown; marked with reddish or blackish brown. Incubated by ♂ and ♀.

Young. Seminidifugous. Downy. Cared for by ♂ and ♀.

Technical diagnosis. Ridgway, Pt. 8, 1919:673.

Classification. Peters, **2,** 1934:309–312.

References. Bent, No. 113:1–28; Witherby, **5,** 1941:122–142; Murphy, **2:** 1006–1040; Alexander, 140–145.

M. E. Pryor, The avifauna of Haswell Island, Antarctica. In *Antarctic Birds Studies,* Am. Geophys. Union No. 1686, 1968:79–81.

G. P. Dement'ev *et al., Birds of the Soviet Union,* Vol. 3, Washington, 1969:430–462.

Synonym: Catharactidae.

LARIDAE

Gull Family (82 species)

Figure 66. Great Black-backed Gull (*Larus marinus*). Total length 28 in. (712 mm).

Physical characteristics. Length—203 to 762 mm (8 to 30 in.). Plumage—typically gray and white, with black on wing tips (most gulls); some spp. all white; others white, with black on upper parts, or largely dark gray or black; many spp. have black cap (terns) or hood (gulls) in breeding plumage. Some spp. (terns) are crested. Bill and feet brightly colored in many spp. Bill slender to rather heavy; medium to rather long; sharply pointed (terns) to rather blunt and slightly hooked. Wings long and pointed; tail medium to long, square (rarely graduated) to deeply forked. Legs very short to medium; feet webbed; hallux small (vestigial in some). Sexes alike.

Range. Worldwide. Habitat—seacoasts, large rivers and lakes, or even ponds and marshes. Migratory exc. in low latitudes.

Habits. Gregarious. Aquatic. Flight strong (gulls soar). Many spp. dive for food from wing. Voice—harsh screams or squawks.

Food. Fish, crustacea, mollusca, insects, carrion; berries and other vegetable matter.

Breeding. Nest variable—from no material to bulky mass of vegetation; on ground, on cliffs, in trees, on floating vegetation, or even in burrows (*Larosterna*).

    Eggs. 1 to 4; almost white to deep olive-buff; heavily spotted and speckled. Incubated by ♂ and ♀.

    Young. Seminidicolous. Downy. Cared for by ♂ and ♀.

Technical diagnosis. Witherby, **5**, 1941:1.

Classification. Peters, **2**, 1934:312–349.

References. Bent, No. 113:29–310; Witherby, **5**, 1941:1–122; Murphy, **2**:1040–1169; Alexander, 70–136.

    N. G. Smith, Evolution of some arctic gulls (*Larus*): an experimental study of isolating mechanisms, *Ornithol. Monogr.* No. 4, 1966.

    S. Ali and S. D. Ripley, *Handbook of the Birds of India and Pakistan,* Vol. 3, Bombay, 1969:22–74.

    G. P. Dement'ev *et al., Birds of the Soviet Union,* Vol. 3, Washington, 1969:462–677.

    D. F. Parmelee and S. J. Maxson, The Antarctic terns of Anvers Island, *Living Bird,* **1974**:233–250.

Synonymy: The Sternidae are included.

RYNCHOPIDAE

Skimmer Family (3 species)

Figure 67. Black Skimmer (*Rynchops nigra*). Total length 20 in. (508 mm).

Physical characteristics. Length—368 to 508 mm (14.5 to 20 in.). Plumage—black, slate, or brownish above; white or whitish below. Bill red, orange, or yellow; feet red. Bill long, mandibles laterally compressed to thin blades, lower mandible much longer than upper. Wings very long and pointed; tail short and forked. Legs very short; feet small and slightly webbed; claws slender and sharp. Sexes alike, but ♀ smaller.

Range. Atlantic coast of N. America from Long Island southward; coasts and large rivers of C. America and most of S. America; tropical Africa; larger rivers of India, Burma, Indochina. Habitat—ocean shores, large rivers and lakes. Migratory in part.

Habits. Gregarious. In part crepuscular and nocturnal. Flight dextrous and swift but not strong. Feed by flying over the water, ploughing the surface with knifelike lower mandible. Voice—loud raucous cries, a "bark"; softer courtship notes.

Food. Small fish, shrimps, other small crustacea.

Breeding. Colonial. Nest a shallow, unlined depression in sand.

    Eggs. 3 to 5; bluish, buff, greenish, or pale salmon; blotched and streaked with brown. Incubated by ♀ (assisted by ♂?).

    Young. Nidicolous. Downy. Fed by ♂ and ♀.

Technical diagnosis. Ridgway, Pt. 8, 1919:449–457.

Classification. Peters, **2**, 1934:349–350.

References. Bent, No. 113:310–318; Murphy, **2**:1169–1178; Bannerman, **2**, 1931:xxiv–xxv, 280–284.

    W. Stone, *Bird Studies at Old Cape May,* Vol. 2, Philadelphia, 1937:598–608.

    R. L. Zusi, Structural adaptations of the head and neck in the Black Skimmer, *Rynchops nigra* Linnaeus, *Publ. Nuttall Ornithol. Club* No. 3, 1962.

    S. Ali and S. D. Ripley, *Handbook of the Birds of India and Pakistan,* Vol. 3, Bombay, 1969:74–76.

Synonym: Rhynchopidae.

ALCIDAE

Auk Family (22 species)

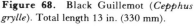

Figure 68. Black Guillemot (*Cepphus grylle*). Total length 13 in. (330 mm).

Physical characteristics. Length—165 to 762 mm (6.5 to 30 in.). Plumage—black, brownish black, or gray; patterned above with rusty or buff in 2 spp.; white below in most spp. (but chin and throat of breeding plumage black in many). A white wing patch, ornamental head plumes, and/or brightly colored bill, mouth, and feet in some spp. Bill short and stout to moderately long and slender, laterally compressed, elaborately sculptured in some spp. Head large; neck short; body heavy and compact. Wings small (with extremely short secondaries); tail short. Legs short and set far back; feet webbed; hallux absent or vestigial; claws strong. Sexes alike.

Range. Northern Pacific, n. Atlantic, and Arctic oceans and their coasts. Some spp. migratory.

Habits. Gregarious, esp. in breeding season. Pelagic. Flight direct and rapid but not usually sustained (Great Auk was flightless). Dive from surface for food. Use wings to swim under water. Voice—grunting, moaning, yelping, or sibilant notes.

Food. Fish, crustacea, mollusca, worms, algae.

Breeding. Colonial. Nest on rock ledges, in rock crevices, in burrows; no nest material (or but scanty lining).

> *Eggs.* 1 or 2; white to deep buff or light green or blue; immaculate or lightly to heavily spotted and scrawled with brown and/or black. Incubated by ♂ and ♀, or by ♀.
>
> *Young.* Seminidicolous. Downy. Cared for by ♂ and ♀.

Technical diagnosis. Witherby, **5**, 1941:141–142.

Classification. Peters, **2**, 1934:350–359.

References. Bent, No. 107:82–224; Witherby, **5**, 1941:142–173; Alexander, 160–177.

> R. W. Storer, A comparison of variation, behavior, and evolution in the sea bird genera *Uria* and *Cepphus, Univ. Calif. Publ. Zool.,* **52**, 1952:121–222.
>
> F. Richardson, Breeding biology of the Rhinoceros Auklet on Protection Island, Washington, *Condor,* **63**, 1961:456–473.
>
> S. G. Sealy, Breeding biology of the Horned Puffin on St. Lawrence Island, Bering Sea . . . , *Pacific Sci.,* **27**, 1973:99–119.

Synonym: Uriidae.

PTEROCLIDAE

Sandgrouse Family

(16 species)

Figure 69. Tibetan Sandgrouse (*Syrrhaptes tibetanus*). Total length 16 in. (406 mm).

Physical characteristics.   Length—228 to 406 mm (9 to 16 in.). Plumage—dense; sandy buff, reddish brown, or gray above, barred, spotted, or mottled in cryptic patterns; face and breast boldly marked with yellow, chestnut, vinaceous, or white and black; belly black or dark brown in some spp.; tail barred. Bill short and conical. Neck short. Wings long and pointed; tail medium and wedge-shaped to very long and acuminate, the tail coverts very long. Legs very short; tarsi (and toes in *Syrrhaptes*) feathered; toes medium to very short; claws short and thick. Sexes unlike (the ♀ smaller and more spotted and mottled).

Range.   Portugal, Spain, s. France (erupting irregularly into n.w. Europe and Great Britain), Africa, s. Asia. Habitat—open, sandy plains and deserts, thin brush, or (rarely) forest. Some spp. migratory.

Habits.   Gregarious. Terrestrial. Flight very swift. Fly considerable distances in flocks daily at regular hours to drink. Some spp. swim readily. Voice—whistles; clucking and croaking sounds.

Food.   Seeds, berries, buds; small insects.

Breeding.   Nest a scrape in ground, with little or no lining.

  Eggs.   2 to 3; buffish white, pale cinnamon, grayish, or greenish; with brown, red-brown, lavender, and/or gray markings. Incubated by ♂ and ♀.

  Young.   Nidifugous. Downy. Cared for by ♂ and ♀.

Technical diagnosis.   Witherby, **4**, 1940:147.

Classification.   Peters, **3**, 1937:3–10.

References.   Bannerman, **2**, 1931:xxvi–xxvii, 285–303, col. pls. 10–11.

  E. C. S. Baker, *The Game-birds of India, Burma and Ceylon,* Vol. 2, London, 1921:235–323.

  C. W. Mackworth-Praed and C. H. B. Grant, *Birds of the Southern Third of Africa,* London, 1962:371–377.

  G. L. Maclean, Field studies of the sandgrouse of the Kalahari Desert, *Living Bird,* **1968:**209–235.

  S. Ali and S. D. Ripley, *Handbook of the Birds of India and Pakistan,* Vol. 3, Bombay, 1969:76–92.

Synonym: Eremialectoridae.

COLUMBIDAE

Pigeon Family (295 species ±)

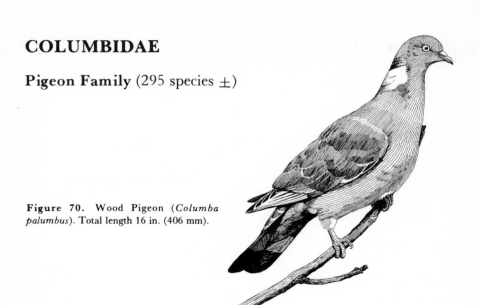

Figure 70. Wood Pigeon (*Columba palumbus*). Total length 16 in. (406 mm).

Physical characteristics. Length—152 to 838 mm (6 to 33 in.). Plumage—soft and very dense; wide range of colors; pale shades of gray or brown; rich shades of reddish brown, green, yellow, purple, etc.; and, in many spp., metallic reflections, esp. on neck and wings. Scaled or barred patterns frequent. A few spp. crested. Bill slender to stout and of medium length, with bare cere (rarely with a fleshy knob). Head small; neck rather short. Body compact. Wings short to long; tail short to long, truncate to pointed. Legs very short to fairly long. Sexes alike in most spp.

Range. Worldwide exc. n. N. America, n. Asia, s. S. America, and many oceanic islands. Many spp. migratory.

Habits. Solitary to highly gregarious. Arboreal or terrestrial. Flight strong in most spp. Drink by immersing bill and sucking. Young fed at first with "pigeon's milk," produced in crop of parents. Voice—variety of cooing or booming notes; also hissing, whistling, or guttural sounds. Occasionally make loud clapping with wings.

Food. Seeds, fruit, acorns, etc.; a few spp. also eat insects, worms, snails.

Breeding. Sometimes colonial. Nest usually a simple platform of sticks, placed on cliff (or building) ledges or in trees; or nest in tree cavities, in burrows, or on ground.

    Eggs. 1 or 2 (rarely 3); white or (rarely) buff. Incubated by ♂ and ♀.

    Young. Nidicolous. With little or no down at hatching. Cared for by ♂ and ♀.

Technical diagnosis. Ridgway, Pt. 7, 1916:275–279.

Classification. D. Goodwin, *Pigeons and Doves of the World,* British Museum of Natural History, 1967.

References. Bent, No. 162:353–458; Witherby, **4,** 1940:129–146.

    C. W. Mackworth-Praed and C. H. B. Grant, *Birds of the Southern Third of Africa,* London, 1962:377–399.

    G. P. Dement'ev *et al., Birds of the Soviet Union,* Vol. 2, Washington, 1968:1–82.

    P. Slater *et al., A Field Guide to Australian Birds,* Wynnewood, Pa., 1971:340–352.

    J. E. duPont, *Philippine Birds,* Greenville, Del., 1971:114–139.

Synonymy: The Didunculidae, Gouridae, Peristeridae, Turturidae, and Treronidae are included.

PSITTACIDAE

Parrot Family (315 species)

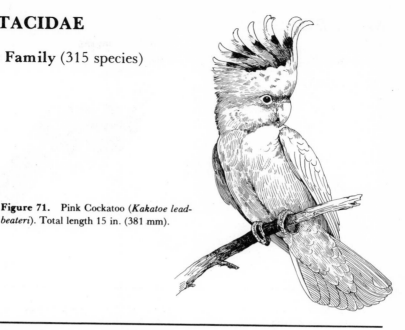

Figure 71. Pink Cockatoo (*Kakatoe lead-beateri*). Total length 15 in. (381 mm).

Physical characteristics. Length—83 to 991 mm (3.25 to 39 in.). Plumage—sparse; hard and glossy; usually brightly colored, with green commonly predominant. Bill short, stout, and strongly hooked. Head and neck short. Body compact. Wings strong; tail short to very long (racquet-shaped central feathers in *Prioniturus*; spine-tipped feathers in *Micropsitta*). Legs short; feet zygodactyl. Sexes usually alike.

Range. Southern Hemisphere (exc. remote oceanic islands and the s. tip of Africa); tropical and subtropical parts of N. Hemisphere (a few species originally ranged to about 40° N. lat. in N. America and 35° in s.e. Asia). Most species nonmigratory.

Habits. Many species gregarious. Typically arboreal. Flight usually strong. Use beak in climbing. Manipulate food with feet. Voice—noisy; often remarkable mimics (in captivity only).

Food. Fruit, nuts, grains; nectar, fungi, and other vegetable matter.

Breeding. Nest in unlined holes in trees, in termite nests, on rocks, or on banks. Several Australasian species nest on the ground, and *Myiopsitta* of S. America builds colonial nests of twigs in branches of trees.

Eggs. 1 to 12; white. Incubated by ♂ and ♀, or by ♀ alone.

Young. Nidicolous. Naked at first, then downy. Cared for by ♂ and ♀.

Technical diagnosis. Ridgway, Pt. 7, 1916:103–106.

Classification. Peters, **3,** 1937:141–273.

References. Bannerman, **2,** 1931:387–414.

W. R. Eastman, Jr., and A. C. Hunt, *The Parrots of Australia,* Sydney, 1966, 194 pp.

A. Wetmore, *The Birds of the Republic of Panamá,* 2, *Smithson. Misc. Collect.,* **150,** 1968:63–107.

J. M. Forshaw, *Parrots of the World,* Doubleday, 1973, 584 pp.

Synonyms: Stringopidae, Nestoridae, Cacatuidae, Loriidae, Cyclopsittacidae.

MUSOPHAGIDAE

Touraco Family
(19 species)

Figure 72. Red-tipped Crested Touraco (*Tauraco macrorhynchus verreauxi*). Total length 17 in. (432 mm).

Physical characteristics. Length—368 to 712 mm (14.5 to 28 in.). Plumage—glossy greens, blues, browns, with violet, crimson, and yellow (exc. *Crinifer,* which is ashy gray). Conspicuous patch of crimson (turacin pigment) on spread wing (exc. *Corythaeola* and *Crinifer*). Bill brightly colored in many spp. Typically crested. In many spp. skin around eye bare and (usually) red; face bare in 1 sp. Bill serrate; strong, stout, and broad; produced to form frontal shield in *Musophaga* and *Ruwenzorornis*. Nostrils placed near tip of bill in some spp. Wings short and rounded; tail long and broad. Feet semizygodactyl (4th toe reversible). Sexes alike.

Range. Africa (s. of Sahara), exc. Madagascar. Habitat—dense forest and forest edges; woods near water. Nonmigratory, exc. for local movements.

Habits. Solitary or in small bands. Arboreal. Flight direct, but labored, brief, and infrequent. Run along branches and jump across gaps. Climb with dexterity. Voice—loud raucous shrieks and croakings; explosive chatter; a soft coo.

Food. Fruits, seeds, buds; insects, snails.

Breeding. Nest a rough, bulky platform of sticks in a tree.

Eggs. 2 to 3; white or nearly so; immaculate. Incubated by ♂ and ♀.

Young. Nidicolous. Downy. Cared for by ♂ and ♀.

Technical diagnosis. G. E. Shelley, Cat. of Birds in Brit. Mus., **19,** 1891:435.

Classification. Peters, **4,** 1940:3–11.

References. Bannerman, **3,** 1933:xviii–xix, 52–79, col. pls. 1–4.

R. E. Moreau, A contribution to the biology of the Musophagiformes . . . , *Ibis,* **1938:**639–671.

F. J. Jackson, *The Birds of Kenya Colony and the Uganda Protectorate,* Vol. 1, London, 1938:514–531.

J. P. Chapin, The touracos: an African bird family, *Living Bird,* **1963:**57–68.

C. W. Mackworth-Praed and C. H. B. Grant, *Birds of the Southern Third of Africa,* London, 1962:425–436.

CUCULIDAE

Cuckoo Family (127 species)

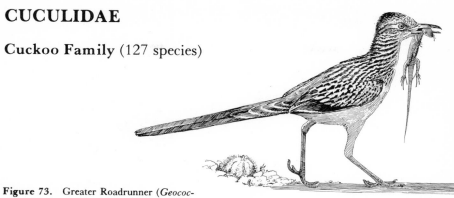

Figure 73. Greater Roadrunner (*Geococcyx californianus*). Total length 23 in. (584 mm).

Physical characteristics. Length—159 to 699 mm (6.25 to 27.5 in.). Plumage—loose-webbed (wiry in some spp.). Brown, olive, gray, or black (purple, blue, or bright iridescent green in a few spp.); uniform or in bold combinations; many spp. streaked or barred, esp. below; tail barred or tipped with white in many spp. A few spp. crested. Bill somewhat curved, rather stout to extremely heavy. Bare, colored orbital skin and/or conspicuous eyelashes in many spp. Wings medium to long; tail medium to extremely long, graduated (forked in *Surniculus*). Legs short (exc. in some terrestrial spp.); feet zygodactyl. Sexes alike in most spp.

Range. Worldwide exc. some oceanic islands and high latitudes of Asia and the Americas. Many spp. migratory.

Habits. Solitary (exc. *Crotophaga* and *Guira*). Most spp. arboreal, a few terrestrial. Flight of some spp. strong; other spp. virtually flightless. Voice—typically loud, unmusical calls, monotonously repeated.

Food. Insects, snails, small vertebrates; fruit.

Breeding. Many spp. parasitic; *Crotophaga* and *Guira* build communal nests; remaining spp. make nests of sticks or grass—shallow saucers to completely domed structures with entrance at side (*Centropus*).

Eggs. 2 to 6; white, blue, green, red, or brown; spotted or immaculate. In nonparasitic spp. incubated by ♂ and ♀.

Young. Nidicolous. Naked or "hairy" at hatching. In nonparasitic spp. cared for by ♂ and ♀.

Technical diagnosis. Witherby, **2**, 1938:296.

Classification. Peters, **4**, 1940:12–76.

References. Bent, No. 176:19–105; Witherby, **2**, 1938:296–308.

H. Friedmann, The parasitic cuckoos of Africa, *Wash. Acad. Sci. Monogr.* No. 1, 1948.

A. Wetmore, *The Birds of the Republic of Panamá*, 2, *Smithson. Misc. Collect.*, **150**, 1968:108–143.

S. Ali and S. D. Ripley, *Handbook of the Birds of India and Pakistan*, Vol. 3, Bombay, 1969:191–247.

H. J. Frith, ed., *Birds in the Australian High Country*, Sydney, 1969:238–252.

TYTONIDAE

Barn Owl Family (11 species)

Figure 74. Barn Owl (*Tyto alba*). Total length 15 in. (381 mm).

Physical characteristics. Length—305 to 534 mm (12 to 21 in.). Plumage—soft; variable in color, with light and dark phases frequent. Above: brown, spotted with white, or grayish brown or orange-buff, vermiculated with gray or brown; below: buff, grayish brown, or white, spotted, barred, or vermiculated with white, gray, or brown; wing and tail barred; face (long and heart-shaped) white or buff. Bill hooked and rather long, but mostly concealed by bristly radiating feathers of the facial discs; a cere at the base. Eyes comparatively small, directed forward. Wings long and rounded; tail short and slightly emarginate. Legs long, the flank feathers elongate; the tarsus completely feathered. Toes strong, sparsely covered with bristles, the outer toe reversible; claws long, sharp, strongly hooked, the middle claw pectinate. Sexes alike or nearly so; ♀ sometimes larger.

Range. Worldwide exc. extreme n. and New Zealand, Hawaiian Ids. (introduced), some islands of Malaysia. Largely nonmigratory.

Habits. Solitary. Nocturnal. Flight noiseless, buoyant but not swift. Prey swallowed whole; the bones, fur, etc., regurgitated later in pellets. Voice—varied; a shrill scream, a snoring sound, chirruping notes, hisses; also a clicking sound made with the beak.

Food. Mammals, birds, insects, crustacea, frogs, fish.

Breeding. Nest in burrows, hollow trees, buildings, etc., or on ground; little or no nest material.
    Eggs. 3 to 11 (usually 4 to 7); white. Incubated by ♀.
    Young. Nidicolous. Downy. Cared for by ♂ and ♀.

Technical diagnosis. Ridgway, Pt. 6, 1914:598–599 (excludes *Phodilus*—see p. 618, footnote).

Classification. Peters, **4**, 1940:77–86 (10 spp.).

References. Bent, No. 170:140–153; Witherby, **2**, 1938:342–347.
    G. P. Dement'ev *et al., Birds of the Soviet Union*, Vol. 1, Washington, 1966:473–477.
    S. Ali and S. D. Ripley, *Handbook of the Birds of India and Pakistan*, Vol. 3, Bombay, 1969:249–255.
    D. G. Smith *et al.,* History and ecology of a colony of Barn Owls in Utah, *Condor*, **76**, 1974:131–136.

Synonyms: Aluconidae, Strigidae (the Typical Owls, now called Strigidae, were formerly called Bubonidae).

STRIGIDAE

Typical Owl Family

(123 species)

Figure 75. Spectacled Owl (*Pulsatrix perspicillata*). Total length 20 in. (508 mm).

Physical characteristics. Length—133 to 686 mm (5.25 to 27 in.). Plumage—long and soft; brownish black, gray, brown, or chestnut (1 sp. white); barred, streaked, or vermiculated with white, buff, or brownish black (a few spp. nearly uniform above). Tail conspicuously barred in many spp. Some spp. have brown ("red") and gray phases. Earlike tufts ("horns") on many spp. Tarsus and (in many spp.) toes completely to slightly feathered. Bill short, strong, and hooked, with cere at base. Eyes very large and directed forward, surrounded by feathered discs. Head large; neck short. Wings broad and rounded, short to rather long; tail short to fairly long. Flank feathers greatly elongated. Legs short to medium; outer toe reversible; claws sharp, strongly hooked. Sexes in most spp. alike in color but ♀ larger.

Range. Worldwide exc. some oceanic islands. Most spp. nonmigratory.

Habits. Solitary. Largely nocturnal or crepuscular. Most spp. arboreal. Flight noiseless, buoyant but not swift. Food swallowed whole; bones, fur, etc., ejected later from the mouth in pellets. Voice—a variety of hooting, quavering, trilling, mewing, or barking sounds; also a loud snapping sound made with the bill.

Food. Mammals, birds, reptiles, amphibians, fish, insects, crabs.

Breeding. Nest in tree cavities; in old nests of eagles, crows, etc.; on the ground; in burrows; and in buildings. Little or no nest material.

Eggs. 1 to 7, rarely up to 10; white. Incubated by ♀, or by ♂ and ♀.

Young. Nidicolous. Downy. Cared for by ♂ and ♀.

Technical diagnosis. Ridgway, Pt. 6, 1914:594–598, 617–618.

Classification. Peters, **4,** 1940:86–174.

References. Bent, No. 170:153–444; Witherby, **2,** 1938:309–342.

C. W. Mackworth-Praed and C. H. B. Grant, *Birds of the Southern Third of Africa,* London, 1962:504–516.

G. P. Dement'ev *et al., Birds of the Soviet Union,* Vol. 1, Washington, 1966:386–473.

A. Wetmore, *The Birds of the Republic of Panamá,* 2, *Smithson. Misc. Collect.,* **150,** 1968:146–185.

S. Ali and S. D. Ripley, *Handbook of the Birds of India and Pakistan,* Vol. 3, 1969:255–317.

Synonyms: Asionidae, Bubonidae.

STEATORNITHIDAE

Oilbird Family (1 species)

Figure 76. Oilbird (*Steatornis caripensis*). Total length 18 in. (457 mm).

Physical characteristics. Length—432 to 482 mm (17 to 19 in.). Plumage—chestnut, bright above, paler below. Marked (exc. on back) with small, angular spots of white edged with black. Narrow dark bars on upper parts, exc. head. Bill short, rather wide-gaped, hooked. Very long rictal bristles. Wings pointed; tail rounded; both long. Legs extremely short; bare; not scaled. Toes long; claws sharp and curved. Sexes alike.

Range. Locally from c. Peru, through Ecuador, Colombia, and Venezuela, to French Guiana; also island of Trinidad. Nonmigratory.

Habits. Gregarious. Nocturnal. Flight strong, noiseless. Hover to feed, picking fruits from trees. Voice—loud squawks, croaks, and shrieks.

Food. Fruits, especially of palms.

Breeding. Colonial nesters (on ledges in caves). Nest a truncated cone of seeds and droppings, with shallow egg cavity.

> *Eggs.* 2 to 4; white. Incubated by ♂ and ♀.
>
> *Young.* Nidicolous. Naked except for sparse down (primarily ventrally) at hatching. Become exceedingly fat (reaching twice the weight of adults) and are used as a source of oil by primitive peoples—hence name of family. Nestling period 90–125 days. Cared for by ♂ and ♀.

Technical diagnosis. Ridgway, Pt. 6, 1914:488–489.

Classification. Peters, **4,** 1940:174.

References. M. A. Carriker, Jr., The cave birds of Trinidad, *Auk,* **48,** 1931:186–194.

> D. R. Griffin, Acoustic orientation in the Oil Bird, *Steatornis, Proc. Natl. Acad. Sci.,* **39,** 1953:884–893 (or—less detailed—*Sci. Am.,* **190,** 1954:78–93).
>
> D. W. Snow, The natural history of the Oilbird, *Steatornis caripensis,* in Trinidad . . . , *Zoologica,* **46,** 1961:27–48.
>
> A. Wetmore, *The Birds of the Republic of Panamá,* 2, *Smithson. Misc. Collect.,* **150,** 1968:186–188.

PODARGIDAE

Frogmouth Family (12 species)

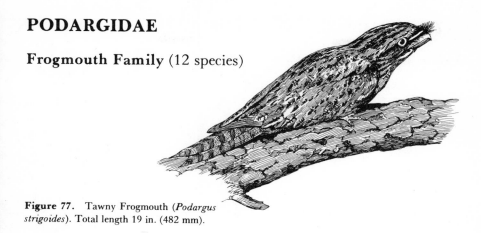

Figure 77. Tawny Frogmouth (*Podargus strigoides*). Total length 19 in. (482 mm).

Physical characteristics.   Length—216 to 534 mm (8.5 to 21 in.). Plumage—soft, silky; brown, tawny, and gray; mottled, vermiculated, barred, or streaked with black and chocolate-brown in cryptic patterns; ear tufts in some spp. Bill broad, flat, and triangular, strongly hooked and extremely wide-gaped; long, bristly feathers at base. Neck short and thick. Wings rounded, of moderate length; tail medium to long. Legs very short; feet small and weak; middle toe elongated. Sexes similar; ♀ more reddish or has red phase.

Range.   Ceylon, s. and n.e. India, Burma, Thailand, Malaysia, Papuan Region, Australia (incl. Tasmania), Solomons, Philippines. Habitat—forest. Migratory in part.

Habits.   Usually in pairs. Nocturnal. Somewhat lethargic. Flight direct, not strong. Characteristically perch lengthwise on branches; when alarmed, assume rigid posture resembling broken branch. Voice—low, hoarse, booming note, loud hiss; *oom-oom* repeated many times.

Food.   Beetles, moths, other insects (usually picked from ground or branches); also mice.

Breeding.   Shallow nest of loosely interwoven sticks or flat pad of birds' own down overlaid with bark, lichen, and moss; placed at fork on horizontal branch.

   Eggs.   1 or 2 (sometimes 3?); white. Incubated by ♂ and ♀.

   Young.   Nidicolous. Downy. Cared for by both ♂ and ♀.

Technical diagnosis.   Baker, *Fauna*, 2nd ed., **4**, 1927:377.

Classification.   Peters, **4**, 1940:175–179.

References.   Baker, *Nidification*, **3**, 1934:492–495.

   B. E. Smythies, *The Birds of Borneo*, London, 1960:273–276.

   H. J. Frith, ed., *Birds in the Australian High Country*, Sydney, 1969:252–254.

   S. Ali and S. D. Ripley, *Handbook of the Birds of India and Pakistan*, Vol. 4, Bombay, 1970:1–4.

   P. Slater *et al., A Field Guide to Australian Birds*, Wynnewood, Pa., 1971:395–397.

NYCTIBIIDAE

Potoo Family (5 species)

Figure 78. Common Potoo (*Nyctibius griseus*). Total length 16 in. (406 mm).

Physical characteristics. Length—406 to 495 mm (16 to 19.5 in.). Plumage—soft; cryptic pattern of gray, buff, blackish brown, and white. Juvenal plumage white, with narrow, dark shaft-streaks. Bill small, narrow, and terminally decurved; projection, or "tooth," on maxillary tomium; mouth very large. No rictal bristles, but loral feathers have bristlelike tips, much elongated and decurved. Wings and tail long. Legs extremely short; toes strong and much flattened below; claws curved (middle claw not pectinate). Sexes alike or nearly so.

Range. Southern Sinaloa and Tamaulipas to s. Brazil and Paraguay; also Jamaica and Hispaniola. Habitat—forested or semiforested areas. Nonmigratory.

Habits. Solitary. Nocturnal. Perch in very upright position. Hawk insects. Voice—"clear, plaintive, far-carrying 'chant' of seven or eight whistled notes," starting high and running down scale (Butler); guttural *ch-r-r* (Scott); mews like cat (Goeldi); quacking notes.

Food. Insects.

Breeding. Lay egg on top of broken tree stub and incubate in stiffly erect posture.

   Eggs. 1; oval; white; lightly and irregularly spotted with violet and brown. Incubated by ♂ and ♀.

   Young. Nidicolous. Downy. Cared for by ♂ and ♀.

Technical diagnosis. Ridgway, Pt. 6, 1914:583–584.

Classification. Peters, **4**, 1940:179–181.

References. Fr. Haverschmidt, Observations on *Nyctibius grandis* in Surinam, *Auk,* **65,** 1948:30–32.

   H. Sick, The voice of the Grand Potoo, *Wilson Bull.,* **65,** 1953:203.

   A. Wetmore, *The Birds of the Republic of Panama,* 2, *Smithson. Misc. Collect.,* **150,** 1968:188–195.

   J. I. Borrero H., A photographic study of the Potoo in Columbia, *Living Bird,* **1970:**257–263.

   A. F. Skutch, Life history of the Common Potoo, *Living Bird,* **1970:**265–280.

AEGOTHELIDAE

Owlet-frogmouth Family (8 species)

Figure 79. Owlet-frogmouth (*Aegotheles insignis*). Total length 12 in. (305 mm).

Physical characteristics. Length—190 to 323 mm (7.5 to 12.75 in.). Plumage—soft; rufous cinnamon, gray, brown (to black), and buff, mottled, spotted, vermiculated, and barred with brown, gray, black, and white. Elongated, erect loral bristles with hairlike barbs. Flank feathers lengthened. Bill small, flat, with large gape; hooked (but not toothed). Wings long and rounded; tail long and wedge-shaped. Legs and feet small and weak; toes long and slender; claws long (middle claw not pectinate). Sexes alike or nearly so.

Range. Australia, Tasmania, New Guinea, New Caledonia; Moluccan, Fergusson, Goodenough, and Aru Ids. Habitat—dense brush and open, wooded country. Nonmigratory.

Habits. Solitary. Nocturnal. Arboreal. Flight noiseless; direct and rapid, but not prolonged. Hawk insects; also feed on ground insects. Spend day in hollow tree or log. Voice—loud hissing, whistling call, churring notes, shrill squeak.

Food. Insects.

Breeding. Nest in hollow tree (sometimes in holes in banks), lined with leaves or unlined.

Eggs. 3 to 5; white; immaculate or striated. Incubation not described.

Young. Nidicolous. Downy. Care of young not described.

Technical diagnosis. Mathews, **7**, 1918:51.

Classification. Peters, **4**, 1940:181–184 (7 spp.).

References. H. Burrell, Owlet-Nightjar nestlings, *Emu,* **13**, 1914:216–217.

D. L. Serventy and H. M. Whittell, *Birds of Western Australia,* Perth, 1967:287.

H. J. Frith, ed., *Birds in the Australian High Country,* Sydney, 1969:255.

CAPRIMULGIDAE

Nightjar Family (67 species)

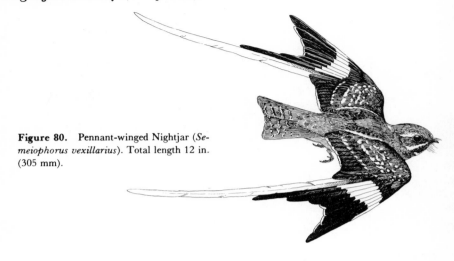

Figure 80. Pennant-winged Nightjar (*Semeiophorus vexillarius*). Total length 12 in. (305 mm).

Physical characteristics. Length—190 to 305 mm (7.5 to 12.0 in.), excl. the extremely elongate tail feathers of 2 spp., which may add as much as 686 mm (27 in.). Plumage—soft; rufous, buff, gray, black, and white, mottled and finely vermiculated in cryptic patterns. Many spp. with barring below and/or with white throat; also areas of white on wings and/or tail (visible only when spread). Some spp. have extremely elongated rectrices or inner primaries; some have tufts of lengthened feathers on head. Bill small and weak, but with very wide gape. Long rictal bristles in many spp. Large eyes. Head large; neck appearing short. Wings and tail medium to long. Legs very short; tarsus feathered in some spp.; toes and claws small, but 3rd (middle) toe long, with pectinate claw. Sexes unlike in most spp.

Range. Worldwide, exc. n. N. America, n. Asia, s. S. America, New Zealand, and most oceanic islands. Some spp. migratory.

Habits. Solitary (some spp. migrate in loose flocks). Largely nocturnal or crepuscular. Flight weak to strong. Hawk insects. Perch lengthwise on branches. Voice—purring, rasping, and whistling notes; loud, 2- to 4-syllable calls.

Food. Insects (rarely small birds).

Breeding. Eggs laid directly on ground (no nest material).

    Eggs. 1 to 2 (very rarely 3); white to pinkish buff; with dark specks and blotches (immaculate in a few spp.). Incubated by ♂ and ♀.

    Young. Seminidicolous. Downy. Cared for by ♂ and ♀.

Technical diagnosis. Witherby, **2**, 1938:251.

Classification. Peters, **4**, 1940:184–220 (70–72 spp.).

References. Bent, No. 176:147–254; Witherby, **2**, 1938:251–262; Bannerman, **3**, 1933:147–178.

    A. Wetmore, *The Birds of the Republic of Panamá*, 2, *Smithson. Misc. Collect.*, **150**, 1968:195–222.

    S. Ali and S. D. Ripley, *Handbook of the Birds of India and Pakistan*, Vol. 4, Bombay, 1970:4–25.

APODIDAE

Swift Family (67 species ±)

Figure 81. White-throated Swift (*Aeronautes saxatalis*). Total length 6.5 in. (165 mm).

Physical characteristics. Length—89 to 228 mm (3.5 to 9 in.). Plumage—bluish or brownish black to grayish brown (1 sp. with chestnut breast and collar). Some spp. have throat, collar, flanks, or rump white (or very pale brown). Bill very small, slightly decurved; gape large. Neck short; body compact. Wings very long and pointed; tail short and truncate (with the feathers in some spp. spine-tipped) to very long and forked. Legs extremely short; many spp. have tarsi (or even toes) feathered. Feet very small but strong; pamprodactyl; hallux reversible in Apodinae; claws strong and curved. Sexes alike.

Range. Worldwide, exc. n. N. America, n. Asia, s. S. America, and some oceanic islands. Many spp. migratory.

Habits. Usually gregarious. Flight very strong and swift—the most aerial of birds; rest only by clinging to cliffs, inner walls of caves, chimneys, hollow trees, etc. Capture all food on the wing. Voice—rasping, twittering notes.

Food. Insects.

Breeding. Sometimes colonial. Nest of a salivary secretion or (in most spp.) of plant fragments glued together with the secretion, ranging from slight bracket to long elaborate tube and placed in cave, cleft, or hollow tree, under overhanging rock or tree limb, or on under surface of palm leaf; or a few grasses and feathers in swallow nest or in old burrow in earth bank.

Eggs. 1 to 6; white. Incubated by ♂ and ♀.

Young. Nidicolous. Naked at hatching (downy later in *Cypseloides* and *Cypsiurus*). Cared for by ♂ and ♀.

Technical diagnosis. Ridgway, Pt. 5, 1911:683–684.

Classification. Peters, **4**, 1940:220–256 (73–81 spp.).

References. Bent, No. 176:254–319; Witherby, **2**, 1938:242–251.

Lord Medway, The antiquity of trade in edible birds' nests, *Fed. Mus. J.* (Kuala Lumpur), **8**, 1963:36–47.

C. T. Collins, Notes on the biology of Chapman's Swift *Chaetura chapmani* (Aves, Apodidae), *Am. Mus. Novit.*, No. 2320, 1968.

A. Wetmore, *The Birds of the Republic of Panamá*, 2, Smithson. Misc. Collect., **150**, 1968:223–247.

S. Ali and S. D. Ripley, *Handbook of the Birds of India and Pakistan,* Vol. 4, Bombay, 1970:25–58.

Synonyms: Micropodidae, Cypselidae, Triopidae.

HEMIPROCNIDAE

Crested-swift Family

(3 species)

Figure 82. Indian Crested-swift (*Hemiprocne coronata*). Total length 8.5 in. (216 mm).

Physical characteristics. Length—165 to 330 mm (6.5 to 13 in.). Plumage—soft; gray or bronze-brown above, with head, wings, and sometimes the back glossed with green or blue; under parts uniform gray or bronze-brown, or with white on abdomen and under tail coverts, white on chin, blue or green gloss on breast. In 2 spp. conspicuous white lines extend from base of bill over the eye and under the cheek, ending in tufts at nape and shoulder. All spp. have a crest and a patch of silky feathers on flanks; all (♂ only) have some chestnut on head. Bill small, flat, and broadly triangular, the gape large. Eye very large. Neck short. Wings extremely long and pointed; tail long and deeply forked, the outer feathers much attenuated. Legs very short; feet weak; toes long and slender (the hallux not reversible). Sexes unlike.

Range. India and Ceylon, Burma, Thailand, Indochina, Philippines; Malaysia, Papuan Region, e. to Solomons. Habitat—open, wooded hillsides and forest clearings. Nonmigratory.

Habits. Somewhat gregarious. Aerial, but perch regularly in trees. Flight strong, rapid, wheeling. Hawk insects. Voice—a loud, screaming cry.

Food. Insects.

Breeding. Nest an extremely small and shallow cup of bark fragments and small feathers plastered with saliva, attached to branch of tree.

Eggs. 1; pale gray; tinged with blue. Incubated by ♂ and ♀.

Young. Nidicolous. Downy. Cared for by ♂ and ♀.

Technical diagnosis. Ridgway, Pt. 5, 1911:683 ("Dendrochelidonidae").

Classification. Peters, **4,** 1940:257–259.

References. E. H. N. Lowther, *A Bird Photographer in India,* London, 1949:1–10, 5 pls.

S. Ali and S. D. Ripley, *Handbook of the Birds of India and Pakistan,* Vol. 4, Bombay, 1970:58–60.

G. M. Henry, *A Guide to the Birds of Ceylon,* New York, 1971:156–157.

Synonyms: Macropterygidae, Dendrochelidonidae.

TROCHILIDAE

Hummingbird Family (319 species)

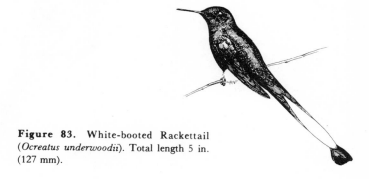

Figure 83. White-booted Rackettail (*Ocreatus underwoodii*). Total length 5 in. (127 mm).

Physical characteristics. Length—63 to 216 mm (2.25 to 8.5 in.). Plumage—mainly green, brown, or black. Some spp. are gray or white below. Areas of brilliant iridescent green, red, blue, purple, or gold (on throat, crown, sides of head, back) in most spp. Some spp. crested. Bill slender and pointed, rather short to extremely long (7–100 mm), straight to strongly decurved (recurved in a few spp.). Wings long and narrow; tail extremely varied—acuminate to deeply forked, the rectrices highly modified (racquet-tipped, etc.) in some spp. Legs very short (covered in some spp. with "muffs" of long downy feathers); feet very small and weak. Sexes unlike in most spp.

Range. South, C., and N. America (exc. extreme n.). Many spp. migratory.

Habits. Solitary. Pugnacious. Arboreal. Flight very swift and agile, with extremely rapid wing beat. Perch but do not walk or hop. Frequently hover, esp. when feeding. ♂ has elaborate display flight. Voice—squeaking or twittering notes; also various sounds made by vibration of flight feathers during courtship display of ♂.

Food. Insects, arachnids; nectar.

Breeding. Some spp. polygamous? Deep, cup-shaped nest of plant down and spider web, saddled on branch, suspended in fork, or (in palms, etc.) fastened to under side of leaf tip.

Eggs. 2 (in some spp. only 1?); white; immaculate. Incubated by ♀ (assisted by ♂ in *Colibri coruscans*).

Young. Nidicolous. With trace of down. Cared for by ♀ (assisted by ♂ in *Colibri coruscans*).

Technical diagnosis. Ridgway, Pt. 5, 1911:300–301.

Classification. Peters, **5,** 1945:1–143 (327 spp).

References. Bent, No. 176:319–472.

C. H. Greenewalt, *Hummingbirds,* New York, 1960.

A. F. Skutch, Life histories of Central American highland birds, *Publ. Nuttall Ornithol. Club,* No. 7, 1967:19–50.

K. A. and V. Grant, *Hummingbirds and Their Flowers,* New York, 1968, 115 pp.

D. W. Snow, The singing assemblies of Little Hermits, *Living Bird,* **1968:**47–55.

A. Wetmore, *The Birds of the Republic of Panama,* 2, *Smithson. Misc. Collect.,* **150,** 1968:247–378.

COLIIDAE

Coly Family (6 species)

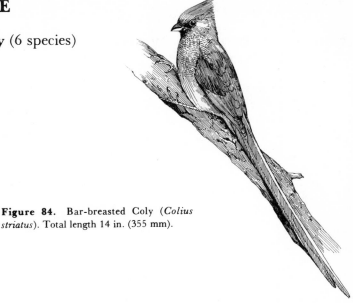

Figure 84. Bar-breasted Coly (*Colius striatus*). Total length 14 in. (355 mm).

Physical characteristics. Length—292 to 355 mm (11.5 to 14 in.), the tail more than twice the length of the body. Plumage—soft, the feathers of head and neck lax; somber brown (barred with dusky) or gray, marked with a blue patch on the nape, with black on the throat, or chestnut on the rump; under parts lighter. All spp. crested. Bill red or black and white in some spp.; feet red. Bare skin around eye red, blue, or gray. Bill short, stout, and curved, fleshy about the nostrils. Neck short; body slender. Wings short and rounded; tail extremely long and slender, sharply graduated, with stiff shafts. Legs short; feet strong; toes very long, the 1st toe reversible; claws long, sharp, and strong. Sexes alike.

Range. Africa s. of the Sahara, exc. Madagascar. Habitat—forest edge and brushland. Nonmigratory.

Habits. Gregarious (sleep hanging together in clusters). Arboreal. Flight rapid and direct, but not sustained. Creep about trees, using bill and feet, the tarsus resting on the branches. Acrobatic; often hang head downward (even when sleeping). Preen each other. Voice—harsh calls, mewing cries, whistling notes.

Food. Fruit, new shoots, leaves, seeds; rarely insects. Some spp. drink by sucking.

Breeding. Nest an open, shallow cup of sticks, bark, and roots, lined with wool, cotton, or leaves; placed in bush or tree.

> *Eggs.* 2 to 7 (usually 2 to 4?); white or creamy white; immaculate or streaked with brown. Incubated by ♂ and ♀.
>
> *Young.* Nidicolous. With sparse down. Cared for (by ♂ and ♀)?

Technical diagnosis. P. L. Sclater, Fam. Coliidae, Wytsman's *Genera Avium,* Pt. 6, 1906:1–2.

Classification. Peters, **5,** 1945:143–146.

References. Bannerman, **3,** 1933:xx–xxi, 139–146.

> V. G. L. Van Someren, Days with birds, *Fieldiana: Zool.,* **38,** 1956:203–208.
>
> C. W. Mackworth-Praed and C. H. B. Grant, *Birds of the Southern Third of Africa,* London, 1962:528–534.
>
> T. J. Cade and L. I. Greenwald, Drinking behavior of mousebirds in the Nambi Desert, southern Africa, *Auk,* **83,** 1966:126–128.

TROGONIDAE

Trogon Family (34 species)

Figure 85. Mexican Trogon (*Trogon mexicanus*). Total length 12 in. (305 mm).

Physical characteristics. Length—228 to 337 mm (9 to 13.25 in.). Plumage—soft and dense. Back, tail coverts, and central rectrices brilliant metallic green (chestnut in a few spp.; coverts red in 1 sp.); head, neck, and chest varied: green, black, gray, blue, violet, chestnut, rose, red; wings and outer rectrices black with white markings; abdomen and lower tail coverts red, yellow, or orange. Upper tail coverts extremely elongate in the Quetzal. Crest or tufts behind eyes in some spp. Brightly colored bare orbital ring. Eyes large. Bill short, broad, usually brightly colored; serrate in some spp. Nostrils and base of bill covered with bristles. Neck short. Wings short, rounded; tail long, broad, truncate or graduated. Legs and feet very small and weak; tarsus feathered; 1st and 2nd toes turned backward. Sexes unlike in most spp.

Range. Tropics. Southern half of Africa, India, and s.e. Asia; Malaysia and the Philippines; Arizona, extreme s. Texas, Mexico, C. America, West Indies, and n. three-quarters of S. America. Habitat—forest. Nonmigratory (exc. for local movements).

Habits. Usually solitary. Strictly arboreal. Flight undulating, rapid but not prolonged. Rarely walk or hop. Feed by darting from perch and snatching insects from air or from foliage, or small fruits from stem. Voice—a variety of simple call notes.

Food. Insects; small fruits; snails, small lizards, frogs.

Breeding. Nest a hollow in decayed stump or tree or in termite or wasp nest; no lining.

> *Eggs.* 2 to 4; white or buff to greenish blue; immaculate. Incubated by ♂ and ♀.
>
> *Young.* Nidicolous. Naked at hatching. Cared for by ♂ and ♀.

Technical diagnosis. Ridgway, Pt. 5, 1911:729–731.

Classification. Peters, 5, 1945:148–164.

References. Bent, No. 176:106–110; Bannerman, 3, 1933:355–360.

> A. F. Skutch, Life history of the Quetzal, *Condor, 46,* 1944:213–235. (Reprinted in *Smithson. Rept. for 1946:*265–293, pls. 1–4.)
>
> A. Wetmore, *The Birds of the Republic of Panamá,* 2, *Smithson. Misc. Collect.,* **150,** 1968:379–419.
>
> S. Ali and S. D. Ripley, *Handbook of the Birds of India and Pakistan,* Vol. 4, Bombay, 1970:60–67.

ALCEDINIDAE

Kingfisher Family (86 species)

Figure 86. Laughing Kookaburra (*Dacelo novaeguinae*). Total length 16 in. (406 mm).

Physical characteristics. Length—102 to 457 mm (4 to 18 in.). Plumage—green, blue, purple, reddish brown, white, usually in contrasting solid areas. Some spp. crested, and some barred (esp. on tail) and/or spotted; many with broad collar and/or pectoral band. Bill and feet of many spp. bright red or yellow. Bill massive, straight (but gonys typically upcurved), and pointed (hooked in 1 sp.; shovel-shaped in *Clytoceyx*). Head large; neck short; body compact. Wings generally short and rounded; tail very short to very long, with central feathers much elongated (and even racquet-tipped). Legs very short; feet (4 or 3 toes) strongly syndactyl. Sexes alike or unlike.

Range. Worldwide (exc. some oceanic islands and extreme n. parts of N. America and Eurasia). Spp. of higher latitudes migratory.

Habits. Solitary. Flight strong and direct but usually not sustained. Watch from exposed perch or hover over water; feed by diving for fish, by catching insects in the air, or by pouncing on large insects or small vertebrates on the ground. Voice—sharp calls or rattling cries.

Food. Fish, crustacea, insects, amphibians, reptiles, or even small birds or mammals.

Breeding. Nest in burrows in banks, in termite nests, or in tree cavities; no nest lining.

Eggs. 2 to 7; white. Incubated by ♂ and ♀.

Young. Nidicolous. Naked or (*Dacelo*) with down on upper parts. Cared for by ♂ and ♀.

Technical diagnosis. Ridgway, Pt. 6, 1914:404–406.

Classification. Peters, **5**, 1945:165–219.

References. Bent, No. 176:11–146; Witherby, **2**, 1938:272–276; Bannerman, **3**, 1933:239–281.

G. P. Dement'ev *et al.*, *Birds of the Soviet Union*, Washington, 1966:567–585.

A. Wetmore, *The Birds of the Republic of Panamá*, 2, *Smithson. Misc. Collect.*, **150**, 1968:420–437.

H. J. Frith, ed., *Birds in the Australian High Country*, Sydney, 1969:260–271.

S. Ali and S. D. Ripley, *Handbook of the Birds of India and Pakistan*, Vol. 4, Bombay, 1970:68–98.

Synonym: Halcyonidae. The Dacelonidae are included.

TODIDAE

Tody Family (5 species)

Figure 87. Cuban Tody (*Todus multi-color*). Total length 3.75 in. (95 mm).

Physical characteristics. Length—89 to 114 mm (3.5 to 4.5 in.). Plumage—uniform bright green above, with lores and forehead yellow in 1 sp.; subauricular region blue or gray; malar stripe and chin white; throat geranium-red; under parts mostly white or washed with green, yellow, pink, gray, flanks pink or (1 sp.) yellow; bill brown and red; feet orange-red. Bill long, straight, flattened, obtusely pointed. Head large; neck short; body compact. Wings short and rounded; tail medium in length, slightly rounded. Legs slender; feet weak, syndactyl; toes long. Sexes alike.

Range. Greater Antilles. Habitat—thickets on hillsides and stream banks. Nonmigratory.

Habits. Usually in pairs. Strictly arboreal. Flight weak. Sedentary exc. for brief, swift darts from perch to seize insects. Very tame. Voice—harsh chattering; chipping notes; a harsh *chreck*; *terp terp terp* (*Todus subulatus*). Wings make whirring rattle in flight.

Food. Insects; rarely minute lizards.

Breeding. Nest in unlined burrow which they excavate in earth bank or even in very slight vertical elevation, as in the side of a rut.

    Eggs. 2 to 5; white. Incubated by ♂ and ♀.

    Young. Nidicolous. Naked (?). Care not described.

Technical diagnosis. Ridgway, Pt. 6, 1914:441.

Classification. Peters, **5,** 1945:220.

References. A. Wetmore and B. H. Swales, The birds of Haiti and the Dominican Republic, *U.S. Natl. Mus. Bull.,* **155,** 1931:283–290.

    J. Bond, Nesting of the Narrow-billed Tody, *Wilson Bull.,* **61,** 1949:188.

    F. J. Rolle, Notes on the nesting burrow and the young of the Puerto Rican Tody (*Todus mexicanus*), *Auk,* **80,** 1963:551.

    J. Bond, *Birds of the West Indies,* London, 1971:141–142.

MOMOTIDAE

Motmot Family (8 species)

Figure 88. Blue-crowned Motmot (*Momotus momota*). Total length 17 in. (432 mm).

Physical characteristics. Length—171 to 502 mm (6.75 to 19.75 in.). Plumage—loose-webbed; green, blue, and brown. Crown, face, or throat boldly marked with black, blue, or brown. Usually a spot in center of breast. Bill large, broad, decurved; usually serrate. Wings short and rounded. Tail long (exc. *Hylomanes*); graduated; racquet-tipped (exc. *Hylomanes, Aspatha,* and 2 subspp. of *Baryphthengus*)—racquet shape develops after growth is complete but is not caused by deliberate action of bird. Legs very short; feet syndactyl. Sexes alike or nearly so.

Range. Neotropical. Forests of Mexico (s. Sonora and Tamaulipas) to Paraguay and n.e. Argentina; also the islands of Cozumel, Trinidad, and Tobago. Most spp. nonmigratory.

Habits. Usually solitary. Perch long in one spot. Twitch tail from side to side in irregular, mechanical manner. Flight undulating. Voice—low-pitched hooting or cooing notes, singly or in series.

Food. Insects, spiders, worms, snails, lizards; also fruit.

Breeding. Nest in a crevice in a rock, or in a hollow at the end of a burrow in a vertical bank or in the level ground; no lining.

Eggs. 3 to 4; white. Incubated by ♂ and ♀.

Young. Nidicolous. Naked. Cared for by ♂ and ♀. Nest not cleaned.

Technical diagnosis. Ridgway, Pt. 6, 1914:450–452.

Classification. Peters, **5,** 1945:221–228.

References. A. F. Skutch, Life history of the Blue-throated Green Motmot, *Auk,* **62,** 1945:489–517, pl. 22.

G. M. Sutton, Blue-crowned Motmot, *Wilson Bull.,* **58,** 1946:frontispiece.

A. F. Skutch, Life history of the Turquoise-browed Motmot, *Auk,* **64,** 1947:201–217, pl. 9.

H. O. Wagner, Observations on the racquet-tips of the Motmot's tail, *Auk,* **67,** 1950:387–389.

A. Wetmore, *The Birds of the Republic of Panamá,* 2, *Smithson. Misc. Collect.,* **150,** 1968:437–455.

H. C. Land, *Birds of Guatemala,* Wynnewood, Pa., 1970:174–177.

MEROPIDAE

Bee-eater Family (24 species)

Figure 89. European Bee-eater (*Merops apiaster*). Total length 11 in. (280 mm).

Physical characteristics. Length—152 to 355 mm (6 to 14 in.). Plumage—soft and compact. Most spp. are green, with broad black line through eye, black wing tips, and bold areas (head, chin, throat, tail coverts) of bright yellow, vermilion, chestnut, blue; 1 sp. black, with turquoise-blue streaks and a red throat; 1 sp. slate-gray and red, with white markings; 2 spp. red, with green head and blue and black markings; 2 spp. largely blue and chestnut. *Nyctiornis* and *Meropogon* have the chin and throat feathers much elongated. Bill long, slender, laterally compressed, and decurved; both mandibles pointed. Wings long and pointed; tail long, with the central pair of rectrices elongated, square (or slightly emarginate), or forked (1 sp.). Lower tibia bare or sparsely feathered; feet syndactyl, rather small and weak; toes slender, short to long; claws slender and acute. Sexes alike or nearly so.

Range. Temperate and tropical parts of Old World. Most spp. migratory, at least locally.

Habits. Most spp. gregarious. Graceful wheeling flight (spectacular evolutions in some spp.). Hawk insects or capture them in short sallies from a perch. Fearless. Voice—musical trills, chirps, and whistles; hoarse chuckling and croaking notes.

Food. Insects, esp. bees and their allies.

Breeding. Colonial in most spp. Nest in burrows, excavated (by ♂ and ♀) very frequently in riverbanks, sometimes in level ground; no lining.

Eggs. 2 to 9; white; immaculate. Incubated by ♂ and ♀.

Young. Nidicolous. Naked at hatching. Cared for by ♂ and ♀.

Technical diagnosis. Witherby, **2**, 1938:262–263.

Classification. Peters, **5**, 1945:229–239.

References. Witherby, **2**, 1938:262–263; Bannerman, **3**, 1933:281–313.

C. W. Mackworth-Praed and C. H. B. Grant, *Birds of the Southern Third of Africa,* London, 1962:468–480.

P. A. Clancey, *The Birds of Natal and Zululand,* Edinburgh, 1964:256–261.

S. Ali and S. D. Ripley, *Handbook of the Birds of India and Pakistan,* Vol. 4, Bombay, 1970:98–113.

CORACIIDAE

Roller Family (16 species)

Figure 90. Indian Roller (*Coracias benghalensis*). Total length 13 in. (330 mm).

Physical characteristics. Length—241 to 457 mm (9.5 to 18 in.). Plumage of typical rollers (Coraciinae)—brightly colored in most spp.: blended shades of blue, bluish green, green, violet, reddish brown; usually unmarked, but streaked below in a few spp. Bill and feet yellow, red, or black. Bill wide, strong, decurved, slightly hooked. Neck short. Wings long. Tail rather long; truncate, emarginate, or deeply forked (with outer feathers lengthened and attenuate or even spatulate). Legs very short; feet strong, the 2nd and 3rd toes united basally. Sexes alike or nearly so. (The 5 spp. of Ground Rollers, the Brachypteraciinae of Madagascar, have mottled, cryptic plumage, longer legs, shorter wings, and more pointed—sometimes very long—tails.)

Range. Africa, Eurasia (exc. n. part), East Indies, Philippines, n. and e. Australia; e. to Solomons. Some spp. migratory.

Habits. Usually solitary. Arboreal. Flight strong and skillful (tumble or roll over during display flight). Fly from perch to perch, but hop on ground. Sit motionless on high, exposed perches. Voice—harsh, loud cries, frequently uttered. (Ground Rollers are mainly terrestrial birds which frequent heavy forest or—*Uratelornis*—sandy brush country.)

Food. Small animals, esp. insects; fruit exceptionally.

Breeding. Nest in holes in trees, in banks, in rock crevices, or in abandoned nests of magpies (*Pica*); little or no lining.

Eggs. 3 to 6; white. Incubated by ♂ and ♀ (or by ♀ only?).

Young. Nidicolous. Naked. Cared for by ♂ and ♀.

Technical diagnosis. Witherby, **2**, 1938:269.

Classification. Peters, **5**, 1945:240–247.

References. Witherby, **2**, 1938:269–272; Bannerman, **3**, 1933:206–221.

A. L. Rand, The distribution and habits of Madagascar birds, *Bull. Am. Mus. Nat. Hist.*, **72**, 1936:416–421 (omit *Leptosomus*).

P. A. Clancey, *The Birds of Natal and Zululand,* Edinburgh, 1964:261–264.

G. P. Dement'ev *et al., Birds of the Soviet Union,* Vol. 1, Washington, 1966:539–553.

S. Ali and S. D. Ripley, *Handbook of the Birds of India and Pakistan,* Vol. 4, Bombay, 1970:113–124.

Synonymy: The Brachypteraciidae are included.

LEPTOSOMATIDAE

Cuckoo-roller Family (1 species)

Figure 91. Cuckoo-roller (*Leptosomus discolor*). Total length 18 in. (457 mm).

Physical characteristics. Length—406 to 457 mm (16 to 18 in.). Plumage—♂ dark plumbeous gray above, with strong metallic green and coppery red reflections; face, throat, and complete collar ashy gray; rest of under parts grayish white; ♀ with gray largely replaced by rufous brown; head and hind neck barred, and whole under parts boldly spotted with black. Short crest in both sexes. Bill stout, decurved, slightly hooked. Head large; neck short. Wings long and pointed; tail long and truncate. Legs extremely short; feet semizygodactyl; toes, exc. hallux, long. ♀ larger than ♂, as well as differently colored.

Range. Madagascar and the nearby Comores Ids. Habitat—forests and brushland. Nonmigratory.

Habits. Somewhat gregarious. Arboreal. Flight strong; perform spectacular aerial evolutions above the forest. Feed largely in treetops. Vociferous. Voice—loud, whistled *wheu* or *wha-ha-ha-ha.*

Food. Large insects, lizards.

Breeding. Polyandrous? Nest in hollows in trees (and in holes in banks?).

Eggs. 3?; white. Incubation not described.

Young. Not described.

Technical diagnosis. R. B. Sharpe, Cat. of Birds in Brit. Mus., **17,** 1892:1.

Classification. Peters, **5,** 1945:239–240.

References. R. B. Sharpe, On the *Coraciidae* of the Ethiopian Region. Subfamily III. Leptosominae, *Ibis,* **1871:**285–289.

H. E. Dresser, *A Monograph of the Coraciidae* . . . , Farnborough, England, 1893:101–108, pls. 26, 27.

A. L. Rand, The distribution and habits of Madagascar birds, *Bull. Am. Mus. Nat. Hist.,* **72,** 1936:417–418.

E. T. Gilliard, *Living Birds of the World,* New York, 1958:247–248.

Synonymy: Sometimes included in the Coraciidae.

UPUPIDAE

Hoopoe Family (1 species)

Figure 92. Hoopoe (*Upupa epops*). Total length 11 in. (280 mm).

Physical characteristics. Length—266 to 305 mm (10.5 to 12 in.). Plumage—pinkish cinnamon to rufous chestnut (paler below), with bands of black, white, and buff on back and wings; tail barred with white. Long, conspicuous, black-tipped crest (with white subterminal area in some races). Bill long and very slender; tongue short. Wings broad and rounded; tail square, moderate in length. Tarsi short, slender, and bare; toes long (3rd and 4th fused at base); claws short. Sexes similar (♀ duller and/or smaller in some cases).

Range. Central and s. Europe, Africa (except c. region), Madagascar, and Asia (exc. n. third) to Korea. Habitat—semiopen country and cultivated clearings. Migratory in parts of range.

Habits. Solitary or in small bands. Terrestrial, but perch and roost in trees and occasionally hawk insects. Flight slow, undulating, and erratic, but efficient in danger. ♂ feeds ♀ in courtship and during breeding. Voice—typically *hoop-hoop* or *poup-poup*; hawing and mewing sounds.

Food. Insects, worms, spiders, etc.

Breeding. Nest a hole in tree, wall, earth bank, or termite nest, sometimes a bulky structure of sticks, etc., but usually without lining.

Eggs. 4 to 12 (usually 4 to 6); pale blue to olive-brown; usually immaculate. Incubated by ♀.

Young. Nidicolous. With sparse down. Cared for by ♂ and ♀.

Technical diagnosis. Witherby, **2,** 1938:266.

Classification. Peters, **5,** 1945:617–620.

References. Witherby, **2,** 1938:266–269; Bannerman, **3,** 1933:222–228.

F. J. Jackson, *The Birds of Kenya Colony and the Uganda Protectorate,* Vol. 2, London, 1938:617–620.

C. J. Skead, A study of the African Hoopoe, *Ibis,* **92,** 1950:434–463.

G. P. Dement'ev *et al., Birds of the Soviet Union,* Vol. 1, Washington, 1966:586–597.

S. Ali and S. D. Ripley, *Handbook of the Birds of India and Pakistan,* Vol. 4, Bombay, 1970:124–129.

PHOENICULIDAE

Woodhoopoe Family (6 species)

Figure 93. Scimitarbill (*Rhinopomastus cyanomelas*). Total length 11 in. (280 mm).

Physical characteristics. Length—222 to 381 mm (8.75 to 15 in.). Plumage—blackish blue, blackish purple, and blackish green with metallic gloss, uniform in color exc. 2 spp., which have white or light brown heads; terminal and subterminal white spots on tail in some; bill brightly colored in some. Bill long, slender, and laterally compressed, almost straight to sickle-curved; tongue short. Wings rounded; tail long and steeply graduated. Tarsus very short, partly feathered in some spp.; toes rather long (esp. hallux), the 3rd and 4th fused basally; claws long and sharply curved. ♀ similar to ♂ but smaller and/or browner in some spp.

Range. Central and s. Africa, exc. Madagascar. Habitat—dense forest, forest edges, wooded grasslands. Nonmigratory (exc. for local movements).

Habits. Solitary or in small bands. Arboreal. Run along trunks and branches; climb (often head downward) with dexterity. Flight brief and infrequent, rather labored. Voice—loud, chattering notes.

Food. Insects, spiders, small fruits, seeds.

Breeding. Nest a hole in a tree.

Eggs. 3 to 5 (usually 3?); pale blue, green, or greenish blue. Incubated by ♀.

Young. Nidicolous. Downy. Cared for by ♂ and ♀.

Technical diagnosis. Bannerman, **3**, 1933:xxvi–xxvii, 228.

Classification. Peters, **5**, 1945:250–253.

References. Bannerman, **3**, 1933:228–239.

F. J. Jackson, *The Birds of Kenya Colony and the Uganda Protectorate*, Vol. 2, London, 1938:620–629.

J. P. Chapin, The birds of the Belgian Congo, II, *Bull. Am. Mus. Nat. Hist.*, **75**, 1939:323–332.

C. W. Mackworth-Praed and C. H. B. Grant, *Birds of the Southern Third of Africa*, London, 1962:499–504.

Synonym: Irrisoridae.

BUCEROTIDAE

Hornbill Family

(45 species)

Figure 94. Great Hornbill (*Buceros bicornis*). Total length 42 in. (1067 mm).

Physical characteristics. Length—381 to 1600 mm (15 to 63 in.). Plumage—loose-webbed and wiry; brown; black, white, and brown; or (typically) black and white. Bill typically red or yellow, very large, curved, variously sculptured, serrated in some spp.; usually with casque on culmen. Many spp. crested. Bare skin about eye and, sometimes, throat brightly colored. Conspicuous eyelashes. Wings strong; tail long. Legs very short (exc. Ground Hornbill, *Bucorvus*); feet broad-soled and syndactyl. Sexes unlike in many spp., alike in others.

Range. Africa, s. of Sahara (exc. Madagascar), tropical Asia, Malaysia, Philippines, and east to Solomons. Nonmigratory.

Habits. Usually in pairs or small flocks. Arboreal (exc. *Bucorvus*). Voice (most spp. vociferous)—harsh calls and loud whistles. Wings very noisy in flight.

Food. Omnivorous; some live principally on fruit.

Breeding. Nest in hollow trees, occasionally in caves. From before egg-laying until fledging of young, ♀ remains in nest and is fed by ♂ (exc. *Bucorvus*); one or both members of pair wall up nest entrance with mud, etc., leaving small feeding aperture.

Eggs. 1 to 6; white. Incubated by ♀.

Young. Nidicolous. Naked. Heel-pads. Fed by ♀ with food brought by ♂; in some small spp. (*Tockus*), ♀ leaves nest when young are half-fledged and assists in gathering food.

Technical diagnosis. Baker, *Fauna*, **4**, 1927:282.

Classification. Peters, **5**, 1945:245–272.

References. Bannerman, **3**, 1933:314–354.

R. E. Moreau and W. M. Moreau, Breeding biology of Silvery-cheeked Hornbill, *Auk*, **58**, 1941:13–27.

G. Ranger, Life of the Crowned Hornbill, *Lophoceros suahelicus australis*, *Ostrich*, **20**, 1949:54–65, 152–167; **21**, 1950:2–13; **22**, 1951:77–93; **23**, 1952:26–36.

S. Ali and S. D. Ripley, *Handbook of the Birds of India and Pakistan*, Vol. 4, Bombay, 1970:129–146.

J. E. duPont, *Philippine Birds*, Greenville, Del., 1971:210–213.

GALBULIDAE

Jacamar Family (15 species)

Figure 95. Rufous-tailed Jacamar
(*Galbula melanogenia*). Total length 9.5 in.
(241 mm).

Physical characteristics. Length—127 to 299 mm (5 to 11.75 in.). Plumage—soft and loose-webbed; usually metallic green or black above, tawny or black below; throat usually white in ♂, buff in ♀. Long, attenuate bill (nearly straight in most spp.), sharply ridged above and below. Wings short; tail usually long and graduated (or acuminate). Feet zygodactyl (3-toed in *Jacamaralcyon*). Sexes unlike.

Range. Mainland from s. Mexico (Veracruz) to s. Brazil, but chiefly Amazon Valley. Habitat—tropical forests. Nonmigratory.

Habits. Solitary. Perch quietly on tree branches for long periods and then make lengthy, elaborate sallies for flying insects. Voice—varied squeaks and trills; ♂ of at least 1 sp. (*Galbula melanogenia*) has a long, rather melodious song.

Food. Insects, esp. butterflies and dragonflies.

Breeding. Nest in hole in bank, excavated by ♂ and ♀.

 Eggs. 3 to 4; white; nearly round. Incubated by ♂ and ♀.

 Young. Nidicolous. Long, white down. Heel-pads. Fed by ♂ and ♀.

Technical diagnosis. Ridgway, Pt. 6, 1914:360.

Classification. Peters, **6**, 1948:3–9.

References. A. F. Skutch, Life-history of the Black-chinned Jacamar, *Auk,* **54,** 1937:135–146.

 A. Wetmore, *The Birds of the Republic of Panamá*, 2, *Smithson. Misc. Collect.,* **150,** 1968:456–467.

 R. Meyer de Schauensee, *A Guide to the Birds of South America,* Wynnewood, Pa., 1970:169–171.

BUCCONIDAE

Puffbird Family (30 species)

Figure 96. White-necked Puffbird (*Notharcus macrorhynchos*). Total length 9.5 in. (241 mm).

Physical characteristics. Length—139 to 317 mm (5.5 to 12.5 in.). Plumage—thick and loose-webbed; brown (or black) and white, sometimes with chestnut or gray; often streaked and spotted; throat often white or pale buff; broad breast band in many spp. Bill (often red or yellow) large and strong, rounded, almost straight to markedly decurved and hooked, with tip of maxilla sometimes bifid. Conspicuous rictal bristles. Wings rounded (exc. *Chelidoptera*); tail medium to long. Legs short; feet zygodactyl. Sexes alike or nearly so.

Range. Mainland from s. Mexico (Oaxaca) to s. Brazil and Paraguay. Habitat—tropical forests. Nonmigratory.

Habits. Usually solitary. Arboreal. Very stolid and sedentary (exc. *Chelidoptera,* which has long, pointed wings and flies strongly). Capture insects on the wing or pick them from ground or trees during brief sallies from a perch. Voice (rarely heard)—low peeps, thin whistles, a high-pitched *tzeeee tzeeee,* a twittering song.

Food. Insects.

Breeding. Nest (sometimes lined with grass or leaves) is excavated by ♂ and ♀—holes in arboreal termite nests or banks or tunnels (up to 1.5 m long) in level ground, with leaves and sticks piled around entrance by some spp.

Eggs. 2 to 3; glossy white. Incubated by ♂ and ♀.

Young. Nidicolous. No down. Fed by ♂ and ♀ (?).

Technical diagnosis. Ridgway, Pt. 6, 1914:370–371.

Classification. Peters, 6, 1948:10–23.

References. G. Hollister and W. Beebe, The secret of the Swallow-winged Puff-bird, *N.Y. Zool. Soc. Bull.,* **30,** 1927:115–119.

A. F. Skutch, Life history notes on Puff-birds, *Wilson Bull.,* **60,** 1948:81–97.

F. Haverschmidt, Notes on the Swallow-wing, *Chelidoptera tenebrosa,* in Surinam, *Condor,* **52,** 1950:74–77.

A. Wetmore, *The Birds of the Republic of Panamá,* 2, *Smithson. Misc. Collect.,* **150,** 1968:467–491.

R. Meyer de Schauensee, *A Guide to the Birds of South America,* Wynnewood, Pa., 1970:171–175.

INDICATORIDAE

Honeyguide Family (11 species)

Figure 97. Greater Honeyguide (*Indicator indicator*). Total length 7.5 in. (190 mm).

Physical characteristics. Length—108 to 203 mm (4.25 to 8 in.). Plumage—brown, olive, and gray above, lighter below; variously streaked, spotted; with small areas of yellow in some spp.; tail marked with white. Bill short; stout and blunt to slender and pointed; nostrils have raised rims. Wings long, pointed; tail somewhat graduated, the outermost rectrix always shorter than the next (*Melichneutes* has lyre-shaped tail). Tarsus rather short; feet zygodactyl; toes strong; claws long, strongly hooked. Sexes unlike in most spp.

Range. Africa s. of Sahara (exc. Madagascar), Himalayas, Burma, Thailand, Malaya, Sumatra, and Borneo. Habitat—forest and brush country. Nonmigratory exc. for local movements.

Habits. Solitary. Arboreal. Flight rapid; direct or undulating. Some spp. lead man (and other mammals) to stores of wild honey. Some spp. hawk insects. Voice—harsh squeak, loud clear whistle, chattering and croaking notes.

Food. Insects (in some spp., largely bees and their larvae), honey, beeswax.

Breeding. Parasitic in all spp. for which data are available. Eggs usually laid in nests of hole- or burrow-nesting birds.

Eggs. Number unknown; white. Incubated by hosts.

Young. Nidicolous. No down. Both mandibles hooked in nestlings of some spp. Cared for by hosts.

Technical diagnosis. H. Friedmann, The Honey-guides, *U.S. Natl. Mus. Bull.,* **208,** 1955:6.

Classification. H. Friedmann, The Honey-guides, *U.S. Natl. Mus. Bull.,* **208,** 1955:6.

References. Bannerman, **3,** 1933:xxxiv–xxxv, 403–423.

J. P. Chapin, The birds of the Belgian Congo, II, *Bull. Am. Mus. Nat. Hist.,* **75,** 1939:535–556.

H. Friedmann, Additional data on brood parasitism in the Honey-guides, *Proc. U.S. Natl. Mus.,* **124,** No. 3648, 1968.

S. Ali and S. D. Ripley, *Handbook of the Birds of India and Pakistan,* Vol. 4, Bombay, 1970:165–167.

RAMPHASTIDAE

Toucan Family (37 species)

Figure 98. Cuvier Toucan (*Ramphastos cuvieri*). Total length 23 in. (584 mm).

Physical characteristics. Length—305 to 610 mm (12 to 24 in.). Plumage—lax; usually brightly colored, with bold contrast (black and white, orange, red, yellow, green, blue). Bare skin around eye. Bill usually bright in color; very large (relatively larger in larger spp.); serrate; with nostrils at extreme base. Tongue very long, narrow, and fringed. Wings short and rounded; tail usually rather long, rounded to extremely graduate. Legs strong; feet zygodactyl. Sexes alike in most spp.

Range. Vera Cruz s. to Brazil, Paraguay, and n. Argentina. Habitat—forests. Altitudinal migration in some spp.

Habits. Rather gregarious. Arboreal. Restless, active. Flight weak. "Mob" birds of prey. Voice (often noisy)—a variety of croaks, shrill calls, and harsh undiversified "songs."

Food. Fruit, large insects, nestlings of smaller birds, lizards, etc.

Breeding. Nest in unlined tree cavity, natural or made by other birds.

Eggs. 2 to 4; white, glossy. Incubated by ♂ and ♀.

Young. Nidicolous. Naked. With heel-pads. Cared for by ♂ and ♀.

Technical diagnosis. Ridgway, Pt. **6,** 1914:327–329; P. R. Lowe, *Ibis,* **1946:**119.

Classification. Peters, **6,** 1948:70–85.

References. A. F. Skutch, Life histories of Central American highland birds, *Publ. Nuttall Ornithol. Club.,* No. 7; 1967:51–59.

A. Wetmore, *The Birds of the Republic of Panama,* 2, *Smithson. Misc. Collect.,* **150,** 1968:504–527.

R. Meyer de Schauensee, *A Guide to the Birds of South America,* Wynnewood, Pa., 1970:177–183.

G. R. Bourne, The Red-billed Toucan in Guyana, *Living Bird,* **1974:**99–126.

Synonym: Rhamphastidae.

CAPITONIDAE

Barbet Family (72 species)

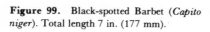

Figure 99. Black-spotted Barbet (*Capito niger*). Total length 7 in. (177 mm).

Physical characteristics. Length—89 to 317 mm (3.5 to 12.5 in.). Plumage—bright green, olive, brown, or black, boldly marked (esp. on head and breast) with solid areas of bright yellow, red, blue, gray, white (many spp. brilliantly multicolored). Some spp. conspicuously spotted; some uniformly dull-colored. Most spp. have tufts of feathers over the nostrils and/or well-developed rictal and chin bristles. Bill large and heavy, somewhat curved, pointed. Head large; body heavy. Wings short to medium, rounded; tail short to medium. Legs short and strong; feet large, zygodactyl. Sexes alike in most spp.

Range. Costa Rica, Panama, and n.w. S. America; Africa (s. of Sahara); India, Burma, Thailand, Indochina, Malaysia, and Philippines. Nonmigratory.

Habits. Usually solitary. Arboreal. Perch long in one spot. Flight weak. Voice—typically harsh single notes indefinitely repeated; also low whistling calls.

Food. Fruit, insects.

Breeding. A few spp. somewhat colonial. Nest in holes which they excavate in trees, or in holes in banks (*Trachyphonus*); no nest lining.

Eggs. 2 to 4; white. Incubated by ♂ and ♀.

Young. Nidicolous. No down. Cared for by ♂ and ♀.

Technical diagnosis. Ridgway, Pt. 6, 1914:310–311. (See also P. R. Lowe, *Ibis,* **1946:**118).

Classification. Peters, **6,** 1948:24–63 (78 spp.).

References. Bannerman, **3,** 1933:xxxiv–xxxv, 361–403.

S. D. Ripley, The Barbets, *Auk,* **62,** 1945:542–563. See also *Auk, 63,* 1946:452–453 (Ripley); 384–388 (R. E. Moreau); 481 (C. M. N. White).

S. Ali and S. D. Ripley, *Handbook of the Birds of India and Pakistan,* Vol. 4, Bombay, 1970:146–165.

R. Meyer de Schauensee, *A Guide to the Birds of South America,* Wynnewood, Pa., 1970:175–177.

Synonym: Megalaemidae.

PICIDAE

Woodpecker Family (208 species)

Figure 100. Imperial Woodpecker (*Campephilus imperialis*). Total length 22 in. (559 mm).

Physical characteristics. Length—89 to 559 mm (3.5 to 22 in.). Plumage—black, white, yellow, red, brown, green. Red or yellow on the head of many spp. Many are barred, spotted, or streaked, esp. below; some are crested. Bill strong; typically straight and chisel-like. Head large; neck slender but very strong. Wings strong, rather rounded; tail rounded or wedge-shaped, the rectrices stiff and pointed (exc. in Picumninae). Legs short; feet 3-toed, or 4-toed and zygodactyl. Sexes unlike in most spp.

Range. Worldwide (exc. the extreme n., Madagascar, the Papuan Region, Australia, and most oceanic islands). Most spp. nonmigratory.

Habits. Solitary. Typically arboreal. Flight strong but not sustained; undulating except in largest spp. Bore into wood (or earth) for food. Cling to tree trunks, bracing with the tail (seldom perch). Voice—loud and harsh in most spp.; some have "laughing" or ringing cries. Drumming with the bill partly replaces voice.

Food. Insects; fruits, nuts; sap of trees.

Breeding. Nest (usually freshly excavated) a tree cavity, hole in bank, or termite nest, without lining. (*Colaptes rupicola* nests colonially.)

Eggs. 2 to 8; glossy, white. Incubated by ♂ and ♀.

Young. Nidicolous. Naked or (rarely) with sparse down. Cared for by ♂ and ♀.

Technical diagnosis. Ridgway, Pt. 6, 1914:5.

Classification. Peters, **6,** 1948:88–232.

References. Bent, No. 174; Witherby, **2,** 1938:276–292; Bannerman, **3,** 1933:423–462.

A. F. Skutch, Life history of the Olivaceous Piculet and related forms, *Ibis,* **90,** 1948:433–449.

G. P. Dement'ev *et al., Birds of the Soviet Union,* Vol. 1, Washington, 1966:598–668.

A. Wetmore, *The Birds of the Republic of Panamá,* 2, *Smithson. Misc. Collect.,* **150,** 1968:527–583.

A. F. Skutch, Life histories of Central American birds, III, *Pacific Coast Avifauna,* No. 35, 1969:419–561.

S. Ali and S. D. Ripley, *Handbook of the Birds of India and Pakistan,* Vol. 4, Bombay, 1970:168–246.

JYNGIDAE

Wryneck Family (2 species)

Figure 101. Eurasian Wryneck (*Jynx torquilla*). Total length 6.5 in. (165 mm).

Physical characteristics. Length—165 to 177 mm (6.5 to 7 in.). Plumage—soft; brown, gray, and black in mottled, cryptic pattern. Under parts paler; throat, breast, and under tail coverts buff (barred with black) or chestnut (unmarked); belly white or pale buff, barred or streaked with black. Bill slender, pointed. Wings rounded; tail rather long, the feathers soft and rounded at tips. Legs short; feet 4-toed, zygodactyl. Sexes alike.

Range. Eurasia (exc. extreme n.) and Africa (exc. c. part). Northern forms migratory.

Habits. Solitary (but small groups form during migration and in winter). Arboreal. Flight slow and undulating. Twisting motions of the neck are responsible for the common name. Remain motionless when alarmed. Obtain food largely from surface of trees. Usually perch across branches in passerine fashion; sometimes also cling to tree trunks. Voice—a shrill *quee-quee-quee* monotonously repeated; a harsh screaming cry; hissing notes.

Food. Insects; fruit rarely.

Breeding. Nest in natural cavity of tree or bank, or in crevice in wall; no nest lining.

    Eggs. 2 to 12; white. Incubated chiefly by ♀.

    Young. Nidicolous. Naked. Heel-pads. Cared for by ♂ and ♀.

Technical diagnosis. Ridgway, Pt. 6, 1914:4.

Classification. Peters, **6**, 1948:86–88.

References. Witherby, **2**, 1938:292–296; Bannerman, **3**, 1933:462–466.

    H. Siewert, Beitrage zur Biologie des Wendelhalses, *Beitr. Fortpflanz. Vögel*, **4**, 1928:47–49.

    J. Bussmann, Beitrage zur Kenntnis der Brutbiologie des Wendehalses (*Jynx torquilla torquilla*), *Arch. Suisses Ornithol.*, **1**, 1941:467–480.

    C. W. Mackworth-Praed and C. H. B. Grant, *Birds of the Southern Third of Africa,* London, 1962:587–589.

    G. P. Dement'ev *et al., Birds of the Soviet Union,* Vol. 1, Washington, 1966:668–674.

    S. Ali and S. D. Ripley, *Handbook of the Birds of India and Pakistan,* Vol. 4, Bombay, 1970:168–171.

Synonym: Yungidae.

EURYLAIMIDAE

Broadbill Family (14 species)

Figure 102. Black-and-Yellow Broadbill (*Eurylaimus ochromalus*). Total length 6 in. (152 mm).

Physical characteristics. Length—127 to 280 mm (5 to 11 in.). Plumage—lax; bright green and blue (or black and pale vinous to crimson), marked with black, white, and yellow or orange; gray and chestnut, marked with blue; or brown and buff (plain or streaked), marked with black, white, and small areas of yellow. Most spp. have white dorsal patch, concealed exc. in flight. Bill (largely covered by short crest in some spp.) broad and flattened, with wide gape; moderate to extremely large and heavy; hooked. Eyes large; head broad; body stout. Wings short to long, rounded; tail very short and square to long, slender, and graduated. Legs short; feet strong, syndactyl; toes (incl. hallux) long; claws long, strongly hooked. Sexes unlike in most spp.

Range. Central and s. Africa; Himalayas of India through s. China, Indochina, and the Malay Peninsula to Sumatra, Java, and Borneo; Philippines. Habitat—forest edges, open wooded country. Nonmigratory.

Habits. Solitary or gregarious. Some spp. crepuscular. Arboreal. Unsuspicious. Lethargic exc. for short flights to catch insects on the wing (exc. *Pseudocalyptomena*). Voice—churring notes; clear whistles.

Food. Insects; fruit, seeds, buds, flowers; frogs, lizards.

Breeding. Nest a large, pear-shaped pendent structure of grass, etc., decorated with streamers of moss, etc., with a porched entrance at side; usually placed over water.

    Eggs. 2 to 8 (usually 3 to 5); white to salmon; immaculate or spotted. Incubated by ♂ and ♀.

    Young. Nidicolous. Naked at hatching. Cared for by ♂ and ♀?

Technical diagnosis. Baker, *Fauna,* 2nd ed., **3,** 1926:459–460.

Classification. Peters, **7,** 1951:3–13.

References. Baker, *Nidification, 3,* 1934:260–271; Bannerman, **4,** 1936:5–12.

    S. Ali and S. D. Ripley, *Handbook of the Birds of India and Pakistan,* Vol. 4, Bombay, 1970:246–250.

    H. Friedmann, The status and habits of Grauer's Broadbill in Uganda (Aves: Eurylaemidae), *Contrib. Sci.,* **176,** 1970.

DENDROCOLAPTIDAE

Woodcreeper Family (48 species)

Figure 103. Ivory-billed Woodcreeper (*Xiphorhynchus flavigaster*). Total length 10.5 in. (266 mm).

Physical characteristics. Length—146 to 368 mm (5.75 to 14.5 in.). Plumage—olive-brown or grayish brown to cinnamon; most spp. streaked (sometimes also barred) or spotted with black, gray, buffy, white, esp. on head, shoulders, and under parts; wings and tail in most spp. rufous. Bill typically strong, laterally compressed, short and straight to very long and curved. Wings rather long, rounded. Tail long, rounded or graduated; the feathers with very strong, rigid, sharp-pointed shafts (the tips curved and abruptly attenuate in some spp.). Legs short; feet and claws strong; anterior toes adherent basally. Sexes alike or nearly so.

Range. Mexico, C. and S. America (exc. extreme s.). Habitat—forest or brushland. Largely nonmigratory.

Habits. Solitary, or in mixed flocks of other spp. Arboreal. Flight strong but not sustained. Climb tree trunks (bracing with tail) in search of food, then fly to base of another tree to repeat the process. Some spp. occasionally feed on ground. Voice—loud, ringing, repetitive songs; a musical trill; harsh alarm notes.

Food. Insects, spiders, amphibians.

Breeding. Nest in tree cavities, natural or made by other birds; lined with bark, leaves, etc.

    Eggs. 2 to 3; white or greenish white; immaculate. Incubated by ♂ and ♀.

    Young. Nidicolous. Downy. Cared for by ♂ and ♀.

Technical diagnosis. Ridgway, Pt. 5, 1911:4, 224–226.

Classification. Peters, 7, 1951:13–57.

References. D. R. Dickey and A. J. van Rossem, The Birds of El Salvador, *Field Mus. Nat. Hist., Zool. Ser.,* **23,** 1938:321–328.

    A. F. Skutch, Life histories of Central American birds, III, *Pacific Coast Avifauna,* No. 35, 1969:374–418.

    R. Meyer de Schauensee, *A Guide to the Birds of South America,* Wynnewood, Pa., 1970:194–200.

    A. Wetmore, *The Birds of the Republic of Panamá,* 3, *Smithson. Misc. Collect.,* **150,** 1972:2–56.

FURNARIIDAE

Ovenbird Family (219 species)

Figure 104. Red Ovenbird (*Furnarius rufus*). Total length 8 in. (203 mm).

Physical characteristics. Length—120 to 280 mm (4.75 to 11 in.). Plumage—very dark brown (sometimes with slate or black), olive-brown to cinnamon, buff, or gray. Most spp. show little pattern, but a few are streaked, spotted, or scaled; some have conspicuous light patch on spread wing; many have contrasting crown and/or throat. Under parts lighter in most spp. (largely white in some), marked with yellow in a few. Some spp. crested. Bill rather slender; very short to long; straight, curved, or (rarely) upturned. Wings short and rounded to rather long and pointed. Tail short to long (extremely long in *Sylviorthorhynchus*), rounded to acuminate, the feathers pointed in many spp. but tips soft in most. Legs short to medium; anterior toes basally adherent. Sexes alike or nearly so.

Range. Southern Mexico, C. and S. America. Habitat—forest to semidesert; seashore, mountain cliffs, and rocky slopes. Most spp. nonmigratory.

Habits. Solitary or gregarious. Terrestrial or arboreal. Flight weak to strong, but not sustained. Many terrestrial forms walk. A few spp. climb tree trunks (some brace with tail). Voice (many spp. noisy)—harsh, scolding notes; loud whistled calls; trilled songs.

Food. Insects and spiders; also (some spp.) seeds, crustacea.

Breeding. Well-lined nest in hole in ground, bank, or rocks, or in natural tree cavity; domed nest of mud on tree, post, or building; domed nest of grass or sticks in reeds or bushes, or on ground.

Eggs. 2 to 5 (rarely 6); white to pale blue or bluish green; immaculate. Incubated by ♂ and ♀.

Young. Nidicolous. Downy. Cared for by ♂ and ♀.

Technical diagnosis. Ridgway, Pt. 5, 1911:4, 157–158.

Classification. Peters, **7**, 1951:58–153.

References. R. Meyer de Schauensee, *A Guide to the Birds of South America,* Wynnewood, Pa., 1970:200–226.

C. Vaurie, *Classification of the Ovenbirds (Furnariidae),* London, 1971.

A. Wetmore, *The Birds of the Republic of Panamá,* 3, *Smithson. Misc. Collect.,* **150**, 1972:56–120.

A. Feduccia, Evolutionary trends in the Neotropical ovenbirds and woodhewers, *Ornithol. Monogr.* No. 13, 1973.

FORMICARIIDAE

Antbird Family (232 species)

Figure 105. Black-faced Antthrush (*Formicarius analis*). Total length 7.5 in. (190 mm).

Physical characteristics. Length—95 to 368 mm (3.75 to 14.5 in.). Plumage—loose-webbed; black, gray, browns (sometimes with white, rarely yellowish), in solid areas of color or (esp. below) streaked or strongly barred; ♀ commonly browner than ♂. Some spp. have bare red or blue orbital skin; some are crested; some have feathers of lower back long and dense, with concealed white or rufous spotting. Bill strong, slightly to strongly hooked. Wings short and rounded; tail short to long. Legs short (arboreal spp.) to long (terrestrial spp.); anterior toes somewhat adherent basally. Sexes unlike in most spp.

Range. Southern Mexico, C. America, and S. America to c. Argentina. Habitat—forests or brushland. Nonmigratory.

Habits. Usually solitary or in pairs. Arboreal or terrestrial. Flight weak. Some spp. accompany ant armies and prey on insects flushed by them. Voice—sharp, often harsh, calls; low, rather melodious songs; or loud, mellow whistling notes.

Food. Insects.

Breeding. Open, cuplike nest, typically semipendent in horizontal fork of a bush or low tree; simple cup on or near ground; covered nest on ground; or lined cavity, usually in a tree.

Eggs. Usually 2; white or buffy; speckled or streaked with brown, red, or black. Incubated by ♂ and ♀.

Young. Nidicolous. Downy (*Formicarius*) or naked. Cared for by ♂ and ♀.

Technical diagnosis. Ridgway, Pt. 5, 1911:8–9.

Classification. Peters, **7**, 1951:153–273.

References. J. Van Tyne, The nest of the Antbird *Gymnopithys bicolor bicolor*, *Univ. Mich. Mus. Zool. Occas. Pap.* No. 491, 1944:1–5.

A. F. Skutch, Life histories of Central American birds, III, *Pacific Coast Avifauna*, No. 35, 1969:164–295.

A. Wetmore, *The Birds of the Republic of Panamá*, 3, *Smithson. Misc. Collect.*, **150**, 1972:120–254.

E. O. Willis, The behavior of Spotted Antbirds, *Ornithol. Monogr.* No. 10, 1972, 162 pp.

Synonymy: Includes *Conopophaga* (formerly in Conopophagidae).

RHINOCRYPTIDAE

Tapaculo Family (26 species)

Figure 106. Chestnut-breasted Turco (*Pteroptochos castaneus*). Total length 10 in. (254 mm).

Physical characteristics. Length—114 to 254 mm (4.5 to 10 in.). Plumage—soft and loose-webbed; brown, gray, or black. Some spp. with areas of reddish brown and/or barring below; 1 sp. conspicuously spotted with white above and below. One sp. crested; 1 with loral plumes. Bill sharp-pointed; rather slender (in small spp.) to stout; culmen flat in *Acropternis*. Body compact. Wings rounded; tail short to rather long. Feet and claws large and strong. Sexes alike or nearly so.

Range. Mountains of Costa Rica, through Panama; w. S. America, and e. across s. Brazil to state of Baía. Habitat—dense forest undergrowth, grassland, semiarid brushy country. Nonmigratory.

Habits. Solitary or in small groups. Terrestrial; most spp. walk rather than hop (forest spp. "creep about like mice"); run with great speed (commonly holding tail erect); scratch like hens for food. Fly rarely. Very secretive. Voice—loud whistles; harsh, barking calls; loud, deep *chirrup*; crowing songs; musical notes in descending scale.

Food. Insects, seeds.

Breeding. Nest of grass, moss, etc., in burrow in bank, in hole in cliff, in abandoned mammal burrow, in hollow trunk, or in crevice between bark and trunk; or domed nest in bush.
Eggs. 2 to 4; white. Incubation not described.
Young. Nidicolous. Downy. Cared for by ♂ and ♀.

Technical diagnosis. Ridgway, Pt. 5, 1911:4–5.

Classification. Peters, **7**, 1951:278–289.

References. C. E. Hellmayr, The birds of Chile, *Field Mus. Nat. Hist., Zool. Ser.,* **19**, 1932:214–230.

J. D. Goodall *et al., Las Aves de Chile,* Vol. 1, Buenos Aires, 1946:267–287.

R. Meyer de Schauensee, *A Guide to the Birds of South America,* Wynnewood, Pa., 1970:259–263.

A. Wetmore, *The Birds of the Republic of Panamá,* 3, *Smithson. Misc. Collect.,* **150**, 1972:254–261.

Synonyms: Pteroptochidae, Hylactidae.

COTINGIDAE

Cotinga Family (73 species)

Figure 107. Umbrellabird (*Cephalopterus ornatus*). Total length 18 in. (457 mm).

Physical characteristics. Length—89 to 457 mm (3.5 to 18 in.). Plumage—in many spp. gray or brown, with little pattern; in others white, grayish white, black, or largely brilliant red, purple, blue, or green. Some spp. with bare skin or long erectile caruncles on head, or with bare gular pouch. *Cephalopterus* has great, umbrella-like crest; *Rupicola,* a strange, laterally compressed one. Bill moderately long and compressed to short and flattened. Wings short and rounded to rather long, with one or more primaries highly modified in many spp.; tail short to rather long (deeply forked in 1 sp.). Legs short; feet large. Sexes alike or unlike.

Range. Extreme s. edge of Arizona and Texas; Mexico, C. America, Jamaica, S. America (exc. s. third). Habitat—forests. Largely nonmigratory.

Habits. Usually solitary. Arboreal. Flight medium to strong. Voice—loud bell-like call; grunting sounds; some musical utterances, others with a "mechanical quality."

Food. Fruit, insects.

Breeding. Nest a lined tree cavity; shallow cup on branch; bulky covered nest on, or suspended from, branch tip; shallow mud nest plastered on steep rocky wall.

Eggs. 1 to 6; white to dark-colored; heavily marked. Incubated by ♀.

Young. Nidicolous. Downy or naked. Cared for by ♂ and ♀.

Technical diagnosis. Ridgway, Pt. 4, 1907:769–771.

Classification. Hellmayr, Pt. 6, 1929:92–246 (exclude *Attila, Casiornis, Laniocera, Rhytipterna*).

References. E. T. Gilliard, On the breeding behavior of the Cock-of-the-rock (Aves, *Rupicola rupicola*), *Bull. Am. Mus. Nat. Hist.,* **124,** Art. 2, 1962.

A. F. Skutch, Life histories of Central American birds, III, *Pacific Coast Avifauna,* No. 35, 1969:10–96.

D. W. Snow, Observations on the Purple-throated Fruit-Crow in Guyana, *Living Bird,* **1971:**5–17.

A. Wetmore, *The Birds of the Republic of Panamá,* 3, *Smithson. Misc. Collect.,* **150,** 1972:261–309.

Synonymy: The Rupicolidae are included.

PIPRIDAE

Manakin Family (56 species)

Figure 108. Yellow-thighed Manakin (*Pipra mentalis*). Total length 4.5 in. (114 mm).

Physical characteristics. Length (exc. for long central rectrices of *Chiroxiphia linearis*)—83 to 159 mm (3.25 to 6.25 in.). Plumage (♂)—usually in a solid color (commonly black) with areas, esp. on crown or throat, of white or of bright, lustrous colors (scarlet, orange, yellow, blue); feathers of throat or crown elongated in some spp. Bill short and broad, slightly hooked. Wings short; tail typically short, with (in a few spp.) some elongated feathers. Legs short; 3rd toe fused at base with 2nd or 4th. Sexes usually different; ♀ typically olive-green.

Range. From s. Mexico to Paraguay. Habitat—humid tropical forests. Nonmigratory.

Habits. Solitary (occasionally form small flocks). Flight rapid and direct. Very striking and varied dance display (♂ ♂ alone or in a group). Voice—short, rather low calls (louder, more spectacular sounds are made by remiges in many spp.).

Food. Fruit, some insects.

Breeding. Frail, semipensile nest in low bushes. Built by ♀.

Eggs. 2; spotted. Incubated by ♀.

Young. Nidicolous. With sparse down. Cared for by ♀.

Technical diagnosis. Ridgway, Pt. 4, 1907:723.

Classification. Hellmayr, Pt. 6, 1929:3–92.

References. A. F. Skutch, Life history of the Yellow-thighed Manakin, *Auk,* **66,** 1949:1–24.

P. Slud, The song and dance of the Long-tailed Manakin, *Chiroxiphia linearis, Auk,* **74,** 1957:333–339.

K. C. Parkes, Intergeneric hybrids in the family Pipridae, *Condor,* **63,** 1961:345–350.

H. Sick, Courtship behavior in the manakins (Pipridae): a review, *Living Bird,* **1967:**5–22.

A. F. Skutch, Life histories of Central American birds, III, *Pacific Coast Avifauna,* No. 35, 1969:97–163.

A. Wetmore, *The Birds of the Republic of Panamá,* 3, *Smithson. Misc. Collect.,* **150,** 1972:309–356.

PHYTOTOMIDAE

Plantcutter Family (3 species)

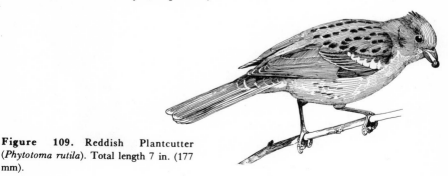

Figure 109. Reddish Plantcutter (*Phytotoma rutila*). Total length 7 in. (177 mm).

Physical characteristics. Length—165 to 177 mm (6.5 to 7 in.). Plumage—gray or brown above, streaked with black; crown and under parts rufous to brick red; wings and tail black, marked with white. Crested. Bill short, heavy, and conical; finely serrated. Body stocky. Wings short, pointed; tail rather long. Legs short; feet large. Sexes unlike.

Range. Western Peru through Chile, Bolivia, and Argentina to about lat. 40°. Habitat—open brush country; cultivated fields and gardens. Migratory in part.

Habits. Usually solitary or in small flocks; sometimes gregarious in nonbreeding season. Flight weak and undulating, not sustained. Voice—loud harsh calls and metallic rasping cries; squeaking and croaking notes.

Food. Fruits, buds, shoots, leaves (also insects?).

Breeding. Round, open nest of twigs lined with fibers, placed in high bushes or in trees.

    Eggs. 2 to 4; bluish green; flecked with black or dark brown. Incubated by ♀.

    Young. Nidicolous. Not otherwise described. Fed by ♂ and ♀.

Technical diagnosis. Ridgway, Pt. 4, 1907:330–331.

Classification. Hellmayr, Pt. 6, 1929:247–250.

References. A. A. Lane, Field-notes on the birds of Chili, *Ibis,* **1897**:35–36.

    R. Barros, La Rara (*Phytotoma rara* Mol.), *An. Zool. Apl.,* **6,** 1919:11–16, pl. 2.

    W. H. Hudson, *Birds of La Plata,* Vol. 1, London, 1920:193–195.

    W. Küchler, Anatomisch Untersuchungen an *Phytotoma rara* Mol., *J. Ornithol.,* **84,** 1936:352–362.

    J. D. Goodall *et al., Las Aves de Chile,* Vol. 1, Buenos Aires, 1946:197–198.

    R. Meyer de Schauensee, *A Guide to the Birds of South America,* Wynnewood, Pa., 1970:327.

TYRANNIDAE

Tyrant-flycatcher Family (374 species)

Figure 110. Eastern Kingbird (*Tyrannus tyrannus*). Total length 8 in. (203 mm).

Physical characteristics. Length—76 to 406 mm (3 to 16 in.). Plumage—typically gray, brown, or olive-green, but some spp. largely black, white, or yellow; a few spp. streaked; many with a partly concealed crown patch of red, yellow, or white. Some spp. crested. Bill extremely varied but usually rather broad, flattened, and slightly hooked; rictal bristles typically well developed. Wings short and rounded to long and pointed, the outer primaries of some spp. attenuate or much shortened; tail in most spp. medium in length and truncate (in a few spp. some tail feathers greatly elongated). Legs and feet small and weak (exc. in terrestrial forms). Sexes alike in most spp.

Range. North America (exc. extreme n.) and S. America. Migratory (exc. most tropical forms).

Habits. Typically solitary and arboreal (in s. S. America some long-legged spp. are terrestrial and, in some cases, gregarious). Many spp. feed by watching from exposed perch, flying out to capture prey in air or on ground. Voice—many distinctive call notes, but song generally not well developed.

Food. Insects; fruit; small mammals, reptiles, amphibians, fish.

Breeding. Cuplike nest in tree or on ground; domed nest in tree; pendent nest with entrance in side or bottom; nest in tree cavity or hole in ground; conical nest attached to reeds; covered nests captured from other spp.

> *Eggs.* 2 to 6; white; spotted and/or streaked, or immaculate. Incubated by ♀, sometimes assisted by ♂.
>
> *Young.* Nidicolous. With down on upper parts. Cared for by ♂ and ♀.

Technical diagnosis. Ridgway, Pt. 4, 1907:335–340.

Classification. Hellmayr, Pt. 5, 1927.

References. Bent, No. 179:11–314.

> A. F. Skutch, Life histories of Central American Birds, II, *Pacific Coast Avifauna,* No. 34, 1960:287–577.
>
> A. F. Skutch, Life histories of Central American highland birds, *Publ. Nuttall Ornithol. Club,* No. 7, 1967:79–102.
>
> R. Meyer de Schauensee, *A Guide to the Birds of South America,* Wynnewood, Pa., 1970:282–326.
>
> A. Wetmore, *The Birds of the Republic of Panamá,* 3, *Smithson. Misc. Collect.,* **150,** 1972:356–601.

Synonymy: Includes *Corythopis* (formerly in Conopophagidae), *Attila, Casiornis, Laniocera,* and *Rhytipterna* (formerly in Cotingidae).

OXYRUNCIDAE

Sharpbill Family (1 species)

Figure 111. Crested Sharpbill (*Oxy-runcus cristatus*). Total length 7 in. (177 mm).

Physical characteristics. Length—165 to 177 mm (6.5 to 7 in.). Plumage—olive-green above; crown brownish black with light barring on forehead and sides of head; a partly concealed median crest of scarlet or orange-red feathers; wings and tail blackish brown, with green edging; under parts white or yellowish white, barred and spotted with brownish black. Bill rather long, straight, and acuminate, with short, fine, bristly feathers at the base. Wings rather long and rounded, the outer primary (at least usually) serrated in ♂; tail moderately long, truncate. Legs short; toes stout and strong; claws acute. Sexes similar, but ♀ may have paler crest.

Range. Costa Rica and Panama; British Guiana; s.e. Brazil and Paraguay. Habitat—humid forest. Nonmigratory?

Habits. Solitary. Flight strong.

Food. Fruit.

Breeding. Nest not described.

Eggs. Not described.

Young. Not described.

Technical diagnosis. Ridgway, Pt. 4, 1907:332.

Classification. Hellmayr, Pt. 6, 1929:1–3; F. M. Chapman, *Am. Mus. Novit.,* No. 1047, 1939.

References. H. L. Clark, Anatomical notes on *Todus, Oxyruncus* and *Spindalis, Auk,* **30,** 1913:402–406.

C. Chubb, *The Birds of British Guiana,* Vol. 2, London, 1921:239–240.

O. Bangs and T. Barbour, Birds from Darien, *Bull. Mus. Comp. Zoöl.,* **65,** 1922:220–221.

F. M. Chapman, The riddle of *Oxyruncus, Am. Mus. Novit.,* No. 1047, 1939.

R. Meyer de Schauensee, *A Guide to the Birds of South America,* Wynnewood, Pa., 1970:326.

A. Wetmore, *The Birds of the Republic of Panamá,* 3, *Smithson. Misc. Collect.,* **150,** 1972:601–605.

Synonym: Oxyrhamphidae.

PITTIDAE

Pitta Family (23 species)

Figure 112. Hooded Pitta (*Pitta sordida*).
Total length 7 in. (177 mm).

Physical characteristics. Length—152 to 280 mm (6 to 11 in.). Plumage—loose-webbed. Most spp. with bright, contrasted coloration; solid patches of scarlet, blue, green, purple, chestnut, tan, white, black. Some spp. heavily barred below, or with white wing patches; some largely dull-colored. Lustrous turquoise-blue on wing coverts and rump of several spp. Some forms with slight bushy crest or with "ear tufts." Bill strong, slightly curved. Neck short; body stout. Wings short and rounded; tail very short. Legs long and strong; feet large. Sexes alike or unlike.

Range. South-central Africa; India, Burma, Thailand, Indochina, s.e. China, s. Japan, East Indies; n. and e. Australia, and e. to Solomons. Habitat—forests or brushland. Some spp. migratory.

Habits. Solitary. Terrestrial, but roost in trees and perch to sing. Flight strong. Hop when on ground. Voice—loud whistling calls; grunting sounds; a whinny.

Food. Insects and other invertebrates; small vertebrates.

Breeding. Loosely constructed, domed nest on ground or in low branches.

Eggs. 2 to 7; white or buffy; speckled and blotched. Incubated by ♂ and ♀.

Young. Nidicolous. Naked. Cared for by ♂ and ♀.

Technical diagnosis. Baker, *Fauna, 3,* 1926:441.

Classification. R. B. Sharpe, *Hand-list of the Genera and Species of Birds,* British Museum, **3,** 1901:179–185 (omit *Mellopitta*).

References. Baker, *Nidification, 3,* 1934:250–260; Bannerman, **4,** 1936:12–17.

D. G. Elliot, *A Monograph of the Pittidae,* London, 1893–1895.

J. Delacour, The first rearing of Pittas in captivity, *Proc. 8th Int. Ornithol. Congr.* **1938:**717–719.

R. E. Moreau, *The Bird Faunas of Africa and Its Islands,* New York, 1966:234–236.

S. Ali and S. D. Ripley, *Handbook of the Birds of India and Pakistan,* Vol. 4, Bombay, 1970:250–257.

714

XENICIDAE

New Zealand Wren Family (4 species)

Figure 113. New Zealand Bushwren (*Xenicus longipes*). Total length 4 in. (102 mm).

Physical characteristics. Length—76 to 102 mm (3 to 4 in.). Plumage—soft; dull green or olive-brown above, darker on head, yellowish on rump; white, gray, or pale purplish brown below; yellowish on sides; wings black, olive-green, or brown, edged with green (barred with yellow in 1 sp.); tail black or olive-green; white superciliary stripe. Bill straight, very slender, and pointed. Wings rather short; tail extremely short, truncate. Legs and toes long and slender, the outer and middle toes joined basally; claws long (esp. on hallux), very acute. Sexes unlike.

Range. New Zealand. Habitat—forest, scrub. Migratory in part.

Habits. Solitary, or in small groups. Mainly arboreal. Flight weak; 1 sp. (now extinct) probably flightless. Very active, running about rocks, tree trunks, and branches in search of insects. Voice—sharp *cheep*; rasping note.

Food. Insects and their larvae; spiders.

Breeding. Nest a rounded structure of leaves, plant fragments, and feathers, with entrance in side, placed in crevice in tree, log, earth bank, or rocks.

  Eggs. 2 to 5; white; immaculate. Incubated by ♂ and ♀.

  Young. Nidicolous. Naked at hatching. Fed by ♂ and ♀.

Technical diagnosis. W. A. Forbes, *Proc. Zool. Soc. London,* **1882**:569–571.

Classification. W. R. B. Oliver, *New Zealand Birds,* Wellington, 1955:447.

References. W. P. Pycraft, Some points in the anatomy of *Acanthidositta chloris,* with some remarks on the systematic position of the genera *Acanthidositta* and *Xenicus, Ibis,* **1905**:603–621.

  J. C. McLean, Field-notes on some of the bush-birds of New Zealand, *Ibis,* **1907**:536–540.

  H. Guthrie-Smith, *Mutton Birds and Other Birds,* London, 1914:123–126.

  W. R. B. Oliver, *New Zealand Birds,* Wellington, 1955:447–457.

  E. G. Turbott, ed., *Buller's Birds of New Zealand,* Honolulu, 1967:60–65.

  R. A. Falla *et al., A Field Guide to the Birds of New Zealand,* Boston, 1967:191–194.

Synonyms: Acanthisittidae, Xenicornithidae, and Traversiidae.

PHILEPITTIDAE

Asity Family (4 species)

Figure 114. Velvet Asity (*Philepitta castanea*). Total length 5.5 in. (139 mm).

Physical characteristics. Length—102 to 165 mm (4 to 6.5 in.). Plumage—soft; the ♂ of 1 sp sooty black, with yellow edge on bend of wing, the other spp. yellow, olive-green, blue above (crown and nape black in ♂ of 1 sp.), yellow or yellowish green below (immaculate or spotted and scaled with darker). Orbital skin bare in ♂, surmounted by bluish or greenish caruncle. Bill moderately long, slender, and slightly curved, to very long, attenuate, sharp-pointed, and strongly curved. Wing of medium length and rounded; tail short and somewhat rounded. Legs and feet large and strong; claws long and acute. Sexes unlike.

Range. Madagascar. Habitat—forest. Nonmigratory.

Habits. Usually solitary or in pairs; sometimes associated with flocks of other spp. Arboreal. Rather torpid. Flight strong but not sustained. Voice (rarely heard)—soft, thrushlike song; a soft hissing note.

Food. Fruit and buds; nectar; insects and spiders.

Breeding. Nest (*Philepitta*) a pear-shaped pendent structure of moss and palm fiber, placed in trees or high bushes.

Eggs. 3; white to bluish white; immaculate. Incubation not described.

Young. Not described.

Technical diagnosis. W. A. Forbes, *Proc. Zool. Soc. London,* **1880**:387–391; D. Amadon, *L'Ois Rev. Franç. Ornithol.* **21,** 1951:59–63. (The spp. are *Philepitta castanea, P. schlegeli, Neodrepanis coruscans, N. hypoxantha.*)

References. A. Milne Edwards and A. Grandidier, Histoire naturelle des oiseaux. In *Histoire physique, naturelle et politique de Madagascar,* Paris, 1876–1885, Vol. 12:288–291, 295–303; Vol. 14: pls. 106–112.

F. Salomonsen, Les Neodrepanis . . . , *L'Ois. Rev. Franç. Ornithol.,* **4,** 1934:1–9.

A. L. Rand, The distribution and habits of Madagascar birds, *Bull. Am. Mus. Nat. Hist.,* **72,** 1936:425–427, 472 (*Neodrepanis*).

Synonym: Paictidae.

MENURIDAE

Lyrebird Family (2 species)

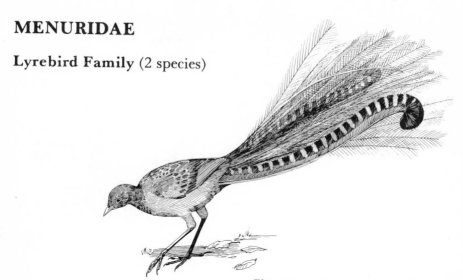

Figure 115. Lyrebird (*Menura superba*).
Total length 38 in. (966 mm).

Physical characteristics. Length—762 to 1016 mm (30 to 40 in.). Plumage—sooty brown above, brownish gray below, rufous brown on chin, throat, and wings. Bare space around eye bluish. Bill elongated, conical, sharp-pointed. Neck long. Wings short and rounded; tail long (elaborate in ♂). Legs and feet large and strong; claws long. Sexes unlike.

Range. Southeastern Australia. Habitat—dense thickets of mountain forests. Nonmigratory.

Habits. Solitary. Terrestrial, but roost in trees. Run rapidly and leap with agility. Fly rarely, but volplane for considerable distances. Build mounds or make scrapes and perform elaborate display dance (♂). Voice—variety of notes and calls. Mimic mechanical sounds as well as voices of other animals.

Food. Mollusca, worms, spiders, insects.

Breeding. Apparently mate for life. Nest a large, roofed structure of sticks and roots lined with bark, roots, and down, placed in hollow stump, under rock ledge, in roots of fallen tree, etc., or (more rarely) up to 60 ft above the ground in a tree.

Eggs. 1; purplish gray; blotched with darker purple and brown. Incubated by ♀.

Young. Nidicolous. Almost naked at hatching; downy later. Cared for by ♀.

Technical diagnosis. Mathews, **7**, 1919:394.

Classification. G. M. Mathews, *Systema Avium Australasianarum,* Pt. 1, London, 1927:425–426.

References. C. Barrett, *Menura*—Australia's Mockingbird, [*N.Y.*] *Zool. Soc. Bull.,* **30,** 1927:207–216.

T. Tregellas, The truth about the Lyrebird, *Emu,* **30,** 1931:243–250, pls. 41–46.

L. H. Smith, *The Lyrebirds of Sherbrooke,* Melbourne, 1951.

N. W. Cayley, *What Bird Is That?,* Sydney, 1966:42–43, col. pl.

H. J. Frith, ed., *Birds in the Australian High Country,* Sydney, 1969:275–278.

ATRICHORNITHIDAE

Scrubbird Family (2 species)

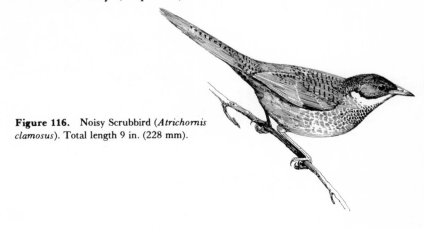

Figure 116. Noisy Scrubbird (*Atrichornis clamosus*). Total length 9 in. (228 mm).

Physical characteristics. Length—165 to 228 mm (6.5 to 9 in.). Plumage—rufous brown above, the feathers finely barred and vermiculated with blackish brown; lighter below (throat and breast white in *clamosus*). Bill rather large. Wings very small; tail long, broad, and slightly graduated. Legs strong; feet large. ♀ smaller than ♂ and somewhat different in coloration.

Range. New S. Wales (*rufescens*); s.w. Australia (*clamosus*). Habitat—dense thickets. Nonmigratory.

Habits. Solitary. Terrestrial. Very active; almost flightless, but run swiftly (holding tail erect). Scratch with feet for food. Very secretive and shy. Voice—notably loud; sharp, shrill, accelerated whistles. Mimic other birds.

Food. Snails' eggs, snails, worms, insects; also seeds.

Breeding. Nest a dome with side entrance, built of grass and leaves, lined with a plaster of wood pulp; placed in a clump of grass.

   Eggs. 2; white or reddish white; with reddish brown markings. Incubated by ♀.

   Young. Nidicolous. With some down on upper parts. Cared for by ♀.

Technical diagnosis. R. B. Sharpe, Cat. of Birds in Brit. Mus., **13**, 1890:659.

Classification. G. M. Mathews, *Systema Avium Australasianarum*, Pt. 2, London, 1930:436.

References. Mathews, **8**, Pt. 1, 1920:22–29, pl. 373.

S. W. Jackson, Second trip to Macpherson Range, South-East Queensland, *Emu,* **20,** 1921:196–203.

H. M. Whittell, The noisy Scrub-bird (*Atrichornis clamosus*), *Emu,* **42,** 1943:217–234.

A. H. Chisholm, The story of the Scrub-birds, *Emu,* **51,** Pt. 2, 1951:89–112, pls. 8–10; Pt. 3, 1952:285–297, pls. 15–16.

N. W. Cayley, *What Bird Is That?,* Sydney, 1966:54–55, col. pl.

D. L. Serventy and H. M. Whittell, *Birds of Western Australia,* Perth, 1967:295–299.

J. Fisher *et al., Wildlife in Danger,* New York, 1969:271.

Synonym: Atrichiidae.

ALAUDIDAE

Lark Family (75 species)

Figure 117. Crested Lark (*Galerida cristata*). Total length 7 in. (177 mm).

Physical characteristics. Length—120 to 228 mm (4.75 to 9 in.). Plumage—typically gray-brown and buff above, marked with dark brown and black in cryptic patterns; paler and less marked below. One sp. all black; others with black areas on head or under parts; many spp. with outer tail feathers white or edged with white; some with crest or "ear tufts." Bill rather long and curved to very short and stout. Wings rather long and, typically, pointed; tail short to medium. Legs short to fairly long; hind claw typically straight, long, and very sharp. Sexes in most spp. alike or nearly so in color, but ♀ smaller.

Range. North America to s. Mexico (Oaxaca); n.w. S. America (mts. of central Colombia); Africa (incl. Madagascar); Eurasia, Philippines, Borneo, Java, Timor, and Australia. Habitat—open, bare areas. Many spp. migratory.

Habits. Often gregarious. Terrestrial. Flight strong in many spp. Walk when on ground. Voice—many spp. have elaborate and beautiful songs, and many a soaring flight song.

Food. Seeds, insects, mollusca.

Breeding. Open or domed nest, almost always built on the ground.

Eggs. 2 to 6; speckled (a few unmarked). Incubated largely or entirely by ♀.

Young. Nidicolous. With thick down, esp. above. Cared for by ♂ and ♀.

Technical diagnosis. Ridgway, Pt. 4, 1907:289.

Classification. Peters, **9**, 1960:3–80.

References. Bent, No. 179:314–371; Witherby, **1**, 1938:163–187; Bannerman, **4**, 1936:xx–xxi, 17–56.

R. Meinertzhagen, Review of the Alaudidae [*Mirafra, Eremopterix*, and *Eremophila* omitted], *Proc. Zool. Soc. London*, **121**, 1951:81–132.

C. W. Mackworth-Praed and C. H. B. Grant, *Birds of the Southern Third of Africa*, London, 1962:605–653.

G. P. Dement'ev *et al.*, *Birds of the Soviet Union*, Vol. 5, Washington, 1970:613–712.

S. Ali and S. D. Ripley, *Handbook of the Birds of India and Pakistan*, Vol. 5, Bombay, 1972:1–48.

HIRUNDINIDAE

Swallow Family (79 species)

Figure 118. Barn Swallow (*Hirundo rustica*). Total length 7 in. (177 mm).

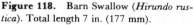

Physical characteristics. Length—95 to 228 mm (3.75 to 9 in.). Plumage—black-brown, dark green, or dark blue (with metallic luster in many spp.); a few spp. strongly streaked, esp. on under parts, and/or with white or buff rump; a number show white areas on spread tail; under parts of many spp. white, chestnut, or gray-brown. Bill short, broad-gaped, and flattened. Neck short; body slender. Wings very long and pointed; tail medium to very long, truncate to deeply forked. Legs very short, tarsi (and even toes) feathered in several spp.; feet very small and weak; front toes more or less united at base. Sexes alike or nearly so in most spp.

Range. Worldwide, exc. extreme n. and some oceanic islands. Migratory.

Habits. Most spp. gregarious. Aerial. Flight very strong and agile. Feed on the wing. Perch, but are barely able to walk. Voice—twittering or squeaking notes; melodious notes or even song in some spp.

Food. Insects; rarely berries.

Breeding. Most spp. colonial or semicolonial. Nest in natural hollows in trees or rocks, or excavate burrows for nests in banks or level ground, or build mud nests (either cup- or retort-shaped).

Eggs. 3 to 7; white; immaculate or speckled. Incubated by ♂ and ♀, or by ♀ alone.

Young. Nidicolous. With some down on upper parts. Cared for by ♂ and ♀.

Technical diagnosis. Ridgway, Pt. 3, 1904:23–24.

Classification. Peters, **9**, 1960:80–129.

References. Bent, No. 179:371–516; Witherby, **2**, 1938:226–241.

R. W. Allen and M. M. Nice, A study of the breeding biology of the Purple Martin (*Progne subis*), *Am. Midl. Nat.*, **47**, 1952:606–665.

W. A. Lunk, The Rough-winged Swallow, *Stelgidopteryx ruficollis* (Vieillot) . . . , *Publ. Nuttall Ornithol. Club*, No. 4, 1962, 155 pp.

P. A. Clancey, *The Birds of Natal and Zululand*, Edinburgh, 1964:299–309.

G. P. Dement'ev *et al., Birds of the Soviet Union*, Vol. 6, Washington, 1968:790–861.

S. Ali and S. D. Ripley, *Handbook of the Birds of India and Pakistan*, Vol. 5, Bombay, 1972:48–77.

CAMPEPHAGIDAE

Cuckoo-shrike Family (70 species)

Figure 119. Great Cuckoo-shrike (*Coracina macei*). Total length 12 in. (305 mm).

Physical characteristics. Length—127 to 355 mm (5 to 14 in.). Plumage—soft, the feathers loosely attached. Bluish or brownish gray, black, blue, red, orange, or yellow, usually in solid areas but strongly barred below in some spp. Under parts white in several spp., chestnut in 2; some spp. have throat and breast gray or black. Tail plain; or tipped or edged with white, yellow, orange, or red. Rump barred or lighter than back in many spp. Bill of medium length, moderately to very heavy, slightly to strongly hooked. Two spp. (*Lobotos*) have orange wattles at gape. Wings medium to long; tail typically graduated (forked in some), long in most spp. Legs short; feet weak to strong. Sexes alike or unlike.

Range. Africa; India to s. and e. China; Japan, Philippines, Malaysia, Papuan Region, Australia; e. to Samoa. Habitat—forests. Largely nonmigratory.

Habits. Often gregarious. Arboreal (1 sp. terrestrial). Flight strong in some spp. but not sustained. Voice (many spp. noisy)—harsh or whistling notes.

Food. Insects, berries.

Breeding. Nest a shallow cup (covered with bark and lichens in many spp.), usually on a horizontal branch.

Eggs. 2 to 4; white, green, or blue; usually speckled and blotched. Incubated by ♂ and ♀.

Young. Nidicolous. Condition at hatching not described. Cared for by ♂ and ♀.

Technical diagnosis. F. J. Jackson, *The Birds of Kenya Colony and the Uganda Protectorate,* Vol. 3, London, 1938:1162–1163.

Classification. Peters, **9,** 1960:167–221.

References. Bannerman, **5,** 1939:xxxii–xxxiii, 303–321, col. pl. 7.

C. W. Mackworth-Praed and C. H. B. Grant, *Birds of Eastern and North Eastern Africa,* London, 1960:555–562.

H. J. Frith, ed., *Birds in the Australian High Country,* Sydney, 1969:288–293.

J. E. duPont, *Philippine Birds,* Greenville, Del., 1971:234–244.

S. Ali and S. D. Ripley, *Handbook of the Birds of India and Pakistan,* Vol. 6, Bombay, 1971:1–47.

Synonym: Campophagidae. The Pericrocotidae are included.

CORVIDAE

Crow Family (103 species)

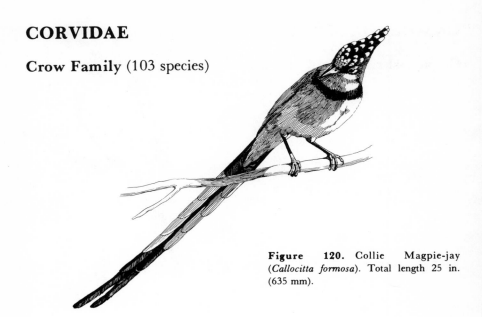

Figure 120. Collie Magpie-jay (*Callocitta formosa*). Total length 25 in. (635 mm).

Physical characteristics. Length—177 to 699 mm (7 to 27.5 in.). The family includes the largest passerine birds. Plumage—black, black and white, or (jays) brightly colored; blue, green, yellow, purple, or brown; usually with large areas of solid color, but wings and tail barred in some. Some have crests, and some have extremely long tails. Bill strong; nostrils usually round, nonoperculate, and shielded by forward-projecting feathers. Wings and tail strong, variable in shape. Tarsi large, strongly scutellated, booted behind. Sexes alike or nearly so.

Range. Worldwide (exc. New Zealand and some oceanic islands). Family best developed in N. Hemisphere. Most spp. nonmigratory.

Habits. Typically gregarious. Flight strong. Bold, aggressive. "Mob" birds of prey. Some walk. Some bury or hide food. Voice—loud, usually harsh calls or croaks; sometimes melodious calls or even songs.

Food. Omnivorous. Often prey on eggs or young of other birds.

Breeding. Usually open nest in trees or on cliffs, but may be covered, or in holes in trees or in ground. Built by ♂ and ♀.

   Eggs. 3 to 10; greenish or white; speckled in most species. Incubated by ♀, or by ♂ and ♀.

   Young. Nidicolous. Down sparse or absent. Fed by ♂ and ♀.

Technical diagnosis. Ridgway, Pt. 3, 1904:252–254.

Classification. Peters, **15,** 1962:204–282.

References. Bent, No. 191:1–322; Witherby, **1,** 1938:7–39.

   C. W. Mackworth-Praed and C. H. B. Grant, *Birds of Eastern and North Eastern Africa,* London, 1960:670–680.

   G. P. Dement'ev *et al., Birds of the Soviet Union,* Vol. 5, Washington, 1970:13–119.

   S. Ali and S. D. Ripley, *Handbook of the Birds of India and Pakistan,* Vol. 5, Bombay, 1972:198–266.

   R. P. Balda and G. C. Bateman, The breeding biology of the Pinon Jay, *Living Bird,* **1972:**5–42.

CRACTICIDAE

Bellmagpie Family (8 species)

Figure 121. New Guinea Forest Butcherbird (*Cracticus cassicus*). Total length 14 in. (355 mm).

Physical characteristics.   Length—260 to 584 mm (10.25 to 23 in.). Plumage—black, or black (or gray) and white; some brown phases. Bill large, very stout, and slightly to strongly hooked in most spp. Head large; body compact. Wings rather short to long and pointed; legs strong, medium to long. Sexes alike or unlike.

Range.   Australia, including Tasmania; New Guinea and adjacent islands. Nonmigratory.

Habits.   Usually gregarious. Arboreal but often feed on ground. Flight strong. Some impale food on thorns or wedge it in forked branches before tearing it apart, or store it in this way. Voice (most spp. vociferous)—loud metallic notes.

Food.   Large insects, small vertebrates; also fruit.

Breeding.   Rather large, open cup of twigs, well lined with grass and rootlets, placed high in a tree.

Eggs.   2 to 5; highly variable: pale blue or green to olive, or pink to reddish brown; usually heavily marked with dark colors. Incubated by ♀. *Gymnorhina t. dorsalis* "more or less promiscuous."

Young.   Nidicolous. Downy (?). Cared for by ♂ and ♀.

Technical diagnosis.   J. A. Leach, *Emu,* **14,** 1914:2–38, pls. 1–3; see also Ridgway, Pt. 3, 1904:253.

Classification.   Peters, **15,** 1962:166–172.

References.   Mathews, 10, Pts. 5–7, 1923:334–434, col. pls. 483–490.

Hugh Wilson, The life history of the Western Magpie (*Gymnorhina dorsalis*), *Emu,* **45,** 1946:233–244, 271–286.

G. M. Storr, Remarks on the Streperidae, *S. Aust. Ornithol.,* **20,** 1952:78–80.

D. Amadon, Further notes on the Cracticidae, *S. Aust. Ornithol.,* **21,** 1953:6–7.

D. L. Serventy and H. M. Whittell, *Birds of Western Australia,* Perth, 1967:413–419.

H. J. Frith, ed., *Birds in the Australian High Country,* Sydney, 1969:457–465.

Synonym: Streperidae.

PTILONORHYNCHIDAE

Bowerbird Family (18 species)

Figure 122. Satin Bowerbird (*Ptilonorhynchus violaceus*). Total length 13 in. (330 mm).

Physical characteristics. Length—228 to 368 mm (9 to 14.5 in.). Plumage—black, gray, brown, green, yellow, orange, lavender—plain or in bold combinations; some spp. spotted. Some spp. have brilliant crest (ranging from several long, narrow crown feathers or a nuchal ruff to a "mane" or cape over most of the back). Bill stout, straight to rather curved, slightly hooked (notched in 3 spp.). Wings short to medium, rounded; tail rather short to long, rounded, truncate, or emarginate. Legs rather short; legs and feet stout. Sexes unlike in most spp.

Range. New Guinea and adjacent islands; n. and e. Australia. Habitat—forests. Nonmigratory.

Habits. Largely solitary. Terrestrial, but nest and feed in trees. Flight swift. In most spp. ♂ ♂ build elaborate "bowers" or "playgrounds" of twigs, decorated with flowers, berries, bits of glass, etc. Voice—variety of ringing calls. Mimic other birds, other animals, and mechanical sounds.

Food. Fruit, berries, seeds, mollusca, insects.

Breeding. Nest a shallow or cup-shaped structure of twigs, sometimes lined with grass or leaves, placed in trees.

Eggs. 1 to 3 (usually 2); white to buff or greenish; spotted and scrawled or immaculate. Incubated by ♀.

Young. Nidicolous. Downy. Cared for by ♀ (assisted by ♂ in some spp.).

Technical diagnosis. C. R. Stonor, *Proc. Zool. Soc. London,* **107B,** 1937:475–490.

Classification. E. T. Gilliard, *Birds of Paradise and Bower Birds,* London, 1969:62–63.

References. A. J. Marshall, *Bower-birds,* London, 1954.

E. T. Gilliard, The courtship behavior of Sanford's Bowerbird (*Archboldia sanfordi*), *Am. Mus. Novit.* No. 1935, 1959.

N. Chaffer, Australia's amazing Bowerbirds, *Natl. Geogr. Mag.,* Dec. 1961:866–873.

H. J. Frith, ed., *Birds in the Australian High Country,* Sydney, 1969:470–472.

Synonymy: Sometimes included in Paradisaeidae.

PARADISAEIDAE

Bird-of-paradise Family
(42 species)

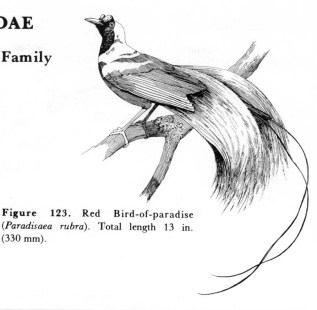

Figure 123. Red Bird-of-paradise (*Paradisaea rubra*). Total length 13 in. (330 mm).

Physical characteristics. Length—139 to 1067 mm (5.5 to 42 in.). Plumage—black with brilliant metallic gloss, or bold combinations of velvet-textured black or brown, and red, yellow, orange, green, blue, purple, white; with spectacular erectile feathers of highly varied and extreme specialization: greatly elongated head or tail wires or plumes; enormous plumed "fans" arising from nape, breast, or flanks. Some spp. have wattles or bare spots on head. Bill medium (rather heavy in some spp.) to long, slender, and sickle-shaped; hooked in some spp. Wings medium, rather rounded; tail short and square to extremely long and graduated. Legs rather short; legs and feet stout. Sexes unlike.

Range. Moluccas; New Guinea and adjacent islands; n. and e. Australia. Habitat—forests. Nonmigratory.

Habits. Largely solitary. Arboreal. Flight slow (swift in a few spp.), not prolonged. In some spp. ♂ ♂ clear vegetation from forest areas for display grounds. ♂ ♂ have spectacular display of erectile plumes, accompanied by elaborate acrobatics. Voice—loud, shrill calls; harsh shrieks; prolonged whistles.

Food. Fruit, berries, seeds, insects, frogs, lizards.

Breeding. Some spp. polygamous? Nest a cup of plant fragments, placed in tree or tree cavity.
    Eggs. 2; pinkish white or brown; longitudinally streaked. Incubated by ♀.
    Young. Nidicolous. Downy or naked. Cared for by ♀, or by ♂ and ♀.

Technical diagnosis. C. R. Stonor, *Proc. Zool. Soc. London,* **107B,** 1937:475–490.

Classification. E. T. Gilliard, *Birds of Paradise and Bower Birds,* London, 1969:62–63.

References. A. L. Rand, On the breeding habits of some Birds of Paradise in the wild, *Am. Mus. Novit.,* No. 993, 1938.

    E. Mayr, Birds of Paradise, *Nat. Hist.,* June 1945.

    [S.] Dillon Ripley, Strange courtship of Birds of Paradise, *Natl. Geogr. Mag.,* **97,** 1950:247–278, 16 col. pls.

    N. W. Cayley, *What Bird Is That?,* Sydney, 1966:13–14, 16.

    E. T. Gilliard and M. Lecroy, Annotated list of birds of the Adelbert Mountains, New Guinea, *Bull. Am. Mus. Nat. Hist.,* **138,** 1967:51–82.

    E. T. Gilliard, *Birds of Paradise and Bower Birds,* London, 1969.

Synonymy: The Epimachidae are included.

GRALLINIDAE

Mudnest-builder Family (4 species)

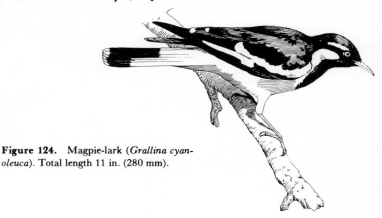

Figure 124. Magpie-lark (*Grallina cyan-oleuca*). Total length 11 in. (280 mm).

Physical characteristics. Length—190 to 502 mm (7.5 to 19.75 in.). Plumage—black, boldly marked with white; or dark bluish and brownish gray, unmarked. Bill very short and stout to long, slender, and curved. Neck short. Wings short and rounded to long and pointed; tail short and truncate to very long and rounded. Legs medium to rather long and stout. Sexes alike or unlike.

Range. Australia and n.w. New Guinea. Some spp. migratory.

Habits. Gregarious. Flight not strong. Jump from branch to branch in trees. Feed largely on the ground. Voice—melodious whistling notes; harsh cries; a plaintive *peewit*.

Food. Insects, seeds.

Breeding. Communal nesting habits (exc. *Grallina*). Nest a deep bowl made of mud and lined with grass and feathers; usually placed on high, horizontal limbs of trees.

    Eggs. 3 to 5; creamy white to pink; marked with brown, purplish brown, or gray. Incubated by ♂ and ♀.

    Young. Nidicolous. With some down. Cared for by ♂ and ♀.

Technical diagnosis. D. Amadon, Australian Mud Nest Builders, *Emu,* **50,** 1950:123–127.

Classification. Peters, **15,** 1962:159–160.

References. H. A. C. Leach, Notes on the White-Winged Chough, *Emu,* **29,** 1929:130–132, pls. 24–25.

    Angus Robinson, Magpie-Larks—A study in behaviour, *Emu,* **46,** 1947:265–281, 382–391; **47,** 1947:11–28, 147–153.

    D. L. Serventy and H. M. Whittell, *Birds of Western Australia,* Perth, 1967:405–407.

    H. J. Frith, ed., *Birds in the Australian High Country,* Sydney, 1969:445–452.

Synonymy: The Corcoraciidae and Struthiididae are included.

PARIDAE

Titmouse Family (64 species)

Figure 125. Green-backed Tit (*Parus monticolus*). Total length 5 in. (127 mm).

Physical characteristics. Length—76 to 203 mm (3 to 8 in.). Plumage—typically long, soft, and thick. In combinations (blended or bold) of gray, brown, yellow, orange, olive-green, gray-blue, wine color, black, white. A few spp. all black or largely white; none spotted, streaked, or barred; but many marked broadly with black on head and under parts. A few spp. crested. Bill rather small; stout in many spp., attenuate and sharp in some. Wings rounded, short to medium; tail very short and truncate to long and graduate. Legs short but strong. Sexes alike in most spp.

Range. North America (exc. extreme n.), s. to Guatemala; Old World (exc. Madagascar, extreme n. Asia, New Guinea, Australia, and Polynesia). Habitat—dense forest to desert brush. Most spp. nonmigratory.

Habits. Rather gregarious. Arboreal. Very restless and active. Flight weak. Voice—chattering or lisping notes; whistled calls.

Food. Insects, other invertebrates; also nuts, seeds.

Breeding. Rather bulky nest of fur, moss, etc., in hole in tree, wall, earth- or rock-bank; or pendent, feltlike pouch with side entrance; or covered nest (with entrance near the top) of feathers, moss, etc., in bush or tree.

Eggs. 4 to 14; white or pinkish white; immaculate to heavily marked. Incubated in most spp. by ♀ alone.

Young. Nidicolous. Downy or naked at hatching. Cared for by ♂ and ♀.

Technical diagnosis. Ridgway, Pt. 3, 1904:375–378.

Classification. Peters, **12,** 1967:52–124 (Aegithalidae, Remizidae, and Paridae).

References. Bent, No. 191:322–460; Witherby, **1,** 1938:244–274; Bannerman, **6,** 1948:xx–xxi, 3–19.

> H. N. Kluijver, The population ecology of the Great Tit, *Parus m. major* L., *Ardea,* **39,** 1951:1–135.

> K. L. Dixon, An ecological analysis of the interbreeding of Crested Titmice in Texas, *Univ. Calif. Publ. Zool.,* **54,** No. 3, 1955:125–206.

> G. P. Dement'ev *et al., Birds of the Soviet Union,* Vol. 5, Washington, 1970:870–941.

Synonymy: The Remizidae (Remizinae) and Aegithalidae (Psaltriparinae) are tentatively included.

CERTHIIDAE

Creeper Family (6 species)

Figure 126. Brown Creeper (*Certhia familiaris*). Total length 5.25 in. (133 mm).

Physical characteristics. Length—120 to 177 mm (4.75 to 7 in.). Plumage—brown to black above (streaked, barred, and spotted with white, buff, or darker brown) and white, gray, or buff below; olive-brown to blackish, streaked below with white; black, spotted with white (1 sp.); or gray above and white below, with large crimson area on wing (1 sp.). Bill rather short to long; slender and laterally compressed; almost straight to strongly decurved. Wings long, rounded or pointed; tail long and graduated, with stiff pointed tips, or short, rounded, and soft. Legs very slender to stout; toes long; claws curved and sharp, very long (esp. on hallux) in most spp. Sexes alike or nearly so.

Range. Holarctic, Africa, India. Some spp. migratory.

Habits. Solitary. Most spp. arboreal. Flight strong. Creep about trunks and branches of trees, over rocks, cliffs, and walls, hunting food in crevices. Voice—soft cheeping call; harsh piping notes; and (in most spp.) a clear sweet song.

Food. Mainly insects and spiders; also seeds.

Breeding. Nest a pad of moss, grass, and bark, lined with hair, feathers, etc., placed in trees under overhanging bark, in rock crevices, etc.; or (*Salpornis*) a lichen-decorated cup of plant fragments placed on horizontal branch.

> *Eggs.* 2 to 9 (usually 5 or 6); white or flesh; spotted with brown. Incubated by ♀ (sometimes assisted by ♂).

> *Young.* Nidicolous. Downy. Fed by ♂ and ♀.

Technical diagnosis. Witherby, **1**, 1938:234.

Classification. Peters, **12**, 1967:150–160.

References. Bent, No. 195:56–79; Witherby, **1**, 1938:234–240; Baker, *Fauna,* **1**, 1922:428–440; Bannerman, **6**, 1948:265–270.

> G. P. Dement'ev *et al., Birds of the Soviet Union,* Vol. 5, Washington, 1970:838–851.

> S. Ali and S. D. Ripley, *Handbook of the Birds of India and Pakistan,* Vol. 9, Bombay, 1973:231–241.

RHABDORNITHIDAE

Philippine Creeper Family (2 species)

Figure 127. Striped-headed Creeper (*Rhabdornis mystacalis*). Total length 5.5 in. (139 mm).

Physical characteristics. Length—139 to 152 mm (5.5 to 6 in.). Plumage—top of head and upper back black streaked with white; lower back, wings, and tail gray with white shaft streaks; lower parts white heavily streaked with black on flanks. Or brown above and white with brown stripes below. Cheeks and sides of neck black, brown, or gray. Iris brown. Bill black, long and strong; tongue brush-tipped. Tail nearly square. Legs and feet long and strong. Sexes unlike.

Range. Philippines. Habitat—deep woods and second growth.

Habits. Feed creeperlike on bark but also feed among flowers like sunbirds (*Arachnothera*); early collectors called this bird a "flowercreeper."

Food. Presumably only insects.

Breeding. Said to nest in holes in trees; nesting habits not otherwise described.

Technical diagnosis. R. C. McGregor, *A Manual of Philippine Birds,* 1909:612–613.

Classification. Peters, **12,** 1967:161–162.

References. J. Delacour and E. Mayr, *Birds of the Philippines,* New York, 1946:220–221.

A. L. Rand, Five new birds from the Philippines, *Fieldiana: Zool.,* **31,** No. 25, 1948:204.

A. L. Rand and D. S. Rabor, Birds of the Philippine Islands . . . , *Fieldiana: Zool.,* **35,** No. 7, 1960:436–437.

J. E. duPont, *Philippine Birds,* Greenville, Del., 1971:258–262.

J. E. duPont and D. S. Rabor, Birds of Dinagat and Siargao, Philippines, *Nemouria,* **10,** 1973:75.

CLIMACTERIDAE

Australian Treecreeper Family (6 species)

Figure 128. Red-browed Treecreeper (*Climacteris erythrops*). Total length 5.75 in. (146 mm).

Physical characteristics. Length—127 to 200 mm (5 to 8 in.). Plumage—brown above, often darker on upper tail coverts (blackish brown tail with black subterminal bar in 1 sp.); white or rusty red eyebrow line and colored wing bar; under parts white to buff, gray, black, or brown, conspicuously streaked with black, white, or brown, either on throat or on breast and flanks. Iris light brown to reddish brown. Bill long, slender, decurved. Tail short, square. Legs and toes long and strong. Sexes unlike.

Range. Australia, New Guinea.

Habits. Solitary. Primarily arboreal, but some feed on the ground. Creep on trunks and larger branches of trees, nearly always working upward on trunk or in spiral around larger limbs. Wing beat rapid; frequently glide. Voice—loud piercing calls; piping or chattering notes; a loud "spink, spink."

Food. Insects, spiders, etc.

Breeding. Nest of hair, feathers, moss, soft bark, and dried grasses, placed in hollow branch or trunk. Built by ?

> *Eggs.* 2 or 3; creamy white, pinkish white, or purplish red; sparingly to profusely spotted with reddish brown, lilac, or dull purple. Incubated by?
>
> *Young.* Nidicolous. Downy? Fed by ♂ and ♀ (?)

Technical diagnosis. A. Keast, Variation and speciation in the genus *Climacteris* Temminck. *Aust. J. Zool.*, **5**, 1957:476–478.

Classification. Peters, **12**, 1967:162–166.

References. A. Keast, Bird speciation on the Australian continent, *Bull. Mus. Comp. Zool.*, **123**, No. 8, 1961:373–375.

N. W. Cayley, *What Bird Is That?*, Sydney, 1966:131–135.

D. L. Serventy and H. M. Whittell, *Birds of Western Australia*, Perth, 1967:369–371.

A. L. Rand and E. T. Gilliard, *Handbook of New Guinea Birds*, New York, 1967:522–523.

H. J. Frith, ed., *Birds in the Australian High Country*, Sydney, 1969:269, 367, 374–378.

H. Frauca, *Australian Bush Birds*, Melbourne, 1971:81.

SITTIDAE

Common Nuthatch Family (22 species)

Figure 129. Red-breasted Nuthatch (*Sitta canadensis*). Total length 4.5 in. (114 mm).

Physical characteristics. Length—95 to 190 mm (3.75 to 7.5 in.). Plumage—gray to blue-gray or blue above; top of head (or forehead) often black or brown; usually a dark eye-line. Under parts unstreaked; light, usually with some brown (pale buff to rich chestnut), esp. posteriorly. Tail usually marked with white. Slender, straight, unnotched bill. Nostrils nonoperculate, partly covered with forward-projecting feathers. Wings rather long, pointed; tail short, truncate. Tarsi short; toes (especially hind toe) long; claws laterally compressed. Sexes alike or nearly so; immatures like adults.

Range. North America and Eurasia exc. extreme n., Malaysia (exc. Celebes), Japan, Formosa, and Philippines. Most spp. nonmigratory.

Habits. Typically solitary (some spp. flock). Arboreal (exc. 2 spp. of "rock nuthatches"). Climb with short, jerky hops—upward or head downward. Flight undulating. Voice—simple call notes; song usually a rhythmic repetition of similar notes.

Food. Insects, nuts, seeds.

Breeding. Nest in lined tree or rock cavity with entrance often reduced to small opening by plaster of pitch or clay (rock nuthatches build a cone-shaped plaster projection extending 6 or 8 in. from rock face).

Eggs. 4 to 12; white; with rufous spots. Incubated by ♀ (sometimes also by ♂?).

Young. Nidicolous. With long, sparse down. Cared for by ♂ and ♀.

Technical diagnosis. Ridgway, Pt. 3, 1904:436–439.

Classification. Peters, **12,** 1967:125–145, 149.

References. Bent, No. 195:1–55; Witherby, **1,** 1938:240–244; Baker, *Nidification, 1,* 1932:89–100.

C. Vaurie, *The Birds of the Palearctic Fauna: Passeriformes,* London, 1959:519–535.

S. Ali, *The Birds of Sikkim,* London, 1962:347–354.

G. P. Dement'ev *et al., Birds of the Soviet Union,* Vol. 5, Washington, 1970:835–838, 851–870.

Synonymy: The Tichodromadinae are included.

NEOSITTIDAE

Australian Nuthatch Family
(3 species)

Figure 130. Orange-winged Treerunner (*Neositta chrysoptera*). Total length 4.25 in. (108 mm).

Physical characteristics. Length—102 to 120 mm (4 to 4.75 in.). Plumage—black, with pink lateral areas on tail (*Daphoenositta*); or gray above, streaked with brown, with crown white or black, rump and under parts white—the latter streaked in some spp.—and outer rectrices tipped with white. All spp. have a broad white or buffy band across the (spread) wing. Bill laterally compressed, slightly hooked, and notched; nostrils exposed, operculate. Bristles about base of bill, characteristic of other families of nuthatches, absent or only slightly developed. Wings long; tail medium. Sexes unlike in some spp.; immature spotted.

Range. New Guinea and Australia. Nonmigratory.

Habits. Gregarious (commonly found in small flocks). Arboreal. Very active, running upward or head downward along tree trunks. Flight undulating. Voice—soft, twittering notes, frequently uttered; "mournful, monotonous cries" also described.

Food. Insects, spiders.

Breeding. Nest (of *Neositta* ["*Sittella*"]), built by ♂ and ♀ of pair (often aided by several other adults), deep, well constructed, cup-shaped (of spider webs and cocoons, or wool and hair, with outer covering of flakes of bark), placed in upright fork of tree, often 50 ft or more above ground.

　Eggs. 3 (rarely 4); grayish white; speckled and blotched with brown and black. Incubated by ♂ and ♀ of pair (assisted by other adults?).

　Young. Nidicolous. Otherwise undescribed. Fed by both parents (and sometimes as many as 4 other adults).

Technical diagnosis. A. L. Rand, *Auk,* **53,** 1936:309.

Classification. Peters, **12,** 1967:145–149.

References. A. J. Campbell, *Nests and Eggs of Australian Birds,* Pt. 1, Sheffield, England, 1901:337–344, 1 pl.

　A. L. Rand, The rediscovery of the Nuthatch *Daphaenositta* with notes on its affinities, *Auk,* **53,** 1936:306–310.

　H. J. Frith, ed., *Birds in the Australian High Country,* Sydney, 1969:378–379.

Synonym: Daphoenosittinae. These birds are sometimes included in the Sittidae.

TIMALIIDAE

Babbler Family (282 species)

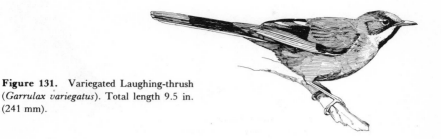

Figure 131. Variegated Laughing-thrush (*Garrulax variegatus*). Total length 9.5 in. (241 mm).

Physical characteristics. Length—89 to 406 mm (3.5 to 16 in.). Plumage—soft, lax, typically long and thick on lower back. Gray, buff, or chestnut (some spp. with much olive-green, black, white, or even yellow and red), usually in solid areas, but some spp. streaked or scaled, esp. below, some with bold markings on head and neck. A few spp. crested. Bill very small and weak to long and sickle-shaped; or short, laterally compressed, and massive; or long and straight; or rather short and swollen; hooked in many spp. Wings short and rounded; tail very short to long, truncate to graduated. Legs and feet strong. Sexes alike or unlike.

Range. Europe, Africa (incl. Madagascar), s. Asia, Malaysia, Papuan Region, Philippines, Australia. Largely nonmigratory.

Habits. Gregarious or solitary. Arboreal (some spp. largely terrestrial). Flight weak. Voice (many spp. very noisy)—harsh calls to rich, musical songs.

Food. Insects and other small animals; fruit.

Breeding. Cuplike, or domed, nest, with side entrance; on ground (or in holes in ground or in bank), in grass, reeds, bushes, trees; or mud nest on rock ledge.

Eggs. 2 to 7; white, green, blue, or pink; immaculate in many spp. Incubated by ♂ and ♀, or by ♀ alone.

Young. Nidicolous. Naked or downy. Cared for by ♂ and ♀.

Technical diagnosis. Baker, *Fauna,* 2nd ed., **1,** 1922:134–135.

Classification. Peters, **10,** 1964:228–429, 431–442.

References. Bannerman, **4,** 1936:88–130, **6,** 1948:113–120.

D. L. Serventy and H. M. Whittell, *Birds of Western Australia,* Perth, 1967:309–312.

J. E. duPont, *Philippine Birds,* Greenville, Del., 1971:262–276.

S. Ali and S. D. Ripley, *Handbook of the Birds of India and Pakistan,* Bombay, Vol. 6, 1971:114–238; Vol. 7, 1972:1–135.

Synonym: Timeliidae. The Cinclosomatidae, Eupetidae, Illadopsidae, Leiotrichidae, Liotrichidae, Orthonycidae, Orthonychinae, Paradoxornithidae, Picathartidae, and Turdoididae are included.

DICRURIDAE

Drongo Family (20 species)

Figure 132. Greater Racket-tailed Drongo (*Dicrurus paradiseus*). Total length 25 in. (635 mm).

Physical characteristics. Length—177 to 635 mm (7 to 25 in.). Plumage—black (typically with a high greenish, bluish, or purplish luster) or gray; in solid colors, but variously marked by ornamental specialization of feather structure; extremely elongate, hairlike feathers in crest; "spangles" on head, chest, and throat feathers; glossy hackles; elaborated tail. White on face or belly in a few forms. Frontal crest in some spp. Eyes red in most spp. Bill stout, somewhat hooked and notched, the culmen arched. Wings long; tail medium to extremely long, truncate (rarely) to deeply forked (in many spp. with outer feathers curved, curled, and/or racquet-tipped). Legs short; toes and claws stout. Sexes alike but ♀ slightly smaller.

Range. Africa (s. of Sahara), incl. Madagascar; India, Burma, Thailand, Indochina, s. and e. China, Philippines, Malaysia, Papuan Region, n. and e. Australia, Solomons. A few forms migratory.

Habits. Solitary. Arboreal. Flight strong but not sustained. Feed chiefly on insects captured on the wing. Very pugnacious. Voice—variety of calls; melodious songs. Mimic other birds.

Food. Insects, nectar.

Breeding. Frail saucerlike nest, usually semipendent in horizontal fork of tree.

> *Eggs.* 2 to 4; white or colored; immaculate or blotched and speckled. Incubated by ♀, sometimes assisted by ♂.
>
> *Young.* Nidicolous. Naked. Cared for by ♂ and ♀.

Technical diagnosis. Baker, *Fauna,* **2,** 1924:352.

Classification. Peters, **15,** 1962:137–157.

References. Baker, *Nidification,* **2,** 1933:316–350; Bannerman, **5,** 1939:321–336.

> C. W. Mackworth-Praed and C. H. B. Grant, *Birds of Eastern and North Eastern Africa,* London, 1960:562–567.
>
> J. E. duPont, *Philippine Birds,* Greenville, Del., 1971:244–248.
>
> S. Ali and S. D. Ripley, *Handbook of the Birds of India and Pakistan,* Vol. 5, Bombay, 1972:113–143.

Synonym: Edoliidae.

ORIOLIDAE

Oriole Family (28 species)

Figure 133. Golden Oriole (*Oriolus oriolus*). Total length 9.5 in. (241 mm).

Physical characteristics. Length—177 to 305 mm (7 to 12 in.). Plumage—largely yellow, olive-green, red, brown, or black; part or all of head black in most spp.; wings dark; tail partly black in most spp. Some spp. (or plumages) heavily streaked, chiefly below. Lores and orbital region feathered in *Oriolus*, bare in *Sphecotheres*. Bill red or blue in *Oriolus*, black in *Sphecotheres*. Bill strong, pointed, and slightly hooked; rather short to long. Wings long and pointed; tail medium to rather long. Legs strong but short. Sexes unlike in most spp.

Range. Africa, Eurasia (exc. n. part), East Indies, Philippines, n. and e. Australia. A few forms migratory.

Habits. Solitary. Arboreal. Flight swift, undulating. Usually wary. Voice—loud, melodious, flutelike notes; harsh alarm notes.

Food. Insects, fruit.

Breeding. Semipendent, cup-shaped nest, usually high in tree; or (*Sphecotheres*) frail, saucer-shaped nest.

> *Eggs.* 2 to 5; white or pinkish (greenish in *Sphecotheres*); strongly marked with brown and black. Incubated by ♀, assisted by ♂.
>
> *Young.* Nidicolous. With some down. Cared for by ♂ and ♀.

Technical diagnosis. Witherby, **1**, 1938:47–48.

Classification. Peters, **15**, 1962:122–137.

References. Baker, *Nidification*, **2**, 1933:498–505; Bannerman, **5**, 1939:450–465.

> G. P. Dement'ev *et al.*, *Birds of the Soviet Union*, Vol. 5, Washington, 1970:143–183.
>
> J. E. duPont, *Philippine Birds*, Greenville, Del., 1971:248–251.
>
> S. Ali and S. D. Ripley, *Handbook of the Birds of India and Pakistan*, Vol. 5, Bombay, 1972:101–113.

Synonymy: The Sphecotheridae are included.

PYCNONOTIDAE

Bulbul Family (119 species)

Figure 134. Chinese Bulbul (*Pycnonotus sinensis*). Total length 7.5 in. (190 mm).

Physical characteristics. Length—139 to 286 mm (5.5 to 11.25 in.). Plumage—soft and long, esp. on lower back. Gray, brown, olive-green, sometimes with yellow, red, white, or black in bold contrast. Some spp. streaked or spotted (esp. below). Many have strong head pattern and/or brightly colored under tail coverts. Tail white-tipped in some spp. Some spp. crested. Hairlike feathers on nape and rictal bristles usually well developed. Bill short to medium in length; slightly curved; slender to stout; in some spp. hooked, or hooked and notched. Neck short. Wings short to medium, rounded; tail medium to long, truncate to rounded (forked in *Hypsipetes*). Legs short. Sexes similar, but ♂ larger in a few spp.

Range. Africa (incl. Madagascar and Mascarene Ids.), s. Asia, Malaysia, Moluccas, and Philippines. Habitat—forests, brush, or gardens. Northern forms migratory.

Habits. Largely gregarious (often in mixed flocks). Mostly arboreal. Restless, agile. Flight not strong. Voice (most spp. garrulous)—a variety of short, loud notes and whistles, harsh or musical; some spp. have well-developed song. Some spp. mimic other birds.

Food. Berries, fruit, insects.

Breeding. Nest rather insubstantial; shallow, cup-shaped, or semipensile; its elevation varying from ground to high in tree.

Eggs. 2 to 4; pink, cream, or white; speckled and blotched. Incubated by ♂ and ♀.

Young. Nidicolous. Without down. Cared for by ♂ and ♀.

Technical diagnosis. J. Delacour, *Zoologica,* **28,** Pt. 1, 1943:17–20.

Classification. Peters, **9,** 1960:221–300.

References. Baker, *Nidification,* **1,** 1932:333–409; Bannerman, **4,** 1936:xxvi–xxvii, 130–198.

J. Delacour, *Birds of Malaysia,* New York, 1947:216–227.

C. W. Mackworth-Praed and C. H. B. Grant, *Birds of the Southern Third of Africa,* London, 1963:33–63.

S. Ali and S. D. Ripley, *Handbook of the Birds of India and Pakistan,* Vol. 6, Bombay, 1971:65–114.

J. E. duPont, *Philippine Birds,* Greenville, Del., 1971:276–286.

Synonym: Brachypodidae. The Tylidae are included.

IRENIDAE

Leafbird Family (14 species)

Figure 135. Orange-bellied Leafbird (*Chloropsis hardwickii*). Total length 7.5 in. (190 mm).

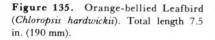

Physical characteristics. Length—120 to 241 mm (4.75 to 9.5 in.). Plumage—bright grass-green and black (or brown), with areas of blue and yellow; or olive-green and black with yellow areas and white markings; or glossy blue and black. Elongated hairlike feathers on nape in some spp. Bill fairly long, somewhat curved, slightly hooked (notched in some spp.). Wings short to medium, rounded; tail short to fairly long, square or slightly rounded, the tail coverts much elongated in some spp. Feet rather small. Sexes unlike.

Range. India, Burma, Thailand, Indochina, s. China, Philippines, Malaysia. Habitat—forests and second growth. Nonmigratory except for local movements.

Habits. Most spp. gregarious. Arboreal. Flight swift. Active. Typically shy and retiring. *Aegithina tiphia* has acrobatic aerial courtship display. Voice—chattering and flutelike notes; shrill musical whistles; loud, clear song. Some spp. mimic other birds.

Food. Mainly fruit; berries, seeds, and buds; insects.

Breeding. Nest a neat, lined cup or loosely constructed shallow saucer, placed in tree (often at great heights) or in scrub.

    Eggs. 2 to 4; gray, cream, or pink; marked with brown. Incubated by ♂ and ♀ (at least in *Aegithina tiphia*).

    Young. Not described? Cared for by ♂ and ♀.

Technical diagnosis. H. C. Oberholser, *J. Wash. Acad. Sci.,* **7,** 1917:538 (*Irena* only).

Classification. Peters, **9,** 1960:300–308.

References. Baker, *Nidification,* **1,** 1932:315–329; **2,** 1933:496–498.

    H. C. Robinson and F. N. Chasen, *The Birds of the Malay Peninsula,* London, Vol. 1, 1927:213–216; Vol. 2, 1928:171–176; Vol. 4, 1939:265–272.

    B. E. Smythies, *The Birds of Borneo,* Edinburgh, 1960:370–374.

    S. Ali and S. D. Ripley, *Handbook of the Birds of India and Pakistan,* Vol. 6, Bombay, 1971:47–65.

    J. E. duPont, *Philippine Birds,* Greenville, Del., 1971:286–290.

Synonym: Aegithinidae. The Phyllornithidae are included.

TROGLODYTIDAE

Wren Family (59 species)

Figure 136. Gray-breasted Woodwren (*Henicorhina leucophrys*). Total length 4.5 in. (114 mm).

Physical characteristics. Length—95 to 222 mm (3.75 to 8.75 in.). Plumage—reddish brown, grayish brown, olive-brown, or blackish brown, often with white or chestnut areas, and typically barred, streaked, and/or spotted with white, buff, dark brown, or black, esp. on wings, tail, and sides. Bill rather slender, medium to long and curved. Wings short and rounded; tail short to long. Legs and feet strong; anterior toes partly adherent; claws long. Sexes alike or nearly so.

Range. N. America (exc. extreme n.), S. America, and (1 sp.) Europe (exc. extreme n.), Asia (exc. extreme n. and extreme s.), and n.w. Africa (Tunis to Morocco). Northern forms migratory.

Habits. Typically solitary; certain tropical spp. somewhat gregarious when not breeding. Flight weak. Very active and inquisitive. Tail commonly carried erect. Voice—song highly developed; in many spp. the ♀ also sings, and antiphonal singing is recorded.

Food. Insects, spiders, etc.

Breeding. Polygamy frequent in at least some spp. Nest in cavities in trees, rocks, buildings, or banks; or build covered nests in grass, reeds, bushes, or trees; ♂ ♂ may make several unlined nests, not used for breeding ("cock nests").

Eggs. 2 to 10; white or brown; speckled or immaculate. Incubated by ♀ (or, in some spp., by ♂ and ♀?).

Young. Nidicolous. With some down on upper parts. Cared for by ♂ and ♀, or by ♀ alone.

Technical diagnosis. Ridgway, Pt. 3, 1904:473.

Classification. Peters, **9**, 1960:379–440.

References. Bent, No. 195:113–295; Witherby, **2**, 1938:213–219.

A. F. Skutch, Life histories of Central American birds, II, *Pacific Coast Avifauna,* No. 34, 1960:116–210.

G. P. Dement'ev *et al., Birds of the Soviet Union,* Vol. 6, Washington, 1968:762–773.

O. S. Pettingill, Jr., Passerine birds of the Falkland Islands . . . , *Living Bird,* **1973:**95–136.

MIMIDAE

Mockingbird Family (31 species)

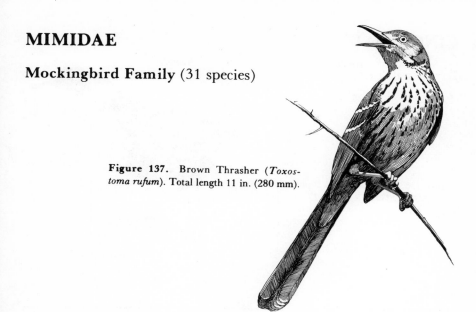

Figure 137. Brown Thrasher (*Toxostoma rufum*). Total length 11 in. (280 mm).

Physical characteristics. Length—203 to 305 mm (8 to 12 in.). Plumage—blue-gray, gray, gray-brown, or reddish brown above. Under parts from paler than back to white, slightly to boldly spotted in many spp. One sp. entirely iridescent black. Wings and tail blackish in some spp. Several spp. have white or white-tipped outer tail feathers and/or white areas or spotting on wings. Iris straw-colored, orange, or red in some spp. Bill strong, medium to long, nearly straight to sharply decurved. Rictal bristles reduced but always present. Wings short and rounded; tail long. Legs rather long. Base of middle toe adherent to outer toe. Sexes alike or nearly so.

Range. North America (from s. Canada), through C. America and the West Indies, to S. America (exc. s. third of Chile and Argentina). Migratory in higher latitudes.

Habits. Solitary or in pairs. Arboreal, but most spp. feed on ground, and some are largely terrestrial. Most spp. make only short flights. Voice—very highly developed song; harsh alarm notes. Some spp. mimic the songs and calls of other birds.

Food. Insects; also fruit, seeds.

Breeding. Bulky, cup-shaped nest in bushes or on ground.

> *Eggs.* 2 to 5; buffy, blue, or green; speckled in most spp. Incubated by ♂ and ♀, or by ♀ alone.

> *Young.* Nidicolous. With some down on upper parts. Cared for by ♂ and ♀.

Technical diagnosis. Ridgway, Pt. 4, 1907:180–181 (omit *Calyptophilus*).

Classification. Peters, **9**, 1960:440–458.

References. Bent, No. 195:295–435.

A. F. Skutch, Life history of the White-breasted Blue Mockingbird, *Condor*, **52**, 1950:220–227.

B. L. Monroe, Jr., A distributional survey of the birds of Honduras, *Ornithol. Monogr.* No. 7, 1968:298–301.

R. I. Bowman and A. Carter, Egg-pecking behavior in Galapagos mockingbirds, *Living Bird*, **1971**:243–270.

J. Bond, *Birds of the West Indies*, London, 1971:166–171.

CINCLIDAE

Dipper Family (4 species)

Figure 138. White-breasted Dipper (*Cinclus cinclus*). Total length 7.25 in. (184 mm).

Physical characteristics. Length—139 to 190 mm (5.5 to 7.5 in.). Plumage—firm and dense, with under coat of down; gray, brown (with chestnut shadings), or black; uniform, mottled with white, or with white areas (head, back, under parts, under side of wings). Bill straight, slender, and much compressed laterally, slightly hooked and notched. Body compact. Wings very short, somewhat pointed, very concave beneath; tail short, square or slightly rounded. Legs and toes long and stout; claws short but strong (the middle claw sometimes slightly pectinate). Sexes alike.

Range. Europe, c. Asia to China and Japan; w. America (from the Yukon) to Tucuman, Argentina. Habitat—chiefly swift mountain streams. Nonmigratory (but some altitudinal shift).

Habits. Solitary. Exclusively aquatic. Flight direct and rapid, usually close over water. Dive. Swim well, often under water. Walk on bottom. Restless; a frequent bobbing motion is responsible for the common name. Voice—rapid chatter; shrill whistle; prolonged deep *zur-rrrr* (in display); elaborate song.

Food. Insects, esp. aquatic larvae; small crustacea, mollusca, and fishes; flatworms; some vegetable matter.

Breeding. Nest large, domed, with side entrance; built of moss and grass, lined with leaves; placed in hollow of wall, bridge, tree roots, etc., or on rock or fallen tree in midst of stream.
Eggs. 3 to 7 (usually 4 or 5); white; immaculate. Incubated by ♀.
Young. Nidicolous. Downy. Fed by ♂ and ♀.

Technical diagnosis. Ridgway, Pt. 3, 1904:675.

Classification. Peters, **9**, 1960:374–379.

References. Bent, No. 195:96–113; Witherby, **2**, 1938:220–226.

W. R. Goodge, Locomotion and other behavior of the dipper, *Condor,* **61**, 1959:4–17.

S. Ali, *The Birds of Sikkim,* London, 1962:296–298.

G. P. Dement'ev *et al., Birds of the Soviet Union,* Vol. 6, Washington, 1968:773–790.

TURDIDAE

Thrush Family (306 species)

Figure 139. Song Thrush (*Turdus ericetorum*). Total length 9 in. (228 mm).

Physical characteristics. Length—114 to 330 mm (4.5 to 13 in.). Plumage—highly glossy in a few spp.; browns, grays, olives, black, blue, usually in soft blended combinations, sometimes with bold contrast (e.g., black with white or bright chestnut), sometimes marked with red, rarely with green or yellow. Most juvenal plumages (and some adult) at least partly spotted or squamate. Bill of medium length, slender to fairly stout. Wings short and rounded to long and pointed; tail very short and truncate to long and graduate (deeply forked in a few spp.). Legs and feet typically stout; legs short in some spp. Most spp. with "booted" tarsus. Sexes alike or unlike.

Range. Worldwide, exc. extreme n., New Zealand, and some oceanic islands. Most spp. migratory.

Habits. Solitary, but some flock in nonbreeding season. Arboreal or terrestrial. Flight weak to strong. Voice—extremely varied; some spp. have highly developed song.

Food. Varied (animal and vegetable).

Breeding. Nest an open cup (rarely domed) in bush or tree or on ground; or nest in tree cavity or among rocks (rarely in hole in ground).

> *Eggs.* 2 to 6; white, greenish or bluish white, olive-green; speckled or (uncommonly) immaculate. Incubated by ♀, or by ♂ and ♀.
>
> *Young.* Nidicolous. Downy. Cared for by ♂ and ♀.

Technical diagnosis. Ridgway, Pt. 4, 1907:1–4.

Classification. Peters, **10**, 1964:13–227 (omit *Zeledonia*).

References. Bent, No. 196:1–330; Witherby, **2**, 1938:104–204.

> A. F. Skutch, Life histories of Central American birds, II, *Pacific Coast Avifauna,* No. 34, 1960:66–115.
>
> A. F. Skutch, Life histories of Central American highland birds, *Publ. Nuttall Ornithol. Club,* No. 7, 1967:109–122.
>
> G. P. Dement'ev *et al., Birds of the Soviet Union,* Vol. 6, Washington, 1968:461–717.
>
> S. Ali and S. D. Ripley, *Handbook of the Birds of India and Pakistan,* Bombay, Vol. 8, 1973:203–270; Vol. 9, 1973:1–134.

Synonymy: The Enicuridae and Myiadestidae are included.

741

SYLVIIDAE

Old-world Warbler Family (395 species ±)

Figure 140. Blackcap (*Sylvia atricapilla*).
Total length 6 in. (152 mm).

Physical characteristics. Length—89 to 292 mm (3.5 to 11.5 in.). Plumage—typically pale, blending shades of brown, gray, or olive-green (some are white or yellow below); with little pattern, but a few spp. are streaked or barred below, or have contrasting head or throat. (The spp. in *Malurus* are glossy blue, chestnut, red, white, black in bold contrasts.) Young not spotted. Bill very small to rather long; slender to stout; straight or curved. Wings of medium length, rounded; tail short to long, often graduated; broad and fan-shaped to vestigial. Legs short to medium; slender to rather stout. Sexes alike in most spp.

Range. North America (exc. extreme n.) s. to c. Brazil and e. Peru; the Old World (exc. extreme n. and some oceanic islands). Many spp. migratory.

Habits. Solitary or gregarious. Arboreal, or frequent brush, reeds, etc. Flight weak. Voice—variety of notes and songs; some spp. have well-developed song.

Food. Insects, spiders, snails.

Breeding. Nest cup-shaped, domed, or placed in a curled leaf—in tree, bush, or marsh vegetation.

> *Eggs.* 3 to 12; white, pink, or buff; speckled in most spp. Incubated by ♂ and ♀, or by ♀ alone.

> *Young.* Nidicolous. With or without down. Cared for by ♂ and ♀.

Technical diagnosis. Ridgway, Pt. 3, 1904:691–693.

Classification. New World: Hellmayr, Pt. 7, 1934:510–514; Peters, **10**, 1964:330–331 (*Chamaea*); 443–455 (*Microbates, Ramphocaenus, Polioptila*). Old World: C. Vaurie, *Birds of the Palearctic Fauna: Passeriformes,* London, 1959:220–315 (Sylviinae).

References. Bent, 196:330–418; Witherby, **1**, 1938:314–320, and **2**, 1938:1–104.

> P. A. Clancey, *The Birds of Natal and Zululand,* Edinburgh, 1964:347–379.

> G. P. Dement'ev *et al., Birds of the Soviet Union,* Vol. 6, Washington, 1968:147–461.

> H. J. Frith, ed., *Birds in the Australian High Country,* Sydney, 1969:301–339.

> J. E. duPont, *Philippine Birds,* Greenville, Del., 1971:303–326.

> S. Ali and S. D. Ripley, *Handbook of the Birds of India and Pakistan,* Vol. 8, Bombay, 1973:1–202.

Synonymy: The Regulidae, Acanthizidae, Polioptilidae, Chamaeidae, Maluridae, and Epthianuridae are included.

MUSCICAPIDAE

Old-world Flycatcher Family (378 species)

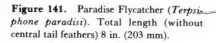

Figure 141. Paradise Flycatcher (*Terpsiphone paradisi*). Total length (without central tail feathers) 8 in. (203 mm).

Physical characteristics. Length—89 to 228 mm (3.5 to 9 in.). *Terpsiphone,* with streamerlike central tail feathers, may reach 533 mm (21 in.). Plumage—extremely diverse in color: from dull gray or brown to brilliant blue, red, chestnut, yellow, black and white, etc.; usually in solid areas or broad bands. Young typically spotted. A few spp. crested; a few with face wattles. Bill from very narrow to very broad and flat, but typically broad. Wings short and rounded to long and rather pointed; tail very short and narrow to extremely long and fanlike (or with very long central feathers). Legs rather short. Sexes alike or unlike.

Range. Old World (exc. extreme n. Asia) and e. in the Pacific to Hawaiian Ids. and Marquesas. Many spp. migratory.

Habits. Solitary. Largely arboreal, but some feed on ground. Many spp. sally from exposed perch to capture prey in air or on ground. Voice—varied; some spp. have a well-developed (but not loud) song.

Food. Insects, spiders.

Breeding. Cup-shaped nest, placed on a branch of bush or tree, in tree cavity, on bank or ledge of rock.

Eggs. 2 to 6; spotted in most spp. Incubated by ♂ and ♀, or by ♀ alone.

Young. Nidicolous. With sparse down. Cared for by ♂ and ♀.

Technical diagnosis. Witherby, **1**, 1938:299.

Classification. C. Vaurie, *Birds of the Palearctic Fauna: Passeriformes,* London, 1959:316–333; Peters, **12**, 1967:3–51 (Pachycephalinae).

References. Witherby, **1**, 1938:299–314; Bannerman, **4**:198–309.

C. W. Mackworth-Praed and C. H. B. Grant, *Birds of Eastern and North Eastern Africa,* London, 1960:153–226, 1101–1102.

G. J. H. Moon, *Refocus on New Zealand Birds,* Wellington, 1967:112–113.

H. C. Officer, *Australian Flycatchers and Their Allies,* Melbourne, 1969, 111 pp., 12 col. pls.

S. Ali and S. D. Ripley, *Handbook of the Birds of India and Pakistan,* Vol. 7, Bombay, 1972:135–229.

A. J. Berger, *Hawaiian Birdlife,* Honolulu, 1972:110–114.

Synonymy: The Falcunculidae and Pachycephalidae are included; *Turnagra capensis* of New Zealand is now considered to be *incertae sedis*.

PRUNELLIDAE

Hedge-sparrow Family (12 species)

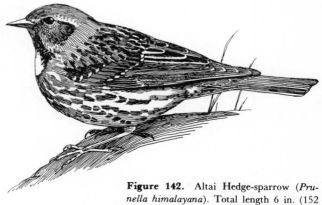

Figure 142. Altai Hedge-sparrow (*Prunella himalayana*). Total length 6 in. (152 mm).

Physical characteristics. Length—127 to 177 mm (5 to 7 in.). Plumage—brown (to black), gray, buffy, and chestnut; streaked above (exc. 1 sp.), plain or streaked below; chin and/or breast in contrasting color (sometimes spotted) in most spp.; light superciliary line in some. Bill medium in length, slender, finely pointed. Wings short and rounded to moderately long and pointed; tail short to moderately long, even or emarginate. Legs and feet strong. Sexes alike or nearly so.

Range. North Africa, Europe, Asia (exc. s. peninsulas). Habitat—brush or barrens. Migratory (only altitudinally in some spp.).

Habits. Most spp. gregarious. Largely terrestrial, feeding chiefly on or near ground, but some spp. perch in trees, esp. to sing. Flight strong and rapid in most spp., but brief and usually low; straight or undulating. On ground walk or hop. Characteristically flick wings and jerk tail. Voice—loud metallic calls; twittering and chattering notes; warbling song.

Food. Insects and other small invertebrates; berries; seeds in winter.

Breeding. Nest a lined open cup of plant fragments, feathers, etc., placed on ground or in low shrubs (rarely at tip of tree branch), or among rocks or in a rock crevice.

> *Eggs.*   2 to 7 (usually 3 or 4); blue; immaculate (rarely spotted). Incubated by ♂ and ♀, or by ♀ alone.

> *Young.* Nidicolous. Downy. Fed by ♂ and ♀.

Technical diagnosis. Witherby, **2,** 1938:205.

Classification. Peters, **10,** 1964:3–12.

References. Bent, No. 197:1–3; Witherby, **2,** 1938:205–213.

C. Vaurie, *The Birds of the Palearctic Fauna: Passeriformes,* London, 1959:208–219.

G. P. Dement'ev *et al., Birds of the Soviet Union,* Vol. 6, Washington, 1968:721–762.

S. Ali and S. D. Ripley, *Handbook of the Birds of India and Pakistan,* Vol. 9, Bombay, 1973:144–162.

Synonym: Accentoridae.

ZOSTEROPIDAE

White-eye Family (81 species)

Figure 143. Gray-breasted White-eye (*Zosterops lateralis*). Total length 5 in. (127 mm).

Physical characteristics. Length—102 to 139 mm (4 to 5.5 in.). Plumage—typically olive-green above, yellow (esp. throat) and white below; with (in most spp.) a conspicuous white eye ring. Some spp. have considerable gray or brown in the plumage (esp. below); others have black crowns. Bill slender, pointed, and slightly curved. Wings rather pointed (with but 9 functional primaries); tail medium in length, truncate. Legs short but strong. Sexes alike.

Range. Africa (s. of Sahara), incl. Madagascar; India, Burma, Thailand, Indochina, e. China, Japan, Philippines, Malaysia, Papuan Region, Australia; e. to Fiji Ids., New Zealand, and Chatham Ids. Largely nonmigratory.

Habits. Occur often in small flocks with other spp. Arboreal. Very active and restless. Voice—a twitter, a musical trill, a melodious warbling song.

Food. Insects, fruit, nectar.

Breeding. Semipendent cuplike nest, in forked twig.

> *Eggs.* 2 to 4; white, pale blue or blue-green; immaculate. Incubated by ♂ and ♀.
> *Young.* Nidicolous. With some down on head. Cared for by ♂ and ♀.

Technical diagnosis. Baker, *Fauna,* **3,** 1926:357.

Classification. Peters, **12,** 1967:289–337.

References. Baker, *Fauna,* **3,** 1926:357–367; Mathews, **11,** Pt. 3, 1923:134–170; Bannerman, **6,** 1948:xxvi–xxvii, 120–137.

> C. W. Mackworth-Praed and C. H. B. Grant, *Birds of Eastern and North Eastern Africa,* London, 1960:724–736.
> B. E. Smythies, *The Birds of Borneo,* London, 1960:484–487.
> N. W. Cayley, *What Bird Is That?,* Sydney, 1966:112–113, 177–178.
> B. Lekagul, *Bird Guide of Thailand,* Bangkok, 1968:234–235.
> J. E. duPont, *Philippine Birds,* Greenville, Del., 1971:399–408.

Synonymy: The Cinnamon White-eye (*Hypocryptadius cinnamomeus*) of the Philippine Islands is considered *incertae sedis.*

MOTACILLIDAE

Pipit Family (54 species)

Figure 144. White Wagtail (*Motacilla alba*). Total length 7.5 in. (190 mm).

Physical characteristics. Length—127 to 222 mm (5 to 8.75 in.). Plumage—black, gray, brown, to olive or yellow, plain or broadly streaked. Tail and often wings edged with white, buff, or yellow. Under parts in most spp. yellow, buff, or black and white, plain or streaked or spotted; with boldly contrasting chin, breast, and/or pectoral band in some spp. Bill medium to long; slender and pointed. Neck short; body slender. Wings medium to long and pointed; tail typically long. Legs medium to long; toes long; hind claw elongated in most spp. Sexes alike or unlike.

Range. Worldwide, exc. extreme n. and some oceanic islands. Many spp. migratory.

Habits. Rather gregarious exc. in breeding season. Mainly terrestrial. Flight strong. Walk when on ground. "Tail-wagging" habit characteristic of most spp. Voice—many sharp call notes; a song, usually simple and repetitive, delivered in many cases on the wing while mounting or while hovering at considerable heights (several hundred feet).

Food. Insects, spiders, mollusca; some vegetable matter.

Breeding. Cuplike, or sometimes domed, nest on ground; or lined nest in cavities of rocks, walls, or trees.

Eggs. 2 to 7; white or colored; typically speckled. Incubated by ♀, or by ♂ and ♀.

Young. Nidicolous. With thick down on upper parts. Cared for by ♂ and ♀, or by ♀ alone.

Technical diagnosis. Ridgway, Pt. 3, 1904:1–2.

Classification. Peters, **9**, 1960:129–167.

References. Bent, No. 197:3–62; Witherby, **1**, 1938:187–233; Bannerman, **4**, 1936:56–87.

S. Smith, *The Yellow Wagtail,* London, 1950.

W. E. Godfrey, *The Birds of Canada,* Ottawa, 1966:307–309.

G. P. Dement'ev *et al., Birds of the Soviet Union,* Vol. 5, Washington, 1970:712–829.

S. Ali and S. D. Ripley, *Handbook of the Birds of India and Pakistan,* Vol. 9, Bombay, 1973:241–298.

BOMBYCILLIDAE

Waxwing Family (3 species)[1]

Figure 145. Bohemian Waxwing (*Bombycilla garrulus*). Total length 8 in. (203 mm).

Physical characteristics. Length—159 to 203 mm (6.25 to 8.0 in.). Plumage—soft and blended; rich fawn to soft dark gray, with chestnut shadings, black throat, and black eye-line; wings slate, marked with white and red or yellow in 2 spp., and often with waxy red tips on the secondaries; tail slate, with subterminal black band, tipped with yellow or red, the under tail coverts chestnut, white, or crimson. Prominent crest. Bill short and thick, slightly notched and hooked; gape broad. Wings long and pointed; tail moderate in length, truncate or slightly rounded, the tail coverts much elongated. Legs very short; toes strong, the middle and outer toes united basally; claws long. Sexes nearly alike.

Range. Subarctic and temperate portions of N. Hemisphere (erratic in local distribution). Habitat—generally coniferous and birch forest. Migratory.

Habits. Gregarious, esp. in nonbreeding season. Arboreal, but often feed on ground. Flight rapid, light, and graceful. Catch insects on wing. Often tame and rather sluggish. Voice—soft, lisping call; high trill; short, loud chatter.

Food. Fruit, esp. berries; insects; flowers.

Breeding. Nest a bulky open cup of twigs, moss, and grass, lined with hair, down, and feathers; usually placed in trees.

> *Eggs.* 3 to 7 (usually 4 or 5); ashy gray or ashy blue; spotted and flecked with dark brown and black. Incubated by ♀ (at least chiefly).
>
> *Young.* Nidicolous. Naked. Fed by ♂ and ♀.

Technical diagnosis. Ridgway, Pt. 3, 1904:103–104.

Classification. Peters, **9**, 1960:369–371, 373.

References. Bent, No. 197:62–102; Witherby, **1**, 1938:296–299.

L. S. Putnam, The life history of the Cedar Waxwing, *Wilson Bull.*, **61**, 1949:141–182.

C. Vaurie, *The Birds of the Palearctic Fauna: Passeriformes,* London, 1959:178–180.

G. P. Dement'ev *et al., Birds of the Soviet Union,* Vol. 6, Washington, 1968:71–80.

S. Ali and S. D. Ripley, *Handbook of the Birds of India and Pakistan,* Vol. 5, Bombay, 1972:266–270.

R. E. Moreau, *The Palaearctic-African Bird Migration Systems,* London, 1972:295.

Synonym: Ampelidae.

---

[1] *Hypocolius ampelinus* of Arabia is now placed in a subfamily (Hypocoliinae) of the Bombycillidae; it also has been given full family status, Hypocoliidae.

PTILOGONATIDAE

Silky-flycatcher Family (4 species)

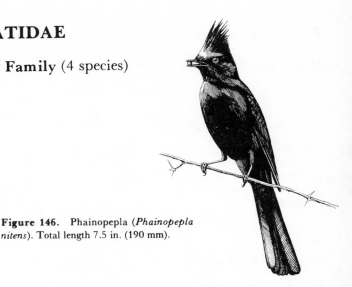

Figure 146. Phainopepla (*Phainopepla nitens*). Total length 7.5 in. (190 mm).

Physical characteristics. Length—184 to 248 mm (7.25 to 9.75 in.). Plumage—silky; never entirely unicolored, but never spotted or streaked, even in young; black, gray, brown, with (in 2 spp.) yellow and olive-yellow markings; large white central areas on wings or tail. Crested. Bill small, rather broad. Wings rather short; tail long. Legs short. Sexes unlike (*Phainoptila*, which does not conform to some of the above, should perhaps not be included in this family).

Range. Southwestern U. S. (s. Utah and c. California) through highlands of Mexico and C. America to w. Panama. Habitat—arid, brushy country. Migratory in part.

Habits. (based mainly on *Phainopepla*). Somewhat gregarious. Arboreal. Shy, active. Capture insects on wing. Voice—weak, warbling song.

Food. Largely vegetable, esp. mistletoe and other berries; also insects.

Breeding. *Phainopepla:* shallow nest in crotches or saddled on limbs of trees, 8 to 20 (sometimes 50) ft up, built chiefly by ♂. *Ptilogonys:* bulky nest (primarily of *Usnea barbata*) in trees, 6 to 60 ft above ground, built by both sexes.

   Eggs. *Phainopepla:* 2 to 3 (sometimes 4); grayish white; speckled profusely with brown and black; incubated by ♂ and ♀. *Ptilogonys:* 2; pale gray; blotched and spotted with dark brown and lilac; incubated by ♀.

   Young. Nidicolous. Long white down in *Phainopepla;* short in *Ptilogonys.* Cared for by ♂ and ♀.

Technical diagnosis. Ridgway, Pt. 3, 1904:113–114.

Classification. Peters, **9,** 1960:371–373, 458.

References. Bent, No. 197:102–114.

   R. J. Newman, A nest of the Mexican Ptilogonys, *Condor,* **52,** 1950:157–158.

   A. F. Skutch, Life history of the Long-tailed Silky-Flycatcher, with notes on related species, *Auk,* **82,** 1965:375–426.

   L. I. Davis, *A Field Guide to the Birds of Mexico and Central America,* Austin, Tex., 1972:180.

   E. P. Edwards, *A Field Guide to the Birds of Mexico,* Sweet Briar, Va., 1972:196.

DULIDAE

Palmchat Family (1 species)

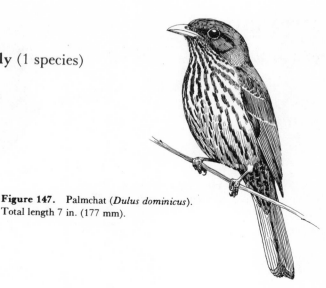

Figure 147. Palmchat (*Dulus dominicus*).
Total length 7 in. (177 mm).

Physical characteristics. Length—about 177 mm (7 in.). Plumage—olive to dark brown above, the feathers of head with dusky centers; the rump and upper tail coverts more greenish; wings and tail edged with yellowish olive; yellowish white below, with broad, dark brown shaft streaks. Bill rather long, heavy, laterally compressed, the upper mandible decurved. Wings medium in length, rounded; tail rather long. Legs and feet stout; toes and claws rather long. Sexes alike.

Range. Hispaniola and Gonave in the West Indies. Nonmigratory.

Habits. Gregarious. Arboreal. Noisy. Voice—a variety of chattering notes, some harsh and some "quite pleasing" (usually uttered in chorus).

Food. Berries, blossoms.

Breeding. Nest a large communal structure of twigs, usually placed in top of palm, with several compartments (lined with shreds of bark, grass, etc.), each having separate outside entrance; or smaller (sometimes single) nests in pines.

Eggs. 2 to 4; white; heavily spotted (in wreath at blunt end) with deep purplish gray. Incubated by?

Young. Nidicolous. Naked at hatching? Cared for by?

Technical diagnosis. Ridgway, Pt. 3, 1904:125.

Classification. Peters, **9**, 1960:373–374.

References. A. Wetmore and B. H. Swales, The birds of Haiti and the Dominican Republic, *U.S. Natl. Mus. Bull.*, **155**, 1931:345–352, pls. 23, 24.

J. Bond, *Birds of the West Indies*, London, 1971:179–180.

ARTAMIDAE

Wood-swallow Family (10 species)

Figure 148. White-browed Wood-swallow (*Artamus superciliosus*). Total length 7.5 in. (190 mm).

Physical characteristics. Length—146 to 203 mm (5.75 to 8 in.). Plumage—soft and fine-textured. Head and upper parts plain black, slate-gray, brownish gray, or brown (the rump or whole back white in a few spp.). Under parts white, vinaceous ashy, gray-brown, brown, or chestnut. Tip of tail and/or margin of primaries white in some spp.; white superciliary line in 1 sp. Bill stout, rather long, slightly curved, rounded in cross section, pointed; gape wide. Neck short; body stout. Wings very long and pointed; tail medium in length, square or emarginate. Legs very short but stout; feet strong. Sexes alike or nearly so.

Range. India, Burma, Thailand, Indochina, s. and s.w. China, East Indies, Philippines, Australia, and e. to Fiji Ids. Habitat—clearings or open country, esp. near water. Some spp. migratory.

Habits. Strongly gregarious. Rest on occasion in clustered masses. Arboreal. Flight gliding, very graceful. Usually feed on the wing. Sit quietly on exposed perch. Voice—twittering notes; harsh cries.

Food. Insects; rarely seeds.

Breeding. Sometimes colonial. Shallow cup-shaped nest in low tree or bush, or on a stump; or nest built in shallow cavity in tree.

> *Eggs.* 2 to 4; white, pinkish white, greenish white, or pale brown; speckled or blotched with brown or purplish brown. Incubated by ♂ and ♀.

> *Young.* Nidicolous. With some down on upper parts. Cared for by ♂ and ♀.

Technical diagnosis. Baker, *Fauna*, **2**, 1924:348.

Classification. Peters, **15**, 1962:160–165.

References. B. E. Smythies, *The Birds of Borneo,* London, 1960:369–370.

D. L. Serventy and H. M. Whittell, *Birds of Western Australia,* Perth, 1967:407–413.

H. J. Frith, ed., *Birds in the Australian High Country,* Sydney, 1969:452–457.

J. E. duPont, *Philippine Birds,* Greenville, Del., 1971:360.

S. Ali and S. D. Ripley, *Handbook of the Birds of India and Pakistan,* Vol. 5, Bombay, 1972:143–146.

HYPOSITTIDAE

Coral-billed Nuthatch Family (1 species)

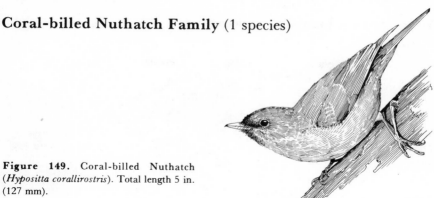

Figure 149. Coral-billed Nuthatch (*Hypositta corallirostris*). Total length 5 in. (127 mm).

Physical characteristics. Length—about 127 mm (5 in.). Plumage—greenish blue. Bill short, coral-red; slightly hooked. Nostrils not operculate (or only slightly so), partly concealed by forward-projecting feathers. Wings rather long and rounded; tail relatively long. Feet syndactyl; hind toe very long. Sexes unlike (♀ duller than ♂ above, brown and white below); immatures like adult ♀.

Range. Humid forest of e. Madagascar. Nonmigratory.

Habits. Behavior creeperlike (described as "climbing silently up one tree trunk, then flying down to climb another trunk"). Sometimes accompany mixed flocks of other forest birds. Voice—undescribed.

Food. Insects.

Breeding. Nest unknown.

   Eggs. Not described.

   Young. Not described.

Technical diagnosis. Ridgway, Pt. 3, 1904:439.

Classification. Peters, **12**, 1967:124.

References. R. Ridgway, Relationships of the Madagascar genus *Hypositta* Newton, *Proc. Biol. Soc. Wash.*, **16**, 1903:125.

   C. E. Hellmayr, Fam. Hyposittidae, Wytsman's *Genera Avium,* Pt. 24, 1913 (with col. pl.).

   A. L. Rand, The distribution and habits of Madagascar birds, *Bull. Am. Mus. Nat. Hist.,* **72,** Art. 5, 1936:468–469.

   O. L. Austin, Jr., *Birds of the World,* New York, 1961:239.

Synonymy: Sometimes included in the Sittidae, but relationships remain unknown; some authors suspect an affinity with the Vangidae.

VANGIDAE

Vanga-shrike Family (12 species)

Figure 150. Hook-billed Vanga (*Vanga curvirostris*). Total length 9.5 in. (241 mm).

Physical characteristics. Length—127 to 311 mm (5 to 12.25 in.). Plumage—black (with green or purple metallic reflections) above and all or largely white below; chestnut and/or gray areas in some spp.; head white in 2 spp. *Oriolia* is all black with metallic reflections; *Leptopterus* is brilliant blue above, white below; *Euryceros* is black and chestnut, with blue bill. Bill typically rather long and robust, hooked and toothed (long, slender, and sickle-shaped in *Falculea*; heavy and casqued in *Euryceros*). Wings short to rather long, rounded; tail moderately long and abruptly truncate in most spp. (rounded in 3 spp.). Legs and feet strong. Sexes alike or unlike.

Range. Madagascar. Habitat—forest and brushy areas. Nonmigratory.

Habits. Most spp. gregarious (often in mixed flocks of other birds). Arboreal. Flight strong in most spp. Fly or jump from branch to branch in search of insects. Voice—shrill whistle; chattering calls; harsh repetitive notes.

Food. Insects; small lizards and amphibians.

Breeding. Nest (*Falculea*) a large, shallow structure of sticks, lined with grass, usually in fork, high in tree.

> *Eggs.* (described in 2 spp.; clutch size not recorded). White or green; spotted with pinkish brown. Incubation not recorded.

> *Young.* Not described.

Technical diagnosis. A. Reichenow, *Die Vögel: Handbuch der Systematischen Ornithologie*, Vol. 2, Stuttgart, 1914:278, 291 ("Vanginae"). (For photograph of skins of 7 spp., see D. Amadon, *Bull. Am. Mus. Nat. Hist.*, **95**, 1950:pl. 13.)

Classification. Peters, **9**, 1960:365–369.

References. A. Milne Edwards and A. Grandidier, *Histoire physique, naturelle et politique de Madagascar*, Vol. 12: *Histoire naturelle des oiseaux*, Paris, Vol. 1, 1882:303–311; Vol. 2, 1885:409–444; Vol. 3, 1879:col. pls. 117 and (various) 156–172.

> A. L. Rand, The distribution and habits of Madagascar birds, *Am. Mus. Nat. Hist. Bull.*, **72**, 1936:460–468.

Synonymy: The Eurycerotidae ("Aerocharidae") and Falculidae are included.

LANIIDAE

Shrike Family (65 species)

Figure 151. Loggerhead Shrike (*Lanius ludovicianus*). Total length 9 in. (228 mm).

Physical characteristics. Length—159 to 368 mm (6.25 to 14.5 in.). Plumage—many spp. gray or brown above and white below, with face and flight feathers boldly patterned in black and white; some African spp. brightly colored (e.g., green and yellow, bright red) or entirely black; a very few spp. with streaked or barred areas. Bill strong, hooked, and in some spp. toothed; head large. Wings medium; tail long and narrow. Legs and feet strong; claws sharp. Sexes alike or unlike.

Range. Africa, Europe, Asia (s. to Timor and New Guinea), and N. America (s. to Tehuantepec). Habitat—open or semiopen country, rarely forest edge. Northern forms migratory.

Habits. Typically solitary. Watch for prey from exposed perches. Bold and aggressive. Flight strong. Most spp. impale prey on thorns. Voice—great variety of notes; many spp. have well-developed song.

Food. Insects; small reptiles, birds, and mammals.

Breeding. Deep bulky nest, usually placed in bushes or trees. Built (varying with the spp.) by ♂, by ♀, or by both.

　Eggs. 2 to 8; spotted. Incubated chiefly by ♀, assisted by ♂ (in the few spp. studied).

　Young. Nidicolous. With some down on upper parts. Fed by ♂ and ♀.

Technical diagnosis. Ridgway, Pt. 3, 1904:232–234.

Classification. Peters, **9**, 1960:309–365 (exclude Prionopinae).

References. Bent, No. 197:114–182; Witherby, **1**, 1938:277–296.

　　C. W. Mackworth-Praed and C. H. B. Grant, *Birds of Eastern and North Eastern Africa*, London, 1960:576–644.

　　P. A. Clancey, *The Birds of Natal and Zululand*, Edinburgh, 1964:403–415.

　　G. P. Dement'ev *et al.*, *Birds of the Soviet Union*, Vol. 6, Washington, 1968:3–67.

　　J. E. duPont, *Philippine Birds*, Greenville, Del., 1971:360–364.

　　S. Ali and S. D. Ripley, *Handbook of the Birds of India and Pakistan*, Vol. 5, Bombay, 1972:78–101.

Synonymy: The Laniariidae and Pityriasidae are included.

PRIONOPIDAE

Wood-shrike Family (9 species)

Figure 152. Black-winged Helmetshrike (*Prionops cristata*). Total length 8.5 in. (216 mm).

Physical characteristics. Length—190 to 254 mm (7.5 to 10 in.). Plumage—soft. Bold patterns of black (or brown) and white, or black and brown; some spp. with areas of gray, chestnut, or yellow. Tail conspicuously marked with white in most spp. Brightly colored fleshy orbital ring (exc. in *Eurocephalus*). Most spp. crested. Bill of medium length, rather stout, hooked. Wings medium to rather long; tail medium to long, rounded. Legs short but strong. Sexes alike or nearly so.

Range. Africa s. of Sahara (exc. Madagascar). Nonmigratory exc. for local movements.

Habits. Gregarious. Largely arboreal but often descend to ground for food; also capture insects on the wing. Flight buoyant, but slow and brief. Voice—harsh, chattering notes; a nasal humming sound. Make snapping sound with bill.

Food. Insects.

Breeding. Some spp. colonial. Cup-shaped nest (of grass, leaves, and rootlets, bound with cobwebs) placed in tree.

    Eggs. 2 to 6 (probably 2 or more ♀ ♀ lay in some nests); white or bluish green; spotted. Incubated by ♂ and ♀; also probably by several associated adults (*Prionops*).

    Young. Nidicolous. Downy. Cared for by ♂ and ♀; also by several associated adults (*Prionops*).

Technical diagnosis. Bannerman, **5**, 1939:337; E. Mayr, *Ibis,* **1943**:216–218.

Classification. Peters, **9**, 1960:309–314 (Prionopinae).

References. Bannerman, **5**, 1939:xxxvi–xxxvii, 337–348.

    F. M. Benson, Field-notes from Nyasaland, *Ostrich,* **17**, 1946:298, 308–314.

    C. W. Mackworth-Praed and C. H. B. Grant, *Birds of Eastern and North Eastern Africa,* London, 1960:567–576, 583–584.

    P. A. Clancey, *The Birds of Natal and Zululand,* Edinburgh, 1964:416–418.

    K. Newman, ed., *Birdlife in Southern Africa,* Johannesburg, 1971:126, 236.

CALLAEIDAE

Wattlebird Family (3 species)

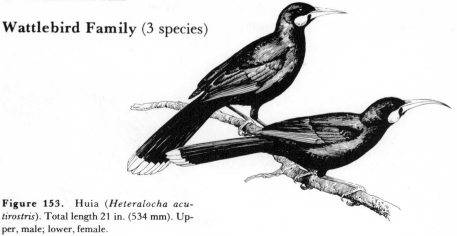

Figure 153. Huia (*Heteralocha acutirostris*). Total length 21 in. (534 mm). Upper, male; lower, female.

Physical characteristics. Length—254 to 534 mm (10 to 21 in.). Plumage—plain blue-gray, brown, or black (back and crissum bright ferruginous in *Creadion,* tail broadly tipped with white in *Heteralocha*). Bill strong; from short and heavy to long and sickle-shaped. Large orange or blue wattles at corner of mouth. Wings weak. Legs and feet large and strong. Sexes alike in plumage but unlike in size, bill shape, or size of wattle. Extreme sexual dimorphism in bill of *Heteralocha.*

Range. New Zealand. Habitat—forests. Nonmigratory.

Habits. Usually found in pairs or small flocks. Very active, but nearly flightless. Progress by long hops from branch to branch or along the ground. Voice—great variety of musical whistling and flutelike sounds; rapid series of harsh notes; mewing.

Food. Insects, fruit, leaves, nectar.

Breeding. Nest of sticks lined with grass and feathers, in crotches or in cavities of trees.

   Eggs. 2 to 3; gray; spotted and blotched with brown. Incubated by ♂ and ♀?

   Young. Nidicolous. Some down on upper parts. Fed by ♂ and ♀.

Technical diagnosis. C. R. Stonor, *Ibis,* 1942:1–18.

Classification. Peters, **15,** 1962:157–159.

References. H. R. McKenzie, Breeding of Kokako, *Notornis,* **4,** 1951:70–76, pls. 13–18.

   W. R. B. Oliver, *New Zealand Birds,* Wellington, 1955:512–524.

   R. A. Falla *et al., A Field Guide to the Birds of New Zealand,* Boston, 1967:233–238.

   G. J. H. Moon, *Refocus on New Zealand Birds,* Wellington, 1967:138–141.

   E. G. Turbott, *Buller's Birds of New Zealand,* Honolulu, 1967:3–17.

Synonymy: The Creadiontidae and Philesturnidae of Oliver are included.

STURNIDAE

Starling Family (111 species)

Figure 154. Rose-colored Starling (*Sturnus roseus*). Total length 9 in. (228 mm).

Physical characteristics. Length—177 to 432 mm (7 to 17 in.). Plumage—silky, the contour feathers, esp. on head and neck, often lanceolate. Dark metallic purples, greens, blues, and bronze (some spp. spotted, esp. in winter plumage, or with chestnut and/or white areas); black (usually glossy), with white, gray, or yellow (rarely crimson) markings; gray (or brown) and buff, streaked below in some spp. Some spp. with bare areas and/or wattles on head; some crested. Bill typically straight and rather long and slender (heavy and/or hooked in some spp.). Wings short and rounded to long and pointed; tail usually short and square, but long and steeply graduated in some spp. Legs and feet strong. Sexes alike or unlike.

Range. Africa, Eurasia (exc. extreme n.), Malaysia, Papuan Region, n.e. Australia, and the islands of Oceania, e. to Tuamoto. Most spp. migratory, at least in part.

Habits. Most spp. highly gregarious. Arboreal or terrestrial. Flight swift and direct (spectacular mass evolutions in many spp.). Typically walk and run on ground. Voice (most spp. garrulous)—variety of harsh, grating notes, musical whistles, and warbles. Many spp. mimic other birds.

Food. Omnivorous: insects, fruit, grain, offal, birds' eggs, crustacea, lizards.

Breeding. Often colonial. Nest a pile of plant fragments in tree cavity, rock crevice, hole in bank, etc.; domed or cup-shaped nest in bush or on ground; 1 sp. builds hanging nest.

Eggs. 2 to 9 (usually 3 or 4); white or blue; spotted or immaculate. Incubated by ♂ and ♀, or by ♀ alone.

Young. Nidicolous. Downy. Fed by ♂ and ♀.

Technical diagnosis. Witherby, **1**, 1938:39.

Classification. Peters, **15**, 1962:75–121.

References. Bent, No. 197:182–222; Witherby, **1**, 1938:39–47; Bannerman, **6**, 1948:41–112.

B. E. Smythies, *The Birds of Borneo,* Edinburgh, 1960:488–491.

J. E. duPont, *Philippine Birds,* Greenville, Del., 1971:364–371.

S. Ali and S. D. Ripley, *Handbook of the Birds of India and Pakistan,* Vol. 5, Bombay, 1972:146–198.

Synonymy: The Eulebetidae ("Graculidae") and Buphagidae are included.

MELIPHAGIDAE

Honeyeater Family (172 species)

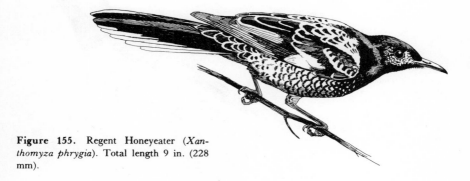

Figure 155. Regent Honeyeater (*Xanthomyza phrygia*). Total length 9 in. (228 mm).

Physical characteristics. Length—102 to 432 mm (4 to 17 in.). Plumage—green, gray, brown, red, or black, more or less solid or in combination. Some spp. streaked, barred, or "scaled"; many have small areas or stripes of yellow or white on head. Many spp. have bare skin on head (often brightly colored) and/or wattles or casque; some have white or yellow plumy tufts on chin, throat, or sides of head. Bill slender, curved, medium to long. Wings rather long and pointed; tail medium to long. Legs strong; short to medium. Sexes alike or unlike.

Range. Bali, Papuan Region, Australia; e. to New Zealand, Samoa, and Hawaii; n. to Marianas; also S. Africa (*Promerops*). Some spp. migratory.

Habits. Rather gregarious. Pugnacious. Largely arboreal. Flight typically undulating. Voice—loud, musical calls; harsh notes; well-developed song in some spp.

Food. Nectar, insects, fruit.

Breeding. A few spp. colonial. Nest in trees or bushes: cuplike, semipendent nest fastened to twigs; abandoned nests of other birds; domed nest with side entrance; open nest of twigs; or (*Moho*) hole in tree.

> *Eggs.* 1 to 4; buff, pink, or white; marked with red or black. Incubated by ♂ and ♀, or by ♀ alone.

> *Young.* Nidicolous. With sparse down. Cared for by ♂ and ♀.

Technical diagnosis. H. Gadow, Cat. of Birds in Brit. Mus., **9**, 1884:127 (omit Zosteropinae).

Classification. Peters, **12**, 1967:338–450.

References. G. F. Broekhuysen, Biology of Cape Sugarbird, *Promerops cafer, Ostrich,* Suppl., 1959:180–226.

> B. E. Smythies, *The Birds of Borneo,* Edinburgh, 1960:468–483.

> G. J. H. Moon, *Refocus on New Zealand Birds,* Wellington, 1967:128–133.

> D. L. Serventy and H. M. Whittell, *Birds of Western Australia,* Perth, 1967:377–397.

> H. J. Frith, ed., *Birds in the Australian High Country,* Sydney, 1969:388–391, 394–423.

Synonym: Melithreptidae. The Promeropidae are included.

NECTARINIIDAE

Sunbird Family (115 species)

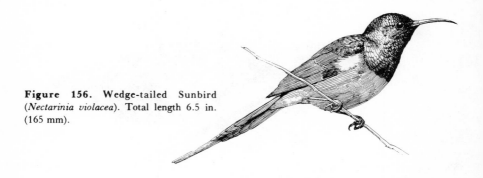

Figure 156. Wedge-tailed Sunbird (*Nectarinia violacea*). Total length 6.5 in. (165 mm).

Physical characteristics. Length—95 to 254 mm (3.75 to 10 in.). Plumage—typically marked by solid areas of red, orange, yellow, brown, black, and metallic green, blue, or purple. Some spp. plain olive-green (rarely streaked). Many spp. have tufts of yellow or orange feathers at sides of breast. Bill long, curved, and very finely serrate; very slender and pointed in most spp. Wings short and rounded; tail short to medium and truncate, to very long and pointed (the central feathers extremely long in some spp.). Legs short; legs and feet strong. Sexes unlike in most spp.

Range. Africa (s. of Sahara), incl. Madagascar; Palestine, s.e. Arabia, s. Persia, India, s.e. Asia (n. to c. China); Philippines, Malaysia, Papuan Region, extreme n.e. Australia (n. Queensland). Most spp. nonmigratory.

Habits. Sometimes gregarious. Arboreal. Flight strong. Very active. Males very pugnacious, esp. in breeding season. Habitually feed at flowers, usually while perched. Voice—sharp, metallic notes; some spp. have well-developed song.

Food. Nectar and small insects and spiders about flowers; fruit.

Breeding. Pouch nest with side entrance (often with porchlike projection), hung from leaves or twigs of tree, from grass clumps, or from roots projecting from bank; or sewn to under side of large leaf.

Eggs. 1 to 3; typically speckled or blotched. Incubated by ♀; rarely by ♂ and ♀.

Young. Nidicolous. With down on upper parts, or naked. Cared for by ♀, or by ♂ and ♀.

Technical diagnosis. J. Delacour, *Zoologica*, **29**, Pt. 1, 1944:17–20.

Classification. Peters, **12**, 1967:208–289.

References. Baker, *Fauna, 3*, 1926:368–412; Bannerman, **6**, 1948:137–264.

B. E. Smythies, *The Birds of Borneo*, Edinburgh, 1960:468–483.

S. Ali, *The Birds of Sikkim*, London, 1962:356–361.

P. A. Clancey, *The Birds of Natal and Zululand*, Edinburgh, 1964:427–437.

C. J. Skead, *The Sunbirds of Southern Africa . . .* , Cape Town, 1967:351 pp.

J. E. duPont, *Philippine Birds*, Greenville, Del., 1971:372–385.

Synonymy: The Chalcopariidae are included.

DICAEIDAE

Flowerpecker Family (58 species)

Figure 157. Fire-breasted Flowerpecker (*Dicaeum ignipectum*). Total length 3.25 in. (83 mm).

Physical characteristics. Length—76 to 190 mm (3 to 7.5 in.). Plumage—♂ typically dark and glossy above, light below, with (in many spp.) solid areas of bright red or yellow on breast (usually central), crown, back, or rump. ♀ typically dull-colored; ♂ also dull-colored in a few spp. Some spp. broadly streaked below. Bill thin, curved, and serrate to stout and not serrate. Neck short. Wings rather long; tail short. Legs short. Sexes alike or unlike.

Range. India, Burma, Thailand, Malaysia, Papuan Region, Indochina, s. China; Philippines, Australia, and e. to Solomons. Nonmigratory.

Habits. Somewhat gregarious. Arboreal, typically frequenting treetops. Flight strong and swift. Voice—variety of sharp, metallic notes; some spp. have a warbling song.

Food. Insects, nectar, fruit (esp. mistletoe berries).

Breeding. Nest a cup or pendent pouch in trees or bushes (*Pardalotus* in holes in trees or in banks).

Eggs. 1 to 3; white; plain or spotted. Incubated by ♀ or by both sexes.

Young. Nidicolous. No down? Cared for by ♂ and ♀.

Technical diagnosis. E. Mayr and D. Amadon, *Am. Mus. Novit.,* No. 1360, 1947:1.

Classification. Peters, **12,** 1967:166–208.

References. Baker, *Fauna,* **3,** 1926:420–440; Baker, *Nidification,* **3,** 1934:237–249.

H. C. Robinson and F. N. Chasen, *The Birds of the Malay Peninsula,* Vol. 4, London, 1939:403–416.

B. E. Smythies, *The Birds of Burma,* Edinburgh, 1953:277–281.

B. E. Smythies, *The Birds of Borneo,* Edinburgh, 1960:461–468.

S. Ali, *The Birds of Sikkim,* London, 1962:354–356.

H. J. Frith, ed., *Birds in the Australian High Country,* Sydney, 1969:372–373, 379–387.

J. E. duPont, *Philippine Birds,* Greenville, Del., 1971:385–399.

G. M. Henry, *A Guide to the Birds of Ceylon,* New York, 1971:113–116.

Synonymy: The Paramythiidae are included.

CYCLARHIDAE

Pepper-shrike Family (2 species)

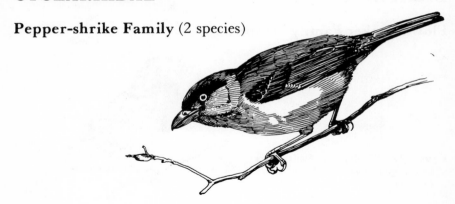

Figure 158. Pale-billed Pepper-shrike (*Cyclarhis gujanensis*). Total length 6.5 in. (165 mm).

Physical characteristics. Length—139 to 177 mm (5.5 to 7 in.). Plumage—loose-webbed; olive-green above; crown gray or brown in some forms. Broad reddish brown stripe along side of head; cheeks gray. Under parts bright greenish yellow to buffy white. Bill heavy, laterally compressed, hooked. Head large; neck short; body heavy. Wings short and rounded; tail medium in length. Legs and feet strong. Sexes alike or nearly so.

Range. Southern Mexico (Veracruz and Puebla), through C. America and S. America to Uruguay. Habitat—brushland or thin forest. Probably nonmigratory.

Habits. Found singly or in pairs. Arboreal. Movements rather deliberate. Flight weak. Voice—harsh scolding notes; sweet warbling song (loud and frequently given).

Food. Insects, fruit.

Breeding. Fragile, loosely constructed, semipendent nest placed in horizontal fork of bush or low branch of tree.

Eggs. 2 to 3; pinkish white; blotched and speckled with brown. Incubated by ♂ and ♀?

Young. Nidicolous. Naked. Cared for by ♂ and ♀.

Technical diagnosis. W. P. Pycraft, *Proc. Zool. Soc. London,* **1907**:378.

Classification. Peters, **14,** 1968:103–108 (Cyclarhinae).

References. C. B. Worth, Nesting of the Pepper-shrike, *Auk,* **55,** 1938:539–540.

G. M. Sutton, *Mexican Birds: First Impressions,* Norman, Okla., 1951:242.

A. F. Skutch, Life histories of Central American highland birds, *Publ. Nuttall Ornithol. Club,* No. 7, 1967:123–129.

B. L. Monroe, Jr., A distributional survey of the birds of Honduras, *Ornithol. Monogr.* No. 7, 1968:313–314.

R. Meyer de Schauensee, *A Guide to the Birds of South America,* Wynnewood, Pa., 1970:349.

E. P. Edwards, *A Field Guide to the Birds of Mexico,* Sweet Briar, Va., 1972:197.

Synonym: Cyclorhidae.

VIREOLANIIDAE

Shrike-vireo Family (3 species)

Figure 159. Chestnut-sided Shrike-vireo (*Vireolanius melitophrys*). Total length 7 in. (177 mm).

Physical characteristics. Length—146 to 184 mm (5.75 to 7.25 in.). Plumage—loose-webbed, silky; parrot-green or olive-green above, with part or all of crown bright blue or gray. Two spp. have sides of head broadly banded with yellow, white, and black (gray in *Vireolanius leucotis*). Under parts greenish yellow (2 spp.) or white; *V. melitophrys* has pectoral band and sides of chestnut. Bill rather stout, hooked. Head large; neck short; body heavy. Wings short and rounded; tail short or medium. Legs short but strong. Sexes alike (2 spp.) or very similar.

Range. Southern Mexico (Veracruz) through C. America and n. S. America to n. Bolivia and c. Brazil (n. Matto Grosso). Habitat—forests. Nonmigratory.

Habits. Solitary. Arboreal (*Vireolanius pulchellus* is a bird of the treetops). Voice—titmouse-like (*V. pulchellus*); "a short ascending whine terminating abruptly" (*V. melitophrys*).

Food. Primarily insects, some fruit; gleaned from leaves, branches, and trunks.

Breeding. Nest semipendent, in horizontal fork of tree.

Eggs. Not described.

Young. Not described.

Technical diagnosis. W. P. Pycraft, *Proc. Zool. Soc. London,* **1907**:378–379, 362 (fig. e).

Classification. Peters, **14,** 1968:108–110 (Vireolaniinae).

References. F. Haverschmidt, *Birds of Surinam,* Edinburgh, 1968:361.

B. L. Monroe, Jr., A distributional survey of the birds of Honduras, *Ornithol. Monogr.* No. 7, 1968:314, 392, 401.

H. C. Land, *Birds of Guatemala,* Wynnewood, Pa., 1970:271–272.

R. Meyer de Schauensee, *A Guide to the Birds of South America,* Wynnewood, Pa., 1970:349–350.

E. P. Edwards, *A Field Guide to the Birds of Mexico,* Sweet Briar, Va., 1972:198.

J. Barlow and R. D. James, Aspects of the biology of the Chestnut-sided Shrike-vireo, *Wilson Bull.,* **87,** 1975:320–334.

VIREONIDAE

Vireo Family (38 species)

Figure 160. Yellow-throated Vireo (*Vireo flavifrons*). Total length 5.5 in. (139 mm).

Physical characteristics. Length—102 to 165 mm (4 to 6.5 in.). Plumage—yellowish or greenish olive, olive-brown, or gray above. Crown in some spp. contrasting in color (olive-green, gray, reddish brown, or, in 1 sp., black); light superciliary line in some. Under parts white or buffy white to canary yellow; or gray. Some spp. have wing bars, but plumage never otherwise barred, streaked, or spotted, even in young. Irides of some spp. white or red. Bill medium to somewhat long, rather thick, slightly hooked. Neck short. Wings long and pointed to short and rounded (the 10th primary short to vestigial); tail medium. Legs rather short; anterior toes basally adherent. Sexes alike or nearly so.

Range. North America (exc. extreme n.), C. America, West Indies, and S. America (exc. s. third). Habitat—forests or brushland. Many spp. migratory.

Habits. Solitary. Arboreal. Typically feed among leaves and twigs, moving about rather deliberately. Often notably tame, esp. at nest. Voice—rather loud, frequently repeated, warbling songs; harsh, scolding notes.

Food. Insects, some fruit.

Breeding. Cuplike, semipendent nest placed in horizontal fork of bush or tree.

Eggs. 2 to 5; white; very lightly speckled. Incubated by ♂ and ♀, or by ♀ alone.

Young. Nidicolous. With some down on upper parts, or naked. Cared for by ♂ and ♀.

Technical diagnosis. Ridgway, Pt. 3, 1904:128–130 (omit *Vireolanius* and *Cyclarhis*).

Classification. Peters, **14**, 1968:103, 110–138.

References. Bent, No. 197:222–379.

G. M. Sutton, Studies of the nesting birds of the Edwin S. George Reserve, Pt. 1, the vireos, *Univ. Mich. Mus. Zool. Misc. Publ.* No. 74, 1949.

L. de K. Lawrence, Nesting life and behaviour of the Red-eyed Vireo, *Can. Field-Nat., 67,* 1953:47–77.

A. F. Skutch, Life histories of Central American birds, II, *Pacific Coast Avifauna,* No. 34, 1960:11–42.

J. W. Graber, Distribution, habitat requirements, and life history of the Black-capped Vireo (*Vireo atricapilla*), *Ecol. Monogr., 31,* 1961:313–336.

A. F. Skutch, Life histories of Central American highland birds, *Publ. Nuttall Ornithol. Club,* No. 7, 1967:130–136.

PLOCEIDAE

Weaverbird Family (267 species)

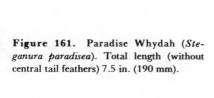

Figure 161. Paradise Whydah (*Steganura paradisea*). Total length (without central tail feathers) 7.5 in. (190 mm).

Physical characteristics. Length—76 to 648 mm (3 to 25.5 in.). Plumage—from brown and gray in cryptic patterns to yellow, red, purple, blue, green, black in bold patterns; many spp. conspicuously spotted or barred with white or black. Bill short (typically rather thick) and sharp-pointed; never long. Wings short and rounded to long and pointed; tail very short to extremely long (very broad, spatulate, or filiform feathers). Legs short. Sexes alike or unlike.

Range. Africa (incl. Madagascar), Eurasia, Malaysia, Papuan Region, Australia, Philippines, Oceania (e. to Samoa). A few spp. are migratory.

Habits. Many spp. gregarious. Arboreal or terrestrial. Flight strong in most spp. Voice—typically simple, chirping notes, often harsh; many spp. have well-developed song.

Food. Vegetable; animal (to a lesser extent).

Breeding. A number of spp. colonial. Monogamous, polygamous, polyandrous, or promiscuous. Nest types extremely diverse: elaborately woven and flask-shaped nest (often with tunnel entrance); or massive covered colonial nest in tree; old or new covered nests of other birds; bulky cup-shaped nest in tree; lined tree cavity or rock crevice. Some spp. (Viduinae) entirely parasitic.

Eggs. 2 to 8; white, green, pale blue, or reddish; speckled or immaculate. Incubated by ♀, or by ♂ and ♀.

Young. Nidicolous. With or without down. Cared for by ♂ and ♀.

Technical diagnosis. P. P. Sushkin, *Bull. Am. Mus. Nat. Hist.,* **57**, 1927:1–32.

Classification. Peters, **14**, 1968:306–397; **15**, 1962:3–75.

References. Bannerman, **6**, 1948:320–352; 7, 1949.

H. Friedmann, The parasitic weaverbirds, *U.S. Natl. Mus. Bull.,* **223**, 1960.

N. E. Collias and E. C. Collias, Evolution of nest-building in the weaverbirds (Ploceidae), *Univ. Calif. Publ. Zool.,* **73**, 1964.

G. P. Dement'ev *et al., Birds of the Soviet Union,* Vol. 5, Washington, 1970:363–445.

Synonyms: Malimbidae, Passeridae. The Bubalornithidae and Estrildidae are included. Data published after the Peters' volumes cast doubt on the classification therein; hence, we elect to retain tentatively an older concept of the group.

ZELEDONIIDAE

Wrenthrush Family (1 species)

Figure 162. Wrenthrush (*Zeledonia coronata*). Total length 4.5 in. (114 mm).

Physical characteristics. Length—114 to 120 mm (4.5 to 4.75 in.). Plumage—soft and lax. Crown rufous orange, bordered laterally with dull black stripes. Nape, back, wings, tail, and flanks dark brownish olive; rest of plumage slate. Bill weak, somewhat flattened. Neck short. Wings and tail short, rounded, and soft. Legs rather long; feet large. Sexes alike.

Range. High mountain peaks (above 5000 ft) of Costa Rica and Panama. Habitat—dense, humid forest. Nonmigratory.

Habits. Solitary. "Continually creeping and hopping about under the masses of half decayed branches, searching for insects and larvae, and would be rarely seen or collected were it not for their song." Voice—"a clear, musical whistle . . . repeats the same note from six to eight times with the same interval between . . . the length of the note and the interval is about the same" (Carriker, 1910).

Food. "Insects and larvae."

Breeding. Domed nest with side entrance, concealed on moss-covered bank; mostly of moss, lined with dead, fine plant materials.

Eggs. 2; white or buffy white; finely speckled with brown. Incubated by ♀ only (?).

Young. Nidicolous. Naked. Cared for by ♀, or by ♂ and ♀.

Technical diagnosis. Ridgway, Pt. 4, 1907:69–70, 885, pl. 2.

Classification. Peters, **10,** 1964:18.

References. W. P. Pycraft, On the systematic position of *Zeledonia coronata* with some observations on the position of the Turdidae, *Ibis,* **1905:**1–24.

P. Slud, The birds of Costa Rica, *Bull. Am. Mus. Nat. Hist.,* **128,** 1964, 430 pp.

C. G. Sibley, The relationships of the "Wren-Thrush," *Zeledonia coronata* Ridgway, *Postilla,* **125,** 1968.

C. G. Sibley, A comparative study of the egg-white proteins of Passerine birds, *Peabody Mus. Nat. Hist. Bull.* No. 32, 1970.

J. H. Hunt, A field study of the Wrenthrush, *Zeledonia coronata, Auk,* **88,** 1971:1–20.

PARULIDAE

American Wood-warbler Family (123 species[1])

Figure 163. Black-throated Gray Warbler (*Dendroica nigrescens*). Total length 5 in. (127 mm).

Physical characteristics. Length—108 to 184 mm (4.25 to 7.25 in.). Plumage—usually olive or gray, often brightly marked with yellow, orange, red, or gray-blue. Bill usually slender and pointed (a few spp. have a rather broad bill and pronounced rictal bristles; others have a rather heavy, tanager-like bill). Primaries reduced to 9. Sexes alike or unlike.

Range. Alaska to Labrador s. through C. America and the West Indies to Uruguay and n. Argentina. North American spp. migratory.

Habits. Typically arboreal; some (*Seiurus*) are terrestrial (and walk rather than hop); others have creeper-like habits; others feed mainly by catching insects on the wing. Voice—a few have highly developed songs.

Food. Insects, fruits, nectar.

Breeding. Nest cup-shaped or domed; placed on ground, in low plants, in shrubs, or in trees, or in holes in trees or banks.

Eggs. 2 to 5 (rarely 6); usually white; marked with brown. Incubated by ♀.

Young. Nidicolous. With scant down, on upper surface; or naked. Cared for by ♂ and ♀.

Technical diagnosis. Ridgway, Pt. 2, 1902:425.

Classification. Peters, **14,** 1968:3–93.

References. Bent, No. 197:379–382 (*Coereba*); No. 203.

    A. F. Skutch, Life histories of Central American birds, *Pacific Coast Avifauna,* No. 31, 1954:339–386, 404, 420.

    L. Griscom *et al., The Warblers of America,* New York, 1957, 356 pp.

    H. Mayfield, The Kirtland's Warbler, *Cranbrook Inst. Sci. Bull.* No. 40, 1960.

    A. F. Skutch, Life histories of Central American highland birds, *Publ. Nuttall Ornithol. Club,* No. 7, 1967:137–164.

    B. Meanley, Natural history of the Swainson's Warbler, *North Am. Fauna,* No. 69, 1971.

Synonyms: Compsothlypidae, Mniotiltidae, Sylvicolidae.

---

[1] Sixteen species (genera *Peucedramus, Xenoligea, Granatellus, Icteria, Conirostrum,* and *Coereba*) are *incertae sedis.*

ICTERIDAE

Troupial Family (91 species)

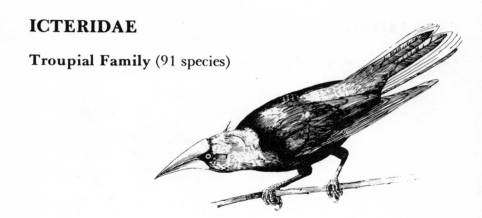

Figure 164. Wagler's Oropendola (*Icterus wagleri*). Total length 14.5 in. (368 mm).

Physical characteristics. Length—171 to 546 mm (6.75 to 21.5 in.). Plumage—uniform black (often with brilliant metallic gloss) or bold combinations of black with brown, chestnut, buff, orange, crimson, yellow; rarely streaked; often with white wing bars. A few spp. with neck ruff; a few with sparse crest. Bill conical; short and massive to moderately long and slender; in larger spp. rather heavy and casqued. Wings (9 primaries) typically long and pointed (short and rounded in some spp.); tail short to long, truncate, rounded, or graduate (plicate in a few). Feet strong. Sexes unlike in most spp.

Range. South America, C. America, West Indies, and N. America (exc. extreme n.). Many spp. migratory.

Habits. Gregarious or solitary. Arboreal or terrestrial. Some spp. walk when on ground. Voice—a variety of calls, harsh and shrill to whistling or flutelike; some spp. have well-developed song.

Food. Insects, seeds, fruit, nectar, fish, amphibians, crustacea, other birds, small mammals.

Breeding. Some spp. colonial, some parasitic, some polygamous or promiscuous. Pensile or semipensile nest; cup-shaped nest in trees, on ground, in aquatic vegetation, in holes or crevices; nests of other birds.

    Eggs. 2 to 7; pale blue or green, gray, buff, white; usually strongly marked. Incubated by ♀.

    Young. Nidicolous. With sparse down or (*Cacicus*) naked. Cared for by ♀, assisted by ♂ in some spp.

Technical diagnosis. Ridgway, Pt. 2, 1902:169–173.

Classification. Peters, **14,** 1968:138–202.

References. A. F. Skutch, Life histories of Central American birds, *Pacific Coast Avifauna*, No. 31, 1954:263–337.

    F. Haverschmidt, *Birds of Surinam,* Edinburgh, 1968:375–388.

    G. H. Orians and G. M. Christman, A comparative study of the behavior of Red-winged, Tricolored, and Yellow-headed blackbirds, *Univ. Calif. Publ. Zool.,* **84,** 1968.

    R. B. Payne, Breeding seasons and reproductive physiology of Tricolored Blackbirds and Redwinged Blackbirds, *Univ. Calif. Publ. Zool.,* **90,** 1969.

    R. Meyer de Schauensee, *A Guide to the Birds of South America,* Wynnewood, Pa., 1970:352–361.

TERSINIDAE

Swallow-tanager Family (1 species)

Figure 165. Swallow-tanager (*Tersina viridis*). Total length 6.25 in. (159 mm).

Physical characteristics. Length—152 to 159 mm (6 to 6.25 in.). Plumage—of ♂ bright turquoise-blue, barred with black on flanks, the face and throat black, crissum and center of belly pure white; of ♀ green, barred with dark green and black on flanks, the crissum and center of belly pale yellow, lightly streaked with gray. Bill short, somewhat flattened, triangular (very broad at base), slightly hooked. Neck short. Wings long; tail medium. Legs very short. Sexes unlike, as described above.

Range. Eastern Panama and S. America (exc. s. third). Migratory in part.

Habits. Gregarious. Arboreal. Flight strong. Feed in part on the wing. Voice—variety of call notes; song little developed.

Food. Fruit, green shoots, insects.

Breeding. Nest of roots and grass in hole in ground, in bank, or in tree. Use abandoned nesting holes of jacamars (Galbulidae) and swallow-winged puffbirds (Bucconidae).

Eggs. 2 to 4 (usually 3); glossy, white; immaculate. Incubated by ♀.

Young. Nidicolous. With sparse down. Cared for by ♀, assisted by ♂.

Technical diagnosis. R. Ridgway, *Proc. U.S. Natl. Mus.,* **18,** 1896:449–450; F. A. Lucas, *Ibid.,* 505–507.

Classification. Peters, **13,** 1970:408–409 (Tersininae).

References. W. E. C. Todd and M. A. Carriker, Jr., The birds of the Santa Marta region of Colombia, *Ann. Carnegie Mus.,* **14,** 1922:438–439.

E. Schaefer, Contribution to the life history of the Swallow-tanager, *Auk,* **70,** 1953:403–460, pls. 11–18.

F. Haverschmidt, *Birds of Surinam,* Edinburgh, 1968:388.

R. Meyer de Schauensee, *A Guide to the Birds of South America,* Wynnewood, Pa., 1970:374.

Synonym: Procniatidae.

THRAUPIDAE

Tanager Family (236 species)

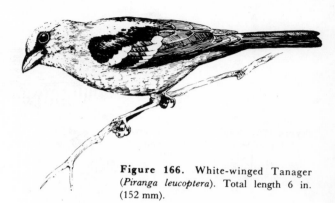

Figure 166. White-winged Tanager (*Piranga leucoptera*). Total length 6 in. (152 mm).

Physical characteristics. Length—76 to 305 mm (3 to 12 in.). Plumage—brightly colored in many spp.; highly glossy in a few; green, yellow, orange, red, purple, blue, brown, black, white; color generally in solid patches with bold contrasting areas, but some spp. have spotted areas. A few spp. crested. Bill short to medium; typically rather conical; commonly hooked or notched; more or less "toothed" in large spp. Wings short to rather long; tail short to medium in most spp.; emarginate, truncate, or rounded (long and graduate in *Cissopis*). Legs short. Sexes alike or unlike.

Range. North America (exc. extreme n.), C. America, West Indies, and S. America (exc. s. third). Habitat—forest or brushland. Some spp. migratory.

Habits. Solitary or (less commonly) gregarious. Arboreal. Flight strong but not sustained. Voice—varied; a few spp. have well-developed song.

Food. Fruit, flowers, insects.

Breeding. Shallow, often loosely constructed, cuplike nest, or domed nest with side entrance; placed in trees or bushes, or in shallow cavity in bank or tree.

    Eggs. 1 to 5; white, greenish, or bluish; marked with brown or black (or, rarely, immaculate). Incubated by ♀.

    Young. Nidicolous. With sparse down. Cared for by ♂ and ♀.

Technical diagnosis. Ridgway, Pt. 2, 1902:1.

Classification. Peters, **13,** 1970:246–408.

References. Bent, No. 211:466–509.

    A. F. Skutch, Life histories of Central American birds, *Pacific Coast Avifauna,* No. 31, 1954:123–261.

    K. W. Prescott, The Scarlet Tanager, *N.J. State Mus. Invest.* No. 2, 1965, 159 pp.

    A. F. Skutch, Life histories of Central American highland birds, *Publ. Nuttall Ornithol. Club,* No. 7, 1967:165–178.

    F. Haverschmidt, *Birds of Surinam,* Edinburgh, 1968:388–407.

    R. Meyer de Schauensee, *A Guide to the Birds of South America,* Wynnewood, Pa., 1970:374–399.

Synonyms: Tanagridae, Tangaridae.

CATAMBLYRHYNCHIDAE

Plush-capped Finch Family (1 species)

Figure 167. Plush-capped Finch (*Catamblyrhynchus diadema*). Total length 5.5 in. (139 mm).

Physical characteristics. Length—about 139 mm (5.5 in.). Plumage—forepart of crown golden, "plushlike," the feathers erect and stiffened in the northern form, *Catamblyrhynchus d. diadema,* softer and recumbent, as well as lighter in color, in the southern form, *C. d. citrinifrons*; hindcrown and nape black. Rest of upper parts bluish gray; under parts a uniform deep chestnut (paler in the immature). Bill short, massive, slightly hooked, higher than broad, flattened and grooved laterally; the lower mandible very thick. Wings short and rounded; tail medium in length, graduated. Legs stout, medium in length; feet strong, with large hindclaw. Sexes similar, but ♀ duller.

Range. Andes of Colombia, Ecuador, and Peru, to c. Bolivia. Nonmigratory?

Habits. The following two statements represent, apparently, the only data on the habits of this family (the first is here translated from the French). "They are met with in isolated pairs or mingled with flocks of other birds" (Jelski, quoted by Taczanowski, *Ornithologie du Pérou,* Vol. 3, Rennes, 1886:25). "We found them [3 ♂ ♂] singly in the higher trees" (W. Goodfellow, *Ibis,* **1901**:473).

Food. Not described.

Breeding. Nest not described.

    Eggs. Not described.

    Young. Not described.

Technical diagnosis. Ridgway, Pt. 1, 1901:19; G. R. Gray, *The Genera of Birds,* Vol. 2, London, 1849:385, col. pl. 93.

Classification. Peters, **13,** 1970:215 (Catamblyrhynchinae).

References. O. L. Austin, Jr., *Birds of the World,* New York, 1961:292, 294.

    R. Meyer de Schauensee, *A Guide to the Birds of South America,* Wynnewood, Pa., 1970:399.

DREPANIDIDAE

Hawaiian Honeycreeper Family (23 species)

Figure 168. Iiwi (*Vestiaria coccinea*).
Total length 6 in. (152 mm).

Physical characteristics. Length—102 to 203 mm (4 to 8 in.). Plumage—olive-green, yellow, orange, red, brown, gray, or black, with little pattern exc. for darker wings and tail; 1 sp. (*Palmeria*) has brightly spotted and streaked plumage and a crest. Bill extremely varied, ranging from very long, slender, and curved to very short, massive, and strongly hooked. Wings pointed (with but 9 functional primaries); tail short to medium, truncate or emarginate. Legs medium to rather short; feet strong. Sexes unlike or alike (but the ♂ larger).

Range. Hawaiian Ids. Nonmigratory.

Habits. Solitary or in small loose flocks. Arboreal. Flight strong and, in some spp., noisy. Voice—loud clear notes and simple trills; some spp. have a "rather sweet" song; discordant and "rusty" in *Vestiaria*.

Food. Nectar, insects, fruit, seeds.

Breeding. Cup-shaped nest placed in trees, treeferns, lava tubes, or grass tussocks.
   Eggs. 2 to 4; white; spotted. Incubated by ♀.
   Young. Nidicolous. With some down on upper parts. Cared for by ♂ and ♀.

Technical diagnosis. D. Amadon, *Bull. Am. Mus. Nat. Hist., 95,* Art. 4, 1950:163–164, pls. 9–12, 15.

Classification. D. Amadon, *Bull. Am. Mus. Nat. Hist., 95,* Art. 4, 1950:163–192.

References. W. Rothschild, *The Avifauna of Laysan and Neighbouring Islands,* London, 1893–1900.

   R. C. L. Perkins, An introduction to the study of the Drepanididae . . . , *Ibis,* **1901**:562–585.

   P. H. Baldwin, Annual cycle, environment and evolution in the Hawaiian honeycreepers . . . , *Univ. Calif. Publ. Zool., 52,* 1953:285–398.

   J. C. Greenway, Jr., Family Drepanididae. In Peters, **14,** 1968:93–103.

   A. J. Berger, *Hawaiian Birdlife,* Honolulu, 1972, 270 pp.

Synonyms: Drepaniidae, Drepanidae.

FRINGILLIDAE

Finch Family (436 species)

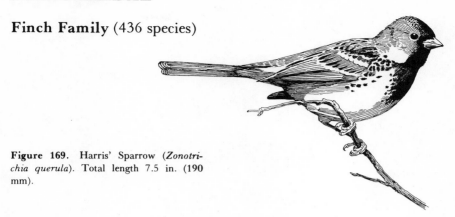

Figure 169. Harris' Sparrow (*Zonotrichia querula*). Total length 7.5 in. (190 mm).

Physical characteristics. Length—95 to 273 mm (3.75 to 10.75 in.). Plumage—from brown and gray in cryptic patterns to combinations of yellow, red, purple, blue, green, black, white, in bold patterns; many spp. with broad stripes on top or sides of head. Bill short and pointed; rather thick to very massive; sometimes hooked; mandibles crossed in *Loxia*. Wings short and rounded to long and rather pointed; tail short to long. Legs typically of medium length. Sexes alike or unlike.

Range. Worldwide (exc. Madagascar, the Papuan Region, Australia, and Oceania). Many spp. migratory.

Habits. Solitary or (esp. in nonbreeding season) gregarious. Terrestrial or arboreal. Flight weak to strong. Voice—extremely varied; many spp. have well-developed song.

Food. Typically seeds; other plant food and some insects.

Breeding. Cup-shaped nest in trees, shrubs, or herbage, or on ground; in rock crevice; in holes in trees; or in covered nests of other birds (*Sicalis*).

    Eggs. 2 to 6 (rarely 8); white, bluish or reddish white, or green; marked or immaculate. Incubated by ♀, or by ♂ and ♀.

    Young. Nidicolous. With some down. Cared for by ♂ and ♀.

Technical diagnosis. Ridgway, Pt. 1, 1901:24.

Classification. Peters, **13**, 1970:3–214, 216–245; **14**, 1968:202–306.

References. Witherby, 1, 1938:51–153; Bannerman, **6**, 1948:270–320; Bent, No. 237, 1968.

    A. F. Skutch, Life histories of Central American birds, *Pacific Coast Avifauna*, No. 31, 1954:19–121.

    R. I. Bowman, Morphological differentiation and adaptation in the Galápagos Finches, *Univ. Calif. Publ. Zool.*, **58**, 1961.

    R. I. Bowman and S. L. Billeb, Blood-eating in a Galápagos Finch, *Living Bird*, **1965**:29–44.

    A. F. Skutch, Life histories of Central American highland birds, *Publ. Nuttall Ornithol. Club*, No. 7, 1967:179–205.

    R. Meyer de Schauensee, *A Guide to the Birds of South America*, Wynnewood, Pa., 1970:400–427.

Synonymy: The Geospizinae, Fringillinae, Richmondeninae, Cardinalinae, Carduelinae, and Emberizinae are included.

Index